PROGRESS IN BRAIN RESEARCH

VOLUME 154

VISUAL PERCEPTION, PART 1

**FUNDAMENTALS OF VISION:
LOW AND MID-LEVEL PROCESSES IN PERCEPTION**

Other volumes in PROGRESS IN BRAIN RESEARCH

PROGRESS IN BRAIN RESEARCH

VOLUME 154

VISUAL PERCEPTION, PART 1

FUNDAMENTALS OF VISION: LOW AND MID-LEVEL PROCESSES IN PERCEPTION

EDITED BY

S. MARTINEZ-CONDE

Department of Neurobiology, Barrow Neurological Institute, Phoenix,
AZ 85013, USA

S.L. MACKNIK

Departments of Neurosurgery and Neurobiology, Barrow Neurological Institute, Phoenix,
AZ 85013, USA

L.M. MARTINEZ

Departamento de Medicina, Facultade de Ciencias da Saúde, Campus de Oza, Universidade da Coruña,
15006, A Coruña, Spain

J.-M. ALONSO

Department of Biological Sciences, State University of New York – Optometry, New York, NY 10036, USA

P.U. TSE

Department of Psychological and Brain Sciences, Dartmouth College, Hanover, NH 03755, USA

ELSEVIER

AMSTERDAM – BOSTON – HEIDELBERG – LONDON – NEW YORK – OXFORD
PARIS – SAN DIEGO – SAN FRANCISCO – SINGAPORE – SYDNEY – TOKYO

Elsevier
Radarweg 29, PO Box 211, 1000 AE Amsterdam, The Netherlands
The Boulevard, Langford Lane, Kidlington, Oxford OX5 1GB, UK

First edition 2006

Library of Congress Cataloging-in-Publication Data
A catalog record for this book is available from the Library of Congress

British Library Cataloguing in Publication Data

European Conference on Visual Perception (28th : 2005 :
A Coruna, Spain)
  Visual Perception
  Part 1: Fundamentals of vision: low and mid-level processes in
  perception. – (Progress in brain research; v. 154)
  1. Vision – Congresses 2. Physiological optics – Congresses
  3. Visual perception – Congresses
  I. Title II. Martinez-Conde, S.
  612.8'4

  ISBN-13: 9780444529664
  ISBN-10: 0444529667

ISBN-13: 978-0-444-52966-4 (this volume; Part 1)
ISBN-10: 0-444-52966-7 (this volume; Part 1)
ISBN-13: 978-0-444-51927-6 (vol. 155; Part 2)
ISBN-10: 0-444-51927-0 (vol. 155; Part 2)
ISBN-13: 978-0-444-80104-3 (series)
ISBN-10: 0-444-80104-9 (series)
ISSN: 0079-6123

For information on all Elsevier publications
visit our website at books.elsevier.com

Printed and bound in The Netherlands

06 07 08 09 10 10 9 8 7 6 5 4 3 2 1

Working together to grow
libraries in developing countries

www.elsevier.com | www.bookaid.org | www.sabre.org

ELSEVIER   BOOK AID
            International   Sabre Foundation

Contents

SECTION III. EYE MOVEMENTS AND PERCEPTION DURING VISUAL FIXATION

SECTION IV.   PERCEPTUAL COMPLETION

SECTION V. FORM, OBJECT AND SHAPE PERCEPTION

List of Contributors

J.-M. Alonso, Department of Biological Sciences, State University of New York—Optometry, 33 West 42nd Street, New York, NY 10036, USA

A. Angelucci, Department of Ophthalmology and Visual Science, Moran Eye Center, University of Utah, 50 North Medical Drive, Salt Lake City, UT 84132, USA

I. Ballesteros-Yáñez, Instituto Cajal (CSIC), Avenida Dr. Arce 37, 28002-Madrid, Spain

N.E. Barraclough, Department of Psychology, University of Hull, Hull HU6 7RX, UK

A. Basole, Department of Neurobiology, Box 3209, Duke University Medical Center, 427C Bryan Research Building, Durham, NC 27710, USA

P.C. Bressloff, Department of Mathematics, University of Utah, 155 South 1400 East, Salt Lake City, UT 84112, USA

I. Bülthoff, Max-Planck Institut für biologische Kybernetik, Spemannstrasse 38, D 72076 Tübingen, Germany

G.P. Caplovitz, Department of Psychological and Brain Sciences, H.B. 6207, Moore Hall, Dartmouth College, Hanover, NH 03755, USA

M. Carrasco, Department of Psychology and Center for Neural Science, New York University, 6 Washington Place, 8th floor, New York, NY 10003, USA

J. De Dios Navarro-López, Department of Physiology, University College London, London WC1E 6BT, UK

P. De Weerd, Neurocognition Group, Psychology Department, University of Maastricht, 6200 MD Maastricht, The Netherlands

J. DeFelipe, Instituto Cajal (CSIC), Avenida Dr. Arce 37, 28002-Madrid, Spain

J.M. Delgado-García, División de Neurociencias, Universidad Pablo de Olavide, Ctra. De Utrera, Km. 1, 41013-Seville, Spain

R. Engbert, Computational Neuroscience, Department of Psychology, University of Potsdam, 14415 Potsdam, Germany

D. Fitzpatrick, Department of Neurobiology, Box 3209, Duke University Medical Center, 427C Bryan Research Building, Durham, NC 27710, USA

J. Gyoba, Department of Psychology, Graduate School of Arts and Letters, Tohoku University, Kawauchi 27-1, Aoba-ku, Sendai 980-8576, Japan

M.C. Inda, Department of Cell Biology, Universidad Complutense, Madrid, Spain

A. Kitaoka, Department of Psychology, Ritsumeikan University, 56-1 Toji-in Kitamachi, Kita-ku, Kyoto 603-8577, Japan

V. Kreft-Kerekes, Department of Neurobiology, Box 3209, Duke University Medical Center, 427C Bryan Research Building, Durham, NC 27710, USA

P. Mamassian, CNRS UMR 8581, LPE Université Paris 5, 71 avenue Edouard Vaillant, 92100 Boulogne-Billancourt, France

L.M. Martinez, Departamento de Medicina, Facultade de Ciencias da Saúde, Campus de Oza, Universidade da Coruña, 15006 La Coruña, Spain

S. Martinez-Conde, Department of Neurobiology, Barrow Neurological Institute, 350 W Thomas Road, Phoenix, AZ 85013, USA

A. Muñoz, Department of Cell Biology, Universidad Complutense, Madrid, Spain

I. Murakami, Department of Life Sciences, University of Tokyo, 3-8-1 Komaba, Meguro-ku, Tokyo 153-8902, Japan

F.N. Newell, Department of Psychology, University of Dublin, Trinity College, Aras an Phiarsaigh, Dublin 2, Ireland

M.W. Oram, School of Psychology, University of St. Andrews, St. Mary's College, South Street, St. Andrews, Fife KY16 9JP, UK

A. Pasupathy, Picower Institute for Learning and Memory, Massachusetts Institute of Technology, 77 Massachusetts Avenue, 46-6241, Cambridge, MA 02139, USA

D.I. Perrett, School of Psychology, University of St. Andrews, St. Mary's College, South Street, St. Andrews, Fife KY16 9JP, UK

K. Sakurai, Department of Psychology, Tohoku Gakuin University, 2-1-1 Tenjinzawa, Izumi-ku, Sendai 981-3193, Japan

C. Stoelzel, Department of Psychology, University of Connecticut, Storrs, CT 06269, USA

P.U. Tse, Department of Psychological and Brain Sciences, H.B. 6207, Moore Hall, Dartmouth College, Hanover, NH 03755, USA

C. Weng, Department of Biological Sciences, State University of New York—Optometry, 33 West 42nd Street, New York, NY 10036, USA

L.E. White, Department of Community and Family Medicine, Physical Therapy Division, Duke University Medical Center, Durham, NC 27710, USA

D. Xiao, School of Psychology, University of St. Andrews, St. Mary's College, South Street, St. Andrews, Fife KY16 9JP, UK

J. Yajeya, Departamento de Fisiología y Farmacología, Facultad de Medicina, Instituto de Neurociencias de Castilla y León, Universidad de Salamanca, Salamanca, Spain

C.-I. Yeh, Department of Psychology, University of Connecticut, Storrs, CT 06269, USA

General introduction

"Visual Perception" is a two-volume series of Progress in Brain Research, based on the symposia presented during the 28th Annual Meeting of the European Conference on Visual Perception (ECVP), the premier transnational conference on visual perception. The conference took place in A Coruña, Spain, in August 2005. The Executive Committee members of ECVP 2005 edited this volume, and the symposia speakers provided the chapters herein.

The general goal of these two volumes is to present the reader with the state-of-the-art in visual perception research, with a special emphasis in the neural substrates of perception. "Visual Perception (Part 1)" generally addresses the initial stages of the visual pathway, and the perceptual aspects than can be explained at early and intermediate levels of visual processing. "Visual Perception (Part 2)" is generally concerned with higher levels of processing along the visual hierarchy, and the resulting percepts. However, this separation is not very strict, and several chapters encompass both early and high-level processes.

The current volume "Visual Perception (Part 1) — Fundamentals of Vision: Low and Mid-Level Processes in Perception" contains 17 chapters, organized into 5 general sections, each addressing one of the main topics in vision research today: "Visual circuits and perception since Ramón y Cajal"; "Recent discoveries on receptive field structure"; "Eye movements and perception during visual fixation"; "Perceptual completion"; and "Form, object and shape perception". Each section includes a short introduction and two to four related chapters. The topics are tackled from a variety of methodological approaches, such as single-neuron recordings, fMRI and optical imaging, psychophysics, eye movement characterization and computational modeling. We hope that the contributions enclosed will provide the reader with a valuable perspective on the current status of vision research, and more importantly, with some insight into future research directions and the discoveries yet to come.

Many people helped to compile this volume. First of all, we thank all the authors for their contributions and enthusiasm. We also thank Shannon Bentz, Xoana Troncoso and Jaime Hoffman, at the Barrow Neurological Institute, for their assistance in obtaining copyright permissions for several of the figures reprinted here. Moreover, Shannon Bentz transcribed Lothar Spillmann's lecture (in "Visual Perception (Part 2)"), and provided general administrative help. Xoana Troncoso was heroic in her effort to help us to meet the submission deadline by collating and packing all the chapters, and preparing the table of contents. We are indebted to Johannes Menzel and Maureen Twaig, at Elsevier, for all their encouragement and assistance; it has been wonderful working with them.

Finally, we thank all the supporting organizations that made the ECVP 2005 conference possible: Ministerio de Educación y Ciencia, International Brain Research Organization, European Office of Aerospace Reseach and Development of the USAF, Consellería de Educación, Industria e Comercio-Xunta de Galicia, Elsevier, Pion Ltd., Universidade da Coruña, Sociedad Española de Neurociencia, SR Research Ltd., Consellería de Sanidade-Xunta de Galicia, Mind Science Foundation, Museos Científicos Coruñeses, Barrow Neurological Institute, Images from Science Exhibition, Concello de A Coruña, Museo Arqueolóxico e Histórico-Castillo de San Antón, Caixanova, Vision Science, Fundación Pedro Barrié de la Maza, and Neurobehavioral Systems.

Susana Martinez-Conde
Executive Chair, European Conference on Visual Perception 2005

On behalf of ECVP 2005's Executive Committee: Stephen Macknik, Luis Martinez, Jose-Manuel Alonso and Peter Tse

SECTION I

Visual Circuits and Perception Since Ramón y Cajal

Introduction

Ramón y Cajal is one of the most distinguished scientists in Spanish history and one of the greatest neuroanatomists of all times. Not surprisingly, any scientific meeting that takes place in Spain rarely happens without making a specific mention of the outstanding contributions of this scientist. The 2005 European Conference on Visual Perception (ECVP) was no exception. The Symposium 'Visual circuits and Perception since Ramón y Cajal' started with an acknowledgement of Cajal's legacy, which was followed by a series of lectures on the study of neural circuitry with modern methods. In 1906, Ramón y Cajal and Camillo Golgi shared the Nobel prize in Medicine and Physiology, while still maintaining completely opposite views on how the brain is organized. Cajal's view prevailed. Cajal defended the idea that neurons were separate entities in the brain (neuron doctrine) that transmitted information from the dendrites and soma to the axon terminal (law of dynamic polarization). This view of the brain, which was revolutionary a century ago, is now one of the basic tenets of neuroscience.

Cajal was not only an outstanding scientist but also an excellent photographer and artist. His beautiful drawings of the neural circuits, and his visionary insights based purely on observations made under a light microscope, still serve as inspiration for modern research in the organization, function and development of the visual system. Cajal spent most of his career studying the circuitry of the brain in different species and different neural systems with the aid of mostly one method — Golgi staining. The philosophy that guided all his work is still valid today — if we want to understand how we perceive, we have to understand in detail the circuitry of the visual pathway. The first symposium of ECVP 2005 provided a brief glimpse at some of the new methods that have emerged 100 years after Cajal and that are currently used to study neural circuitry. The symposium included approaches that extend from modern neuroanatomical and electrophysiological techniques to recent methods of functional magnetic resonance imaging (fMRI) combined with psychophysics. The following chapters review the circuitry of different stages in the visual pathway studied using these different methods.

Jose-Manuel Alonso

Martinez-Conde, Macknik, Martinez, Alonso & Tse (Eds.)
Progress in Brain Research, Vol. 154
ISSN 0079-6123

CHAPTER 1

Retinogeniculate connections: a balancing act between connection specificity and receptive field diversity

J.-M. Alonso[1,2,*], C.-I. Yeh[1,2], C. Weng[1] and C. Stoelzel[2]

[1]*Department of Biological Sciences, SUNY State College of Optometry, 33 West 42nd Street, New York, NY 10036, USA*
[2]*Department of Psychology, University of Connecticut, Storrs, CT 06269, USA*

Abstract: Retinogeniculate connections are one of the most striking examples of connection specificity within the visual pathway. In almost every connection there is one dominant afferent cell per geniculate cell, and both afferent and geniculate cells have very similar receptive fields. The remarkable specificity and strength of retinogeniculate connections have inspired comparisons of the lateral geniculate nucleus (LGN) with a simple relay that connects the retina with the visual cortex. However, because each retinal ganglion cell diverges to innervate multiple cells in the LGN, most geniculate cells must receive additional inputs from other retinal afferents that are not the dominant ones. These additional afferents make weaker connections and their receptive fields are not as perfectly matched with the geniculate target as the dominant afferent. We argue that these 'match imperfections' are important to create receptive field diversity among the cells that represent each point of visual space in the LGN. We propose that the convergence of dominant and weak retinal afferents in the LGN multiplexes the array of retinal ganglion cells by creating receptive fields that have a richer range of positions, sizes and response time courses than those available at the ganglion cell layer of the retina.

Keywords: thalamus; thalamocortical; visual cortex; V1; Y cell; X cell; response latency; simultaneous recording

The cat eye has 160,800 retinal ganglion cells that fit within a retinal area of $450 \, mm^2$ (Illing and Wassle, 1981). One-half of these cells (53–57%) has small receptive fields and is classified as X and a much smaller proportion (2–4%) has larger receptive fields and is classified as Y (Enroth-Cugell and Robson, 1966; Friedlander et al., 1979; Illing and Wassle, 1981). X and Y retinal ganglion cells are the origin of two major functional channels within the cat visual pathway that remain relatively well segregated within the lateral geniculate nucleus

(LGN) (Cleland et al., 1971a, b; Mastronarde, 1992; Usrey et al., 1999). These two major channels have pronounced anatomical differences. For example, the X retinal afferents have very restricted axon terminals ($\sim100 \, \mu m$ diameter) that are confined to a single layer of LGN and connect small geniculate cells. In contrast, the Y axon terminals are twice as large, usually diverge into two different LGN layers (Sur and Sherman, 1982; Sur et al., 1987) and connect geniculate cells with large dendritic trees that tend to cross layer boundaries (Friedlander et al., 1979; Fig. 1).

X and Y retinal ganglion cells diverge at the level of the LGN to connect up to 20 geniculate cells per

*Corresponding author. Tel.: + 1-212-938-5573;
Fax: + 1-212-938-5796; E-mail: jalonso@sunyopt.edu

DOI: 10.1016/S0079-6123(06)54001-4

4

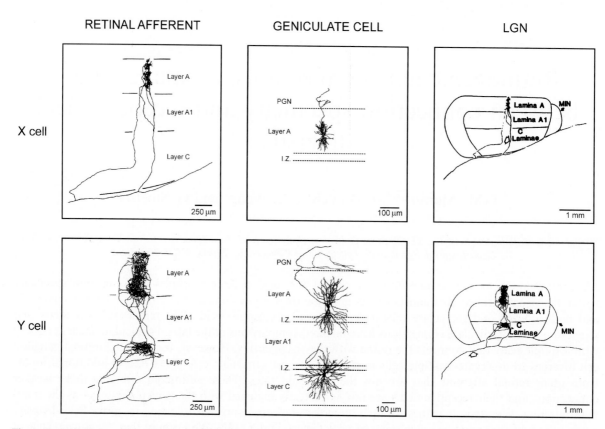

RETINAL AFFERENT   GENICULATE CELL   LGN

Fig. 1. Retinal afferents and geniculate cells. Left: axon terminals from X and Y retinal afferents in LGN. X retinal axons project into a single LGN layer and they are very restricted. Y retinal axons can project into two different LGN layers and are wider. Middle: X and Y geniculate cells. X cells have small dendritic trees that are restricted to a single LGN layer. Y cells have larger dendritic trees that frequently cross layers. Right: the same axon terminals on the left of the figure are shown at a different scale. Reprinted with permission from Sur and Sherman (1982); Copyright 1982 AAAS; left and right: Sur (1988); middle: Friedlander et al. (1981). MIN: medial interlaminar nucleus; PGN: perigeniculate nucleus; I.Z.: interlaminar zone.

retinal afferent (Hamos et al., 1987). This divergence could do much more than just copying the properties of each retinal ganglion cell into the geniculate neurons; it could diversify the spatial and temporal properties of the receptive fields that represent each point of visual space. This receptive field diversity could then be used at the cortical level to maximize the spatiotemporal resolution needed to process visual stimuli. In this review, we illustrate this idea with two different examples. In the first example, we show evidence that a single class of Y retinal afferent can be used to build two different types of Y receptive fields within the LGN. In the second example, we show that geniculate neurons representing the same point of visual space have a rich variety of receptive field sizes

and response latencies that emerge as a consequence of retinogeniculate divergence/convergence.

Retinogeniculate divergence in the Y visual pathway of the cat

Y retinal ganglion cells are a conspicuous minority within the cat retina (2–4%), which is greatly amplified at subsequent stages of the visual pathway. While X retinal ganglion cells diverge, on average, into 1.5 geniculate cells, Y retinal ganglion cells diverge into 9 geniculate cells (X geniculate cells/retinal cells: 120,000/89,000; Y geniculate cells/retinal cells: 60,000/6700; and the Y cells from layer C are not included in this estimate (LeVay and Ferster, 1979; Illing and Wassle, 1981; Peters and Payne, 1993)).

The huge amplification of the Y pathway in the cat is reminiscent of the magnocellular pathway in the primate. In the rhesus monkey, there is little retinogeniculate divergence, probably because there is a limit on how many retinogeniculate connections can be accommodated within the LGN (the primate retina has 1,120,000 parvocellular cells and 128,000 magnocellular cells (see Masland, 2001, for review)). However, as in the cat, magnocellular cells are a minority within the primate retina (~8% of all retinal ganglion cells) and, by connecting to magnocellular geniculate cells, they are able to reach a remarkably large number of cortical neurons — at the cortical representation of the fovea in layer 4C, magnocellular geniculate cells connect about 29 times more cortical cells than parvocellular geniculate cells (Connolly and Van Essen, 1984). Interestingly, neuronal divergence seems to be delayed by one synapse in primate with respect to the cat, as is also the case for the construction of simple receptive fields (Hubel and Wiesel, 2005).

The cat LGN is an excellent model to study the functional consequences of the Y pathway divergence. Unlike the primate, the cat LGN has two main layers that receive Y contralateral input (A and C; A1 receives ipsilateral input) and the retinotopic map of each layer is not excessively folded, making it easier to record from multiple cells with overlapping receptive fields across the different LGN layers. Fig. 2 illustrates the retinotopic map of cat LGN (Fig. 2A) and the response properties of four cells that were simultaneously recorded from different layers. The four cells had on-center receptive fields with slightly different positions and receptive field sizes (Fig. 2B, left). Their response time courses, represented as impulse responses, were also different (receptive fields and impulse responses were obtained with white noise stimuli by reverse correlation (Reid et al., 1997; Yeh et al., 2003)).

As shown in the figure, the Y cells had the largest receptive fields and fastest response time courses within the group. Moreover, the receptive field was larger and the response latency faster for the Y cell from layer C (Y_C, shown in green) than

the Y cell from layer A (Y_A, shown in orange). Simultaneous recordings, like the one shown in Fig. 2, allowed us to compare the response properties from the neighboring Y_A and Y_C cells that had overlapping receptive fields. These measurements demonstrated that, on average, the receptive fields from Y_C cells are 1.8 times larger than those from Y_A cells and the response latencies are 2.5 ms faster ($p < 0.001$, Wilcoxon test).

The differences in receptive field size and response latency between Y cells located in different layers were sometimes as pronounced as the differences between X and Y cells located within the same layer. To quantify these differences, we did simultaneous triplet recordings from the neighboring Y_A, Y_C and X_A cells[1]. Fig. 3, top, shows an example of a triplet recording from three off-center geniculate cells of different types (X_A, Y_A and Y_C). The Y_C cell had the largest receptive field and the fastest response latency and the X cell the smallest receptive field and the slowest response latency. For each cell triplet recorded, we calculated a similarity ratio to compare the differences between the Y_A and Y_C cells with the differences between the Y_A and X_A cells. A ratio higher than 1 indicates that the Y_C cell differed from the Y_A cell more than the Y_A cell differed from the X_A cell. As shown in the histograms at the bottom of Fig. 3, in many cell triplets, the similarity ratio for receptive field size and response latency was higher than 1. Moreover, the mean difference in receptive field size was significantly higher for the Y_C–Y_A cells than that for the Y_A–X_A cells ($p < 0.001$, Wilcoxon test). Y_C and Y_A cells also differed significantly in other properties such as spatial linearity, response transience and contrast sensitivity (Frascella and Lehmkuhle, 1984; Lee et al., 1992; Yeh et al., 2003), and are not illustrated here. These results indicate that Y retinal afferents connect to two

---

[1] The precise retinotopy of LGN and the interelectrode distances used in our experiments strongly suggest that all our recordings came from cells (and not axons) that were located within a cylinder of $< 300 \, \mu m$ in diameter (Sanderson, 1971). Recordings from axons, which were extremely rare in our experiments, had a characteristic spike waveform (Bishop et al., 1962), and could not be maintained for the long periods of time needed for our measurements.

6

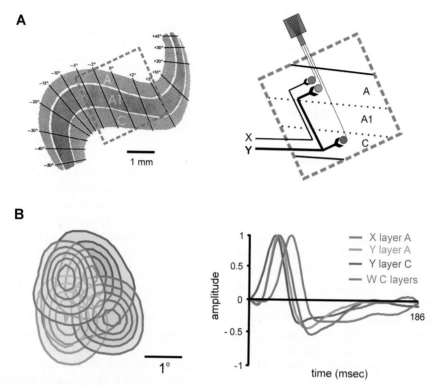

Fig. 2. Simultaneous recordings from four geniculate cells recorded at different layers in the cat LGN. (A) Left: retinotopic map of cat LGN (adapted from Sanderson, 1971). Right: schematic representation of the simultaneous recordings. (B) Left: receptive fields of the four simultaneously recorded geniculate cells mapped with white noise by reverse correlation. The contour lines show responses at 20–100% of the maximum response. Right: impulse responses of the four cells obtained by reverse correlation; the impulse response represents the time course of the receptive field pixel that generated the strongest response. The different cell types are represented in different colors (X cell from layer A, X_A, in blue; Y cell from layer A, Y_A, in orange; Y cell from layer C, Y_C, in green and W cell from the deep C layers in pink). Throughout this review, on-center receptive fields are represented as continuous lines and off-center receptive fields as discontinuous lines. Reprinted with permission from Yeh et al. (2003).

types of Y geniculate cells with significantly different response properties, Y_C and Y_A.

At first sight, this conclusion seems at odds with the idea that retinogeniculate connections are highly specific. If the receptive field of each geniculate neuron resembles very closely the receptive field of the dominant retinal afferent (Cleland et al., 1971a, b; Mastronarde, 1983; Cleland and Lee, 1985; Usrey et al., 1999), it should not be possible to construct two types of Y receptive fields with one type of Y retinal afferent. Certainly, there is no evidence for two types of Y retinal afferents that could match the properties of Y_A and Y_C geniculate receptive fields and almost every Y retinal afferent has been found to diverge in the two layers of the LGN (Sur and Sherman, 1982; Sur et al., 1987).

A better understanding of how Y_A and Y_C receptive fields are generated requires a precise comparison of the response properties from Y_A and Y_C cells that share input from the same retinal afferent. Geniculate neurons that share a common retinal input can be readily identified with cross-correlation analysis because they fire in precise synchrony — their correlogram has a narrow peak of <1 ms width centered at zero (Alonso et al., 1996; Usrey et al., 1998; Yeh et al., 2003). Fig. 4 shows an example of a pair recording from a Y_C cell and a Y_A cell that were tightly correlated (see narrow peak centered at zero in the correlogram, Fig. 4A, bottom). As expected from cells that share a retinal afferent, the receptive fields of the Y_A and Y_C cells were similar in many respects.

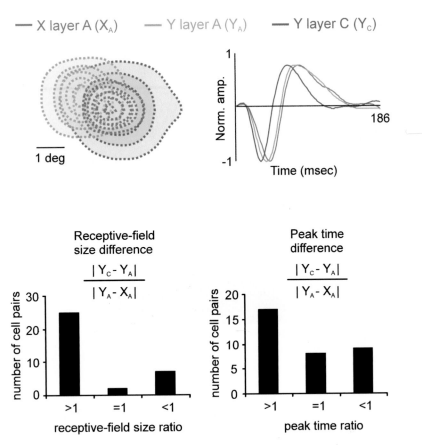

Fig. 3. Comparisons of receptive field size and response latency obtained from triplet recordings of Y_A, Y_C and X_A cells. Top: an example of a triplet recording from three cells with off-center receptive fields. Receptive fields are shown on the left and impulse responses on the right. Bottom: comparisons of receptive field size (left) and response latency (right). An index higher than 1 indicates that the differences between Y_C and Y_A were higher than the differences between Y_A and X_A. An index lower than 1 indicates the opposite. Note that the differences between Y_C and Y_A were frequently higher than those between Y_A and X_A. Reprinted with permission from Yeh et al. (2003).

They were both off-center and they had similar positions in visual space (Fig. 4A, left). On the other hand, the receptive fields showed substantial differences that were reminiscent of the differences between Y_A and Y_C cells described above. For example, the receptive field was larger and the response latency faster for the Y_C cell than those for the Y_A cell (Fig. 4A, top). A similar finding was obtained in recordings from other Y_C–Y_A cell pairs. Y_A and Y_C cells sharing a retinal afferent always had receptive fields of the same sign (e.g., off-center superimposed with off-center) that were highly overlapped ($>80\%$). However, they differed frequently in receptive

field size and response latency, probably owing to the inputs from other retinal afferents that were not shared.

Interestingly, cell synchrony across layers was weaker and more frequently found than cell synchrony within the same layer (when considering only cell pairs with $>80\%$ receptive field overlap). These findings point to a possible mechanism that could allow two types of Y geniculate receptive fields to be constructed with one type of Y retinal afferent. The weaker and more frequent synchrony across layers could be due to a higher divergence of Y retinal afferents within layer C than within layer A. As a consequence of this higher divergence,

8

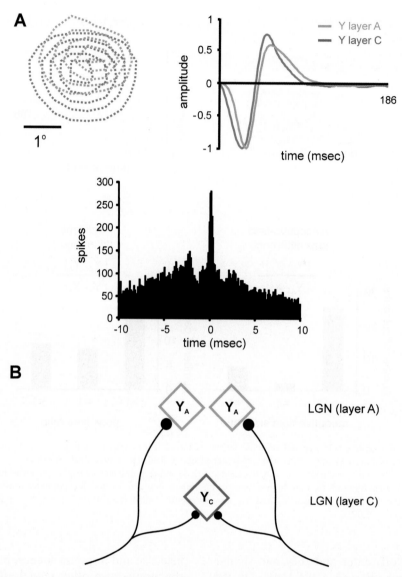

Fig. 4. Different types of Y receptive fields (Y_A and Y_C) are constructed in the LGN with one type of Y retinal afferent. (A) Example of a pair of Y_A and Y_C cells that shared input from the same retinal afferent. The two cells had off-center receptive fields that were well overlapped (top, left). However, the Y_C cell had a slightly larger receptive field and faster response latency (top, right) than the Y_A cell. The correlogram shows, a narrow peak centered at zero indicating that both cells fired in precise synchrony, as is characteristic of cells that share a retinal afferent. (B) Cartoon of a possible neural mechanism to construct two different types of Y receptive fields with a single type of Y retinal afferent. Reprinted with permission from Yeh et al. (2003).

the Y_C geniculate cells would receive input from more retinal afferents than would the Y_A cells (Fig. 4B), and owing to this higher convergence, Y_C cells would have larger receptive fields and faster response latencies than Y_A cells (Fig. 4B; Yeh et al., 2003).

Receptive field properties of geniculate neurons representing the same point of visual space

The differences in the response properties of Y_A and Y_C cells could be an extreme case of a common phenomenon in the LGN: geniculate cells that share input from a common retinal afferent may have substantially different receptive fields owing to weak retinal inputs that are not shared.

The ganglion cell layer of the retina is a thin structure ($<100\,\mu m$ thickness) that can only accommodate a limited number of retinal ganglion cells to cover each point of visual space ($\sim$30 X cells and 5 Y cells in central retina; Peichl and Wassle, 1979). Reaching such coverage factors is particularly challenging at the area centralis, where receptive fields are the smallest and therefore, cell density has to be the highest (6500 X cells/mm^2 and 200 Y cells/mm^2 at the area centralis compared with 80 X cells/mm^2 and 3 Y cells/mm^2 at the far periphery; Peichl and Wassle, 1979). The limited space to fit all these retinal ganglion cells has functional consequences: the receptive fields of neighboring cells of a given type (e.g., X or Y) have to be separated by at least half a receptive field center within most of the

retina (Wassle et al., 1981a, b; Mastronarde, 1983; Meister et al., 1995).

In the cat, the limitation in physical space is somewhat alleviated once the retinal ganglion cells leave the eye. Fig. 5A shows the receptive fields of four neighboring geniculate cells that were simultaneously recorded within layer A of the LGN. The four cells had well-overlapped receptive fields of the same sign (off-center). Furthermore, unlike the retina, the receptive field overlap was almost complete among three cells of the same type (Y cell). Moreover, although the three Y cells showed precise synchronous firing indicating that they shared input from the same retinal afferent, their receptive field sizes (Fig. 5A) and response latencies (Fig. 5B) were substantially different. Interestingly, there was a correlation between the receptive field size and response latency (Fig. 5C), suggesting that both properties may be generated by a common

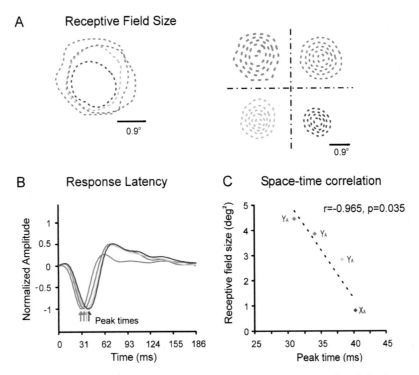

Fig. 5. Receptive field properties of neighboring geniculate neurons that represent the same point of visual space. (A) Receptive fields from four off-center geniculate cells that were simultaneously recorded. The receptive fields of the four neurons have very similar positions, but they differed substantially in size. The receptive fields are shown as contour plots on the right and superimposed on the left (only the 20% contour is shown on the left for clarity). (B) The four neurons also differed in their response latencies, as illustrated by the impulse responses obtained by reverse correlation. (C) There was a strong correlation between receptive field size and response latency: the larger the receptive field, the faster the response to visual stimuli. Reprinted with permission from Weng et al. (2005).

mechanism. Receptive field size and response latency could both be determined by the number of retinal afferents that converge onto a given geniculate cell — more convergent afferents will lead to larger receptive fields and faster responses.

Recordings like the one shown in Fig. 5 demonstrate a surprising diversity of receptive field positions, sizes and response latencies among neighboring neurons within the LGN. This receptive field diversity could provide the cortex with a richer representation of space and time than the one available at the retina.

Multiplexing the receptive field properties of the retinal ganglion cells

The connections from the retina to the LGN are among the strongest and the most specific connections within the visual pathway. One retinal axon can provide more than 100 synapses to the same geniculate cell (Hamos et al., 1987; Chen and Regehr, 2000), a number which is at least 10 times larger than the number of synapses provided by a geniculate axon to a cortical cell (Freund et al., 1985). Moreover, each geniculate cell receives highly specific input from one dominant afferent, whose receptive field is very similar to the geniculate receptive field (Cleland et al., 1971a, b; Cleland and Lee, 1985; Mastronarde, 1992; Usrey et al., 1999).

In addition to the dominant afferent, there are other weak retinal inputs that converge at the same geniculate cell, but whose receptive fields are not a 'near-perfect match' as is the case with the dominant afferent (Cleland et al., 1971a; Mastronarde, 1992; Usrey et al., 1999). The functional significance of these weaker inputs remains unclear. A reasonable possibility is that the weak inputs are remnants of the pruning process during development (Sur et al., 1984; Hamos et al., 1987; Chen and Regehr, 2000). These developmental mistakes (Garraghty et al., 1985) could explain the existence of a few geniculate cells that receive mixed X and Y inputs and have intermediate X/Y properties (Cleland et al., 1971b; Mastronarde, 1992; Usrey et al., 1999). The idea of a developmental error is attractive since most retinogeniculate cells are

known to be highly specific of cell type: most X retinal ganglion cells connect to X geniculate cells and most Y retinal ganglion cells connect to Y geniculate cells (Cleland et al., 1971a, b; Cleland and Lee, 1985; Mastronarde, 1992; Usrey et al., 1999). However, our simultaneous recordings from neighboring geniculate cells suggest an alternative interpretation. The weak retinal inputs could be important to interpolate the spatiotemporal receptive fields of the retinal ganglion cells into a more continuous representation of visual space and time (see also Mastronarde, 1992, for a similar idea). Fig. 6 illustrates this idea with a cartoon. Representative examples of receptive fields from neighboring neurons recorded within the cat retina (taken from Mastronarde, 1983) and cat LGN (taken from Weng et al., 2005) are shown at the top, and a possible mechanism for the coverage transformation at the bottom. The bottom-left of the cartoon shows the receptive fields of three retinal ganglion cells, illustrated as Gaussian curves (shown in red, green, and blue) and the bottom-right, the geniculate receptive fields that result from combining the retinal inputs. The LGN Gaussian at

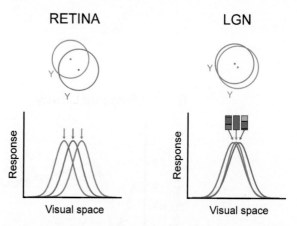

Fig. 6. Multiplexing the receptive field positions of retinal inputs in the LGN. Top: the cartoon illustrates the receptive fields of two neighboring Y cells in the cat retina (based on Mastronarde, 1983) and two neighboring Y cells in the cat LGN (based on Weng et al., 2005). Bottom left: the receptive fields of three Y cells in the retina are represented as Gaussian curves in three different colors. Bottom right: the combination of the three retinal inputs yields LGN receptive fields that can sample stimuli at a finer spatial resolution than in the retina. The bar graphs on the top illustrate the relative weights of the retinal inputs that were used to generate the LGN Gaussians.

the center is an exact copy of the green retinal Gaussian; it represents a geniculate cell that receives only one retinal input. The LGN Gaussians on the sides are obtained from a weighted sum of the green retinal afferent (that contribute 52% of the total input) and the weaker red and blue afferents (that contribute 40% and 8%). The input percentages used in the cartoon are consistent with the synaptic weights estimated from counts of retinogeniculate synapses (Hamos et al., 1987) and retinogeniculate correlations measured in pair recordings from retinal and geniculate cells (Cleland et al., 1971a, b; Mastronarde, 1992; Usrey et al., 1999). The weaker the additional retinal inputs, the closer the receptive field positions within the LGN.

It is estimated that 8–50% of geniculate cells receives input from just one retinal afferent (Cleland et al., 1971a; Cleland and Lee, 1985; Hamos et al., 1987; Mastronarde, 1992). These one-input geniculate cells could be the carriers of a nearly exact copy of the retinal receptive field array (position, size, and response time-course) to the cortex. The rest of the geniculate cells are dominated by one retinal input, but they also receive input from additional afferents (Cleland et al., 1971a, b; Hamos et al., 1987; Mastronarde, 1992; Usrey et al., 1999). These multiple-input geniculate cells could carry spatiotemporal interpolations that are heavily based on the receptive field of each dominant afferent. Notice that although the cartoon (Fig. 6) shows retinal and geniculate Gaussians with identical widths, the geniculate Gaussians should be narrower (Cleland et al., 1971a; Cleland and Lee, 1985) because center–surround interactions are stronger in the LGN than in the retina (Hubel and Wiesel, 1961; Singer and Creutzfeldt, 1970; Levick et al., 1972; Singer et al., 1972; Usrey et al., 1999). This increase in surround strength in the LGN could be important to reduce the overlap among the geniculate Gaussians shown in Fig. 6.

Multiplexing retinal inputs could increase the range of receptive field positions in the LGN, and also the sizes and response latencies. A continuous representation of response latencies in the LGN could be obtained by a weighted sum of the impulse responses from the retinal afferents equivalent to the one illustrated in Fig. 6 for visual space.

Impulse responses are slower at the borders than at the middle of the retinal receptive field center. Therefore, the combined inputs from dominant afferents (retinal center superimposed with geniculate center) and weak afferents (retinal border superimposed with geniculate center) could provide the basis to generate a continuous range of response latencies within the LGN (see Mastronarde (1992) for good examples of the receptive field relation of a geniculate cell with their multiple retinal inputs).

The idea of using interpolation to improve spatial acuity has been proposed decades ago (Barlow, 1979; Crick et al., 1981; Fahle and Poggio, 1981) and is usually associated with some type of cortical computation (however, see Barlow, 1979). However, the properties of retinogeniculate divergence (Sur and Sherman, 1982; Hamos et al., 1987; Sur et al., 1987; Alonso et al., 1996; Yeh et al., 2003) strongly suggest that spatiotemporal interpolation could already be taking place at the level of the LGN, at least in the cat. In that sense, the LGN could serve an important function: to multiplex the receptive field array of retinal ganglion cells and create, by interpolation, a diverse representation of space and time that can be used by the cortex to process visual stimuli more precisely.

Acknowledgments

The research was supported by the NIH (EY 05253) and The Research Foundation at the University of Connecticut and the State University of New York.

References

Alonso, J.M., Usrey, W.M. and Reid, R.C. (1996) Precisely correlated firing in cells of the lateral geniculate nucleus. Nature, 383: 815–819.

Barlow, H.B. (1979) Reconstructing the visual image in space and time. Nature, 279: 189–190.

Bishop, P.O., Burke, W. and Davis, R. (1962) The interpretation of the extracellular response of single lateral geniculate cells. J. Physiol. (Paris), 162: 451–472.

Chen, C. and Regehr, W.G. (2000) Developmental remodeling of the retinogeniculate synapse. Neuron, 28: 955–966.

Cleland, B.G., Dubin, M.W. and Levick, W.R. (1971a) Simultaneous recording of input and output of lateral geniculate neurons. Nat. New Biol., 231: 191–192.

Cleland, B.G., Dubin, M.W. and Levick, W.R. (1971b) Sustained and transient neurons in the cat's retina and lateral geniculate nucleus. J. Physiol., 217: 473–496.

Cleland, B.G. and Lee, B.B. (1985) A comparison of visual responses of cat lateral geniculate nucleus neurons with those of ganglion cells afferent to them. J. Physiol., 369: 249–268.

Connolly, M. and Van Essen, D. (1984) The representation of the visual field in parvicellular and magnocellular layers of the lateral geniculate nucleus in the macaque monkey. J. Comp. Neurol., 226: 544–564.

Crick, F.H.C., Marr, D.C. and Poggio, T. (1981) An information-processing approach to understanding the visual cortex. In: Francis, O.S., Frederic, W., George, A. and Stephen, G.D. (Eds.), The Organization of the Cerebral Cortex. MIT Press, Cambridge, MA.

Enroth-Cugell, C. and Robson, J.G. (1966) The contrast sensitivity of retinal ganglion cells of the cat. J. Physiol. (London), 187: 517–552.

Fahle, M. and Poggio, T. (1981) Visual hyperacuity:spatiotemporal interpolation in human vision. Proc. R. Soc. Lond. B Biol. Sci., 213: 451–477.

Frascella, J. and Lehmkuhle, S. (1984) A comparison between Y-cells in A-laminae and lamina C of cat dorsal lateral geniculate nucleus. J. Neurophysiol., 52: 911–920.

Freund, T.F., Martin, K.A., Somogyi, P. and Whitteridge, D. (1985) Innervation of cat visual areas 17 and 18 by physiologically identified X- and Y- type thalamic afferents, II. Identification of postsynaptic targets by GABA immunocytochemistry and Golgi impregnation. J. Comp. Neurol., 242: 275–291.

Friedlander, M.J., Lin, C.S. and Sherman, S.M. (1979) Structure of physiologically identified X and Y cells in the cat's lateral geniculate nucleus. Science, 204: 1114–1117.

Friedlander, M.J., Lin, C.S., Stanford, L.R. and Sherman, S.M. (1981) Morphology of functionally identified neurons in lateral geniculate nucleus of the cat. J. Neurophysiol., 46: 80–129.

Garraghty, P.E., Salinger, W.L. and Macavoy, M.G. (1985) The development of cell size in the dorsal lateral geniculate nucleus of monocularly paralyzed cats. Brain Res., 353: 99–106.

Hamos, J.E., Van Horn, S.C., Raczkowski, D. and Sherman, S.M. (1987) Synaptic circuits involving an individual retinogeniculate axon in the cat. J. Comp. Neurol., 259: 165–192.

Hubel, D.H. and Wiesel, T.N. (1961) Integrative action in the cat's lateral geniculate body. J. Physiol., 155: 385–398.

Hubel, D.H. and Wiesel, T.N. (2005) Brain and Visual Perception. Oxford University Press, New York.

Illing, R.B. and Wassle, H. (1981) The retinal projection to the thalamus in the cat: a quantitative investigation and a comparison with the retinotectal pathway. J. Comp. Neurol., 202: 265–285.

Lee, D., Lee, C. and Malpeli, J.G. (1992) Acuity-sensitivity trade-offs of X and Y cells in the cat lateral geniculate complex: role of the medial interlaminar nucleus in scotopic vision. J. Neurophysiol., 68: 1235–1247.

LeVay, S. and Ferster, D. (1979) Proportion of interneurons in the cat's lateral geniculate nucleus. Brain Res., 164: 304–308.

Levick, W.R., Cleland, B.G. and Dubin, M.W. (1972) Lateral geniculate neurons of cat: retinal inputs and physiology. Invest. Ophthalmol., 11: 302–311.

Masland, R.H. (2001) The fundamental plan of the retina. Nat. Neurosci., 4: 877–886.

Mastronarde, D.N. (1983) Correlated firing of cat retinal ganglion cells, I. Spontaneously active inputs to X- and Y-cells. J. Neurophysiol., 49: 303–324.

Mastronarde, D.N. (1992) Nonlagged relay cells and interneurons in the cat lateral geniculate nucleus: receptive-field properties and retinal inputs. Vis. Neurosci., 8: 407–441.

Meister, M., Lagnado, L. and Baylor, D.A. (1995) Concerted signaling by retinal ganglion cells. Science, 270: 1207–1210.

Peichl, L. and Wassle, H. (1979) Size, scatter and coverage of ganglion cell receptive field centres in the cat retina. J. Physiol, 291: 117–141.

Peters, A. and Payne, B.R. (1993) Numerical relationships between geniculocortical afferents and pyramidal cell modules in cat primary visual cortex. Cereb. Cortex, 3: 69–78.

Reid, R.C., Victor, J.D. and Shapley, R.M. (1997) The use of m-sequences in the analysis of visual neurons: linear receptive field properties. Vis. Neurosci., 14: 1015–1027.

Sanderson, K.J. (1971) The projection of the visual field to the lateral geniculate and medial interlaminar nuclei in the cat. J. Comp. Neurol., 143: 101–108.

Singer, W. and Creutzfeldt, O.D. (1970) Reciprocal lateral inhibition of on- and off-center neurons in the lateral geniculate body of the cat. Exp. Brain Res., 10: 311–330.

Singer, W., Poppel, E. and Creutzfeldt, O. (1972) Inhibitory interaction in the cat's lateral geniculate nucleus. Exp. Brain Res., 14: 210–226.

Sur, M. (1988) Development and plasticity of retinal X and Y axon terminations in the cat's lateral geniculate nucleus. Brain Behav. Evol., 31: 243–251.

Sur, M., Esguerra, M., Garraghty, P.E., Kritzer, M.F. and Sherman, S.M. (1987) Morphology of physiologically identified retinogeniculate X- and Y-axons in the cat. J. Neurophysiol., 58: 1–32.

Sur, M. and Sherman, S.M. (1982) Retinogeniculate terminations in cats: morphological differences between X and Y cell axons. Science, 218: 389.

Sur, M., Weller, R.E. and Sherman, S.M. (1984) Development of X- and Y-cell retinogeniculate terminations in kittens. Nature, 310: 246–249.

Usrey, W.M., Reppas, J.B. and Reid, R.C. (1998) Paired-spike interactions and synaptic efficacy of retinal inputs to the thalamus. Nature, 395: 384–387.

Usrey, W.M., Reppas, J.B. and Reid, R.C. (1999) Specificity and strength of retinogeniculate connections. J. Neurophysiol., 82: 3527–3540.

Wassle, H., Boycott, B.B. and Illing, R.B. (1981a) Morphology and mosaic of on- and off-beta cells in the cat retina and some functional considerations. Proc. R. Soc. Lond. B Biol. Sci., 212: 177–195.

Wassle, H., Peichl, L. and Boycott, B.B. (1981b) Morphology and topography of on- and off-alpha cells in the cat retina. Proc. R. Soc. Lond. B Biol. Sci., 212: 157–175.

Weng, C., Yeh, C.I., Stoelzel, C.R. and Alonso, J.M. (2005) Receptive field size and response latency are correlated within the cat visual thalamus. J. Neurophysiol., 93: 3537–3547.

Yeh, C.I., Stoelzel, C.R. and Alonso, J.M. (2003) Two different types of Y cells in the cat lateral geniculate nucleus. J. Neurophysiol., 90: 1852–1864.

Martinez-Conde, Macknik, Martinez, Alonso & Tse (Eds.)
Progress in Brain Research, Vol. 154
ISSN 0079-6123

CHAPTER 2

Double-bouquet cells in the monkey and human cerebral cortex with special reference to areas 17 and 18

Javier DeFelipe[1,*], Inmaculada Ballesteros-Yáñez[1], Maria Carmen Inda[1,2] and Alberto Muñoz[1,2]

[1]*Cajal Institute (CSIC), Avenida Dr. Arce 37, 28002-Madrid, Spain*
[2]*Department of Cell Biology, Faculty of Biology, Universidad Complutense, Madrid, Spain*

Abstract: The detailed microanatomical study of the human cerebral cortex began in 1899 with the experiments of Santiago Ramón y Cajal, who applied the Golgi method to define the structure of the visual, motor, auditory and olfactory cortex. In the first article of this series, he described a special type of interneuron in the visual cortex capable of exerting its influence in the vertical dimension. These neurons are now more commonly referred to as double-bouquet cells (DBCs). The DBCs are readily distinguished owing to their characteristic axons that give rise to tightly interwoven bundles of long, vertically oriented axonal collaterals resembling a horsetail (DBC horsetail). Nevertheless, the most striking characteristic of these neurons is that they are so numerous and regularly distributed that the DBC horsetails form a microcolumnar structure. In addition, DBCs establish hundreds of inhibitory synapses within a very narrow column of cortical tissue. These features have generated considerable interest in DBCs over recent years, principally among those researchers interested in the analysis of cortical circuits. In the present chapter, we shall discuss the morphology, synaptic connections and neurochemical features of DBCs that have been defined through the study of these cells in different cortical areas and species. We will mainly consider the immunocytochemical studies of DBCs that have been carried out in the visual cortex (areas 17 and 18) of human and macaque monkey. We will see that there are important differences in the morphology, number and distribution of DBC horsetails between areas 17 and 18 in the primate. This suggests important differences in the microcolumnar organization between these areas, the functional significance of which awaits detailed correlative physiological and microanatomical studies.

Keywords: neocortex; interneurons; visual cortex; GABA; inhibition; circuits; calbindin; minicolumns

Introduction

The detailed study of the microanatomy of the primate visual cortex began with the studies of Santiago Ramón y Cajal in 1899. Using the Golgi method, he commenced a series of studies on the comparative structures of different functional regions of the human cerebral cortex (Cajal, 1899a–c, 1900, 1901; DeFelipe and Jones, 1988). The aim of Cajal and other authors at that time was to determine whether it was possible to explain functional specialization through structural specialization:

> [...] for example, if an organizational detail is exclusively found or is particularly exaggerated in the visual cortex, we will be justified in suspecting that it has

*Corresponding author. Tel.: +34-1-585-4735;
Fax: +34-1-585-4754; E-mail: defelipe@cajal.csic.es

DOI: 10.1016/S0079-6123(06)54002-6

15

something to do with [cerebral visual function]. Conversely, if an anatomical detail is repeated similarly in all cortical regions, we will be justified in assuming that it is devoid of a specific functional significance and instead is of a more general [significance]
(Cajal, 1899b).

Cajal examined the visual cortex layer by layer (Fig. 1), producing beautiful and accurate drawings to illustrate the neuronal components of each layer and their possible connections (Fig. 2). The first article of these series of studies was a preliminary report that appeared in the *Revista Ibero-Americana de Ciencias Médicas* (Cajal, 1899a). In this article he described two new types of aspiny interneurons: a giant cell with a horizontal axon, which according to Cajal would be similar to the special cells of layer I or the Cajal–Retzius cells; and a small fusiform bitufted cell with a characteristic axon that is formed of small bundles, comparable to locks of hair [that were] so long that they extend through almost the whole thickness of the gray matter. This bitufted cell type was considered by Cajal as a special type of interneuron capable of exerting its function in the vertical dimension:

> In some places, it can be seen that the small bundles of threads are applied to the [apical dendrites] and somata of a series of vertical pyramids, from which we think it very probable that the cells referred to are a special category of cells with a short axon, whose role would be to associate pyramids resident in different layers in the vertical direction
> (Cajal, 1899a; DeFelipe and Jones, 1988).

However, instead of giving a different name to this particular neuronal type, Cajal used the term bitufted cell (*células bipenachadas* in Spanish, *double-bouquet* cell in French) to describe neurons with different dendritic and axonal morphologies (Cajal, 1899a–c, 1900, 1901, 1909/1911; DeFelipe, 2002). After the studies of Cajal, these cells were virtually ignored until interest in the Golgi method resurged with the analysis of cortical microanatomy that was carried out in the 1960s and 1970s by

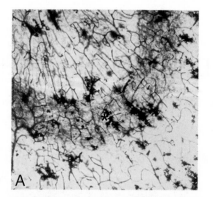

Fig. 1. Photomicrographs of Cajal's Golgi-stained preparations, labeled child, Gennari (A) and Gennari, Golgi, child (B), showing the dense plexus of Golgi-labeled elements (asterisk) in the middle layers of the visual cortex. Arrows in (B) indicate an oblique afferent fiber. Scale bar: 170 μm in (A); 60 μm in (B). From DeFelipe and Jones (1988).

a number of investigators (Sholl, 1956; Colonnier, 1966; Marin-Padilla, 1969; Szentágothai, 1969; Scheibel and Scheibel, 1970; Valverde, 1970; Lund, 1973; Jones, 1975). Because these cells did not have

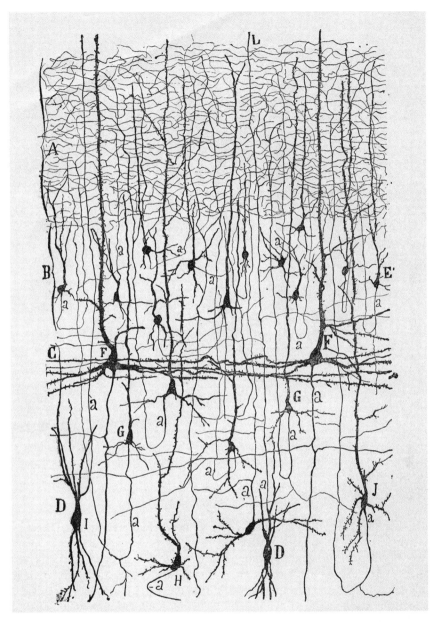

Fig. 2. One of the first drawings of Cajal (1899a) showing Golgi-impregnated elements in the human cerebral cortex. The legend says: The deep layers of the visual cortex of the cerebrum of a child of 27 days. A, layer of the stellate cells; B, layer of the granules; C, layer of the giant cells; D, layer of the polymorphic cells; E, granule cell with ascending axon; F, giant pyramid; G, small pyramid with ascending axon; H, J, other cells with similar axon; I, giant cell with axon destined to the molecular layer; a, axon. Copyright *Herederos de Santiago Ramón y Cajal*.

a particular name, they received different names, which generated some confusion in the literature. For example, Szentágothai referred to them as cells with horsetail-shaped axons, Jones as type 3 neurons, and Valverde as cells with axons forming vertical bundles (Szentágothai, 1973, 1975; Jones, 1975; Valverde, 1978, 1985), while along with other authors, we have preferred to apply the

18

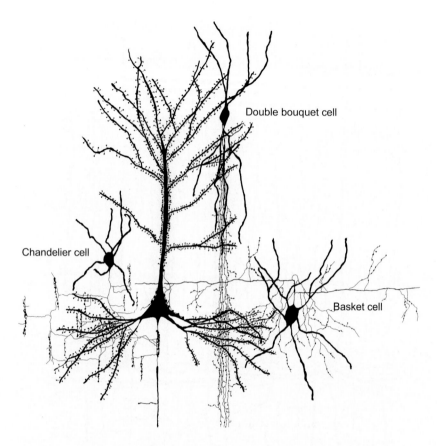

Double bouquet cell

Chandelier cell

Basket cell

Fig. 3. Drawing to illustrate the synaptic relationships between the DBCs, chandelier, and large basket cells with the pyramidal cells. Morphologically and chemically, these cells constitute the best-characterized types of aspiny interneurons. Inset is a schematic diagram to illustrate the synaptic connections between the three nonpyramidal cells and the pyramidal cell. Note that each type of nonpyramidal neuron innervates a different region of the pyramidal cell. From DeFelipe and Fariñas (1992).

French term double-bouquet cell (DBC) specifically to those neurons whose axons form such vertical bundles, irrespective of their somato-dendritic morphology (DeFelipe, 2002). The axonal arbors of DBCs are generally termed DBC axonal bundles or horsetails owing to their resemblance to a horse's tail. Together with basket cells and chandelier cells, these neurons are currently considered as the three main types of inhibitory GABAergic interneurons that innervate pyramidal cells in the neocortex (Fig. 3). In this chapter, we shall discuss features of DBCs that have been identified in the study of these cells in different cortical areas and species. In particular, we will focus on the morphology and distribution of DBCs in the visual cortex (areas 17 and 18) of the human and macaque monkey.

General characteristics of DBCs

Using the Golgi method, DBCs have been identified in layers II and III of different areas of the cat, monkey, and human cortex (reviewed in Peters and Regidor, 1981; Fairén et al., 1984; DeFelipe, 2002). At present, DBCs are defined as interneurons with a somato-dendritic morphology that is either multipolar or bitufted, and whose axons give rise to tightly interwoven bundles of long vertically orientated axonal collaterals resembling a horsetail (DBC horsetail) that descends from layer II or the upper half of layer III to layer V or layer VI. These collaterals bear numerous varicose dilations, short side branches, and club-like bouton appendages. However, because of the inconsistency of the Golgi method and the general difficulties in staining DBCs, little was known about their detailed distribution and neurochemical characteristics. The introduction of immunocytochemistry for the calcium-binding proteins calbindin, parvalbumin and calretinin (Celio et al., 1986; Celio, 1990, reviewed in Baimbridge et al., 1992; Andressen et al., 1993; DeFelipe, 1997) represented an important step in the study of the distribution and biochemical characteristics of interneurons. In particular, it was found that DBCs were among those neurons that were consistently immunolabeled for calbindin (CB) (DeFelipe et al., 1989; Hendry et al., 1989). The advantage of immunocytochemical staining over either the Golgi method or intracellular labeling is that instead of labeling only occasional DBCs, CB immunocytochemistry labeling is far more widespread and homogeneous (Figs. 4 and 5). In addition, CB immunocytochemistry permitted the density and distribution of DBC horsetails to be examined as well as facilitating the analysis of the synaptic connections of large populations of DBCs (see below). Furthermore, their neurochemical characteristics could also be defined by double-labeling immunocytochemical techniques (DeFelipe et al., 1989, 1990, 1999; Hendry et al.,

1989; Del Rio and DeFelipe, 1995, 1997; Peters and Sethares, 1997; Ballesteros-Yáñez et al., 2005).

Neurochemical characteristics of DBCs

Using double-labeling immunocytochemical techniques as well as correlative light and electron microscopy to examine the synaptic connections of neurochemically defined neurons, DBCs can be considered as interneurons containing GABA and CB, although subpopulations of DBCs have also been shown to express calretinin. The peptides somatostatin and tachykinin are also found in certain cortical areas and species (DeFelipe et al., 1999). In contrast, these cells are never labeled for other peptides or parvalbumin, nor do they appear to express markers found in nitric oxide- and tyrosine hydroxylase-expressing neurons (Somogyi et al., 1981; DeFelipe et al., 1989, 1990, 1999; De Lima and Morrison, 1989; DeFelipe and Jones, 1992; Del Rio and DeFelipe, 1995, 1997; DeFelipe, 1997; Benavides-Piccione and DeFelipe, 2003). Therefore, there are specific neurochemical subtypes of DBCs, even though there is little or no information available regarding the possible variation in the neurochemical characteristics of DBCs in different cortical areas and species.

Distribution of DBCs: microcolumnar structure

One of the unique features of DBCs is that while the vast majority of axons from other cortical neurons surpass the size limit for the minicolumn (vertical cylinder of tissue with a diameter of approximately 25–50 µm, basically defined by the space occupied by the small vertical aggregate of pyramidal cells), DBC horsetails fit well within these limits. However, the most striking characteristic of these neurons is that they are so numerous and regularly distributed that the DBC horsetails themselves form a microcolumnar structure (Fig. 4). This microcolumnar organization has been demonstrated in the visual, somatosensorial, auditory and temporal cortex of the macaque monkey as well as in the human prefrontal, motor, somatosensory, temporal and visual cortex (DeFelipe et al., 1990, 1999; Peters and Sethares, 1997; Del Rio and DeFelipe, 1997;

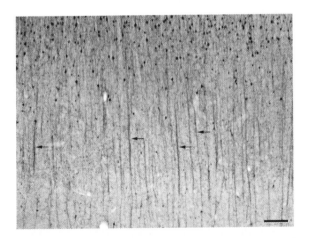

Fig. 4. Low-power photomicrograph of a CB-immunostained section from the human temporal cortex (area 22) showing the distribution of CB-immunostained cell bodies and DBC horsetails. Note the large number and the regular distribution of DBC horsetails. Arrows indicate some DBC horsetails. Scale bar: 140 µm.

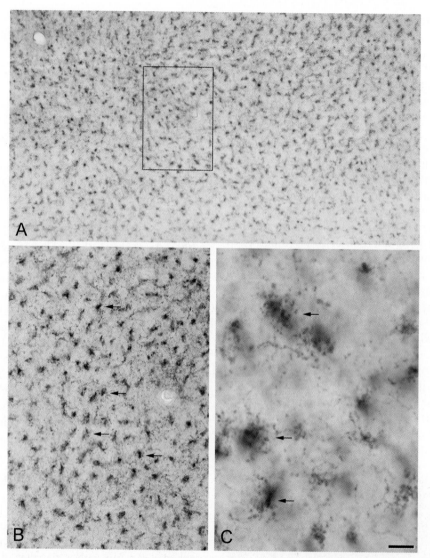

Fig. 5. (A) Low-power photomicrograph from a tangential section taken at the level of layer III of the human secondary visual cortex (area 18) immunostained for CB, illustrating the regular spacing of DBC horsetails. (B) Higher magnification of the area boxed in (A) in a different focal plane. (C) Higher magnification of (B). Arrows indicate some of the tangentially sectioned DBC horsetails. Scale bar: 100 μm for (A); 55 μm for (B) and 14 μm for (C).

Ballesteros-Yáñez et al., 2005). For example, it has been shown that in the macaque primary visual and somatosensory cortex, a mean of 10 DBC horsetails was found in a field of 10,000 μm^2 from tangential sections through layer III immunostained for CB (range from 7 to 15). The center-to-center spacing of these cells was 15–30 μm and the mean width of the cross-sectioned DBC axonal arborizations was

9 μm (from 5 to 15 μm). The number of axonal collaterals that made up each DBC axon varied depending on the layer examined, ranging from as few as 3 in the deepest part of the axons' course to as many as 10 or 15 in the upper part (DeFelipe et al., 1990; see also Fig. 5 in Ballesteros-Yáñez et al., 2005). A similar distribution has been found in the human cerebral cortex. For example, a mean

number of 12 DBC axons can be found in a field of 10,000 μm² in the temporal cortex, with a mean diameter of 12 μm (range 5–20 μm) and a mean center-to-center spacing of 30 μm (Del Rio and DeFelipe, 1995, 1997). The homogeneous distribution of CB-immunostained DBC horsetails in a tangential section of layer III from area 18 of the human visual cortex is illustrated in Fig. 5. In addition, two morphological types of DBC horsetails have been observed, which in the human cerebral cortex show the following morphometric values (Ballesteros-Yáñez et al., 2005): the complex type (type I: Fig. 6a) with a mean thickness of 8.98 ± 3.27 μm and 5.9 ± 1.6 axon collaterals; and the simple type (type II: Fig. 6b) with a mean axon thickness of 3.37 ± 1.12 μm and 3.9 ± 1.0 axon collaterals.

The relationship between DBC horsetails and bundles of myelinated axons

The distribution of DBC horsetails is remarkably similar to the distribution of bundles of myelinated fibers or radial fasciculi that originate from pyramidal cells and that form small vertical aggregates or minicolumns (Del Rio and DeFelipe, 1997; Peters and Sethares, 1997; for a review, see DeFelipe, 2005). Recently, we examined the relationship between bundles of myelinated fibers and DBC horsetails, using dual immunocytochemistry for the myelin basic protein and CB. In these double-labeled sections, it was clear that DBC horsetails were intermingled with bundles of myelinated axons in all the cortical areas examined (Fig. 7).

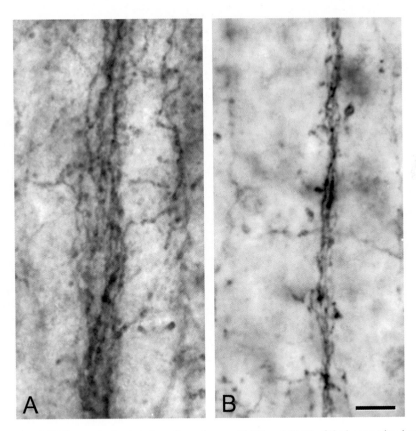

Fig. 6. Photomicrographs of CB-immunostained sections through areas 18 (A) and 17 (B) of the human visual cortex, illustrating the differences in thickness and number of axon collaterals between type I (A) and type II (B) DBC horsetails. In general, the axon arbor of DBCs is more complex in area 18 (type I) than in area 17 (type II, see Fig. 10). Scale bar: 11 μm.

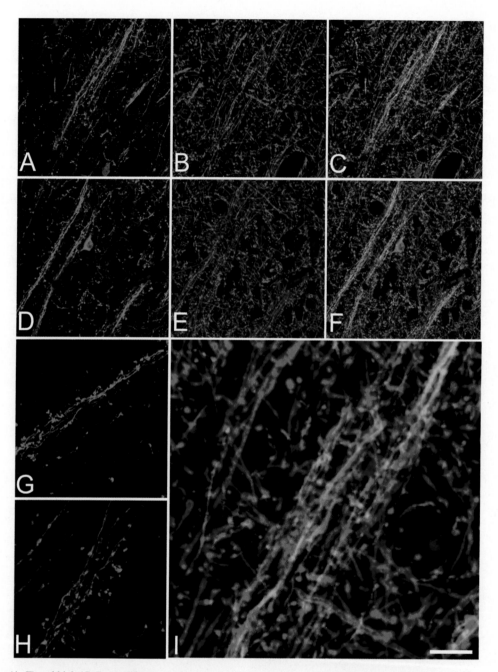

Fig. 7. Low (A–F) and high (G–I) magnifications of serial confocal images from the same microscopic field in a single section of layer III from area 18 (A–C, G and I) and area 17 (D–F and H) immunostained for CB (A, D, G and H; green) or for the basic myelin protein (B and E; red). (C) and (F) were obtained by combining images (A–B) and (D–E), respectively. (I) is a higher magnification of (C). Note the overlap of both type I (A–C) and type II (D–F) CB-immunostained DBC horsetails with myelinated axonal bundles. Images (A–C) and (D–F) were obtained from a stack of 1 optical image with 1.57 μm thickness. (G): A stack of five optical images separated by 0.51 μm in the z-axis; total: 4 μm. (H): A stack of eight optical images separated by 0.52 μm in the z-axis; total: 4 μm. Scale bar: 70 μm in (A–F); 50 μm in (G, H); 18 μm in (I). From Ballesteros-Yáñez et al. (2005).

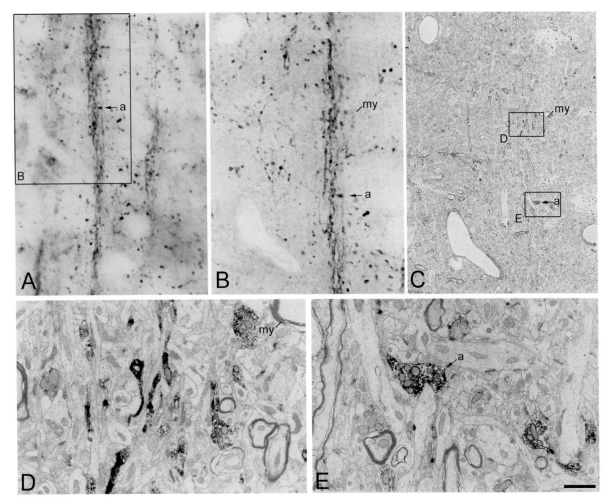

Fig. 8. Correlative light (A, B) and electron micrographs (C–E) of tachykinin-immunoreactive DBC horsetails in the monkey primary auditory cortex. (A): Photomicrograph of DBC horsetails embedded for electron microscopy. (B): A semithin plastic section of one of the tachykinin-ir DBC horsetails illustrated in (A) (boxed area), showing the same axon terminal (a). (C): Electron micrograph after sectioning the semithin section illustrated in (B), showing the same tachykinin-ir DBC horsetail. *a* and *my* indicate the same axon terminal and the myelinated axon shown in (B). (D, E): Higher power electron micrographs of the boxed areas indicated as (D, E) in panel (C), respectively. *a* is the same axon terminal as in panels (A–C). Scale bar: 14 μm for (A); 8 μm for (B); 10 μm for (C); 1.1 μm for (D); 1.2 μm for (E). From DeFelipe et al. (1990).

Indeed, disregarding some exceptions, there appears to be one DBC horsetail per minicolumn (see the section on Distribution of DBC horsetails in areas 17 and 18).

Synaptic connections of DBCs

In several areas of the monkey cerebral cortex (visual, somatosensory, temporal and auditory), it has been shown that DBC axons form symmetrical synapses with small dendritic shafts (57–62%) and spines (38–43%) (Somogyi and Cowey, 1981; DeFelipe et al., 1989, 1990; De Lima and Morrison, 1989) (Figs. 8 and 9). Furthermore, it is relatively frequent that the DBC axon terminals establish synapses with two or more postsynaptic elements (multiple synapses). For example, the proportion of multiple synapses formed by DBC horsetails has been estimated to be approximately 16% in the

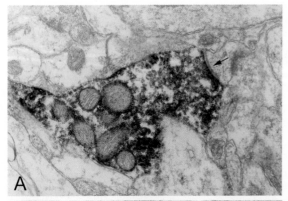

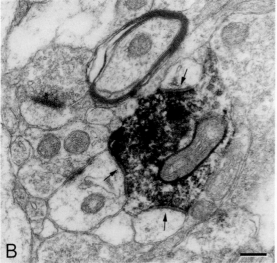

Fig. 9. (A): Higher power electron micrograph of the axon terminal *a* (illustrated in Fig. 8) establishing a symmetrical synapse (arrow) with a dendritic spine. (B): An axon terminal from the tachykinin-ir DBC horsetail illustrated in Fig. 8, which forms symmetrical synapses (arrows) with three different postsynaptic elements (probably dendritic spines). Scale bar: 0.27 μm for (A) and 0.25 μm for (B). From DeFelipe et al. (1990).

primary auditory and somatosensory cortex of the macaque monkey (DeFelipe et al., 1989, 1990; Fig. 9B). Moreover, DBC axons in the human cerebral cortex are virtually identical and establish a microcolumnar structure similar to that found in the monkey. Indeed, with regard to their morphology, distribution and the proportion of synapses, DBC axons establish symmetrical synapses with small dendritic shafts (59%) and spines (41%) in the human temporal cortex (Del Rio and DeFelipe,

1995). Thus, DBCs are likely to participate in similar synaptic circuits in both monkeys and humans (Table 1). The origin of the dendritic shafts that are postsynaptic to DBC axons is unknown. However, the postsynaptic spines belong to pyramidal cells and possibly to spiny stellate cells, which are the two types of spiny neurons found in the cerebral cortex.

Origin of the postsynaptic dendritic shafts of DBCs
In principle, we can assume that the dendritic shafts of both pyramidal cells (excluding their apical dendrites) and interneurons are postsynaptic to DBCs. However, the dendritic shafts of those interneurons with vertically oriented dendritic arbors can be mostly disregarded. Owing to the very narrow extension of the DBC horsetail arbor, the cell body of a presumptive postsynaptic interneuron with vertical dendrites should lie within the axonal arborization of the DBC horsetails, a circumstance that as far as we know has yet to be observed. Thus, the postsynaptic dendritic shafts arise from dendrites crossing the DBC horsetail axonal arbor. Collateral dendritic branches of apical and basal pyramidal cells and the dendrites of some multipolar interneurons, like large basket cells, can run for several hundred micrometers in the monkey and human neocortex. These observations suggest that the synapses of a given DBC horsetail are not restricted to the dendrites of the neurons in the adjacent minicolumn, but rather that they may also establish synapses with dendrites belonging to other surrounding minicolumns, which cross the trajectory of the DBC horsetail.

Origin of the postsynaptic dendritic spines of DBCs
The apical dendrites of pyramidal cells do not lie within the axonal arborization of the DBC horsetails. Hence, it was proposed that the postsynaptic dendritic spines of DBCs arise from collateral branches of apical and basal pyramidal cell dendrites as well as from spiny stellate cells that the DBC horsetails may encounter in their trajectory through the mid-cortical layers (DeFelipe et al., 1989, 1990; Del Rio and DeFelipe, 1995). Furthermore, pyramidal neurons are far more numerous than spiny

Table 1. Synaptic connections of DBCs in the monkey and human neocortex

Staining	Species	Cortical region and references	Number of cells[a]	Number of axon terminals forming synapses	Postsynaptic elements[b]
Golgi	Monkey	Visual (area 17) Somogyi and Cowey (1981)	1	35	60% shafts, 40% spines
CB	Monkey	Somatic sensory (areas 3a, 1) DeFelipe et al. (1989)	7	237	62% shafts, 38% spines
SOM	Monkey	Superior temporal gyrus De Lima and Morrison (1989)	Not specified	64	63% shafts, 37% spines
SOM	Monkey	Inferior temporal gyrus De Lima and Morrison (1989)	Not specified	36	61% shafts, 39% spines
TK	Monkey	Primary auditory DeFelipe et al. (1990)	9	277	57% shafts, 43% spines
CB	Monkey	Primary visual Peters and Sethares (1997)	Not specified	125	68% shafts, 28% spines, 4% somata
CB	Human	Middle temporal gyrus Del Rio and DeFelipe (1995)	2	66	59% shafts, 41% spines

[a]With reference to individual DBC horsetails. [b]The vast majority of dendritic shafts are of small caliber (1–2 μm in diameter). Only 3% (DeFelipe et al., 1989) or 8% (Peters and Sethares, 1997) are apical dendritic shafts. The percentage of postsynaptic elements varies from bundle to bundle with a range of 37–45% of synapses on spines and 63–55% on shafts (DeFelipe et al., 1990).

stellate cells and, thus, the vast majority of dendritic spines clearly arise from pyramidal neurons. As a result, dendritic spines of pyramidal neurons are one of the major targets of DBC. In addition, these spines establish additional asymmetrical synapses with excitatory axons (DeFelipe et al., 1989; Del Rio and DeFelipe, 1997). Because DBCs are very abundant and establish hundreds of inhibitory synapses within a very narrow column of cortical tissue, it is likely that many spines will form symmetrical synapses. This contrasts with the classic view that the majority of dendritic spines (80–90%) only form one asymmetrical excitatory synapse. Indeed, when spines establish synapses with two separate axon terminals, it is rare that this second synapse is symmetrical (reviewed in DeFelipe and Fariñas, 1992). However, pyramidal cells display thousands of spines and they are much more numerous than DBCs. Therefore, it is conceivable that in a random examination of dendritic spines, those spines that are not innervated by these neurons would be principally included. All of these observations led us to propose that DBCs form synapses with a special type of spine capable of forming synapses with both excitatory and inhibitory axon terminals. Furthermore, these must be particularly abundant in the side branches of apical and basal dendrites of pyramidal cells (Del Rio and DeFelipe, 1995), at least in the primate cerebral cortex. Indeed, each DBC horsetail can form several hundreds of synapses within a narrow column of tissue and, thus, they are considered to be a key element in the microcolumnar organization of the cerebral cortex acting on groups of pyramidal cells located in different layers in the minicolumns (DeFelipe et al., 1989, 1990, 1999; Favorov and Kelly, 1994a, b; Del Rio and DeFelipe, 1995; DeFelipe, 1997, 2002, 2005; Jones, 2000; Ballesteros-Yáñez et al., 2005).

DBC horsetails in areas 17 and 18 of the macaque monkey and human

In general, the density and morphology of DBCs vary in different areas of the macaque and human neocortex (DeFelipe et al., 1999; Ballesteros-Yáñez et al., 2005). What follows is the description of DBC horsetails in areas 17 and 18 of both the macaque and human (Table 2).

Table 2. Summary of the morphological characteristics, the density of DBC horsetails and the density of CB-ir neurons in areas 17 and 18 of the human visual cortex

Parameter	Area 18	Area 17
Axon thickness[a]	10.2 ± 2.7 ($n = 51$)	3.3 ± 1.1 ($n = 59$)**
Number of axon collaterals[b]	5.9 ± 1.7 ($n = 30$)	3.9 ± 1.0 ($n = 30$)*
Density of DBC horsetails[c]	13.7 ± 2.9	14.4 ± 2.0
Density of CB-ir somata[d]	74.3 ± 13.7	46.5 ± 9.3**

Source: Data from Ballesteros-Yáñez et al. (2005).
[a]Thickness of axon arborization of DBC horsetails in micrometer (mean $\pm$ S.D.). [b]Number of axon collaterals of the axonal arborization of DBCs (mean $\pm$ S.D.). [c]Number of axons that cross a line of 500 μm in layer III parallel to the pial surface (mean $\pm$ S.D.). [d]Number of CB-ir somata in 200,000 μm^2 in layers II–III (mean $\pm$ S.D.) **$p = 0.0001$; *$p = 0.012$.

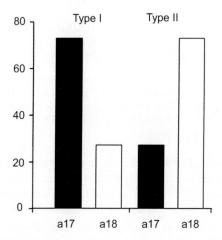

Fig. 10. Percentages of type I and type II DBC horsetails in areas 17 and 18 of the human visual cortex. From Ballesteros-Yáñez et al. (2005).

Morphology of DBC horsetails in areas 17 and 18

There are differences in the morphology of DBC horsetails in area 17 when compared to area 18. First, both complex (type I) and simple (type II) DBC horsetails are found in areas 17 and 18, although the most common type found in area 18 is type I, while in area 17, type II axons are the most abundant (Fig. 10).

Furthermore, there are more short side branches and club-like bouton appendages of the axonal collaterals in area 18 than in area 17, giving DBC horsetails a spiny appearance in area 18. In addition, DBC horsetails can be classified into another two morphological types: short and long. The short

type is the most frequent in area 17 where they run from layer II to layer IVB, while the long type runs from layer III to layers V–VI and are the most frequent type found in area 18 (Fig. 11). Assuming that the percentage of multiple synapses made by the single axon terminals in the DBC horsetails (see the section on Synaptic connections of DBCs) is similar in both the complex and simple type of DBC horsetails, the greater complexity of DBC horsetails in area 18 suggests that each single DBC horsetail establishes more synapses. This regional specialization of DBCs is likely to have an important impact on the connectivity of minicolumns.

DBC horsetail density in areas 17 and 18

The density of DBC horsetails has been estimated in coronal sections of the human cerebral cortex immunostained for CB. This was achieved by counting the number of DBC horsetails crossing a 500 μm long line in the middle of layer III, parallel to the pial surface (Ballesteros-Yáñez et al., 2005). We found that the density of CB-immunoreactive (-ir) DBC horsetails was similar in areas 18 and 17 (13.6 ± 2.9 and $14.4 \pm 2/500$ μm, respectively). We also examined the possible correlation between the density of DBC horsetails and the density of CB-ir somata in layers II–IIIA. The density of CB-ir somata in layers II–IIIA was significantly lower in area 17 (46.5 ± 9.3 somata per 200,000 μm^2) than in area 18 (74.35 ± 13.6). Therefore, the additional CB-ir neurons in area 18 probably represent other cell types.

27

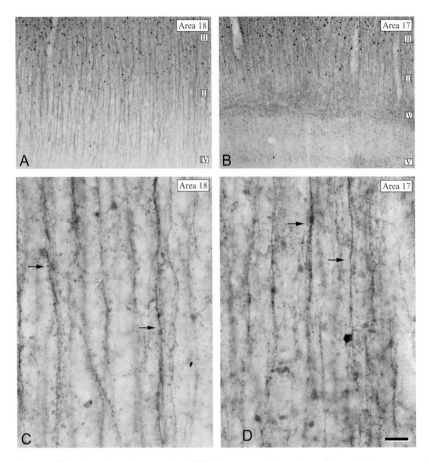

Fig. 11. (A, B) are low-magnification photomicrographs of CB-immunostained sections from the human secondary (area 18) and primary (area 17) visual areas. These show the distribution of CB-ir cell bodies as well as the large number and regular distribution of DBC horsetails in both areas. Note that DBC horsetails are shorter in area 17 than in area 18. (C) and (D) are higher magnifications of (A) and (B), respectively. Arrows indicate some DBC horsetails. Scale bar: 150 µm in (A, B) and 32 µm in (C, D).

Distribution of DBC horsetails in areas 17 and 18

In addition to the aforementioned differences in morphology (Table 2), the most remarkable characteristic that distinguishes DBC horsetails in areas 17 and 18 is related to their distribution. In both areas, CB-ir DBC horsetails are regularly distributed and numerous (Fig. 11). However, while practically the entire extent of area 18 is populated by DBC horsetails, in area 17, DBC horsetails are not present throughout the whole area, but rather are located in groups that occupy cortical segments of a few hundred to several thousand micrometers in width. This differential distribution of DBCs is most dramatic at the border between these two areas, where there were very few or no CB-ir DBC horsetails in area 17, but there were many in area 18 (Figs. 12 and 13; see also Fig. 1 in DeFelipe et al., 1999). A similar uneven distribution of CB-immunostaining in area 17 has also been described in layer III of the macaque monkey, where a higher density of CB-ir elements were found around cytochrome-oxidase-rich puffs in layer III (Celio, 1986; Van Brederode et al., 1990; Hendry and Carder, 1993; Blümcke et al., 1994; Carder et al., 1996). However, when we compared CB and cytochrome-oxidase immunoreactivity in serial sections from the macaque, the distribution of DBC horsetails was not related to the pattern of cytochrome-oxidase staining in layers II and III (unpublished observations; see also Peters and Sethares, 1997).

28

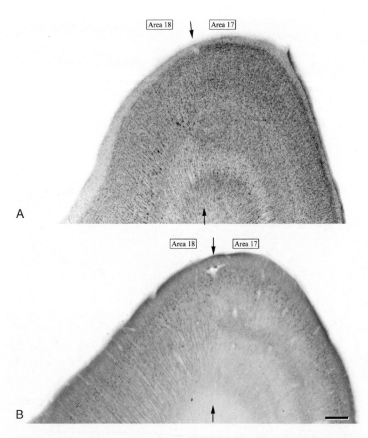

Fig. 12. (A, B): Photomicrographs of two adjacent sections through the 17/18 border (indicated by arrows), one stained with thionin (A) and the other processed for CB-immunocytochemistry to illustrate the differences in the pattern of CB-immunostaining between areas 17 and 18. Scale bar: 270 μm.

Are all pyramidal cells of the minicolumns innervated by DBCs?

In sections stained for both the myelin basic protein and CB, it is clear that DBC horsetails are intermingled with bundles of myelinated axons in all the cortical areas of the human cerebral cortex examined (Ballesteros-Yáñez et al., 2005). As such, it generally appears that there is one DBC horsetail per minicolumn. Whether all pyramidal neurons of the minicolumn or just a fraction of them are innervated by the corresponding DBC horsetail is unknown. However, not all minicolumns are associated with DBC horsetails. For example, adjacent to the numerous consecutive minicolumns that are typically associated with DBC horsetails, one, two or more consecutive minicolumns may not be associated with DBC horsetails. This dissociation of DBC horsetails and minicolumns is the most

evident at the border between areas 17 and 18 where as described above, relatively few DBC horsetails are observed. Of course, the absence of immunostaining may indicate the lack of expression and not the absence of a given type of neuron. Therefore, the lack of labeling of DBC horsetails in certain regions or the lack of association of some minicolumns with DBC horsetails can be interpreted in three ways:

- It is possible that DBCs are present in the whole neocortex and that each minicolumn is associated with a DBC horsetail. However, these neurons may be chemically heterogeneous giving rise to an uneven staining. Nevertheless, it should be emphasized that the expression or lack of a given peptide or calcium-binding protein in a particular type of

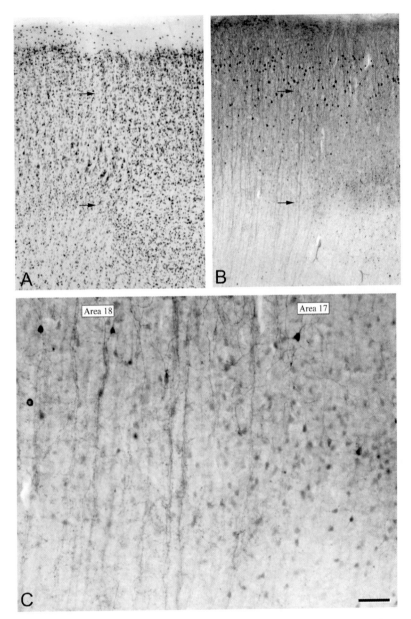

Fig. 13. (A, B): High-power photomicrographs of Figs. 12A and B, respectively, showing numerous CB-immunostained DBC horse-tails in area 18, but not in area 17 at the 17/18 border (indicated by arrows). (C): Higher magnification of (B). Scale bar: 185 μm for (A, B), and 55 μm for (C).

neuron has important functional consequences (e.g., Kawaguchi and Kubota, 1997; Gupta et al., 2000; Galarreta and Hestrin, 2002; Blatow et al., 2003; Monyer and Markram, 2004; Toledo-Rodriguez et al., 2004). Therefore, the variation in the expression found between DBC horsetails in different cortical areas or species probably represents significant differences in their cortical circuits.

• Collateral branches of pyramidal cell and apical and basal dendrites can run for several hundred micrometers across the neocortex in a horizontal or oblique direction. Thus, in spite of its very narrow vertical extension, a given DBC horsetail may innervate the dendrites of pyramidal cells located in distant minicolumns that are not associated with adjacent DBC horsetails. In this way, some pyramidal neurons could be contacted by more proximal DBCs than others.

- It is also possible that DBCs are not present within certain cortical regions and thus, certain minicolumns are not postsynaptic to DBCs, particularly in area 17 adjacent to area 18. In addition, DBC horsetails have been found in a variety of nonprimate species, but not in rodents (mouse, rat), lagomorphs (rabbit), or artiodactyls (goat), where there are also CB-positive or other neurochemically identified vertically oriented axons. However, in these species, such cells did not present an axonal arbor typical of DBCs or horsetails (Ballesteros-Yáñez et al., 2005). Furthermore, DBC horsetails have been identified in the frontal, parietal, and occipital regions of carnivores (cat, cheetah, lion and dog), although they are mainly restricted to the occipital cortex, and they rarely exist in the parietal and frontal cortical regions studied (Ballesteros-Yáñez et al., 2005). We also found that there are much fewer of DBC horsetails in carnivores than in humans and monkeys. Thus, DBCs do not appear to be a fundamental or basic type of neocortical neuron, but rather a species-specific neuronal type. Indeed, this cell type is particularly prevalent in the primate neocortex and its presence might reflect species-specific programs of GABAergic neurogenesis (see DeFelipe, 2002, 2005) and references therein). This phenomenon raises the question of whether the postsynaptic sites in the minicolumns without DBC horsetails are covered by other types of inhibitory interneurons that supplant the function of the DBCs. Alternatively, these postsynaptic sites may remain vacant.

In conclusion, the differences in the morphology, number, and distribution of CB-positive DBC horsetails in areas 17 and 18 of the primate suggest important differences in the microcolumnar organization between these areas. The significance of these differences in terms of the physiological parameters of individual pyramidal cells and in information processing within the minicolumns of areas 17 and 18 must await detailed correlative physiological, neurochemical and microanatomical studies.

Abbreviations

DBCs	double-bouquet cells
DBC horsetails	double-bouquet cell horsetails
CB	calbindin
-ir	immunoreactive
SOM	somatostatin
TK	tachykinin

Acknowledgments

This work was supported by the Spanish Ministry of Education and Science (Grant nos. BFI2003-02745 and BFI2003-01018 and a research fellowship to Inmaculada Ballesteros-Yáñez, AP 2001-0671) and the *Comunidad de Madrid* (Grant no. 08.5/0027/2001).

References

Andressen, C., Blümcke, I. and Celio, M.R. (1993) Calcium-binding proteins: selective markers of nerve cells. Cell Tissue Res., 271(2): 181–208.

Baimbridge, K.G., Celio, M.R. and Rogers, J.H. (1992) Calcium-binding proteins in the nervous system. Trends Neurosci., 15(8): 303–308.

Ballesteros-Yáñez, I., Munoz, A., Contreras, J., Gonzalez, J., Rodriguez-Veiga, E. and DeFelipe, J. (2005) Double bouquet cell in the human cerebral cortex and a comparison with other mammals. J. Comp. Neurol., 486(4): 344–360.

Benavides-Piccione, R. and DeFelipe, J. (2003) Different populations of tyrosine-hydroxylase-immunoreactive neurons defined by differential expression of nitric oxide synthase in the human temporal cortex. Cereb. Cortex, 13: 297–307.

Blatow, M., Rozov, A., Katona, I., Hormuzdi, S.G., Meyer, A.H., Whittington, M.A., Caputi, A. and Monyer, H. (2003) A novel network of multipolar bursting interneurons generates theta frequency oscillations in neocortex. Neuron, 38(5): 805–817.

Blümcke, I., Weruaga, E., Kasas, S., Hendrickson, A.E. and Celio, M.R. (1994) Discrete reduction patterns of parvalbumin and calbindin D-28k immunoreactivity in the dorsal lateral geniculate nucleus and the striate cortex of adult macaque monkeys after monocular enucleation. Vis. Neurosci., 11(1): 1–11.

Cajal, S.R. (1899a) Apuntes para el estudio estructural de la corteza visual del cerebro humano. Rev. Ibero-Americana Cienc. Méd., 1: 1–14.

Cajal, S.R. (1899b) Estudios sobre la corteza cerebral humana I: corteza visual. Rev. Trim. Micrográf. Madrid, 4: 1–63.

Cajal, S.R. (1899c) Estudios sobre la corteza cerebral humana II: estructura de la corteza motriz del hombre y mamíferos superiores. Rev. Trim. Micrográf. Madrid, 4: 117–200.

Cajal, S.R. (1900) Estudios sobre la corteza cerebral humana III: estructura de la corteza acústica. Rev. Trim. Micrográf. Madrid, 5: 129–183.

Cajal, S.R. (1901) Estudios sobre la corteza cerebral humana IV: estructura de la corteza cerebral olfativa del hombre y mamíferos. Trab. Lab. Invest. Biol. Univ. Madrid, 1: 1–140.

Cajal, S.R. (1909,1911) Histologie du système nerveux de l'homme et des vertébrés. Paris, Maloine.

Carder, R.K., Leclerc, S.S. and Hendry, S.H. (1996) Regulation of calcium-binding protein immunoreactivity in GABA neurons of macaque primary visual cortex. Cereb. Cortex, 6(2): 271–287.

Celio, M.R. (1986) Parvalbumin in most gamma-aminobutyric acid-containing neurons of the rat cerebral cortex. Science, 231(4741): 995–997.

Celio, M.R. (1990) Calbindin D-28k and parvalbumin in the rat nervous system. Neuroscience, 35(2): 375–475.

Celio, M.R., Scharer, L., Morrison, J.H., Norman, A.W. and Bloom, F.E. (1986) Calbindin immunoreactivity alternates with cytochrome c-oxidase-rich zones in some layers of the primate visual cortex. Nature, 323(6090): 715–717.

Colonnier, M. (1966) The structural design of the neocortex. In: Eccles, J.C. (Ed.), Brain and Conscious Experience. Berlin, Springer, pp. 1–23.

DeFelipe, J. (1997) Types of neurons, synaptic connections and chemical characteristics of cells immunoreactive for calbindin-D28K, parvalbumin and calretinin in the neocortex. J. Chem. Neuroanat., 14(1): 1–19.

DeFelipe, J. (2002) Cortical interneurons: from Cajal to 2001. Prog. Brain Res., 136: 215–238.

DeFelipe, J. (2005) Reflections on the structure of the cortical minicolumn. In: Casanova, M.F. (Ed.), Neocortical Modularity and the Cell Minicolumn. New York, Nova Science Publishers, pp. 57–91.

DeFelipe, J. and Fariñas, I. (1992) The pyramidal neuron of the cerebral cortex: morphological and chemical characteristics of the synaptic inputs. Prog. Neurobiol., 39(6): 563–607.

DeFelipe, J., Gonzalez-Albo, M.C., Del Rio, M.R. and Elston, G.N. (1999) Distribution and patterns of connectivity of interneurons containing calbindin, calretinin, and parvalbumin in visual areas of the occipital and temporal lobes of the macaque monkey. J. Comp. Neurol., 412(3): 515–526.

DeFelipe, J., Hendry, S.H., Hashikawa, T., Molinari, M. and Jones, E.G. (1990) A microcolumnar structure of monkey cerebral cortex revealed by immunocytochemical studies of double bouquet cell axons. Neuroscience, 37(3): 655–673.

DeFelipe, J., Hendry, S.H. and Jones, E.G. (1989) Synapses of double bouquet cells in monkey cerebral cortex visualized by calbindin immunoreactivity. Brain Res., 503(1): 49–54.

DeFelipe, J. and Jones, E.G. (1988) Cajal on the Cerebral Cortex. New York, Oxford University Press.

DeFelipe, J. and Jones, E.G. (1992) High-resolution light and electron microscopic immunocytochemistry of colocalized GABA and calbindin D-28k in somata and double bouquet cell axons of monkey somatosensory cortex. Eur. J. Neurosci., 4(1): 46–60.

De Lima, A.D. and Morrison, J.H. (1989) Ultrastructural analysis of somatostatin-immunoreactive neurons and synapses in the temporal and occipital cortex of the macaque monkey. J. Comp. Neurol., 283(2): 212–227.

Del Rio, M.R. and DeFelipe, J. (1995) A light and electron microscopic study of calbindin D-28k immunoreactive double bouquet cells in the human temporal cortex. Brain Res., 690(1): 133–140.

Del Rio, M.R. and DeFelipe, J. (1997) Double bouquet cell axons in the human temporal neocortex: relationship to bundles of myelinated axons and colocalization of calretinin and calbindin D-28k immunoreactivities. J. Chem. Neuroanat., 13(4): 243–251.

Fairén, A., DeFelipe, J. and Regidor, J. (1984) Nonpyramidal neurons. General account. Cerebral cortex. In: Peters, A. and Jones, E.G. (Eds.), Components of the Cerebral Cortex. New York, Plenum Publishing Corporation, pp. 201–253.

Favorov, O.V. and Kelly, D.G. (1994a) Minicolumnar organization within somatosensory cortical segregates: I. Development of afferent connections. Cereb. Cortex, 4(4): 408–427.

Favorov, O.V. and Kelly, D.G. (1994b) Minicolumnar organization within somatosensory cortical segregates: II. Emergent functional properties. Cereb. Cortex, 4(4): 428–442.

Galarreta, M. and Hestrin, S. (2002) Electrical and chemical synapses among parvalbumin fast-spiking GABAergic interneurons in adult mouse neocortex. Proc. Natl. Acad. Sci. USA, 99(19): 12438–12443.

Gupta, A., Wang, Y. and Markram, H. (2000) Organizing principles for a diversity of GABAergic interneurons and synapses in the neocortex. Science, 287(5451): 273–278.

Hendry, S.H. and Carder, R.K. (1993) Neurochemical compartmentation of monkey and human visual cortex: similarities and variations in calbindin immunoreactivity across species. Vis. Neurosci., 10(6): 1109–1120.

Hendry, S.H., Jones, E.G., Emson, P.C., Lawson, D.E., Heizmann, C.W. and Streit, P. (1989) Two classes of cortical GABA neurons defined by differential calcium binding protein immunoreactivities. Exp. Brain Res., 76(2): 467–472.

Jones, E.G. (1975) Varieties and distribution of non-pyramidal cells in the somatic sensory cortex of the squirrel monkey. J. Comp Neurol., 160(2): 205–267.

Jones, E.G. (2000) Microcolumns in the cerebral cortex. Proc. Natl. Acad. Sci. USA, 97(10): 5019–5021.

Kawaguchi, Y. and Kubota, Y. (1997) GABAergic cell subtypes and their synaptic connections in rat frontal cortex. Cereb. Cortex, 7(6): 476–486.

Lund, J.S. (1973) Organization of neurons in the visual cortex, area 17, of the monkey (*Macaca mulatta*). J. Comp. Neurol., 147(4): 455–496.

Marin-Padilla, M. (1969) Origin of the pericellular baskets of the pyramidal cells of the human motor cortex: a Golgi study. Brain Res., 14(3): 633–646.

Monyer, H. and Markram, H. (2004) Interneuron diversity series: molecular and genetic tools to study GABAergic interneuron diversity and function. Trends Neurosci., 27(2): 90–97.

Peters, A. and Regidor, J. (1981) A reassessment of the forms of nonpyramidal neurons in area 17 of cat visual cortex. J. Comp. Neurol., 203: 685–716.

Peters, A. and Sethares, C. (1997) The organization of double bouquet cells in monkey striate cortex. J. Neurocytol., 26(12): 779–797.

Scheibel, M.E. and Scheibel, A.B. (1970) The rapid Golgi method: Indian summer or renaissance? In: Nauta, W.J.H. and Ebbesson, S.O.E. (Eds.), Contemporary Research Methods in Neuroanatomy. New York, Springer, pp. 1–11.

Sholl, D.A. (1956) The Organization of the Cerebral Cortex. London, Methuen.

Somogyi, P. and Cowey, A. (1981) Combined Golgi and electron microscopic study on the synapses formed by double bouquet cells in the visual cortex of the cat and monkey. J. Comp. Neurol., 195(4): 547–566.

Somogyi, P., Cowey, A., Halasz, N. and Freund, T.F. (1981) Vertical organization of neurons accumulating 3H-GABA in visual cortex of rhesus monkey. Nature, 294(5843): 761–763.

Szentágothai, J. (1969) Architecture of the cerebral cortex. In: Jasper, H., Ward, A.A. and Pope, A. (Eds.), Basic Mechanisms of the Epilepsies. Boston, Little Brown, pp. 13–28.

Szentágothai, J. (1973) Synaptology of the visual cortex. In: Jung, R. (Ed.), Handbook of Sensory Physiology, Vol. VII/3, Central Visual Information, Part B. Springer, Berlin, pp. 269–324.

Szentágothai, J. (1975) The 'module-concept' in cerebral cortex architecture. Brain Res., 95(2–3): 475–496.

Toledo-Rodriguez, M., Blumenfeld, B., Wu, C., Luo, J., Attali, B., Goodman, P. and Markram, H. (2004) Correlation maps allow neuronal electrical properties to be predicted from single-cell gene expression profiles in rat neocortex. Cereb. Cortex, 14(12): 1310–1327.

Valverde, F. (1970) The Golgi method: A tool for comparative structural analysis. In: Nauta, W.J.H. and Ebbesson, S.O.E. (Eds.), Contemporary Research Methods in Neuroanatomy. New York, Springer, pp. 11–31.

Valverde, F. (1978) The organization of area 18 in the monkey. A Golgi study. Anat. Embryol. (Berl.), 154(3): 305–334.

Valverde, F. (1985) The organizing principles of the primary visual cortex in the monkey. In: Peters, A. and Jones, E.G. (Eds.) Cerebral Cortex. Visual Cortex, Vol. 3. New York, Plenum Press, pp. 207–257.

Van Brederode, J.F., Mulligan, K.A. and Hendrickson, A.E. (1990) Calcium-binding proteins as markers for subpopulations of GABAergic neurons in monkey striate cortex. J. Comp. Neurol., 298(1): 1–22.

Martinez-Conde, Macknik, Martinez, Alonso & Tse (Eds.)
Progress in Brain Research, Vol. 154
ISSN 0079-6123

CHAPTER 3

Covert attention increases contrast sensitivity: psychophysical, neurophysiological and neuroimaging studies

Marisa Carrasco

Department of Psychology & Center for Neural Science, New York University, 6 Washington Pl. 8th floor, New York, NY 10003, USA

Abstract: This chapter focuses on the effect of covert spatial attention on contrast sensitivity, a basic visual dimension where the best mechanistic understanding of attention has been achieved. I discuss how models of contrast sensitivity, as well as the confluence of psychophysical, single-unit recording, and neuroimaging studies, suggest that attention increases contrast sensitivity via contrast gain, an effect akin to a change in the physical contrast stimulus. I suggest possible research directions and ways to strengthen the interaction among different levels of analysis to further our understanding of visual attention.

Keywords: visual attention; early vision; contrast sensitivity; psychophysics; neurophysiology; neuroimaging

Our understanding of visual attention has advanced significantly over the last two decades thanks to a number of factors: psychophysics research on humans has systematically characterized distinct attentional systems, and single-unit neurophysiological research has made possible the recording of neuronal responses in monkeys under attention-demanding tasks. The coupling of the results from these two approaches, as well as the findings emerging from combining fMRI (functional magnetic resonance imaging) and psychophysics, have begun to provide a mechanistic characterization of this fundamental process, which lies at the crossroads of perception and cognition.

This chapter focuses on the effect of covert spatial attention on contrast sensitivity, a basic visual dimension where the best mechanistic understanding of attention has been achieved. This is due to the existence of models of contrast sensitivity, as well as

to the confluence of psychophysical, single-unit recording, and neuroimaging studies, all indicating that attention increases contrast sensitivity. Growing evidence supports the idea that this effect is mediated by contrast gain, an effect akin to a change in the physical contrast stimulus.

In the first section, I introduce the construct of selective attention, and discuss the idea that it arises from the high bioenergetic cost of cortical computation and the brain's limited capacity to process information. Then I provide an overview of the two systems of covert attention — transient (exogenous) and sustained (endogenous) — and of the mechanisms that underlie attentional effects — signal enhancement and external noise reduction.

The second section deals with the psychophysical effects of transient and sustained attention on contrast sensitivity. After introducing some ways in which attention is manipulated in psychophysical

DOI: 10.1016/S0079-6123(06)54003-8

experiments, I discuss studies of transient attention indicating that contrast sensitivity is increased at the attended location across the contrast sensitivity function and the contrast psychometric function. Conversely, compared to a neutral condition, contrast sensitivity is decreased at the unattended location. I then document how the effect of transient attention on appearance is consistent with its effects on performance: apparent contrast increases at the attended location and decreases at the unattended location. At the end of the psychophysics section, I discuss a study comparing the effects of transient and sustained attention on contrast sensitivity; specifically with regard to the mechanism of signal enhancement and the contrast gain and response gain functions.

The third section presents neurophysiological studies of visual attention. Single-unit recording studies in the monkey have provided detailed, quantitative descriptions of how attention alters visual cortical neuron responses. I provide an overview of the studies showing that attentional facilitation and attentional selection may come about by increasing contrast sensitivity in extrastriate cortex in a way comparable to increasing stimulus contrast. In addition, I discuss parallels between contrast and attentional effects at the neuronal level, which advance our understanding of how effects of attention may come about.

In the fourth section, I discuss a human fMRI study that provides a retinotopic neuronal correlate for the effects of transient attention on contrast sensitivity with a concomitant behavioral effect. This study illustrates how neuroimaging studies, in particular fMRI, offer an intermediate level of analysis between psychophysics and single-unit studies.

To conclude, I discuss how models of contrast sensitivity, as well as the confluence of psychophysical, single-unit recording, and neuroimaging studies, suggest that attention increases contrast sensitivity via contrast gain, i.e., in such a way that its effect is indistinguishable from a change in stimulus contrast. Finally, I offer some thoughts regarding possible research directions and ways to strengthen the interaction among different levels of analysis to further our understanding of visual attention.

Selective attention

Limited resources

Each time we open our eyes we are confronted with an overwhelming amount of information. Despite this fact, we have the clear impression of understanding what we see. This requires selecting relevant information out of the irrelevant noise, selecting the wheat from the chaff. In *Funes el Memorioso* [Funes the Memoirist], Borges suggests that forgetting is what enables remembering and thinking; in perception, ignoring irrelevant information is what makes it possible for us to attend and interpret the important part of what we see. Attention often turns looking into seeing.

Attention allows us to select a certain location or aspect of the visual scene and to prioritize its processing. The limits on our capacity to absorb visual information are severe. They are imposed by the high-energy cost of the neuronal activity involved in cortical computation (Lennie, 2003). Neuronal activity accounts for much of the metabolic cost of brain activity, and this cost largely depends on the rate at which neurons produce spikes (Attwell and Laughlin, 2001). The high bioenergetic cost of firing pressures the visual system to use representational codes that rely on very few active neurons (Barlow, 1972). As only a small fraction of the machinery can be engaged concurrently, energy resources must be allocated flexibly according to task demand. Given that the amount of overall energy consumption available to the brain is constant, the average discharge rate in active neurons will determine the number of neurons that can be active at any time. The bioenergetic limitations provide a neurophysiological basis for the idea that selective attention arises from the brain's limited capacity to process information (Lennie, 2003).

As an encoding mechanism, attention helps the visual system to optimize the use of valuable processing resources. It does so by enhancing the representation of the relevant locations or features while diminishing the representation of the less relevant locations or aspects of our visual environment. The processing of sensory input is enhanced by knowledge and assumptions of the world, by the behavioral state of the organism, and by the (sudden) appearance of possibly relevant information in the environment.

Throughout the 19th and early 20th centuries, scientists such as Wundt, Fechner, James, and Helmholtz proposed that attention plays an important role in perception. It is necessary for effortful visual processing, and may be the 'glue' that binds simple visual features into an object. In the 1980s and 1990s, cognitive psychologists developed experimental paradigms to investigate what attention does and which perceptual processes it affects (Neisser, 1967; Posner, 1980; Treisman and Gelade, 1980). Over the last decade, cognitive neuroscientists have investigated the effects of attention on perception using three different methodological approaches. The physiological brain systems that underlie attention have been explored using two different methodological approaches. One has enabled studying how and where attention modulates neuronal responses by using single-unit recording; this method yields a precise estimate of local activity, but largely ignores behavioral consequences. The second approach has employed brain scanners (fMRI systems) to study the human brain while engaged in attentional tasks. This has enabled the identification of many of the cortical and subcortical brain areas involved in attention, and these experiments have yielded insights into the global structure of the brain architecture employed in selectively processing information. A third approach has focused on behavior; researchers have used cognitive and psychophysical techniques to explore what attention does and what perceptual processes it affects. More recently, they have started to investigate the mechanisms of visual attention, including how visual attention modulates the spatial and temporal sensitivity of early filters, and how it influences the selection of stimuli of interest, and its interaction with eye movements (Baldassi, Burr, Carrasco, Eckstein & Verghese, 2004).

Recent studies show that attention affects early visual processes such as contrast discrimination, orientation discrimination, and texture segmentation — which until recently were considered to be preattentive. Electrophysiological studies have established that neural activity increases at attended locations and decreases at unattended locations. Consequently, we can now infer that attention helps manage energy consumption. Usually we think of the need to selectively process information in cluttered displays with different colors and shapes (i.e., in 'Where's Waldo'-like displays). However, psychophysical evidence shows that even with very simple displays, attention is involved in distributing resources across the visual field. Because of bioenergetic limitations, the allocation of additional resources to an attended location implies a withdrawal of resources from unattended locations. Indeed, we have recently published a study showing that when only two stimuli are present in a display, compared to a neutral attentional state, attention enhances the signal at the attended location, but impairs it at the unattended location (Pestilli and Carrasco, 2005).

Systems of covert attention: transient and sustained

Attention can be allocated by moving one's eyes towards a location, or by attending to an area in the periphery without actually directing one's gaze toward it. This peripheral deployment of attention, known as covert attention, aids us in monitoring the environment, and can inform subsequent eye movements (Posner, 1980). Many human psychophysical studies as well as monkey single-unit recording studies have likened attention to increasing visual salience.

A growing body of behavioral evidence demonstrates that there are two systems of covert attention, which deal with facilitation and selection of information: 'sustained' (endogenous) and 'transient' (exogenous). The former corresponds to our ability to monitor information at a given location at will; the latter corresponds to an automatic, involuntary orienting response to a location where sudden stimulation has occurred. Experimentally, these systems can be differentially engaged by using distinct cues. Symbolic cues direct sustained attention in a goal- or conceptually- driven fashion in about 300 ms, whereas peripheral cues grab attention in a stimulus-driven, automatic manner in about 100 ms. Whereas the shifts of attention by sustained cues appear to be under conscious control, it is extremely hard for observers to ignore transient cues (Nakayama and Mackeben, 1989; Cheal and Lyon, 1991; Yantis, 1996; Giordano et al., 2003). This involuntary transient shift occurs even when the cues are uninformative or may impair performance (Yeshurun and Carrasco, 1998, 2000; Yeshurun, 2004; Pestilli and Carrasco, 2005).

Transient and sustained attentions show some common perceptual effects (Hikosaka et al., 1993; Suzuki and Cavanagh, 1997), but some differences in the mechanisms mediating increased contrast sensitivity have been reported (Lu and Dosher, 2000; Ling and Carrasco, 2006). Of interest, these systems have different temporal characteristics and degrees of automaticity (Nakayama and Mackeben, 1989; Cheal and Lyon, 1991; Yantis, 1996), which suggest that these systems may have evolved for different purposes and at different times — the transient system may be phylogenetically older. There is no consensus as to whether common neurophysiological substrates underlie sustained and transient attention. On the one hand, all single-cell recording studies have manipulated sustained attention; on the other hand, some fMRI studies have found no difference in the brain networks mediating these systems (Peelen et al., 2004); others have reported differences. For example, sustained attention is cortical in nature, but transient attention also activates subcortical processing (Robinson and Kertzman, 1995; Zackon et al., 1999), and partially segregated networks mediate the preparatory control signals of sustained and transient attention. Sustained attention is mediated by a feedback mechanism involving delayed reentrant feedback from frontal and parietal areas (e.g., Martinez et al., 1999; Kanwisher and Wojciulik, 2000; Kastner and Ungerleider, 2000; Corbetta and Shulman, 2002).

Mechanisms of covert attention: signal enhancement and external noise reduction

Although it is well established that covert attention improves performance in various visual tasks (e.g., Morgan et al., 1998; Lu and Dosher, 1998, 2000; Carrasco et al., 2000, 2001, 2002, 2004a,b; Baldassi and Burr, 2000; Baldassi and Verghese, 2002; Blanco and Soto, 2002; Cameron et al., 2002; Solomon, 2004), the nature of the attentional mechanisms, and the stages and levels of processing at which they modulate visual activity are not yet well understood. Explanations of how attention improves perception range from proposals maintaining that the deployment of attention changes observers' decision criteria and reduces spatial uncertainty (Davis et al., 1983; Sperling and Dosher, 1986; Kinchla, 1992; Palmer, 1994; Shiu and Pashler, 1994; Nachmias, 2002), to proposals asserting that attention actually improves sensitivity by reducing external noise (Lu and Dosher, 1998; Morgan et al., 1998; Baldassi and Burr, 2000; Dosher and Lu, 2000; Cameron et al., 2004) or by enhancing the signal (Bashinski and Bacharach, 1980; Carrasco et al., 2000, 2002; Dosher and Lu, 2000; Cameron et al., 2002; Ling and Carrasco, 2006).

The external noise reduction hypothesis maintains that attention selects information by diminishing the impact of stimuli that are outside its focus. Noise-limited models incorporate internal noise arising from such sources as spatial and temporal uncertainty of targets and distracters, as well as external noise resulting from distracters and masks. Several studies have attributed attentional facilitation to reduction of external noise, either because a near-threshold target presented alone could be confused with empty locations (spatial uncertainty) or because a suprathreshold target could be confused with suprathreshold distracters. According to these models, performance decreases as spatial uncertainty and the number of distracters increase, because the noise they introduce can be confused with the target signal (Shiu and Pashler, 1994; Solomon et al., 1997; Morgan et al., 1998; Baldassi and Burr, 2000; Dosher and Lu, 2000). Presumably, precues allow observers to monitor only the relevant location(s) instead of all possible ones. This reduction of statistical noise with respect to the target location is also known as reduction of spatial uncertainty. According to external noise reduction, attention affects performance in a given area by actively suppressing the strength of representation for areas outside its locus. Some studies report that attentional effects emerge when distracters appear with the target (distracter exclusion), but not when the target is presented alone, and are more pronounced as the number of distracters increases (Palmer, 1994; Shiu and Pashler, 1994, 1995; Eckstein and Whiting, 1996; Foley and Schwarz, 1998; Verghese, 2001; Cameron et al., 2004). These studies assert that attention allows us to exclude distracters that differ along some relevant dimension from the signal by narrowing a filter that processes the stimulus.

The signal enhancement hypothesis proposes that attention directly improves the quality of the stimulus representation of the signal within the locus of attention enhancement (Bashinski and Bacharach, 1980; Luck et al., 1996; Muller et al., 1998; Lu and Dosher, 1998; Carrasco et al., 2000, 2002; Cameron et al., 2002; Ling and Carrasco, 2006). In my lab, we have conducted a series of studies to evaluate whether signal enhancement (or internal noise) occurs in addition to external noise reduction. An attentional benefit can be attributed with certainty to signal enhancement only when all the factors that according to the external noise reduction model, are responsible for the attentional effects are eliminated. Presenting a suprathreshold target alone, without added external noise such as distracters or local or multiple masks, and eliminating spatial uncertainty, have allowed us to conclude that transient attention can increase contrast sensitivity (Carrasco et al., 2000; Cameron et al., 2002; Ling and Carrasco, 2006) and spatial resolution (Yeshurun and Carrasco, 1999; Carrasco et al., 2002) via signal enhancement (for a review, see Carrasco, 2005). However, it is reasonable to assume that attentional effects in visual tasks reflect a combination of mechanisms such as signal enhancement, external noise reduction, and decisional factors. Indeed, under some experimental conditions it has been shown that signal enhancement and noise reduction mechanisms coexist (e.g., Lu and Dosher, 2000; Carrasco et al., 2004a,b; Pestilli and Carrasco, 2005).

Neurophysiological (e.g., Luck et al., 1997; Reynolds et al., 1999, 2000; Martinez-Trujillo and Treue, 2002; Reynolds and Chelazzi, 2004), psychophysical (Carrasco et al., 2000; Carrasco and McElree, 2001; Cameron et al., 2002, 2004; Talgar

et al., 2004) and neuroimaging (Pinsk et al., 2004; Liu et al., 2005) studies indicate that both mechanisms affect the processing of visual stimuli. Single-cell studies show that attention can alter the responses of V1 neurons and can result in stronger and more selective responses in both V4 and MT neurons (Motter, 1994; Desimone and Duncan, 1995; McAdams and Maunsell, 1999; Reynolds and Desimone, 1999; Treue and Martinez-Trujillo, 1999). Likewise, signal enhancement is reflected in brain-imaging studies showing that attentional modulation is accompanied by stronger stimulus-evoked brain activity, as measured by scalp potential (see review by Hillyard and Anllo-Vento, 1998) and fMRI in both striate and extrastriate visual areas (e.g., Gandhi et al., 1999; Martinez et al., 1999; Pessoa et al., 2003; Yantis and Serences, 2003; Liu et al., 2005). All these studies support the psychophysical finding that attention affects the quality of sensory representation.

Psychophysical studies

Effects of transient attention on early vision

Much research has focused on the time course and degree of automaticity of the allocation of sustained and transient attention. However, less is known about the ways in which these systems, in particular sustained attention, affect fundamental visual dimensions. In past research, my laboratory has been particularly interested in characterizing the effects of transient attention on early visual processes. Given that transient attention highlights salient changes in the environment, its default, heuristic-like operation may be to enhance the quality of the signal and to reduce the external noise, enabling one to react accurately and quickly in most instances.

Indeed, we have found that transient attention affects spatial and temporal aspects of vision in remarkable ways. Compared to a neutral condition, it enhances contrast sensitivity (Carrasco et al., 2000; Cameron et al., 2002; Ling and Carrasco, 2006; Pestilli and Carrasco, 2005) and apparent contrast (Carrasco et al., 2004a,b) at the attended location, and decreases sensitivity (Pestilli and Carrasco,

2005) and apparent contrast (Carrasco et al., 2004a,b) at the unattended location. Transient attention also enhances spatial resolution (Yeshurun and Carrasco, 1998, 1999, 2000; Carrasco et al., 2002), and apparent spatial frequency (Gobell and Carrasco, 2005). In addition to improving discriminability, transient attention also speeds up information accrual (Carrasco and McElree, 2001; Carrasco et al., 2004a,b, 2006).

By improving discriminability, transient attention enables us to selectively extract relevant information in a noisy environment; by accelerating processing, it enables us to extract this information efficiently in a dynamic environment, before potentially interfering stimuli occur. However, purportedly because of its automatic fashion, transient attention does not always result in improved performance. It causes enhanced contrast sensitivity and spatial resolution; even when doing so leads to deviations from veridical perception (Carrasco et al., 2004; Gobell and Carrasco, 2005), makes us more prone to perceive an illusion (Santella and Carrasco, 2003), or impairs performance (Yeshurun and Carrasco, 1998, 2000; Talgar and Carrasco, 2002; Yeshurun, 2004).

Using fMRI, we have demonstrated a retinotopically specific neural correlate in striate and extrastriate areas for the enhanced contrast sensitivity engendered by transient attention (Liu et al., 2005). The attentional effect increases along the hierarchy of visual areas, from V1 to V4. Because attention can boost the signal by increasing the effective stimulus contrast via contrast gain (Reynolds et al., 2000; Carrasco et al., 2000, 2004a,b; Martinez-Trujillo and Treue, 2002; Cameron et al., 2002; Ling and Carrasco, 2006), its effect would be more pronounced in extrastriate than striate areas, where the contrast response functions get steeper, due to areal summation across progressively larger receptive fields in higher areas (Sclar et al., 1990). Thus, a feedforward mechanism in which attentional modulation accumulates across sequential levels of processing can underlie the transient attention gradient.

Manipulations of spatial covert attention

To interpret the psychophysical results reported here, some methodological issues need to be clarified upfront. First, to investigate attention, it is best to keep the task and stimuli constant across conditions and to explicitly manipulate attention, rather than to infer its role (unfortunately, this has often not been the case in attention studies). We compare performance in conditions where attention is deliberately directed to a given location (attended condition) with performance when attention is distributed across the display (neutral or control condition), and in some cases, with performance in conditions where attention is directed to another location (unattended condition).

In cued trials, attention is directed to the target location via either a transient or a sustained cue. To effectively manipulate transient attention and to prevent forward spatial masking, the transient cue is presented ~100 ms before the display onset, adjacent to the location of the upcoming stimulus. In contrast, sustained cues typically appear at the display center ~300 ms before stimulus onset (e.g., Jonides, 1981; Muller and Rabbitt, 1989; Nakayama and Mackeben, 1989; Cheal and Lyon, 1991; Yantis, 1996). Because ~200–250 ms are needed for goal-directed saccades to occur (Mayfrank et al., 1987), the stimulus-onset-asynchrony (SOA) for the sustained cue may allow observers to make an eye movement toward the cued location. Thus, observers' eyes are monitored to ensure that central fixation is maintained throughout each trial.

In the neutral trials, a small disk appears in the center of the display (central neutral cue) or several small bars appear at all possible target locations (distributed neutral cue), or lines encompass the whole display (distributed neutral cue), indicating that the target is equally likely to occur at any possible location. We have found that performance is comparable with these neutral cues. The performance difference between a single peripheral cue and a distributed neutral cue is comparable to the difference between a single peripheral cue and a central-neutral cue in a letter identification task contingent on contrast sensitivity (Talgar et al., 2004), an acuity task (Cameron et al., 2002), and a temporal resolution task (Yeshurun, 2004). All cues indicate display onset, but only the transient or sustained cue provides information, with a given probability, about the location of the upcoming target.

The following are some critical methodological issues to be considered when using spatial cues to test for sensory effects of attention: Spatial cues should convey only information that is orthogonal to the task, e.g., in a discrimination task they could indicate probable target location but not the correct response (e.g., Carrasco and Yeshurun, 1998). Many experiments manipulate sustained attention in detection tasks with cues indicating that a certain location has a given probability of containing the target (e.g., Posner, 1980). Although a high probability encourages observers to direct their attention to a particular location, it is hard to determine whether the enhanced detection is due to facilitation of information coding at that location, to probability matching, or to a decision mechanism, i.e., the higher probability encourages observers to assign more weight to information extracted from that probability location (Kinchla, 1992). By using a two-alternative-forced-choice (2AFC) in which the observers discriminate stimuli preceded by a cue (e.g., the orientation of a stimulus: left vs. right; Fig. 1), even when the cue is 100% valid in terms of location, it conveys no information as to the correct response. Thus, we can assess whether a cueing effect reflects changes in sensory (d'), rather than decisional (criterion), processes. A second critical factor is that of spatial uncertainty. According to noise-limited models, performance decreases as spatial uncertainty increases, because the empty locations introduce noise that can be confused with the target signal. For instance, a spatial uncertainty effect is present for low-contrast pedestals but not for high-contrast pedestals (Foley and Schwarz, 1998). Uncertainty about the target location produces a more noticeable degradation at low than at high performance levels (Pelli, 1985; Eckstein and Whiting, 1996), and uncertainty is larger for less discriminable stimuli (Nachmias and Kocher, 1970; Cohn, 1981; Pelli, 1985). Thus, uncertainty models predict that the precueing effect would be greater for low-contrast stimuli and when localization performance is poor (e.g., Pelli, 1985; Eckstein and Whiting, 1996; Solomon et al., 1997; Palmer et al., 2000; Carrasco et al., 2000, 2002).

In some studies, we have explored the conditions for which the effect of attention can be attributed to

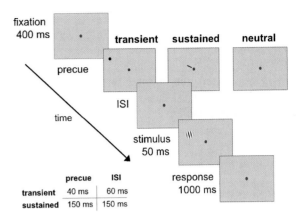

Fig. 1. Sequence of events in a given trial. Observers perform a 2AFC orientation discrimination task on a tilted target Gabor patch, which appears at one of eight isoeccentric locations. The target is preceded by a sustained cue (instructing observers to deploy their attention to the upcoming target location), a transient cue (reflexively capturing attention to the upcoming target location), or a neutral cue (baseline). The timing (precue and interstimulus interval (ISI)) for sustained and transient conditions differs (along with their respective neutral conditions), in order to maximize the effectiveness of the cues (Ling and Carrasco, 2005, Fig. 2).

signal enhancement. To do so, it is necessary to ensure that a performance benefit occurs under conditions that exclude all variables that the external noise reduction models hold to be responsible for the attentional effect. That is, the target should be suprathreshold (to reduce spatial uncertainty) and presented alone, without distracters and local or multiple masks (Lu and Dosher, 1998, 2000; Carrasco et al., 2000, 2002; Cameron et al., 2002; Golla et al., 2004; Ling and Carrasco, 2006).

Many of the studies I describe in this chapter involve an orientation discrimination task because this dimension has been well characterized both psychophysically and neurophysiologically, and a link between these two levels of analysis has been well established (Regan and Beverley, 1985; De Valois and De Valois, 1988; Graham, 1989; Ringach et al., 1997). In addition, we use orientation discrimination to assess the effect of attention on stimulus contrast because performance on this task improves with increasing contrast (Nachmias, 1967; Skottun et al., 1987; Lu and Dosher, 1998; Cameron et al., 2002), and because fMRI response increases monotonically with stimulus contrast (Boynton et al., 1999). Moreover, the shared nonlinearity between the contrast response function and the magnitude of the attentional modulation across different areas of the dorsal and ventral visual pathways indicate a close link between attentional mechanisms and the mechanisms responsible for contrast encoding (Martinez-Trujillo and Treue, 2005; Reynolds, 2005).

Transient attention increases contrast sensitivity

Transient attention increases sensitivity across the contrast sensitivity function

A number of psychophysical studies have shown that in the presence of competing stimuli contrast sensitivity for the attended stimulus is enhanced (Solomon et al., 1997; Lee et al., 1997, 1999; Foley and Schwartz, 1998). We assessed whether attention increases sensitivity in a wide range of spatial frequencies, spanning the contrast sensitivity function. To evaluate whether increased contrast could be mediated by signal enhancement, we explored if this effect also emerges when a suprathreshold target is presented alone (Carrasco et al., 2000).

We compared the stimulus contrast necessary for observers to perform an orientation discrimination task at a given performance level when the target location was preceded by a peripheral cue appearing adjacent to the target location, and

when it is preceded by a neutral cue appearing at fixation, which indicates that the target is equally likely to occur at any of the eight isoeccentric locations. We assessed the effect of transient attention across a wide range of spatial frequencies and found that it increases sensitivity across the contrast sensitivity function (Fig. 2a). Less contrast was necessary to attain the same performance level when a transient cue preceded the Gabor than when a neutral cue did (Fig. 2b). The results are consistent with a signal enhancement mechanism. The display did not contain any added external noise; there were no distracters, or local or global masks, which according to the external noise reduction model are responsible for attentional effects (e.g., Davis et al., 1983; Solomon et al., 1997; Morgan et al., 1998; Dosher and Lu, 2000; Lu and Dosher, 2000; Baldassi and Burr, 2000; Nachmias, 2002).

We found that a signal detection model (SDT) of external noise reduction could account for the cueing benefit in an easy discrimination task (e.g., vertical vs. horizontal Gabor patches). However, such a model could not account for this benefit when location uncertainty was reduced, either by increasing overall performance level, increasing stimulus contrast to enable fine discriminations of slightly tilted suprathreshold stimuli, or presenting a local postmask. An SDT model that incorporates

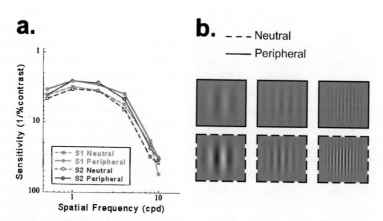

Fig. 2. (a) Data for two individual observers (CPT and YY) illustrating that for a target of constant contrast, precueing the target location enhances sensitivity across the contrast sensitivity function (CSF; Carrasco et al., 2000, Fig. 3). (b) The stimulus contrast necessary to attain the same performance level for a range of spatial frequencies is lower when the target location is precued by a peripheral cue (bottom squares) than by a neutral cue (top squares). The contrast differences depicted in the Gabor patches are based on data reported by Carrasco et al. (2000).

intrinsic uncertainty (the observers' inability to perfectly use information about the elements' spatial or temporal positions, sizes, or spatial frequencies) revealed that the cueing effect exceeded that predicted by uncertainty reduction. Thus, the cueing effect could not be explained by the mere reduction of location uncertainty. Given that the attentional benefits occurred under conditions that exclude all variables predicted by the external noise reduction model, the results support the signal enhancement model of attention. The finding that transient attention operates via signal enhancement under low-noise conditions has been corroborated using the external noise plus attention paradigm (Lu and Dosher, 1998, 2000).

Transient covert attention enhances letter identification without affecting channel tuning

To explore how the enhancement of contrast sensitivity at the attended location comes about we investigated whether covert attention affects the tuning of a spatial frequency channel (Talgar et al., 2004) (see Fig. 3). We chose a task that isolates a spatial frequency channel that mediates the identification of broadband stimuli. A broadband stimulus could be seen through channels with various tunings, allowing us to test for shifts of peak frequency of the channel as a result of directing covert attention. Given that observers have multiple independent channels with various peak frequencies, one would expect a broadband stimulus such as a letter to activate many channels. However, using a critical-band-masking paradigm with unfiltered letters, the same filter tuning is found for detection of narrowband gratings and identification of broadband letters (Solomon and Pelli, 1994). Critical-band masking of letters allows us to test the effects of covert attention on a single spatial frequency channel using a broadband stimulus.

In auditory detection tasks observers are able to switch channels to avoid noise and attain a lower threshold than they would without switching channels, a process termed off-frequency listening (Patterson and Nimmo-Smith, 1980). Correspondingly, in a visual task observers might be able to 'switch channels' to use the noise-free part of the spectrum to reduce their thresholds. When narrow-band noise is superimposed on a broadband stimulus (e.g., a letter), an ideal observer could use the noise-free region of the signal spectrum to perform perfectly.

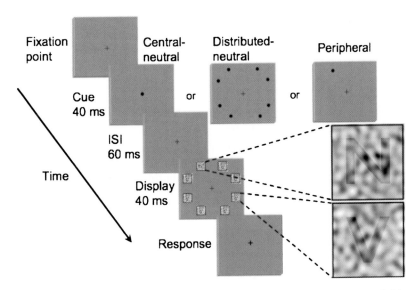

Fig. 3. A schematic representation of a trial sequence. In one third of the blocks, the target was preceded by a central–neutral cue (a dot in the center of the display), in another third by a distributed–neutral cue (a dot adjacent to each of the eight possible target locations), and in the remaining block by a peripheral cue (a single dot adjacent to the actual target location). Note that the eight noise patches were outlined in black to demarcate the locations (Talgar et al., 2004; Fig. 1).

To assess whether transient covert attention affects the spatial frequency tuning of a single channel, we used a task that isolates a single spatial frequency channel which mediates the identification of broadband stimuli (e.g., letters), in conjunction with the critical-band masking paradigm (Solomon and Pelli, 1994). In particular, we investigated the following two hypotheses:

First, covert attention shifts the peak frequency of the channel. Studies dealing with acuity and hyperacuity tasks (Yeshurun and Carrasco, 1999; Carrasco et al., 2002) as well as with texture segmentation tasks (Yeshurun and Carrasco, 1998, 2000; Talgar and Carrasco, 2002) have supported the hypothesis that attention increases spatial resolution at the attended location. Hence, we hypothesized that the peak frequency of the channel may shift to higher spatial frequencies when the low portion of the letter spectrum is masked (high-pass noise), and to lower spatial frequencies when the higher portion of the letter spectrum is masked (low-pass noise).

Second, covert attention alters the channel bandwidth, making it better matched to the signal. There is no consensus as to whether attention increases the selectivity of the neuronal response. Some have reported that attention narrows the tuning for orientation and color of neurons in V4 (Spitzer et al., 1988; Reynolds and Desimone, 1999; Reynolds et al., 2000), whereas others have found an increased gain but unchanged tuning for orientation in area V4 (McAdams and Maunsell, 1999), and for direction of motion in areas MT/MST (Treue and Martinez-Trujillo, 1999). Increased contrast sensitivity for a grating of a given frequency could be mediated by a narrowing of the channel tuning, but increased sensitivity for a broadband stimulus such as a letter would arise from a widening of the bandwidth. In general, better matching the channel to the noise-normalized signal would increase sensitivity.

To investigate these two hypotheses, we used critical-band masking of letters (Solomon and Pelli, 1994) and tested the effects of covert attention on a single spatial frequency channel using a broadband stimulus. The target letter (N, Z, or X; presented in low- or high-pass noise with different cut-off frequencies) followed the transient cue at 1 of 8 locations. Distracter letters (V's) occupied the remaining locations. All stimuli appeared at isoeccentric non-cardinal locations for which contrast sensitivity is similar (Carrasco et al., 2001; Cameron et al., 2002). We measured the energy threshold elevation for each observer at each of the low- and high-pass cut-off noise frequencies with both a peripheral and a neutral cue.

To quantify the attentional benefit, we used two control conditions. The central-neutral cue appeared at the center of the display. To test for the possibility that this cue reduces the extent of the attentional spread by attracting attention to its location, away from the peripheral target locations (Pashler, 1998), we also employed a distributed-neutral cue presented at all possible target locations. By simultaneously stimulating the detectors at all candidate locations, the distributed-neutral cue should also reduce uncertainty as well as any differences in the onset time of activation in response to the central-neutral and the peripheral cues.

We derived the power gain of the inferred filter from the threshold energy elevation at each noise cut-off frequency, by assuming a parabola-shaped filter

$$[\log G(f) = b_0 + b_1 \log f + b_2(\log f)^2].$$

The low- and high-pass noises are additive if their sum leads to a threshold energy elevation that is equivalent to the sum of threshold energy elevations yielded by each noise alone. If observers exhibit channel switching and utilize the noise-free part of the signal spectrum to perform the task, noise additivity would be violated (Majaj et al., 2002). We assume E to be linearly related to the total power passed through the channel filter mediating letter identification (Solomon and Pelli, 1994; Majaj et al., 2002):

$$E = E_0 + a \int_0^\infty 2\pi f G(f) N(f) df,$$

where E_0 is the threshold at 0 noise, N the noise spectrum, f the spatial frequency, and G the power gain of the channel. We estimate its parameters by maximum likelihood methods. The ratio of the E obtained in the peripheral- and neutral-cue conditions is computed to quantify the attention effect.

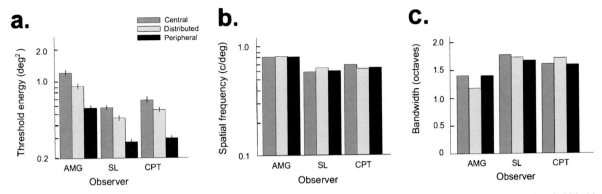

Fig. 4. Transient attention decreases threshold (a), but does not alter a channel's peak spatial frequency (b) or its bandwidth (c) (Talgar et al., 2004; Adapted with permission from Talgar et al., 2004, Figs. 3, 4, and 5.)

We found that directing attention to the target location reduces energy threshold by a factor of 2 (Fig. 4a). The magnitude of the effect is consistent with neurophysiological findings, indicating that attention increases the effective contrast of the attended stimulus by a factor of 1.5 (e.g., Reynolds et al., 2000; Martinez-Trujillo and Treue, 2002; Reynolds and Desimone, 2003). Contrary to our hypotheses, there is no change in the tuning of the channel mediating this task, as assessed by the peak channel frequency (Fig. 4b) and the channel bandwidth (Fig. 4c) in each condition for each observer. The channel characteristics are remarkably stable; neither center frequency nor bandwidth was affected. The absence of channel switching makes it clear that transient covert attention does not induce observers to perform this task in a flexible way. Recently, we have reported that sustained attention yields the same pattern of results. It also increases contrast sensitivity in this task, without affecting the channel's center frequency or bandwidth (Pestilli et al., 2004). Lu and Dosher (2004) have corroborated these results.

Transient attention increases sensitivity across the contrast psychometric function
Two types of gain control mechanisms have been considered in neural responses to luminance-modulated stimuli — contrast gain and response gain (Sclar et al., 1989; Fig. 5). The signature of contrast gain is a shift in the contrast response function to the left. In the case of attention, this reflects

a decrease in the contrast required for the neuron to respond at the same level as in a neutral condition. The signature of a response gain is an increase in firing rate proportional with stimulus intensity. Some have supported a contrast gain model (Reynolds et al., 2000; Martinez-Trujillo and Treue, 2002), but others have reported findings consistent with a response gain model (McAdams and Maunsell, 1999; Treue and Martinez-Trujillo, 1999). How do attentional changes at the neural level affect the psychophysical contrast response functions?

We have examined the effect of transient attention across a range of performance levels, from subthreshold to suprathreshold, when the target was presented alone at 1 of 8 isoeccentric locations (Cameron et al., 2002). We found that transient attention decreased the threshold of the psychometric function for contrast sensitivity in this orientation discrimination task (Fig. 6a). The results were consistent with a contrast gain mechanism; the effect of attention was more pronounced within the dynamic range. However, the high asymptotic level for the neutral condition may have precluded the emergence of response gain.

To assess the role of spatial uncertainty in the precue effect, we conducted two control experiments. First, we made the discrimination task harder by decreasing the tilt of the targets from 15 to 4°. Observers required higher stimulus contrasts to perform this discrimination task, and this in turn diminished spatial uncertainty. Even though the target contrast was higher, an

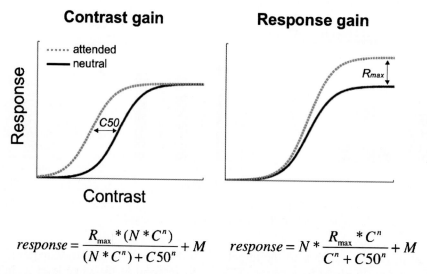

Fig. 5. Possible effects of attention on the contrast response function. The left panel depicts a contrast gain model for attention. Contrast gain predicts an increase in sensitivity that is a function of stimulus intensity, and is characterized by a leftward threshold (C50) shift in the contrast response function. The dashed curve represents the signature curve shift brought about by attentional contrast gain; the shape of the function does not change, but shifts leftward — boosting the effective contrast of the stimulus. In the right panel, the dashed curve (attended) represents the effects of attention according to response gain models. Response gain predicts an increase in firing rate, which is characterized by a change in the shape of the curve — in slope and asymptote (R_{max}). C50, threshold; R_{max}, asymptote, n, slope, C, contrast level, N, attentional modulation, M, response at lowest stimulus intensity.

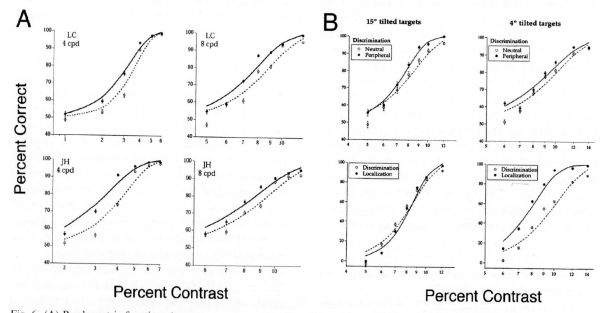

Fig. 6. (A) Psychometric functions (percent correct as a function of target contrast) for two of the spatial frequencies used (4 and 8 cpd), for two observers. Neutral precued condition is represented with open symbols and dotted lines; peripheral precue condition is represented with filled symbols and solid lines. Attention shifts the psychometric function to the left, and in some instances, makes the slope shallower (adapted from Cameron et al.,2002, Fig. 3). (B) The peripheral cue increases contrast sensitivity throughout the psychometric function of contrast sensitivity to the same extent in an orientation discrimination task (top panels) for stimuli that differ in spatial uncertainty, 4° vs. 15° tilted (bottom panels), as assessed by localization performance (Cameron et al., 2002, Fig. 10).

attentional effect of similar magnitude was observed (Fig. 6b). In addition, to directly assess the ease with which observers can localize the stimulus, we also performed a localization task. When the target was tilted 15°, discrimination and localization performance were tightly coupled. However, when the targets were tilted 4°, performance on the localization task was much better than performance on the discrimination task. Notwithstanding the superior localization performance on the 4° discrimination task, the attentional effect was comparable for both orientation conditions. Importantly, at contrasts that yielded perfect localization, there was still an attentional effect in the discrimination tasks. Thus, given that we used suprathreshold stimuli, excluded all sources of added external noise (distracters, local and global masks) and showed experimentally that spatial uncertainty cannot explain this decrease in threshold, the observed attentional benefit is consistent with a signal enhancement mechanism.

Transient attention increases contrast at the cued location and decreases it at the uncued location

It had been proposed that very few neurons can be concurrently engaged, but this proposition only recently became tractable and has now been systematically evaluated. The calculations are astonishing — the cost of a single spike is high and severely limits (possibly to about 1%) the number of neurons that can be (substantially) active concurrently (Lennie, 2003). The limited energy expenditure that the brain can afford necessitates machinery for the system to allocate energy according to task demand. This limited capacity entails selective attention, which enables us to process effectively vast amounts of visual information by selecting relevant information from noise. In this study we investigated the possibility that covert attention helps to control the expenditure of cortical computation by trading contrast sensitivity across attended and unattended areas of the visual field, even with impoverished displays and simple tasks. Specifically, we assessed contrast sensitivity at both cued and uncued locations (Pestilli and Carrasco, 2005).

There is consensus that attention improves performance at the attended location, but there is less agreement regarding the fate of information that is not directly attended, i.e., outside the focus of attention (Eriksen and Hoffman, 1974; Rock and Gutman, 1981; Kinchla, 1992). Although most hypotheses regarding the distribution of attention in the visual field assume that information outside the attended area is not processed, many studies have shown that information beyond the focus of attention affects performance, indicating that it is processed to a certain degree (Carrasco and McElree, 2001; Carrasco et al., 2004a,b; Cameron et al., 2004).

When manipulating attention, a cue is considered valid when it indicates the target location, and it is considered invalid when it indicates a nontarget location. Although assessing the effects of attention by comparing performance in the valid and invalid conditions is useful for distinguishing between sensitivity-based and decisional-based explanations of the cueing effect, this comparison cannot determine whether such an effect is due to an enhanced signal at the cued location, a diminished signal at the uncued location, or both. To pinpoint the source of the attentional effect, it is necessary to compare performance in both the valid and invalid conditions with a neutral condition, in which the cue does not indicate a stimulus location but only the timing of the display onset (Hawkins et al., 1990; Luck et al., 1994; Carrasco and Yeshurun, 1998).

We evaluated the effect of transient attention on contrast sensitivity at both the attended and unattended locations. As discussed above, at the attended area transient attention increases sensitivity in an orientation discrimination task with an informative cue, i.e., when the cue indicates target location but not its orientation (Lu and Dosher, 1998; Carrasco et al., 2000; Cameron et al., 2002). When a peripheral cue is always valid in terms of location, however, some of its effect could be due to a conceptually driven, voluntary component of attention. To eliminate this possible contamination, we ensured cue unpredictability by cueing the target only 50% of the time, and by asking observers to report the orientation of the stimulus indicated by a response cue (a line displayed after stimuli offset). Indeed, observers could have entirely disregarded the cue and based their responses only on the information accumulated during stimulus presentation and still attained the same overall performance level. The use of the nonpredictive cue and the response cue enabled us to isolate the purely automatic orienting of attention. Given that the transient peripheral cue is thought to be automatic (Yantis and Jonides, 1984; Jonides and Yantis, 1988), even an uninformative cue (which indicates neither target location nor orientation) should exert an effect on performance.

Previous studies have examined the effect of attention on contrast sensitivity at parafoveal locations (e.g., Lee et al., 1997; Lu and Dosher, 1998, 2000; Cameron et al., 2002; Solomon, 2004). We investigated the effects of transient attention at both parafoveal and peripheral locations to assess whether the benefit and cost varied as a function of the distance between the attended and unattended stimuli. Observers were asked to discriminate the orientation of 1 of 2 Gabor patches simultaneously presented left and right of fixation (at either 4 or 9° of eccentricity). Contrast sensitivity was measured at the cued (valid cue) and uncued (invalid cue) locations, and compared with the contrast sensitivity obtained at the same locations when the target was preceded by a cue presented at fixation (neutral cue). Based on models of signal enhancement, which propose that attention directly improves the quality of the stimulus representation (Bashinski and Bacharach, 1980; Lu and Dosher, 1998; Muller et al., 1998; Carrasco et al., 2000; Cameron et al., 2002), we hypothesized that sensitivity would be increased at the cued location. Based on models of distracter exclusion, which propose that attention allows us to exclude distracters from the signal by narrowing the filter processing the stimulus (Davis et al., 1983; Palmer, 1994; Solomon et al., 1997; Foley and Schwarz, 1998; Morgan et al., 1998; Baldassi and Burr, 2000), we hypothesized that sensitivity will be reduced at the uncued location.

Following a peripheral or a central-neutral transient cue, two slightly tilted Gabor patches were simultaneously presented to the left and right of fixation (Fig. 7). A response cue was presented after the Gabors, indicating to the observer for which Gabor the orientation was to be reported,

thus defining valid and invalid trials (cue location and response-cue match and do not match, respectively). We estimated contrast thresholds under each attention condition at each eccentricity. Usually, with invalid cue trials attention is diverted away from the target location at stimulus onset, but observers have information regarding the target location because its identity differs from the distracter. However, in this study, observers did not know where the target was, and they had to process the identity of the stimuli presented at both locations to perform the task (Fig. 7).

To quantify the magnitude of the attentional effect, we calculated the ratio of the contrast sensitivity (1/median threshold) for valid vs. neutral cue, and invalid vs. neutral cue at both eccentricities. No difference between the two conditions would yield a ratio equal to 1. A benefit in contrast sensitivity is indicated by values >1; a cost by values <1. All observers followed the same pattern of responses: values >1 for the valid:neutral ratio (benefit) and values <1 for the invalid:neutral ratio (cost). Figure 8 (left panel) shows the values for one observer.

The data for individual observers were consistent with the overall frequency distributions. The histograms represent the threshold values obtained in each cue condition at each eccentricity. Although the absolute contrast threshold and the spread of the distribution varied across observers, the valid cue (blue histograms) improved performance and the invalid cue (red histograms) impaired performance with respect to the neutral cue for each individual observer at both eccentricities. Fig. 8 (right panel) illustrates the frequency distribution for the same observer. The same pattern of results, and of comparable magnitude, was obtained at both parafoveal and peripheral locations.

Results from all observers indicate that despite the fact that they were told that the cue was uninformative as to the target location and orientation, and despite the simplicity of the display, there is a performance trade-off: the cue increases sensitivity at the cued location (benefit) and impairs it at the uncued location (cost), as compared to the neutral condition. This indicates that information at the attended location is processed to a greater degree than in the neutral condition, and that

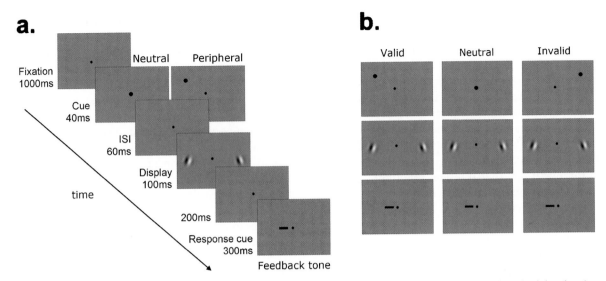

Fig. 7. (a) *A trial sequence.* Following a fixation point, a cue appeared either above one of the two Gabor locations (peripheral cue) or at fixation (neutral cue). After an ISI, two Gabors were simultaneously presented (randomly oriented to the left or to the right) on the horizontal meridian. Then a response cue appeared at fixation to indicate the target Gabor for which the observer had to report the orientation. On one third of the trials the response cue pointed to a precued Gabor. On another third of the trials it pointed to the Gabor that was not precued. In the remaining trials the precue was presented in the center of the screen and the response cue was equally likely to indicate the Gabor to the right or to the left of fixation. (b) *Examples of types of trials.* In a valid trial the locations indicated by the peripheral cue and by the response cue matched. In an invalid trial the locations indicated by the peripheral cue and by the response cue did not match. In a neutral trial the cue was presented at fixation and the response cue indicated the left Gabor in half of the trials and the right Gabor in the other half (Pestilli and Carrasco, 2005, Fig. 1).

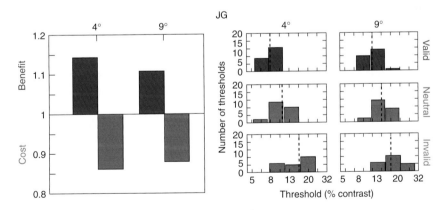

Fig. 8. The *left panel* depicts the ratios of the medians of the sensitivity (1/median threshold) for one observer for 4° (left two bars) and 9° (right two bars) eccentricity. A ratio >1 indicates a benefit of allocating attention to the target location (valid:neutral ratio: sensitivity in the valid condition is higher than sensitivity in the neutral condition). A ratio <1 indicates a cost of allocating attention to the nontarget location (invalid:neutral ratio: sensitivity in the invalid condition is lower than sensitivity in the neutral condition). The black vertical lines reflect the overall attentional effect, i.e., the valid:invalid ratio. The right panel contains six histograms representing the thresholds obtained for the same observer in each cue condition at 4° and 9° eccentricity. Blue histograms represent the threshold obtained for the valid condition; black histograms represent the neutral cue condition; red histograms represent the invalid cue condition. Black vertical lines indicate the median values. (Adapted from Pestilli and Carrasco, 2005, Fig. 2.)

information processed outside of the focus of attention is processed to a lesser degree. Given that for an ideal observer the uninformative cue would not reduce uncertainty, this finding supports sensitivity-based explanations, i.e., signal enhancement at the cued location — the sensory representation of the relevant stimuli is boosted — and distracter exclusion at the uncued location — the influence of the stimuli outside the attentional focus is reduced.

By illustrating that transient attention can help in managing the overall bioenergetic expenditure across the attended and unattended locations of the visual field, this study provides evidence for the notion that transient attention directs observers' attention to the cued location in an automatic fashion (Muller and Rabbitt, 1989; Nakayama and Mackeben, 1989; Cheal and Lyon, 1991; Yantis, 1996).

Transient attention increases apparent contrast
From recent psychophysical and neurophysiological evidence indicating that covert attention increases contrast sensitivity, one might infer that attention changes contrast appearance. But does attention alter appearance? Whether attention can actually affect the perceived intensity of a stimulus has been a matter of debate dating back to the founding fathers of experimental psychology and psychophysics — Helmholtz, James, and Fechner (Helmholtz, 1866/1911; James, 1890). Surprisingly, very little direct empirical evidence has been brought to bear on the issue (Tsal et al., 1994; Prinzmetal et al., 1997, 1998), and a number of methodological concerns limit the conclusions we can draw from these studies (Carrasco et al., 2004a,b; Luck, 2004; Treue, 2004; Gobell and Carrasco, 2005).

To directly investigate this issue, Carrasco et al., (2004a,b) implemented a novel paradigm that enables us to assess the effects of spatial cueing on appearance and to test subjective contrast. This paradigm allows one to objectively assess observers' subjective experience while circumventing methodological limitations of previous studies, and to address other questions about phenomenological experience, making it possible to study subjective

experience more objectively and rigorously (Luck, 2004; Treue, 2004).

Observers were briefly presented with either a peripheral or neutral cue, followed by two Gabor patches (tilted to the left or right) to the left and right of fixation (Fig. 9). The contrast of one of the Gabors was presented at a fixed contrast (standard), whereas the other varied in contrast randomly from a range of values around the standard (test patch). The orientation of each Gabor was chosen randomly. We manipulated transient attention with an uninformative peripheral cue. We asked the observers: what is the orientation of the stimulus that is higher in contrast? These instructions emphasized the orientation judgment, when in fact we were interested in their contrast judgments; i.e., the orientation discrimination task served as a 'cover story' task, which de-emphasized the fact that we were interested in the observers' subjective experience.

The results showed that transient attention significantly increased perceived contrast (Fig. 10). When a Gabor was peripherally cued, the point of subjective equality (PSE) was shifted — the apparent contrast of the stimulus for which transient attention had been drawn to was higher than when attention was not drawn there. That is to say, when observers attend to a stimulus, they perceive it to be of significantly higher contrast than when they perceive the same stimulus without attention.

We conducted multiple control experiments to rule out alternative accounts of these findings: (1) We increased the temporal separation between the cue onset and the display onset from 120 ms, the optimal time for transient attention, to 500 ms, when transient attention is no longer active. Consistent with the quick decay of transient attention to the cued location, this manipulation yielded no contrast enhancement of the cued stimulus, i.e., there is no appearance effect (data not shown). This result shows that observers were not biased to report the orientation of a cued stimulus per se. (2) When observers are asked to report the orientation of the Gabor of *lower* contrast, they select the cued stimulus less often if it is of the same contrast as the uncued stimulus (data not shown). This result is consistent with the enhanced apparent contrast of the cued stimulus observed in the main experiment. This control rules out the possibility

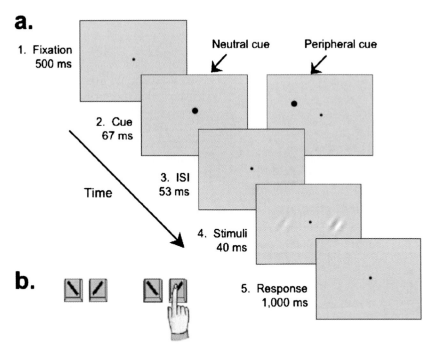

Fig. 9. (a) Sequence of events in a single trial. Each trial began with a fixation point followed by a brief neutral or peripheral cue. The peripheral cue had equal probability of appearing on the left- or right-hand side, and was not predictive of the stimulus contrast or orientation. The timing of this sequence maximized the effect of transient attention and precluded eye movements. (b) Task. Observers performed a 2 × 2 forced choice (2 × 2 AFC) task: they were asked to indicate the orientation (left vs. right) for the stimulus that appeared higher in contrast. In this trial, they would report the orientation for the stimulus on the right. (Carrasco et al., 2004a,b, Fig. 1.)

that observers report the orientation of a cued stimulus more often simply because they find its orientation easier to judge or are subject to some type of cue bias.

This study provides evidence for a contrast gain model (Reynolds et al., 1999, 2000) in which attention allows for greater neuronal sensitivity, suggesting that attention changes the strength of a stimulus by enhancing its effective contrast or salience. It is as if attention boosts the actual stimulus contrast. The finding that the cue not only enhanced the cued stimulus' appearance but also improved the observers' performance supports the hypothesis that the increased saliency at the target location seems to be the basis of perceptual judgments. Many have considered the saliency map to be the basis of perceptual judgments and a tool for directing gaze to potential relevant locations of the visual environment (Itti and Koch, 2001; Treue, 2004; Gobell et al., 2004; Itti, 2005; Zhaoping, 2005).

Sustained attention and contrast sensitivity

Single-unit studies have evaluated the effects of attention on the contrast response function by manipulating sustained attention (Reynolds et al., 2000; Martinez-Trujillo and Treue, 2002). To evaluate the similarity of the transient and sustained systems of attention, it is important to characterize their effects on early vision, and to investigate whether the same mechanism(s) can underlie such effects. Recently, Ling and Carrasco (2006) obtained contrast psychometric functions for both sustained and transient attention to further bridge the gap between neurophysiological and psychophysical results. We systematically compared sustained and transient covert attention using the same task, stimuli, and observers. We tested whether a signal enhancement mechanism underlies both types of attention. Moreover, we investigated the neural model underlying signal enhancement by measuring the psychometric functions for both sustained and

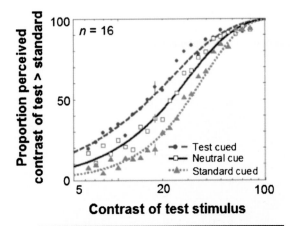

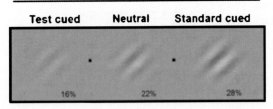

Fig. 10. Attention alters appearance. Top panel: Appearance psychometric function. Percentage of responses in which observers reported the contrast of the test patch as higher than the standard, plotted as a function of the physical contrast of the test patch. Data are shown for the neutral and peripheral conditions (test cued and standard cued). The standard was 22% contrast and that is the contrast at which the test and standard stimuli attained subjective equality (50%). Bottom panel: Effect of covert attention on apparent contrast. If you were looking at one of the two fixation points (black dots), and the grating to the left of that fixation point was cued, the stimuli at both sides of fixation would appear to have the same contrast. A cued 16% contrast grating appears as if it were 22% contrast, and a cued 22% contrast grating appears as if it were 28% contrast. (Adapted from Carrasco et al., 2004a,b, Figs. 4 and 5.)

transient attention to assess whether they have similar or different effects on the contrast response function.

As mentioned above, two types of gain control mechanisms have been considered in neural responses to luminance-modulated stimuli — contrast gain and response gain (Sclar et al., 1989; see Fig. 5). We had provided evidence in support of a contrast gain mechanism for transient attention (Cameron et al., 2002). However, in that study, performance asymptoted close to 100%, leaving little room at the higher contrasts for a possible test of response gain. Neurophysiological studies of sustained attention

that have evaluated these two mechanisms have avoided levels at which neural saturation occurs. Similarly, to properly compare contrast gain and response gain psychophysically, the psychometric functions should arise from a demanding task that ensures that performance on the neutral baseline condition does not asymptote at 100%, leaving room to test for response gain.

Observers performed a 2AFC orientation discrimination task on a slightly tilted Gabor patch. We first established the contrast range required to measure the full extent of the psychometric function with an asymptote that occurs at a performance level that allows room for benefit. We used the method of constant stimuli to measure performance as a function of target contrast in the neutral, transient, and sustained cue conditions. In each trial, a Gabor is presented in 1 of 8 possible isoeccentric locations. The cues (sustained and its neutral control vs. transient and its neutral control) are constant throughout a block, but the spatial frequency and contrast levels are randomized within each block.

Using a nested hierarchical model, for each observer we estimated the probability that the same Weibull distribution can describe the data sets for both cue conditions (sustained vs. its neutral control; transient vs. its neutral control), as opposed to two separate distributions. Additionally, to test the models of response gain vs. contrast gain, we fit the data to their respective models, along with a hybrid model of both response and contrast gain, and compare likelihoods to assess which model describes the data better. Whereas response gain predicts an increasing effect of attention with contrast (a multiplicative effect across the psychometric function), contrast gain predicts only a shift in sensitivity with attention (an overall additive effect independent of stimulus intensity). The hybrid model predicts both a shift in sensitivity as well a multiplicatively increasing effect of attention.

Results indicate that whereas sustained attention operates via contrast gain (Fig. 11; top panel; characterized by a shift in threshold), transient attention operates via a mixture of contrast and response gain (bottom panel; characterized by an effect even at high-contrasts asymptotic levels; Ling and Carrasco, 2006). An uncertainty reduction model of attention

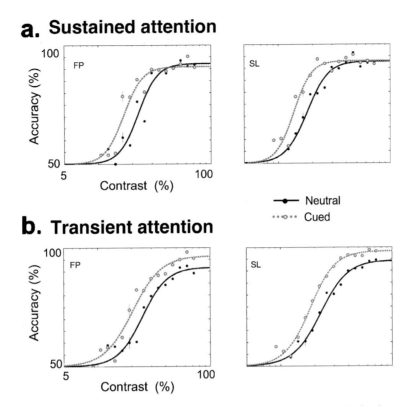

Fig. 11. Psychometric functions for sustained and transient attention. The solid line represents the fits for the neutral condition, and the dashed line represents the fits for the precued. (a) Sustained attention consistently shifted the function to the left, having little impact on its shape, but increasing contrast sensitivity. (b) Transient attention consistently led to an elevation in asymptote, and the fits suggest a decrease in contrast threshold as well. Error bars correspond to mean ± 1 standard error. (Adapted from Ling and Carrasco, 2005, Fig. 3.)

would predict that the attention effect should be most prominent with low-contrast stimuli (where uncertainty is greatest and performance would benefit most from uncertainty reduction) and decrease with increasing stimulus contrast (where uncertainty is diminished and performance would not benefit from uncertainty reduction). However, this was not the case in this study. Moreover, different signature responses across the psychometric function emerged notwithstanding the fact that the reduction of location uncertainty is the same in both cases.

Using the external noise paradigm, Lu and Dosher (1998, 2000) reported that transient covert attention seems to operate via both signal enhancement and external noise reduction. They showed that transient attention increases contrast sensitivity in conditions of low noise, indicative of signal enhancement, and also improves

performance in high-noise conditions, indicative of external noise reduction. However, they have attributed sustained attention effects only to an external noise reduction mechanism (Dosher and Lu, 2000a,b; Lu and Dosher, 2000, 2002).

With regard to transient attention, these and previous findings are in agreement; under low external noise conditions, it operates via signal enhancement. However, the results for sustained attention are inconsistent with those reported previously by Dosher and Lu. The most relevant difference that could help reconcile the discrepancy lies in the amount of time observers were given to deploy their sustained attention. The SOA in their studies was 150 ms because it has been reported that this time was enough for experienced observers to deploy sustained attention (Cheal and Lyon, 1991). Perhaps this short timing precluded emergence of the signal enhancement mechanism. It is possible that the observers who failed to show an effect were not trained optimally to deploy sustained attention within the allotted time; had they had longer time to deploy sustained attention an effect could have emerged.

In a sustained attention task, using a dual-task paradigm in which observers performed tasks under conditions of full or poor attention, evidence for pure response gain has been reported (Morrone et al., 2004). However, a subsequent psychophysical study suggested that dual task, sustained attention may operate via a hybrid model, involving both contrast gain and response gain (Huang and Dobkins, 2005). Whereas the dual-task paradigm has some advantages, such as eliminating location uncertainty reduction as an alternative explanation, it has disadvantages that may have hampered their conclusions. Dual-task paradigms do not control the deployment of attention very well and make it hard to isolate the source of possible processing differences (e.g., Sperling and Dosher, 1986; Pashler, 1998). The difference with the present results may be due to the way in which attention was manipulated. First, in dual-task paradigms, attention is not directed to a specific location, but the amount of resources being spread to all locations is manipulated. Second, to manipulate attention those authors withdrew attention from the target, whereas we directed attention toward the target.

This study systematically compared sustained and transient covert attention using the same task, stimuli, and observers. On the one hand, both types of attention had a similar effect on performance; they increased contrast sensitivity under zero-noise conditions (the display contained nothing to be suppressed, since there was no added external noise). Hence, we conclude that both attentional systems can be mediated by a signal enhancement mechanism. Furthermore, because this effect occurred even with very high-contrast stimuli, it cannot be explained by uncertainty reduction. On the other hand, sustained and transient attention had different effects on the contrast response function. Sustained attention enhances contrast sensitivity strictly via contrast gain, whereas, in addition to contrast gain, transient attention revealed response gain.

Neurophysiological studies of attentional modulation of apparent stimulus contrast: attentional facilitation and selection

The development of techniques to record the activity of neurons in awake-behaving animals has enabled researchers to probe the biological foundations of sustained attention. Single-unit recording studies in the monkey have provided detailed, quantitative descriptions of how attention alters visual cortical neuron responses.

A number of neurophysiological studies have shown that directing attention to a stimulus increases neuronal sensitivity, so that neurons respond to an attended stimulus much as they would were its luminance increased. It is possible to relate these findings to studies in anesthetized cats and monkeys documenting how luminance contrast affects neuronal responses. The same models explaining contrast-dependent changes in neuronal response can account for contrast-dependent modulation of the competitive interactions observed when multiple stimuli appear within a neuron's receptive field (for reviews see Reynolds and Chelazzi, 2004; Martinez-Trujillo and Treue, 2005; Reynolds, 2005).

With regard to attentional facilitation, consistent with psychophysical findings, single-unit recording studies have found that spatial attention can enhance responses evoked by a single stimulus appearing alone in a neuron's receptive field (e.g., Motter, 1993; Ito and Gilbert, 1999; McAdams and Maunsell, 1999; Reynolds et al., 2000). Reynolds et al. (2000) assessed the effects of sustained attention on contrast sensitivity when a single stimulus appeared in a neuron's receptive field. The monkey's task was to detect a target grating that could appear at an unpredictable time at the cued location. The target's luminance contrast was randomly selected to ensure that the monkey had to attend continually to the target location. The contrast response function (CRF) summarizes the way in which changes in stimulus contrast are translated into changes in neuronal firing rate via a nonlinear sigmoid function (Fig. 5). Consistent with a contrast gain, in V4, an extrastriate visual area at an intermediate stage of the ventral processing stream, attention shifts the CRF horizontally with the most pronounced changes occurring at its dynamic range (steepest region). When the grating stimulus appearing in the neuron's receptive field was below the contrast response threshold (5% and 10% contrast), it fails to elicit a response, when unattended. However, when the monkey attended to its location in the RF the same 10% contrast elicits the neuron to respond. Attention does not alter the neuronal response when the stimulus is above saturated contrast. Across a population of V4 neurons, the greatest increments in firing rate were observed at contrasts in the dynamic range of each neuron's CRF (Fig. 12). The finding that similar results were found for preferred and poor stimuli indicates that the lack of attentional effect at high contrast did not reflect an absolute firing rate limit; instead, it reflected a leftward shift in the contrast response function.

Under the conditions of this experiment, for a cell to reliably detect an unattended stimulus, its contrast needed to be 50% higher than that of the attended stimulus; i.e., attention was equivalent to about 50% increase in contrast (Reynolds et al., 2000). This value has been corroborated by other studies that have also quantified spatial attention in units of luminance contrast, including studies in MT (Martinez-Trujillo and Treue, 2002) and in V4 (Reynolds and Desimone, 2003), whose estimates were 50% and 56%, respectively. As mentioned above, this effect of attention is indistinguishable from a change in stimulus contrast (see also Maunsell and McAdams, 2000).

Given our limited ability to process information, it is also crucial to understand how attentional selection of behavioral relevant stimuli from among competing distracters (Wolfe, 1994; Palmer et al., 2000; Carrasco and McElree, 2001; Verghese, 2001; Cameron et al., 2004) may be instantiated at a neural level. Neuronal recordings within the extrastriate cortex have revealed a direct neural correlate of attentional selection. Moran and Desimone (1985) were the first to show that the firing rate is determined primarily by the task-relevant stimulus. This seminal study showed that when two stimuli are presented within the receptive field, the neuron's response to the pair is greater when the monkey is asked to identify the stimulus corresponding to the neuron's preferred color and orientation than when asked to identify the

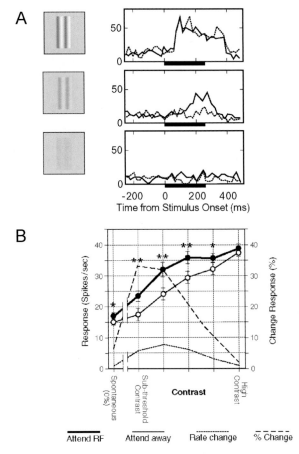

Fig. 12. Response of an example neuron from area V4 as a function of attention and stimulus contrast. (A) The contrast of the stimulus in the receptive filed increased from 5% (bottom panel) to 10% (middle panel) to 80% (top panel). The monkey had to detect a grating at the attended location. On each trial, attention was directed to either the location of the stimulus inside of the receptive field (solid line) or a location far away from the receptive field (dotted line). Attention reduced the contrast threshold to elicit a response (middle panel), but did not affect the response at saturation contrast (top panel). (B) Averages responses of V4 neurons while the monkey attends to the location (thick line) or away (thin line) of the receptive field (thin line). The horizontal line depicts the five different contrast values of the gratings presented inside the RF, which spanned the dynamic range of the neuron. The dashed and dotted lines show percentage and absolute difference in firing rate, respectively, across the two attention conditions, as a function of contrast. (Adapted with permission from Reynolds et al., 2000.)

nonpreferred stimulus. Several labs have replicated this observation that the attentional modulation depends on the similarity between the attended stimulus properties and the sensory preferences of the neuron, both in the ventral (Chelazzi et al., 1993, 1998, 2001; Reynolds et al., 1999; Sheinberg and Logothetis, 2001; Reynolds and Desimone, 2003) and in the dorsal (Treue and Maunsell, 1996; Treue and Martinez-Trujillo, 1999) streams.

For instance, Reynolds et al. (1999) found that in V4 the response to a pair of stimuli lies between the responses elicited by either stimulus alone, the preferred and the nonpreferred. When the monkey attends to the preferred stimulus the response to the pair increases so that it approaches the high response level elicited when it is presented by itself; conversely, when the monkey attends to the non-preferred stimulus the response is reduced so that it approaches the low response elicited when it is presented by itself (Fig. 13). In short, attending to the preferred stimulus increases the response to the pair but attending to the poor stimulus reduces the

response evoked by the pair. This results in increased saliency of the attended stimulus representation and a corresponding suppression of the neuronal representation of unattended stimuli.

Similar results have been obtained for a variety of stimuli. Martinez-Trujillo and Treue (2002) presented two pairs of random dot patterns, one inside the RF and one outside the RF. In each pair, the potential target moved in the null direction and the distracter moved in the preferred direction. They recorded the cell's responses to various contrast levels of the distracter patterns, moving in the preferred direction, in two attentional conditions, when the target was the stimulus outside or inside the receptive field (Fig. 14, top panel). Both sets of functions, attending inside and outside the RF were fit with a sigmoid function (Fig. 5). Attending to the null pattern inside the RF strongly increased threshold

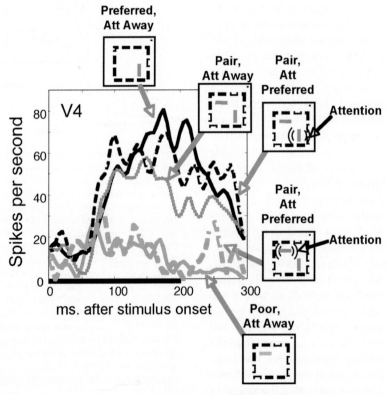

Fig. 13. Effect of attention to one of two stimuli in the RF of a V4 neuron. Each function shows the response of a single V4 neuron averaged over repeated stimulus presentations, under different stimulus and attentional conditions (indicated by the icon). When the preferred stimulus is attended the response for the stimulus pair increases approaching the level of response elicited when the preferred stimulus is presented alone; conversely, when the poor stimulus is attended the level of activity is reduced, approaching the level of activity elicited when the poor stimulus is presented alone. (Adapted with permission from Reynolds et al., 1999.)

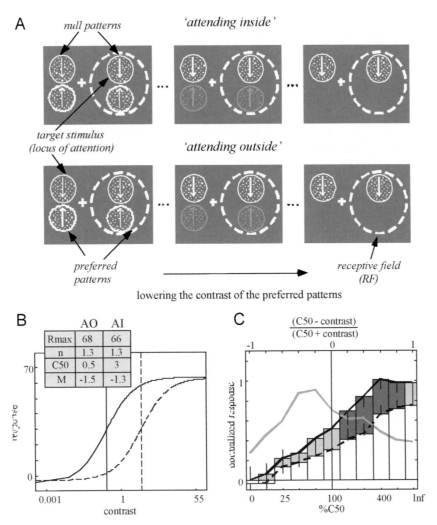

Fig. 14. (A) Two pairs of random dot patterns appeared simultaneously on the screen, one inside and the other outside the cell's receptive field. Each pair consisted of one preferred and one null pattern. The monkeys always attended to the null pattern, inside the receptive field (top row) or outside the receptive field (bottom row). From left to right the panels illustrate decreasing luminance value of the preferred patterns leading to a decrease in response. (B) Average responses of one MT neuron to different contrast levels in the attending outside and attending inside conditions. The vertical lines indicate the C50 value for each curve and the tables shows the values of four parameters (see Fig. 5 for R_{max}, C50, n, and M) (C) Average normalized responses after aligning the contrast response functions in all units to their respective C50 values in the attending outside condition. (Adapted with permission from Martinez-Trujillo and Treue, 2002.)

with only a slight change in asymptote (Fig. 14, bottom panels). This indicates that the response suppression to a distracter was stronger when the stimulus had intermediate contrast. The slight change in asymptote occurred despite the fact that the sampled cells were capable of much higher firing rates. This study demonstrates that the magnitude of attentional modulation depends on the stimulus

contrast of the unattended stimuli inside the receptive field. Attention had a stronger effect on responses of direction-selective MT neurons when a distracter presented in their receptive field had intermediate, rather than low or high contrast. This result is consistent with contrast gain. Martinez-Trujillo and Treue (2002) have suggested that the attentional effects observed in MT may result from the modulation of input gain, which would be similar to a change in stimulus contrast; that the attentional effects vary nonlinearly as a function of stimulus contrast indicates that these effects are not simple multiplications of a cell's response.

Likewise, Chelazzi et al. (2001) obtained a similar pattern of results while recording in area V4 of monkeys performing a visual search task involving objects (faces and houses). The finding that attending to the preferred stimulus increases the response to the pair but attending to the poor stimulus reduces the response evoked by the pair, provides support for attentional models positing that response suppression mediates the selection of one stimulus and the inhibition of the other (e.g., Desimone and Duncan, 1995; Ferrera and Lisberger, 1995; Lee et al., 1999).

The common nonlinearity between the CRF and the magnitude of attentional modulation across different areas of the dorsal and ventral pathways indicate a tight link between attentional mechanisms and the mechanisms responsible for contrast encoding. Indeed, it has been proposed that because this tight link does not seem to exist for the encoding mechanisms of other stimulus properties (e.g., motion coherence), attention can be considered a mechanism aimed at modulating stimulus saliency while leaving other stimulus properties relatively unchanged (Martinez-Trujillo and Treue, 2002).

A similar relative enhancement or suppression of stimulus representations can be obtained when varying stimulus contrast. Given that the role of contrast in modulating the visual response properties of neurons has been well documented, and that response suppression plays an important role in the models developed to account for these modulations, to better understand the role that attention may play in selecting stimuli, it is important to consider the role of suppression in the visual cortex (Reynolds and Chelazzi, 2004; Reynolds, 2005).

Contrast changes in the visual stimulus yield multiplicative changes in neuronal responses, similar to those evoked by attention (e.g., Tolhurst, 1973; Sclar and Freeman, 1982; Treue and Martinez-Trujillo, 1999). The similarity between the effects of attention and the effects of varying stimulus contrast could

indicate that these two processes are closely related and probably use similar mechanisms to modulate neuronal responses. Were this the case, one would expect attentional and contrast modulation of neuronal responses to share the same properties (Martinez-Trujillo and Treue, 2005; Reynolds, 2005).

It is known that contrast modulates neuronal responses in the following ways:

(1) Cortical neuronal responses typically saturate as contrast increases, and this saturation-firing rate is stimulus-dependent. The dynamic range is larger and the saturation point is higher for a stimulus of the neuron's preferred orientation than for a stimulus of poor but excitatory orientation, and no response is elicited by a nonpreferred stimulus. Thus, increasing contrast leads to a multiplicative increase in the response (Sclar and Freeman, 1982; Fig. 15A). Correspondingly, consistent with the idea of attentional facilitation, attention causes a leftward shift in the contrast response function by increasing the effective contrast of a stimulus (Reynolds et al., 2000; Fig. 12).

(2) Increasing the stimulus contrast characteristically results in a multiplicative increase in the neurons' tuning curves for properties such as motion and orientation (Sclar and Freeman, 1982; Fig. 15B). The multiplicative effect of contrast on the orientation-tuning curve is due to the fact that contrast-response functions derived for any orientations can be related to each other by a gain factor, as is the case in Fig. 12. Similarly, spatial attention causes a multiplicative increase in the orientation-tuning curve of neurons in area V4, without otherwise altering its shape. This increase in the gain of the orientation-tuning curve enables neuronal signals to better distinguish the orientation of the stimulus (McAdams and Maunsell, 1999). Likewise, attention multiplicatively increases the direction of motion-tuning curves of neurons in area MT (Treue and Martinez-Trujillo, 1999).

(3) When two spatially superimposed gratings of different orientation appear simultaneously

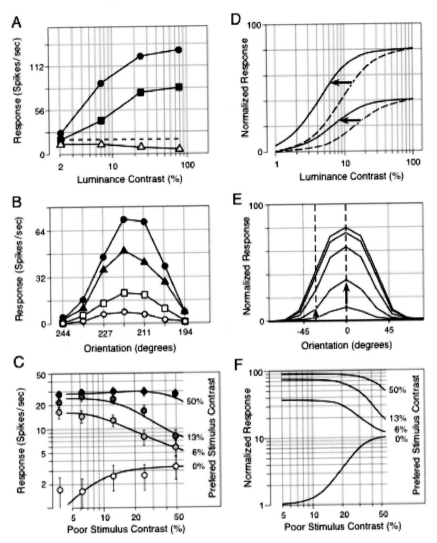

Fig 15. Contrast-dependent response modulations. (A) Contrast–response functions for a stimulus of the neuron's preferred orientation (upper line), a poor but excitatory orientation (middle line), and the null orientation (bottom line). (Adapted with permission from Sclar and Freeman, 1982.) (B) Orientation tuning curves of a second neuron, stimulus contrast varied from 10% (empty circles) to 80% (filled circles). (Adapted with permission from Sclar and Freeman (1982.) (C) Responses of a neuron recorded in area V1 of an anesthetized macaque. Two superimposed gratings appeared within the receptive field: one grating was of optimal orientation, the other was of a suboptimal orientation; both gratings varied from 0% to 50% contrast. (Adapted with permission from Carandini et al., 1997.) (D–F) The contrast gain model can account for these contrast-dependent response modulations. (See text; adapted with permission from Reynolds, 2005.)

within a receptive field in V1, increasing the contrast of one of them results in an increased or decreased response depending on the neuron's selectivity for the two stimuli. In general, increasing the contrast of the preferred grating increases the response to the pair; conversely, increasing the contrast of the nonpreferred grating decreases the response to the pair. However, the highest contrast preferred stimulus seems to be virtually immune to the suppressive effect of the nonpreferred stimulus (Carandini et al., 1997;

Fig. 15C). Correspondingly, this pattern of results is observed when attention is directed to one of two stimuli in the receptive field; attending to one of them will either increase or decrease the response, depending on the cell's relative preference for the two stimuli — the attended stimulus dominates the neuronal response (e.g., Luck et al., 1997; Reynolds et al., 1999; Treue and Martinez-Trujillo, 1999; Chelazzi et al., 2001; Martinez-Trujillo and Treue, 2002; Reynolds and Desimone, 2003).

(4) Similar modulations of the contrast-dependent response occur in V4 when two stimuli appear at separate locations in the visual field. On the one hand, a 5% contrast-poor stimulus has no measurable effect on the neuronal responses of the preferred stimulus but becomes increasingly suppressive as contrast increases. On the other, suppression is diminished if the preferred stimulus is elevated in contrast (Reynolds and Desimone, 2003; Fig. 16). Correspondingly, psychophysical studies show that attention increases contrast sensitivity (Pestilli and Carrasco, 2005; Fig. 8) and apparent contrast (Carrasco et al., 2004a,b; Fig. 9) at the attended location while reducing both contrast sensitivity and apparent contrast at the unattended location.

Based on these findings that facilitation is observed when attention is directed to a single stimulus appearing alone within the receptive field, and that when two stimuli appear within a neuron's receptive field, the neuronal response is dominated by the stimulus that is task relevant, Reynolds et al. (1999) have proposed the contrast gain model of attention. The linking hypothesis is that attention operates by multiplying the effective contrast of the behaviorally relevant stimulus or, equivalently, increases the neuron's contrast sensitivity. This model is mathematically related to models that account for the contrast-dependent effects described above. As would occur with an increase in the stimulus, contrast attention is assumed to lead to increases in the strength of the excitatory and inhibitory inputs activated by the attended stimulus (Reynolds et al., 1999; Fig. 12).

This effect results in a shift of the contrast response function to the left, just as in neurophysiological (Reynolds et al., 2000; Martinez-Trujillo and Treue, 2002) and psychophysical (Dosher and Lu, 2000; Cameron et al., 2002; Carrasco et al., 2004a,b; Ling and Carrasco, 2006; Figs. 6, 10, and 11) studies. See predictions of contrast gain model Fig. 15D. Also, as attention shifts contrast, its effect on the tuning curve is predicted to be the same as an increase in contrast: to cause a multiplicative increase in the tuning curve (McAdams and Maunsell, 1999; Treue and Martinez-Trujillo, 1999). See predictions of contrast gain model Fig. 15E. Moreover, as attention shifts contrast, its effect on the neuronal response depends on whether attention is directed toward the preferred or the nonpreferred stimulus (Reynolds et al., 2000; Chelazzi et al., 2001, Martinez-Trujillo and Treue, 2002). See predictions of contrast gain model in Fig. 15F.

Although it is not the focus of this chapter, it is important to mention that with respect to sustained attention, modulations of responses in the visual cortex occur as a result of feedback from areas like the lateral intraparietal area (LIP) and the frontal eye fields (FEF). At LIP, elevated responses are associated with increased contrast sensitivity at the behavioral level. At FEF, microstimulation causes spatially localized increases in sensitivity both at the behavioral level and in visual cortical neurons, which mimic the effect of spatial attention at the behavioral and the neuronal level (reviewed in Chelazzi and Reynolds, 2004).

Transient attention enhances perceptual performance and fMRI response in human visual cortex

Studies on brain mechanisms of attention have mostly examined sustained attention, and some of them have characterized its effects on stimulus processing in the visual cortex. For instance, in single-unit recording studies, researchers have learned that sustained attention can reduce external noise by reducing the influence of unattended stimuli (Moran and Desimone, 1985; Luck et al., 1997) and that it can also boost the signal by

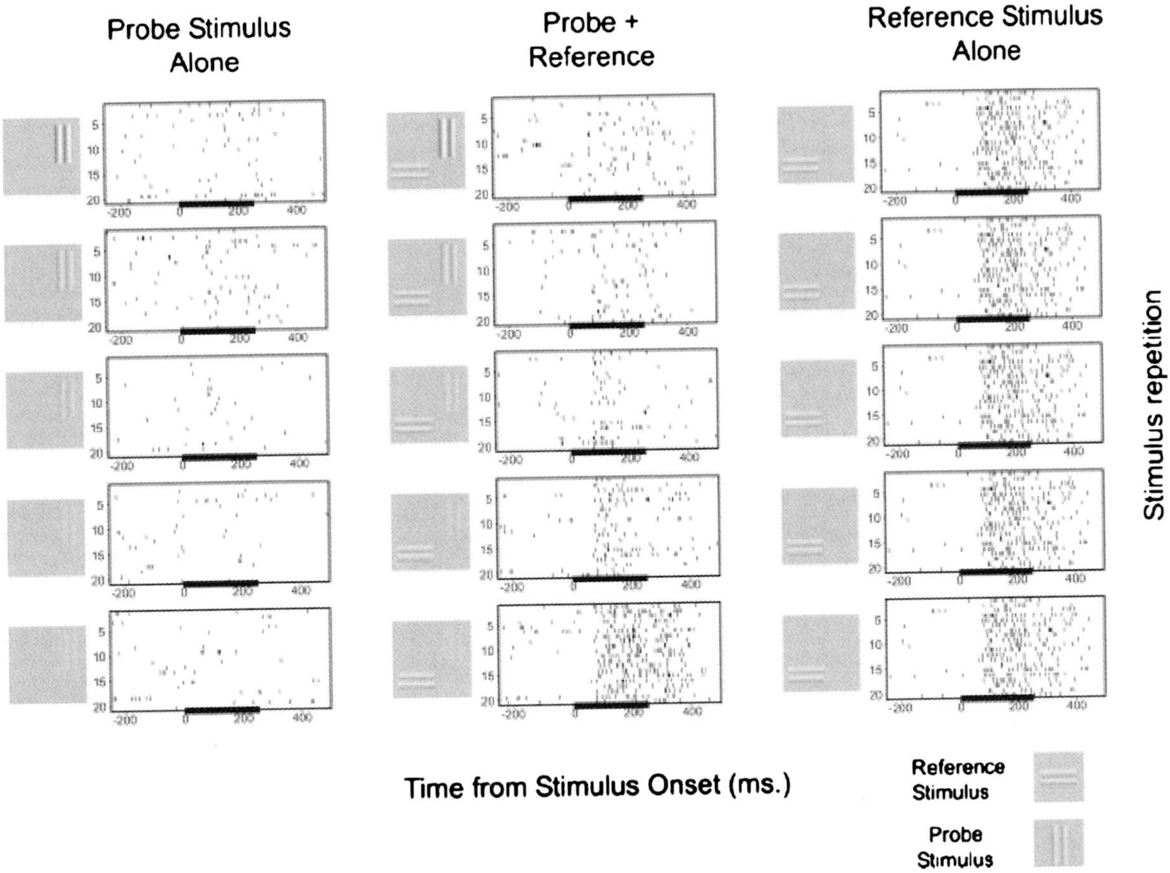

Probe Stimulus Alone — **Probe + Reference** — **Reference Stimulus Alone**

Time from Stimulus Onset (ms.)

Stimulus repetition

Reference Stimulus

Probe Stimulus

Fig. 16. Increasing the contrast of a poor stimulus at one location suppresses the response elicited by a fixed contrast preferred stimulus at a second location in the receptive field of a V4 neuron. The poor stimulus' contrast (left column) increased from 5 to 80%, and did not elicit a clear response at any contrast. The preferred stimulus was fixed in contrast (right column). For the response to the pair (middle column), at low contrast, the poor stimulus had no measurable effect on the response to the preferred stimulus, but as its contrast increased (moving up the column) it became increasingly suppressive. (Adapted with permission from Reynolds and Desimone, 2003.)

increasing the effective stimulus contrast (Reynolds et al., 2000; Martinez-Trujillo and Treue, 2002). Correspondingly, human electrophysiological studies have provided evidence that attention can increase sensory gain (Johannes et al., 1995; Hillyard and Anllo-Vento, 1998), and neuroimaging studies have shown attentional modulation of neural activity in many visual areas (Kanwisher and Wojciulik, 2000; Kastner and Ungerleider, 2000), including the primary visual cortex (Brefczynski and DeYoe, 1999; Gandhi et al., 1999; Martinez et al., 1999; Somers et al., 1999).

Less is known about the neural mechanism for transient attention and its effects on stimulus processing. Psychophysical findings demonstrating that transient attention increases contrast sensitivity (e.g., Lu and Dosher, 1998; Carrasco et al., 2000; Cameron et al., 2002; Ling and Carrasco, 2006) suggest that transient attention should enhance neural activity in early stages of visual processing. We tested this hypothesis by measuring brain activity in early visual areas using rapid event-related fMRI in conjunction with a peripheral cueing paradigm to manipulate transient attention (Liu et al., 2005). Participants discriminated the orientation of one of two gratings preceded or followed by a nonpredictive peripheral cue.

A number of previous neuroimaging studies in humans have examined the control mechanism of attentional capture in the frontoparietal network (reviewed in Corbetta and Shulman, 2002), but those studies have not addressed the effects of transient attention on the stimulus representation in the visual cortex. This is perhaps due to a potential measurement difficulty with the peripheral cueing paradigm used to manipulate transient attention. In this paradigm, a cue is briefly presented in the periphery and quickly followed by a stimulus nearby; the cue draws attention to the location of the upcoming stimulus. Because the spatiotemporal separation between the cue and stimulus is relatively small compared to the spatiotemporal resolution of imaging techniques, it is difficult to differentiate the sensory response to the cue and attentional modulation of the stimulus-evoked response and thus to rule out an explanation based on sensory summation.

We circumvented this methodological limitation with two innovations in our experimental design, involving a spatial and a temporal manipulation that complement each other. First, to anatomically separate the cue and stimulus responses we presented them above and below the horizontal meridian, respectively (Fig. 17a). Early retinotopic areas (V1, V2, and V3) form quadrant representations of the visual field (Horton and Hoyt, 1991) such that the cue and stimulus would activate the ventral and dorsal partition of the visual cortex, respectively. Because V1 has a contiguous hemifield representation, we determined the distance necessary to separate the cue and stimulus activity within V1. However, given that the hemifield representation and larger receptive fields of higher visual areas (e.g., V3a and hV4) are likely to give rise to overlapping activations of the cue and stimulus, and given subthreshold activation, imperfect image coregistration and surface reconstruction, it may not be possible to completely isolate the cortical locations activated by the cue and stimulus. Thus, in addition to the spatial control, we took advantage of the sluggishness of the hemodynamic response and evaluated the effect of postcue trials to control for the sensory effect of the cue. A postcue trial was identical to a precue trial, except that the temporal order of the cue and stimulus were reversed. The

two trial types had identical amounts of visual stimulation, but the postcue does not elicit transient attention. Because of the sluggishness of the hemodynamic response compared to the cue-stimulus interstimulus-interval (50 ms), a sensory response from a region that responded to both the cue and stimulus could not differentiate the order of the two. Thus, any differential effects between the precue and postcue conditions cannot be attributed to purely sensory summation of the hemodynamic response to the cue and stimulus, and must represent attentional modulation.

We presented two Gabor patches simultaneously in the periphery, one vertical and one tilted. Participants were asked to discriminate the orientation of the tilted Gabor (target); the vertical Gabor was a distracter. We used orientation discrimination to assess the effect of attention on stimulus contrast because performance on this task improves with increasing contrast (Nachmias, 1967; Skottun et al., 1987; Lu and Dosher, 1998;

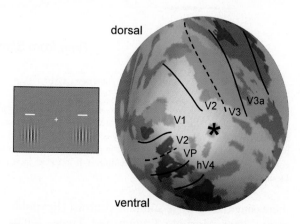

Fig. 17. The diagram on the left illustrates the locations of the cue and the Gabor stimulus, which were presented in alternating blocks. Shown on the right are data from the right hemisphere of one participant, viewed on an inflated surface representation of the posterior occipital cortex. Light and dark gray depict gyral and sulcal surfaces, respectively. Brain activity associated with the cue and the Gabor stimulus are shown in blue and green maps, respectively. Black lines indicate the borders of early visual areas defined by the retinotopic mapping procedure (solid line: vertical meridian, dashed line: horizontal meridian). The asterisk indicates the foveal confluence where borders between areas cannot be resolved. The activation of the cue and the Gabor did not overlap in V1, V2, and V3. Activity started to overlap in V3a and hV4, as they contain a hemifield representation (Liu et al., , 2005, Fig. 2.)

Cameron et al., 2002) and because fMRI response increases monotonically with stimulus contrast (Boynton et al., 1999). Each Gabor was either preceded or followed by a cue that was either valid or invalid. As mentioned above, the terms 'valid' and 'invalid' refer to whether the cue and target appeared on the same or on opposite sides, respectively. In fact, the cue was not predictive of either the location of the target (50% validity) or of its orientation. Participants were explicitly told that the cue was completely uninformative regarding both target location and orientation and that there was no benefit whatsoever in using the cue to perform the task (Carrasco et al., 2004a,b; Gobell and Carrasco, 2005).

In each experimental session, we localized cortical regions responding to the target stimuli and performed region-of-interest analyses on the fMRI signal in early visual areas V1, V2, V3, V3a, and hV4. Representative results from the localizer scan are shown in Fig. 17b on an inflated right hemisphere. The blue and green maps depict activations for the cue and the Gabor stimulus, respectively. Consistent with the known retinotopic organization of early visual areas, the cue and the Gabor stimulus largely activated ventral and dorsal regions of visual cortex, respectively. The separation between cue and Gabor activity was evident in V1 even though the dorsal and ventral representations are contiguous in that area. Activations for the cue and the Gabor remained separate in dorsal V2 and V3, whereas they started to overlap in higher areas such as V3a and hV4, which was expected as these areas contain a hemifield representation (Tootell et al., 1997; Wade et al., 2002).

Discrimination accuracy and reaction time (RT) were computed for each participant in each condition, and the group average is shown in Fig. 18a. The valid precue condition produced the highest accuracy — the accuracy of valid precue condition was higher than that of the invalid precue, valid postcue and invalid postcue conditions; accuracy did not differ for these three conditions. Correspondingly, the valid precue also yielded the shortest responses — RT in the valid precue condition was faster than in the invalid precue, which in turn was faster than for the valid postcue, with

the invalid postcue being the slowest. This pattern of results indicates that there was no speed-accuracy tradeoff across different conditions.

The group-averaged estimates of the fMRI response for contralateral targets and distracters are shown in Fig. 18b. A given trial always contained one target (tilted Gabor) and one distracter (vertical Gabor) in opposite hemifields. The activity level of different conditions did not differ significantly in V1, whereas activity for the valid precue condition was higher than for the other conditions in extrastriate visual areas. We tested for the possible effect of sensory summation of the cue and target, we compared valid postcue vs. invalid precue or vs. invalid postcue conditions. These comparisons yielded no significant effects in any visual area. All observers followed the same pattern of results. If a mere summation of the sensory response of the cue and target were responsible for the differences among the experimental conditions, we should have observed similar levels of fMRI response for the precue and postcue trials. The finding that activity was higher for the valid precue than for the valid postcue conditions allows us to rule out the possibility that the enhanced fMRI signal was due to low-level sensory effects of the cue.

Compared to control conditions, precueing the target location improved performance and produced a larger fMRI response in corresponding retinotopic areas. This enhancement progressively increased from striate to extrastriate areas. Control conditions indicated that enhanced fMRI response was not due to sensory summation of cue and target signals. Thus, an uninformative precue increases both perceptual performance and the concomitant stimulus-evoked activity in early visual areas. These results provide evidence regarding the retinotopically specific neural correlate for the effects of transient attention on early vision.

To further quantify the effect of transient attention, we calculated an attention modulation index, similar to that used in single-unit physiology (Treue and Maunsell, 1996). A large attentional effect leads to an AMI value close to 1, and a small effect leads to an AMI value close to 0. The attention modulation index increased gradually from V1 to extrastriate visual areas (from about 0.05 to about

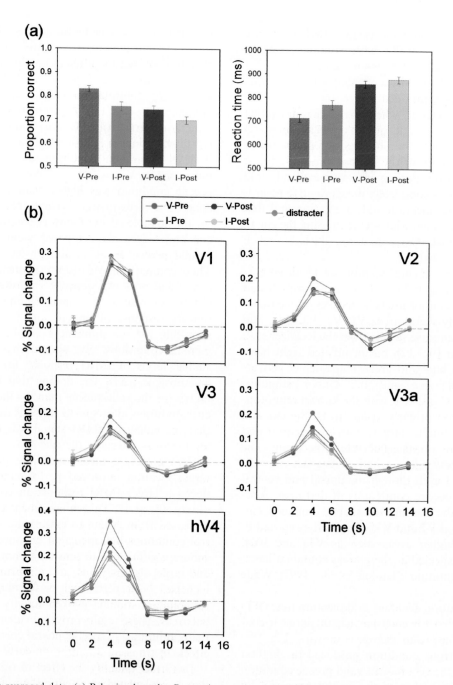

Fig. 18. Group-averaged data. (a) Behavioral results. Proportion correct (left) and reaction time (right) are shown for the four cue conditions (V-Pre: valid precue, I-Pre: invalid precue, V-Post: valid postcue, I-Post: invalid postcue). Error bars are 1 s.e.m. (b) Imaging results. Mean fMRI responses across participants for each cue condition and distracter are shown for each visual area. Response was obtained from the dorsal (V1, V2, V3 and V3a) and ventral (hV4) representations of the target (the green areas in Fig. 15). The average standard error of all time-points along a curve is shown as the error bar on the first time-point. (Adapted with permission from Liu et al., 2005, Fig. 3.)

0.3). Larger attentional effects in higher visual areas have also been found in studies of sustained attention (e.g., Kastner et al., 1999; Maunsell and Cook, 2002). Such a pattern is consistent with top-down modulation from frontal and parietal areas feeding back to the visual cortex, with diminishing effects in earlier visual areas. However, the attentional gradient could also be due to a feed-forward mechanism in which attentional modulation accumulates across sequential levels of processing. Whereas it has been established that sustained attention, a conceptually driven mechanism, is mediated by a feedback mechanism (Desimone and Duncan, 1995; Kanwisher and Wojciulik, 2000; Kastner and Ungerleider, 2000; Schroeder et al., 2001; Corbetta and Shulman, 2002), a feed-forward model seems more likely in the case of transient attention, a stimulus-driven mechanism. Such a feed-forward model could be implemented by steeper CRFs in extrastriate than striate areas. This higher sensitivity in extrastriate areas is due to areal summation across progressively larger receptive fields in higher areas (Sclar et al., 1990). Given that attention can boost the signal by increasing the effective stimulus contrast (Reynolds et al., 2000; Martinez-Trujillo and Treue, 2002) via contrast gain (Reynolds et al., 2000; Cameron et al., 2002; Martinez-Trujillo and Treue, 2002; Carrasco et al., 2004a,b; Ling and Carrasco, 2006), its effect would be more pronounced in areas with steeper CRFs.

To conclude, these results show that transient attention increases neural activity at the retinotopic locations of the subsequent target stimulus. The spatial and temporal parameters used enabled us to rule out a pure sensory explanation of this effect. We demonstrated that a nonpredictive peripheral cue increased both behavioral performance and retinotopic-specific neural response to a subsequent stimulus. Our results are the first to provide a retinotopically specific neural correlate for the *effects* of transient attention on early vision with a concomitant behavioral effect. The increased fMRI response in visual cortex brought about by transient attention provides the neural correlate of enhanced behavioral performance in an early visual task — enhanced contrast sensitivity in orientation discrimination. Previous research in single-unit physiology (Reynolds et al., 2000;

Martinez-Trujillo and Treue, 2002) and human psychophysics (e.g., Lu and Dosher, 1998; Lee et al., 1999; Carrasco et al., 2000, 2004a,b; Cameron et al., 2002; Ling and Carrasco, 2006; Pestilli and Carrasco, 2005) indicate that covert attention increases contrast sensitivity. By supplying evidence from an intermediate scale of analysis — neuroimaging — this study narrows the gap between single-unit physiology and human psychophysics of attention.

Conclusion

As remarkable as the human visual and cognitive systems may be, inevitably we are still limited by both bandwidth and processing power. There is a fixed amount of overall energy consumption available to the brain, and the cost of cortical computation is high. Attention is crucial in optimizing the systems' limited resources.

This chapter has focused on the effects of spatial attention on contrast sensitivity, for which the best mechanistic understanding of visual attention has been achieved due to the confluence of psychophysical, electrophysiological, and neuroimaging studies.

I illustrated how psychophysical studies allow us to probe the human visual system. Specifically, I discussed studies showing that attention enhances contrast sensitivity, and how these studies allow us to characterize the underlying mechanisms, namely external noise reduction and signal enhancement. It is reasonable to assume that attentional modulation may reflect a combination of mechanisms, such as reduction of external noise, reduction of spatial uncertainty, and signal enhancement (e.g., Carrasco et al., 2000, 2002; Dosher and Lu, 2000; Pestilli and Carrasco, 2005). Many of the studies conducted in my lab have been designed to isolate and evaluate the existence of signal enhancement; they indicate that increased contrast sensitivity and increased spatial resolution can be mediated by signal enhancement (Carrasco, 2005). Psychophysical studies characterizing the effects of transient and sustained attention have shown that covert attention increases the signal across the contrast sensitivity function (Carrasco et al., 2000) and across the contrast psychometric function (Cameron et al., 2002). Both sustained and transient attention can be mediated by signal enhancement, as revealed by the finding that the increased contrast sensitivity emerges under conditions of zero-external noise (Ling and Carrasco, 2006).

We have shown that the attentional effect exceeds the effect predicted by reduction of location uncertainty. For instance, although location uncertainty is greater at low- than at high-performance levels, the magnitude of the attentional benefit is similar regardless of the likelihood of observers confusing the target with blank locations. Attention increases sensitivity throughout the psychometric function of contrast sensitivity to the same extent for stimuli that differ in spatial uncertainty (Cameron et al., 2002; Ling and Carrasco, 2006) and even when localization performance indicates that there is no uncertainty with regard to the target location (Carrasco et al., 2000). In addition, the presence of a local postmask, which reduces location uncertainty, does not affect the magnitude of the attentional benefit (Carrasco et al., 2000).

To explore how the enhancement of contrast sensitivity at the attended location comes about we investigated whether covert attention affects the tuning of a spatial frequency channel. Attention halved the contrast threshold necessary for letter identification. However, we found no change in the tuning of the channel mediating letter identification: covert attention did not affect the peak frequency of the channel or the channel bandwidth (Talgar et al., 2004; Pestilli et al., 2004).

Investigating whether the enhancement of contrast sensitivity at the attended location has a concomitant cost at other locations, we found that compared to a neutral condition, an uninformative peripheral precue improves discrimination performance at the cued location and impairs it at the uncued location. This was the case despite the simplicity of the display and despite the fact that observers knew the cue was uninformative, they were explicitly told that the cues contained no information regarding either the location or the orientation of the target (Pestilli and Carrasco, 2005). The presence of a benefit and a cost reflects the bioenergetic limitations of the system. These changes are consistent with the idea that attention elicits two types of mechanisms: signal enhancement — the sensory representation of the relevant stimuli is boosted — and external noise reduction — the influence of the stimuli outside the attentional focus is reduced. In addition, this pattern of results confirms the stimulus-driven, automatic nature of transient attention. Similar results have been reported for contrast appearance (Carrasco et al., 2004a,b), spatial frequency appearance (Gobell and Carrasco, 2005), accuracy and temporal dynamics of visual search (Giordano et al., 2004), and for accuracy of letter identification (Luck and Thomas, 1999).

Pestilli and Carrasco (2005) documented the effect of transient attention on *performance* in an orientation discrimination task. The effect of transient attention on *apparent contrast* is remarkably consistent: compared to a neutral cue, apparent contrast is increased at the cued location and decreased at the other location (Carrasco et al., 2004a,b). This appearance study has been considered a crucial step in completing a chain of findings that provide insights with regard to the immediate perceptual consequences of attention (Treue, 2004). This chain is composed of neurophysiological results indicating that: varying contrast levels create multiplicatively scaled tuning curves (e.g., Sclar and Freeman, 1982); attention similarly scales neural responses (McAdams and Maunsell, 1999; Treue and Martinez-Trujillo, 1999); attention influences contrast gain mechanisms (Di Russo et al., 2001; Cameron et al., 2002; Ling and Carrasco, 2006); and that attentional modulation and changes in stimulus contrast create identical and therefore indistinguishable modulation of firing rates (Reynolds et al., 2000; Treue, 2001; Martinez-Trujillo and Treue, 2002).

As mentioned above, the high bioenergetic cost of firing entails the visual system to use neural coding that relies on very few active neurons (Barlow, 1972). For many perceptual aspects, e.g., to distinguish figure and ground, it is advantageous for the system to enhance contrast in an economic fashion. Treue (2004) has pointed out that much like the center-surround organization of visual receptive fields that serves to enhance the perceived contrast of luminance edges, attention is another tool providing an organism with an optimized representation of the sensory input that emphasizes relevant details, even at the expense of a faithful representation of the sensory input. Indeed, many human psychophysical studies (e.g., Itti and Koch, 2001; Itti, 2005; Zhaoping, 2005) as well as monkey single-unit recording studies (e.g., Reynolds and Desimone, 2003; Treue, 2004) have likened attention to increasing visual salience.

Both sustained and transient attention can increase contrast sensitivity by increasing the signal; however, these attentional systems have different effects on the CRF: sustained attention enhances contrast sensitivity strictly by contrast gain, whereas transient attention does so by a mixture of contrast gain and response gain (Ling and Carrasco, 2006). Our psychophysical findings for sustained attention are consistent with single-cell studies showing that the increased sensitivity brought about by sustained attention is mediated by contrast gain (e.g., Reynolds et al., 2000, Martinez-Trujillo and Treue, 2002). Obviously, comparisons between psychophysical and neurophysiological results need to be made with caution. Whereas the results of psychophysical studies presumably represent the response of the entire visual system, neurometric response functions are based on the response of single neurons or groups of neurons confined to particular regions of the visual system. Moreover, to date, there are no studies of single-unit recordings dealing with transient attention. Nevertheless, the link between psychometric and neurometric findings is tenable; for simple visual tasks such as motion discrimination, responses from single-unit recordings in MT are capable of accounting for behavioral psychometric functions (Britten et al., 1992).

There are several ways in which the link of psychometric and neurometric functions can be strengthened. First, it would be ideal that while characterizing single-unit activity, neurophysiological studies would index behavioral effects. Second, it would be ideal to implement a paradigm that enables the investigation of the effects of transient attention in awake-behaving monkey to develop a system's model of this stimulus-driven attentional system. The lack of single-cell studies of transient attention is probably due to the fact that it is hard to disentangle the effect of transient attention from a sensory cue effect. As mentioned above, we have been able to overcome such limitations and isolate the effect of transient attention in an fMRI study (Liu et al., 2005). Although the methodological challenges and the possible way to overcome them differ, meeting this challenge would significantly advance the field.

A third way to fortify the link of psychometric and neurometric functions is to conduct neuroimaging studies, in particular fMRI, as they provide an intermediate level of analysis capable of indexing retinotopic activity. In my opinion, the usefulness of fMRI studies of attention in narrowing the gap between psychophysical and electrophysiological studies depends on our understanding of the behavioral task performed during imaging, and the degree to which these studies can provide a neural correlate for the effects of attention on vision with a concomitant behavioral effect.

A fourth way is to take more seriously the idea of including biological constraints in the modeling of attention and in the generation of psychophysical experiments. For instance, in this chapter, I discussed how the wealth of knowledge regarding contrast-dependent changes in neuronal response could account for contrast dependent modulation of the competitive interaction observed when multiple stimuli appear within a neuron's receptive field. We could implement psychophysical paradigms to exploit all aspects of this parallel.

Although not the topic of this chapter, it is worth mentioning that attention speeds information processing (Carrasco and McElree, 2001; Carrasco et al., 2004a,b), but the neural basis of this effect is unknown. We know that speed of processing increases with stimulus contrast (Albrecht, 1995; Carandini et al., 1997). We also know that the effect of attention on contrast sensitivity is akin to increasing stimulus contrast (e.g., Reynolds et al., 2000; Carrasco et al., 2004a,b). However, increasing stimulus contrast seems to accelerate information processing to a lesser degree than the speeding of processing time brought about by attention. It remains to be explored, psychophysically and neurophysiologically, to what degree the effect of attention on contrast may mediate its effect on the speed of information processing.

To close, our understanding of visual attention has been advanced by the integration of different levels of analysis and methodologies. In this chapter, it has been illustrated how combining knowledge gathered from single-unit neurophysiology, psychophysics, and neuroimaging techniques, proves useful to understanding the way in which attention increases contrast sensitivity, in particular, and how attention alters perception.

Acknowledgments

We thank the past and present lab members, in particular Leslie Cameron, Sam Ling, Taosheng Liu, Cigdem Penpeci-Talgar, and Franco Pestilli, coauthors in the psychophysical and neuroimaging research described here, to Stuart Fuller, Sam Ling, Taosheng Liu, and Franco Pestilli, for helpful comments on this manuscript, to Jun fukukura for editorial assistance, and to John Reynolds and Stephan Treue for allowing us to reprint figures from their papers.

References

Albrecht, D.G. (1995) Visual cortex neurons in monkey and cat: effect of contrast on the spatial and temporal phase transfer functions. Vis. Neurosci., 12: 1191–1210.

Attwell, D. and Laughlin, S.B. (2001) An energy budget for signaling in the grey matter of the brain. J. Cereb. Blood Flow. Metab., 21: 1133–1145.

Baldassi, S. and Burr, D.C. (2000) Feature-based integration of orientation signals in visual search. Vision Res., 40(10–12): 1293–1300.

Baldassi, S., Burr, D.C., Carrasco, M., Eckstein, M. and Verghese, P. (2004) Visual attention. Vision Res., 44: 1189–1191.

Baldassi, S. and Verghese, P. (2002) Comparing integration rules in visual search. J. Vis., 2(8): 559–570.

Barlow, H.B. (1972) Single units and sensation: a neuron doctrine for perceptual psychology? Perception, 1: 371–394.

Bashinski, H.S. and Bacharach, V.R. (1980) Enhancement of perceptual sensitivity as the result of selectively attending to spatial locations. Percept. Psychophys., 28(3): 241–248.

Blanco, M.J. and Soto, D. (2002) Effects of spatial attention on detection and identification of oriented lines. Acta Psychol. (Amst),, 109(2): 195–212.

Boynton, G.M., Demb, J.B., Glover, G.H. and Heeger, D.J. (1999) Neuronal basis of contrast discrimination. Vision Res., 39(2): 257–269.

Brefczynski, J.A. and DeYoe, E.A. (1999) A physiological correlate of the 'spotlight' of visual attention. Nat. Neurosci., 2(4): 370–374.

Britten, K., Shadlen, M.N., Newsome, W.T. and Movshon, J.A. (1992) The analysis of visual motion: a comparison of neuronal and psychophysical performance. J. Neurosci., 12: 4745–4767.

Cameron, E.L., Tai, J.C. and Carrasco, M. (2002) Covert attention affects the psychometric function of contrast sensitivity. Vision Res., 42(8): 949–967.

Cameron, E.L., Tai, J.C., Eckstein, M.P. and Carrasco, M. (2004) Signal detection theory applied to three visual search tasks: identification, yes/no detection and localization. Spat. Vis., 17(4–5): 295–325.

Carandini, M., Heeger, D.J. and Movshon, J.A. (1997) Linearity and normalization in simple cells of the macaque primary visual cortex. J. Neurosci., 17: 8621–8644.

Carrasco, M. (2005) Transient covert attention increases contrast sensitivity and spatial resolution: support for signal enhancement. In: Itti, L., Rees, G. and Tsotsos, J. (Eds.), Neurobiology of Attention. Elsevier, San Diego, pp. 442–447.

Carrasco, M., Giordano, A.M. and McElree, B. (2004a) Temporal performance fields: visual and attentional factors. Vision Res., 44(11): 1351–1356.

Carrasco, M., Ling, S. and Read, S. (2004b) Attention alters appearance. Nat. Neurosci., 7(3): 308–313.

Carrasco, M. and McElree, B. (2001) Covert attention accelerates the rate of visual information processing. Proc. Natl. Acad. Sci. USA, 98(9): 5363–5367.

Carrasco, M., Penpeci-Talgar, C. and Eckstein, M.P. (2000) Spatial covert attention increases contrast sensitivity across the CSF: support for signal enhancement. Vision Res., 40: 1203–1215.

Carrasco, M., Talgar, C.P. and Cameron, E.L. (2001) Characterizing visual performance fields: effects of transient covert attention, spatial frequency, eccentricity, task and set size. Spat. Vis., 15(1): 61–75.

Carrasco, M., Williams, P.E. and Yeshurun, Y. (2002) Covert attention increases spatial resolution with or without masks: support for signal enhancement. J. Vis., 2: 467–479.

Carrasco, M. and Yeshurun, Y. (1998) The contribution of covert attention to the set-size and eccentricity effects in visual search. J. Exp. Psychol. Hum. Percept. Perform., 24(2): 673–692.

Cheal, M. and Lyon, D. (1991) Central and peripheral precuing of forced-choice discrimination. Q. J. Exp. Psychol. A, 43A(4): 859–880.

Chelazzi, L., Miller, E.K., Duncan, J. and Desimone, R. (2001) Responses of neurons in macaque area V4 during memory-guided visual search. Cereb. Cortex, 11: 761–772.

Chelazzi, L., Duncan, J., Miller, E.K. and Desimone, R. (1998) Responses of neurons in inferior temporal cortex during memory-guided visual search. J. Neurophysiol., 80: 2918–2940.

Chelazzi, L., Miller, E.K., Duncan, J. and Desimone, R. (1993) A neural basis for visual search in inferior temporal cortex. Nature, 363: 345–347.

Cohn, T.E. (1981) Absolute threshold: analysis in terms of uncertainty. J. Opt. Soc. Am., 71(6): 783–785.

Corbetta, M. and Shulman, G.L. (2002) Control of goal-directed and stimulus-driven attention in the brain. Nat. Rev. Neurosci., 3(3): 201–215.

Davis, E.T., Kramer, P. and Graham, N. (1983) Uncertainty about spatial frequency, spatial position, or contrast of visual patterns. Percept. Psychophys., 33(1): 20–28.

Desimone, R. and Duncan, J. (1995) Neural mechanisms of selective visual attention. Annu. Rev. Neurosci., 18: 193–222.

De Valois, R.L. and De Valois, K.K. (1988) Spatial Vision. New York, Oxford University Press.

Di Russo, F., Spinelli, D. and Morrone, M.C. (2001) Automatic gain control contrast mechanisms are modulated by attention in humans: evidence from visual evoked potentials. Vision Res., 41: 2435–2447.

Dosher, B.A. and Lu, Z.L. (2000a) Mechanisms of perceptual attention in precuing of location. Vision Res., 40: 1269–1292.

Dosher, B.A. and Lu, Z.L. (2000b) Noise exclusion in spatial attention. Psychol. Sci., 11: 139–146.

Eckstein, M.P. and Whiting, J.S. (1996) Visual signal detection in structured backgrounds. I. Effect of number of possible spatial locations and signal contrast. J. Opt. Soc. Am., 13(9): 1777–1787.

Eriksen, C.W. and Hoffman, J.E. (1974) Selective attention: noise suppression or signal enhancement? Bull. Psychon. Soc., 4(6): 587–589.

Ferrera, V.P. and Lisberger, S.G. (1995) Attention and target selection for small pursuit eye movements. J. Neurosci., 15: 7472–7484.

Foley, J.M. and Schwartz, W. (1998) Spatial attention: effect of position uncertainty and number of distractor patterns on the threshold-versus-contrast function for contrast discrimination. J. Opt. Soc. Am., 15(5): 1036–1047.

Gandhi, S.P., Heeger, D.J. and Boynton, G.M. (1999) Spatial attention affects brain activity in human primary visual cortex. Proc. Natl. Acad. Sci. USA, 96(6): 3314–3319.

Giordano, A.M., McElree, B. and Carrasco, M. (2003) On the Automaticity of Transient Attention. Annual Meeting of Psychonomic Society, Vancouver, Canada.

Giordano, A.M., McElree, B. and Carrasco, M. (2004) On the automaticity and flexibility of covert attention. http://journalofvision.org/4/8/524/

Gobell, J.L. and Carrasco, M. (2005) Attention alters the appearance of spatial frequency and gap size. Psychol. Sci., 16: 644–651.

Gobell, J.L., Tseng, C.H. and Sperling, G. (2004) The spatial distribution of visual attention. Vision Res., 44: 1273–1296.

Golla, H., Ignashchenkova, A., Haarmeier, T. and Their, P. (2004) Improvement of visual acuity by spatial cueing: a comparative study in human and non-human primates. Vision Res., 44(13): 1589–1600.

Graham, N. (1989) Visual Pattern Analyzers. New York, Oxford University Press.

Hawkins, H.L., Hillyard, S.A., Luck, S.J., Mouloua, M., Downing, C.J. and Woodward, D.P. (1990) Visual attention modulates signal detectability. J. Exp. Psychol. Hum. Percept. Perform., 16(4): 802–811.

Helmholtz, H.V. (1866/1911) Treatise on Physiological Optics. Rochester, Continuum.

Hikosaka, O., Miyauchi, S. and Shimojo, S. (1993) Focal visual attention produces illusory temporal order and motion sensation. Vision Res., 33: 1219–1240.

Hillyard, S.A. and Anllo-Vento, L. (1998) Event-related brain potentials in the study of visual selective attention. Proc. Natl. Acad. Sci. USA, 95(3): 781–787.

Horton, J.C. and Hoyt, W.F. (1991) The representation of the visual field in human striate cortex. a revision of the classic Holmes map. Arch. Ophthal., 109(6): 816–824.

Huang, L. and Dobkins, K.R. (2005) Attentional effects on contrast discrimination in humans: evidence for both contrast gain and response gain. Vision Res., 45: 1201–1212.

Ito, M. and Gilbert, C.D. (1999) Attention modulates contextual influences in the primary visual cortex of alert monkeys. Neuron, 22: 593–604.

Itti, L. (2005) Models of bottom-up attention and saliency. In: Itti, L., Rees, G. and Tsotsos, J. (Eds.), Neurobiology of Attention. Elsevier, San Diego, pp. 576–582.

Itti, L. and Koch, C. (2001) Computational modeling of visual attention. Nat. Rev. Neurosci., 2: 194–203.

James, W. (1890) The Principles of Psychology. New York, Henry Holt.

Johannes, S., Munte, T.F., Heinze, H.J. and Mangun, G.R. (1995) Luminance and spatial attention effects on early visual processing. Brain Res. Cogn. Brain Res., 2: 189–205.

Jonides, J. (1981) Voluntary vs. automatic control of the mind's eye's movement. In: Long, J.B. and Baddeley, A. (Eds.), Attention and Performance IX. Hillsdale, NJ, Erlbaum, pp. 187–204.

Jonides, J. and Yantis, S. (1988) Uniqueness of abrupt visual onset in capturing attention. Percept. Psychophys., 43(4): 346–354.

Kanwisher, N. and Wojciulik, E. (2000) Visual attention: insights from brain imaging. Nat. Rev. Neurosci., 1: 91–100.

Kastner, S., Pinsk, M.A., De Weerd, P., Desimone, R. and Ungerleider, L.G. (1999) Increased activity in human visual cortex during directed attention in the absence of visual stimulation. Neuron, 22(4): 751–761.

Kastner, S. and Ungerleider, L.G. (2000) Mechanisms of visual attention in the human cortex. Annu. Rev. Neurosci., 23: 315–341.

Kinchla, R.A. (1992) Attention. Annu. Rev. Psychol., 43: 711–742.

Lee, D.K., Itti, L., Koch, C. and Braun, J. (1999) Attention activates winner-take-all competition among visual filters. Nat. Neurosci., 2(4): 375–381.

Lee, D.K., Koch, C. and Braun, J. (1997) Spatial vision thresholds in the near absence of attention. Vision Res., 37(17): 2409–2418.

Lennie, P. (2003) The cost of cortical computation. Curr. Biol., 13(6): 493–497.

Ling, S. and Carrasco, M. (2006) Sustained and transient covert attention enhance the signal via different contrast response functions. Vision Res., In press.

Liu, T., Pestilli, F. and Carrasco, M. (2005) Transient attention enhances perceptual performance and FMRI response in human visual cortex. Neuron, 45(3): 469–477.

Lu, Z.L. and Dosher, B.A. (1998) External noise distinguishes attention mechanisms. Vision Res., 38(9): 1183–1198.

Lu, Z.L. and Dosher, B.A. (2000) Spatial attention: different mechanisms for central and peripheral temporal precues? J. Exp. Psychol. Hum. Percept. Perform., 26(5): 1534–1548.

Lu, Z.L. and Dosher, B.A. (2004) Spatial attention excludes external noise without changing the spatial frequency tuning of the perceptual template. J. Vis., 4: 955–966.

Lu, Z.L., Lesmes, L.A. and Dosher, B.A. (2002) Spatial attention excludes external noise at the target location. J. Vis., 3: 312–323.

Luck, S.J. (2004) Understanding awareness: one step closer. Nat. Neurosci., 7: 208–209.

Luck, S.J., Chelazzi, L., Hillyard, S.A. and Desimone, R. (1997) Neural mechanisms of spatial selective attention in areas V1, V2, and V4 of macaque visual cortex. J. Neurophysiol., 77: 24–42.

Luck, S.J., Hillyard, S.A., Mouloua, M. and Hawkins, H.L. (1996) Mechanisms of visual-spatial attention: resource allocation or uncertainty reduction? J. Exp. Psychol. Hum. Percept. Perform., 22(3): 725–737.

Luck, S.J., Hillyard, S.A., Mouloua, M., Woldorff, M.G., Clark, V.P. and Hawkins, H.L. (1994) Effects of spatial cuing on luminance detectability: psychophysical and electrophysiological evidence for early selection. J. Exp. Psychol. Hum. Percept. Perform., 20(4): 887–904.

Luck, S.J. and Thomas, S.J. (1999) What variety of attention is automatically captured by peripheral cues? Percept. Psychophys., 61(7): 1424–1435.

Majaj, N.J., Pelli, D.G., Kurshan, P. and Palomares, M. (2002) The role of spatial frequency channels in letter identification. Vision Res., 42(9): 1165–1184.

Martinez, A., Anllo-Vento, L., Sereno, M.I., Frank, L.R., Buxton, R.B., Dubowitz, D.J., Wong, E.C., Hinrichs, H., Heinze, H.J. and Hillyard, S.A. (1999) Involvement of striate and extrastriate visual cortical areas in spatial attention. Nat. Neurosci., 2(4): 364–369.

Martinez-Trujillo, J. and Treue, S. (2002) Attentional modulation strength in cortical area MT depends on stimulus contrast. Neuron, 35(2): 365–370.

Martinez-Trujillo, J. and Treue, S. (2005) Attentional modulation of apparent stimulus contrast. In: Itti, L., Rees, G. and Tsotsos, J. (Eds.), Neurobiology of Attention. Elsevier, San Diego, p. 428.

Maunsell, J.H. and Cook, E.P. (2002) The role of attention in visual processing. Philos. Trans. R. Soc. London B, 357(1424): 1063–1072.

Maunsell, J.H. and McAdams, C.J. (2000) Effects of attention on neural response properties in visual cerebral cortex. In: Gazzaniga, M.S. (Ed.), The New Cognitive Neurosciences (2nd ed.). MIT Press, Cambridge, pp. 315–324.

Mayfrank, L., Kimmig, H. and Fischer, B. (1987) The role of attention in the preparation of visually guided saccadic eye movements in man. In: O'Regan, J.K. and Levy-Schoen, A. (Eds.), Eye Movements: From Physiology to Cognition. North-Holland, New York, pp. 37–45.

McAdams, C.J. and Maunsell, J.H. (1999) Effects of attention on orientation-tuning functions of single neurons in macaque cortical area V4. J. Neurosci., 19(1): 431–441.

Moran, J. and Desimone, R. (1985) Selective attention gates visual processing in the extrastriate cortex. Science, 229: 782–784.

Morgan, M.J., Ward, R.M. and Castet, E. (1998) Visual search for a tilted target: tests of spatial uncertainty models. Q. J. Exp. Psychol. A,, 51A(2): 347–370.

Morrone, M.C., Denti, V. and Spinelli, D. (2004) Different attentional resources modulate the gain mechanisms for color and luminance contrast. Vision Res., 44: 1389–1401.

Motter, B.C. (1993) Focal attention produces spatially selective processing in visual cortical areas V1, V2, and V4 in the presence of competing stimuli. J. Neurophysiol., 70: 909–919.

Motter, B.C. (1994) Neural correlates of attentive selection for color or luminance in extrastriate area V4. J. Neurosci., 14(4): 2178–2189.

Muller, H.J. and Rabbitt, P.M. (1989) Reflexive and voluntary orienting of visual attention: time course of activation and resistance to interruption. J. Exp. Psychol. Hum. Percept. Perform., 15(2): 315–330.

Muller, M.M., Picton, T.W., Valdes-Sosa, P., Riera, J., Teder-Salejarvi, W.A. and Hillyard, S.A. (1998) Effect of spatial selective attention on the steady-state visual evoked potential in the 20–28 Hz range. Brain Res. Cogn. Brain Res., 6: 249–261.

Nachmias, J. (1967) Effect of exposure duration on visual contrast sensitivity with square-wave gratings. J. Opt. Soc. Am., 57(3): 421–427.

Nachmias, J. (2002) Contrast discrimination with and without spatial uncertainty. Vision Res., 42(1): 41–48.

Nachmias, J. and Kocher, E.C. (1970) Visual detection and discrimination of luminance increments. J. Opt. Soc. Am., 60(3): 382–389.

Nakayama, K. and Mackeben, M. (1989) Sustained and transient components of focal visual attention. Vision Res., 29(11): 1631–1647.

Neisser, U. (1967) Cognitive psychology. Englewood Cliffs, NJ, Prentice-Hall.

Palmer, J. (1994) Set-size effects in visual search: the effect of attention is independent of the stimulus for simple tasks. Vision Res., 34(13): 1703–1721.

Palmer, J., Verghese, P. and Pavel, M. (2000) The psychophysics of visual search. Vision Res., 40(10–12): 1227–1268.

Pashler, H. (1998) The Psychology of Attention. MIT Press, Cambridge.

Patterson, R.D. and Nimmo-Smith, I. (1980) Off-frequency listening and auditory-filter asymmetry. J. Acoust. Soc. Am., 67: 229–245.

Peelen, M.V., Heslenfeld, D.J. and Theeuwes, J. (2004) Endogenous and exogenous attention shifts are mediated by the same large-scale neural network. Neuroimage, 22(2): 822–830.

Pelli, D.G. (1985) Uncertainty explains many aspects of visual contrast detection and discrimination. J. Opt. Soc. Am., 2(9): 1508–1532.

Pessoa, L., Kastner, S. and Ungerleider, L.G. (2003) Neuroimaging studies of attention: from modulation of sensory processing to top-down control. J. Neuroscience, 23: 3990–3998.

Pestilli, F. and Carrasco, M. (2005) Attention enhances contrast sensitivity at cued and impairs it at uncued locations. Vision Res., 45(14): 1867–1875.

Pestilli, F., Talgar, C.P. and Carrasco, M. (2004). Sustained attention enhances letter identification without affecting channel tuning. http://journalofvision.org/4/8/524/

Pinsk, M.A., Doniger, G.M. and Kastner, S. (2004) Push–pull mechanism of selective attention in human extrastriate cortex. J. Neurophysiol., 92(1): 622–629.

Posner, M.I. (1980) Orienting of attention. Q. J. Exp. Psychol. A, 32(1): 3–25.

Prinzmetal, W., Amiri, H., Allen, K. and Edwards, T. (1998) Phenomenology of attention: I. Color, location, orientation, and spatial frequency. J. Exp. Psychol. Hum. Percept. Perform., 24: 261–282.

Prinzmetal, W., Nwachuku, I., Bodanski, L., Blumenfeld, L. and Shimizu, N. (1997) The phenomenology of attention: 2. Brightness and contrast. Conscious. Cogn., 6: 372–412.

Regan, D. and Beverley, K.I. (1985) Postadaptation orientation discrimination. J. Opt. Soc. Am., 2: 147–155.

Reynolds, J.H. (2005) Visual cortical circuits and spatial attention. In: Itti, L., Rees, G. and Tsotsos, J. (Eds.), Neurobiology of Attention. Elsevier, San Diego, pp. 42–49.

Reynolds, J.H. and Chelazzi, L. (2004) Attentional modulation of visual processing. Annu. Rev. Neurosci., 27: 611–647.

Reynolds, J.H., Chelazzi, L. and Desimone, R. (1999) Competitive mechanisms subserve attention in macaque areas V2 and V4. J. Neurosci., 29: 1736–1753.

Reynolds, J.H. and Desimone, R. (1999) The role of neural mechanisms of attention in solving the binding problem. Neuron, 24: 19–29.

Reynolds, J.H. and Desimone, R. (2003) Interacting roles of attention and visual salience in V4. Neuron, 37: 853–863.

Reynolds, J.H., Pasternak, T. and Desimone, R. (2000) Attention increases sensitivity of V4 neurons. Neuron, 26(3): 703–714.

Ringach, D.L., Hawken, M.J. and Shapley, R. (1997) Dynamics of orientation tuning in macaque primary visual cortex. Nature, 387: 281–284.

Robinson, D.L. and Kertzman, C. (1995) Covert orienting of attention in macaques. III. Contributions of the superior colliculus. J. Neurophysiol., 74: 713–721.

Rock, I. and Gutman, D. (1981) The effect of inattention on form perception. J. Exp. Psychol. Hum. Percept. Perform., 7: 275–285.

Santella, D. and Carrasco, M. (2003) Perceptual consequences of temporal disparities in the visual field: the case of the line motion illusion (Abstract). J. Vis., 3(9): 752a.

Schroeder, C.E., Mehta, A.D. and Foxe, J.J. (2001) Determinants and mechanisms of attentional control over cortical neural processing. Front. Biosci., 6: 672–684.

Sclar, G. and Freeman, R.D. (1982) Orientation selectivity in the cat's striate cortex is invariant with stimulus contrast. Exp. Brain Res., 46(3): 457–461.

Sclar, G., Lennie, P. and Depriest, D.D. (1989) Contrast adaptation in striate cortex of macaque. Vision Res., 29(7): 747–755.

Sclar, G., Maunsell, J.H. and Lennie, P. (1990) Coding of image contrast in central visual pathways of the macaque monkey. Vision Res., 30(1): 1–10.

Shiu, L.P. and Pashler, H. (1994) Neglible effect of spatial precuing on identification of single digits. J. Exp. Psychol. Hum. Percept. Perform., 20(5): 1037–1054.

Shiu, L.P. and Pashler, H. (1995) Spatial attention and vernier acuity. Vision Res., 35(3): 337–343.

Skottun, B.C., Bradley, A., Sclar, G., Ohzawa, I. and Freeman, R.D. (1987) The effects of contrast on visual orientation and spatial frequency discrimination: a comparison of single cells and behavior. J. Neurophysiol., 57(3): 773–786.

Solomon, J.A. (2004) The effect of spatial cues on visual sensitivity. Vision Res., 44(12): 1209–1216.

Solomon, J.A., Lavie, N. and Morgan, M.J. (1997) Contrast discrimination function: spatial cuing effects. J. Opt. Soc. Am., 14(9): 2443–2448.

Solomon, J.A. and Pelli, D.G. (1994) The visual filter mediating letter identification. Nature, 369(6479): 395–397.

Somers, D.C., Dale, A.M., Seiffert, A.E. and Tootell, R.B.H. (1999) Functional MRI reveals spatially specific attentional modulation in human primary visual cortex. Proc. Natl. Acad. Sci. USA, 96(4): 1663–1668.

Sperling, G. and Dosher, B.A. (1986) Strategy and optimization in human information processing. In: Boff, K.R., Kaufman, L. and Thomas, J.P. (Eds.), Handbook of Perception and Human Performance. Wiley, New York, pp. 1–65.

Spitzer, H., Desimone, R. and Moran, J. (1988) Increased attention enhances both behavioral and neuronal performance. Science, 240: 338–340.

Suzuki, S. and Cavanagh, P. (1997) Focused attention distorts visual space: an attentional repulsion effect. J. Exp. Psychol. Hum. Percept. Perform., 23: 443–463.

Talgar, C.P. and Carrasco, M. (2002) Vertical meridian asymmetry in spatial resolution: visual and attentional factors. Psychon. Bull. Rev., 9: 714–722.

Talgar, C.P., Pelli, D.G. and Carrasco, M. (2004) Covert attention enhances letter identification without affecting channel tuning. J. Vis., 4(1): 23–32.

Tolhurst, D.J. (1973) Separate channels for the analysis of the shape and the movement of a moving visual stimulus. J. Physiology, 231: 385–402.

Tootell, R.B., Mendola, J.D., Hadjikhani, N.K., Ledden, P.J., Liu, A.K., Reppas, J.B., Sereno, M.I. and Dale, A.M. (1997) Functional analysis of V3A and related areas in human visual cortex. J. Neuroscience, 17(18): 7060–7078.

Treisman, A.M. and Gelade, G. (1980) A feature-integration theory of attention. Cognit. Psychol., 12(1): 97–136.

Treue, S. (2001) Neural correlates of attention in primate visual cortex. Trends Neurosci., 24(5): 295–300.

Treue, S. (2004) Perceptual enhancement of contrast by attention. Trends Cogn. Sci., 8(10): 435–437.

Treue, S. and Martinez-Trujillo, J.C. (1999) Feature-based attention influences motion processing gain in macaque visual cortex. Nature, 399(6736): 575–579.

Treue, S. and Maunsell, J.H. (1996) Attentional modulation of visual motion processing in cortical areas MT and MST. Nature, 382(6591): 539–541.

Tsal, Y., Shalev, L., Zakay, D. and Lubow, R.E. (1994) Attention reduces perceived brightness contrast. Q. J. Exp. Psychol. A, 47A: 865–893.

Verghese, P. (2001) Visual search and attention: a signal detection theory approach. Neuron, 31(4): 523–535.

Wade, A.R., Brewer, A.A., Rieger, J.W. and Wandell, B.A. (2002) Functional measurements of human ventral occipital cortex: retinotopy and colour. Philos. Trans. R. Soc. London B, 357(1424): 963–973.

Wolfe, J.M. (1994) Guided search: 2.0: a revised model of visual search. Psychon. Bull. Rev., 1: 202–238.

Yantis, S. (1996) Attentional capture in vision. In: Kramer, A.F. and Coles, G.H. (Eds.), Converging Operations in the Study of Visual Selective Attention. American Psychological Association, Washington, DC, pp. 45–76.

Yantis, S. and Jonides, J. (1984) Abrupt visual onsets and selective attention: evidence from visual search. J. Exp. Psychol. Hum. Percept. Perform., 10(5): 601–621.

Yantis, S. and Serences, J.T. (2003) Cortical mechanisms of space-based and object-based attentional control. Curr. Opin. Neurobiol., 13(2): 187–193.

Yeshurun, Y. (2004) Isoluminant stimuli and red background attenuate transient spatial attention on temporal resolution. Vision Res., 44(12): 1375–1387.

Yeshurun, Y. and Carrasco, M. (1998) Attention improves or impairs visual perception by enhancing spatial resolution. Nature, 396: 72–75.

Yeshurun, Y. and Carrasco, M. (1999) Spatial attention improves performance in spatial resolution tasks. Vision Res., 39(2): 293–306.

Yeshurun, Y. and Carrasco, M. (2000) The locus of attentional effects in texture segmentation. Nat. Neurosci., 3(6): 622–627.

Zackon, D.H., Casson, E.J., Zafar, A., Stelmach, L. and Racette, L. (1999) The temporal order judgment paradigm: subcortical attentional contribution under exogenous and endogenous cueing conditions. Neuropsychologia, 37(5): 511–520.

Zhaoping, L. (2005) The primary visual cortex creates a bottom-up saliency map. In: Itti, L., Rees, G. and Tsotsos, J. (Eds.), Neurobiology of Attention. Elsevier, San Diego, pp. 570–575.

Recent Discoveries in Receptive Field Structure

Introduction

The visual system must achieve a dauntingly complex task: creating an internal representation of the external world. How does this neural system accomplish the job? A parsimonious explanation proposes that visual information is analyzed in a series of sequential steps starting in the retina and continuing along the multiple visual cortical areas. As a result, the information captured by the approximately 105 million photoreceptors in each eye is continuously rearranged in a complex combination of points and lines of different orientations and curvatures that are defined by differences in local contrast, color, relative timing, depth and/or movement. Ultimately, these elementary features of the image are integrated into the perception of each individual object in the visual scene. The exact mechanisms underlying this process are largely unknown and represent one of the most fascinating challenges of systems neuroscience.

The primary visual cortex (V1), in particular, is a key region in the visual pathway because all information about the visual scene is funneled through V1 to reach higher order cortical areas, where a more complex abstraction of the visual world emerges. In addition, some may argue that responses in V1 reflect, for the first time, not only the physical attributes of the visual stimuli (orientation, direction of motion, color and so on), but also its perceptual experience. Thus, V1 neurons and their receptive fields form the building blocks, the essential elements used by the more advanced stations in the visual hierarchy to generate our visual perception. Unlike cells in the lateral geniculate nucleus (LGN) of the thalamus that supply them, V1 neurons show a great variety of receptive field structures, as Martinez points out in his following chapter. This functional diversity emerges from the specific computations performed by a widespread and distributed synaptic network consisting of feed-forward inputs, local and long-range horizontal connections and feedback influences from other visual cortical areas.

Currently, we have accumulated a wealth of information about the intrinsic properties of V1 cells and how they respond to simple visual stimuli such as bars, gratings and textures. This information has been used to propose various taxonomies of visual cortical neurons and to advance contrasting views and theoretical models of cortical function. Yet, after more than 40 years of intense study, the identity of the circuits that generate each V1 receptive field type, with their distinct functional response properties and even their specific roles in visual processing, are still a matter of intense debate. In the first chapter of this section, Martinez discusses recent data showing how receptive field structure changes according to laminar location in the primary visual cortex. He argues that simple cells are an exclusive feature of the thalamorecipient layers of V1, layer 4 and upper layer 6. Within these layers he shows that a synaptic network consisting of feed-forward inputs from the thalamus and two different sources of intracortical inhibition (simple tuned and complex untuned) contribute to generate the simple receptive field and contrast invariant orientation tuning. Circuits outside layer 4, on the other hand, change the synaptic structure of the complex receptive field and orientation tuning with each step of cortical integration. The first chapter illustrates how experimental designs combining simultaneously anatomy with physiology are very powerful at resolving the synaptic mechanisms that generate distinct functional response properties at different positions within the cortical microcircuit.

Another interesting link between each element of the V1 synaptic network and a specific component of V1 receptive fields is put forward in Angelucci and Bresloff's chapter. They argue that the classical receptive field and extra-classical surround result from the interaction of all three sets of connections (feedforward, lateral and feedback) operating at different spatial scales. According to their experimental and theoretical data, feedforward connections are mainly responsible for the center of V1 receptive fields (see also Martinez's chapter), while lateral and feedback connections modulate responses at the center and cooperate to generate the "near" and "far" surround.

Another important theme of research on visual cortical function is the organization of V1 neurons in functional maps that span the entire primary visual cortex and represents ocular dominance and specific stimulus features such as orientation, direction of motion and spatial frequency. The traditional view holds that the different maps are spatially organized to guarantee full coverage of all stimulus attributes and ocular dominance at each position in visual space. Thus, specific combinations of stimulus features are mapped to individual cortical sites in the form of a place code. However, this view has been difficult to reconcile with the results showing that the spatial properties

and the speed of a visual stimulus can modify, for example, the tuning for other stimulus features such as the direction of motion. The chapter by Basole et al. suggests that this, and other conflicting experimental and theoretical results are best explained by an alternative organization of V1 neurons in a single map of spatiotemporal energy.

The last chapter in this section deals with a rather different topic: the principles underlying the analysis of complex visual stimuli in higher order neurons; in particular, how information about form in static figures influences motion perception. In their chapter, Barraclough et al. illustrate how cells in the macaque monkey superior temporal sulcus discriminate articulated postures implying motion from standing postures, an ability that correlates to sensitivity to motion type for the same neurons. The results of Barraclough et al. are in agreement with a new, feed-forward model of biological motion proposed by Giese and Poggio (cited in the chapter by Barraclough et al.), and suggest that this information about body posture and articulation (form) would strongly influence the activity of neurons downstream coding specific motion patterns.

Luis M. Martinez

Martinez-Conde, Macknik, Martinez, Alonso & Tse (Eds.)
Progress in Brain Research, Vol. 154
ISSN 0079-6123

CHAPTER 4

The generation of receptive-field structure in cat primary visual cortex

L.M. Martinez*

Departamento de Medicina, Facultade de Ciencias da Saude, Campus de Oza, Universidade da Coruña, 15006 La Coruña, Spain

Abstract: Cells in primary visual cortex show a remarkable variety of receptive-field structures. In spite of the extensive experimental and theoretical effort over the past 50 years, it has been difficult to establish how this diversity of functional-response properties emerges in the cortex. One of the reasons is that while functional studies in the early visual pathway have been usually carried out in vivo with extracellular recording techniques, investigations about the precise structure of the cortical network have mainly been conducted in vitro. Thus, the link between structure and function has rarely been explicitly established, remaining a well-known controversial issue. In this chapter, I review recent data that simultaneously combines anatomy with physiology at the intracellular level; trying to understand how the primary visual cortex transforms the information it receives from the thalamus to generate receptive-field structure, contrast-invariant orientation tuning and other functional-response properties.

Keywords: primary visual cortex; receptive field; orientation selectivity; simple cell; complex cell; cortical microcircuit.

Introduction

Receptive field is, nowadays, an elusive concept that cannot be easily defined (Bair, 2005; Hirsch and Martinez, 2006). It was introduced almost 70 years ago in the visual system when Hartline (1938) gave this name to the region of the retina where a change in light brightness modifies the firing rate of a retinal ganglion cell. Thus, in its original description, the term receptive field basically represented a spatial location in sensory, in this case retinal, coordinates. Kuffler (1953), Barlow (1953) and Hubel and Wiesel (1961) later demonstrated the receptive fields of retinal ganglion cells, and their targets in the lateral geniculate nucleus

of the thalamus (LGN), are roughly circular in shape and comprise two concentric subregions known as the center and the surround (Fig. 1a). These subregions have opposite preferences for stimulus contrasts such that On cells are excited by bright spots shone in the center or by dark annuli in the surround; Off cells respond in a reciprocal manner (Kuffler, 1953; Hubel and Wiesel, 1961). As well, the center and surround have a mutually antagonistic relationship because stimuli of the reverse contrast evoke push–pull responses within each of the two subregions — where bright light excites and dark stimuli inhibit (Kuffler, 1953; Hubel and Wiesel, 1961). Remarkably, just the geometry of center and surround and suppressive interactions between them explain why retinal and thalamic cells remain largely indifferent to uniform patterns of illumination, while they are

*Corresponding author. Tel.: +34-981-167000 ext 5864;
Fax: +34-981-167155; E-mail: lmo@neuralcorrelate.com

DOI: 10.1016/S0079-6123(06)54004-X

73

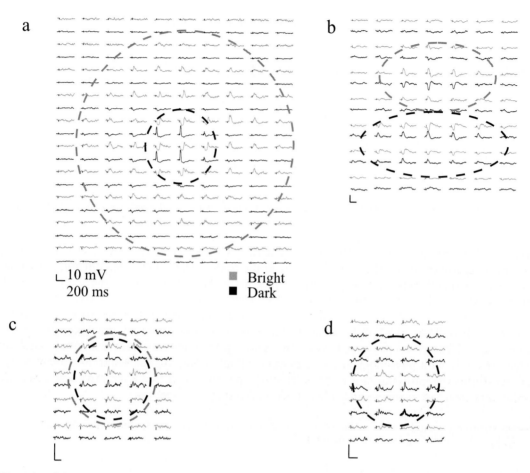

a

b

c

d

∟10 mV
200 ms

■ Bright
■ Dark

Fig. 1. Examples of the main types of receptive fields recorded in the visual thalamus and cortex. The receptive fields are shown as arrays of trace pairs in which each position in the stimulus grid is represented by averages of the corresponding responses to dark (black traces) and bright (gray traces) squares. Gray dashed lines code for On and black dashed lines code for Off subregions. For the relay cell in the thalamus (a) and the simple cell in layer 4, (b) stimuli of the reverse contrast evoked responses of the opposite sign (push–pull) in each subregion. For the complex cell in layer 4 (**c**), both bright and dark stimuli evoked excitation (push–push) in overlapping regions of visual space. Dark but not bright squares (push-null) excited the complex cell in layer 6 (d). The stimulus was flashed for 31 or 47 ms (vertical lines indicates stimulus onset); square size was 0.85° or 1.7° and grid spacing was 0.85°. Adapted from Martinez et al. (2005) and Hirsch and Martinez (2006).

very responsive to local changes in stimulus contrast (Kuffler, 1953; Hubel and Wiesel, 1961; Barlow and Levick, 1976). However, to fully describe the functional properties of a visual neuron, the temporal aspects of the response should also be taken into account (Saul and Humphrey, 1990; DeAngelis et al., 1995). For example, retinal and thalamic cells can be further classified as transient or sustained, depending on whether the cell produces only brief responses after a long-lasting stimulus is switched on or kept firing for most of the duration of the stimulus presentation (Cleland et al, 1971). Furthermore, most thalamic receptive fields exhibit some degree of space–time inseparability (Cai et al., 1997). As a result, there is a progressive reduction in size of the receptive-field center as the surround response, delayed from that of the center, builds up over time. Finally, in recent years a new level of complexity has been introduced to the definition of the receptive field. Many observations have demonstrated that spike responses which evoked from the receptive field of visual neurons can be further modulated by stimulation of a surrounding region that has been called the non-classical, or extra-classical, receptive field (Hubel and Wiesel, 1968; Allman et al., 1985; Fitzpatrick, 2000; Angelucci et al., 2002; Cavanaugh et al., 2002; Bair, 2005).

The receptive-field concept has thus evolved over the years to represent the shape of a spatiotemporal filter that ultimately determines how visual input is analyzed by each element of the visual pathway (Barlow, 1953; Kuffler, 1953; Hubel and Wiesel, 1959; Wiesel, 1959; Hubel and Wiesel, 1961; Hubel and Wiesel, 1962; Hubel and Wiesel, 1968; Levick et al., 1972; Enroth-Cugell and Lennie, 1975; Barlow and Levick, 1976; Gilbert, 1977; Henry, 1977; Movshon et al., 1978a, b, c; Bullier and Henry, 1979a, b; Victor and Shapley, 1979; Toyama et al., 1981; Dean and Tolhurst, 1983; Wiesel and Gilbert, 1983; Enroth-Cugell and Robson, 1984; Martin and Whitteridge, 1984; Cleland and Lee, 1985; Baker and Cynader, 1986; Heggelund, 1986; Jones and Palmer, 1987; Ferster, 1988; Chapman et al., 1991; Emerson et al., 1992; DeAngelis et al., 1993a, b; Volgushev et al., 1993; McLean and Palmer, 1994; Pei et al., 1994; DeAngelis et al., 1995; Reid and Alonso, 1995; Fitzpatrick, 1996; Bosking et al., 1997; Cai et al., 1997; Ohzawa and Freeman, 1997; Borg-Graham et al., 1998; Debanne et al., 1998; Hirsch et al., 1998a, b; Troyer et al., 1998; Murthy and Humphrey, 1999; Usrey et al., 1999; Ferster and Miller, 2000; Hirsch et al., 2000; Lampl et al., 2001; Martinez and Alonso, 2001; Wielaard et al., 2001; Abbott and Chance, 2002; Hirsch et al., 2002; Kagan et al., 2002; Martinez et al., 2002; Mechler and Ringach, 2002; Troyer et al., 2002; Chisum et al., 2003; Hirsch, 2003; Hirsch et al., 2003; Lauritzen and Miller, 2003; Martinez and Alonso, 2003; Monier et al., 2003; Usrey et al., 2003; Van Hooser et al., 2003; Douglas and Martin, 2004; Mooser et al., 2004; Priebe et al., 2004; Ringach, 2004; Bair, 2005; Martinez et al., 2005; Mata and Ringach, 2005; Hirsch and Martinez, 2006). The view of the receptive field as a filter conveys two ideas that are central to our understanding of sensory systems in general. First, not all available sensory information is processed by the brain; in fact, most of it is actually discarded. Second, each stage along a sensory pathway is in charge of analyzing a

distinct aspect of its sensory input; for example, changes in local contrast by retinal ganglion cells. As a result, redundant information is discarded early on while the relevant structure of the visual image is preserved and transmitted to more advanced stations of processing.

In this chapter, we summarize current research in receptive fields in primary visual cortex (V1). First, we concentrate on the circuits that are responsible for the generation of the various classes of cortical cells: simple and complex. Second, we review recent data suggesting that a purely feedforward circuit can achieve contrast invariant orientation tuning at the first stage of cortical processing. Third, we discuss how successive stages of the cortical microcircuit modify the synaptic structure of complex receptive fields and orientation tuning. Understanding how such cortical circuits are built may ultimately help explain orientation selectivity in other species, like primates, where it develops fully at intracortical stages of processing rather than at the thalamocortical level (Hubel and Wiesel., 1968; Bosking et al., 1997; Ringach et al., 1997, 2002, 2003; Chisum et al., 2003; Mooser et al., 2004).

Receptive fields in primary visual cortex

When Hubel and Wiesel first recorded from the primary visual cortex they used the same simple stimuli that had been so successful in mapping the spatial structure of retinal and thalamic receptive fields (Hubel and Wiesel, 1959, 1962). The cortex, however, turn out to have a larger variety of receptive-fields structures. Attending to their functional response properties, Hubel and Wiesel classified cortical cells into two main categories: simple cells and complex cells. The simple cell category included the population of neurons that reminded them of presynaptic cells in the thalamus, for their receptive fields were divided into On and Off subregions that had an antagonistic effect on one another (Hubel and Wiesel, 1962; Hirsch and Martinez, 2006) (Fig. 1b). Unlike the subcortical concentric arrangement, however, the On and Off subregions of simple cells are elongated and lay side by side. What made this observation so

76

interesting was that the new geometric configuration correlated with the emergence of neural sensitivity to stimulus orientation. In the retina and thalamus, a bar of any orientation drove cells vigorously. Cortical cells, on the other hand, responded briskly to an oriented stimulus aligned with the long axis of sign matched subregions, but fired less vigorously, if at all, to stimuli rotated away from the preferred angle. Complex cells were also orientation selective. Unlike simple cells, however, they formed a much more diverse population and were defined by exclusion. The main classification criteria for a complex receptive field was that they lacked spatially discrete subregions (Figs. 1c, d) and their preferred orientation and responses to variously shaped stimuli could not be predicted from a spatial map of their receptive fields.

The hierarchical model of receptive-field construction

Inspired by the comparison of functional response properties in the thalamus and primary visual cortex, Hubel and Wiesel (1962) introduced the idea of a hierarchical organization of receptive-field structures. According to their model, the elongated subregions of simple cells are constructed from the convergent input of On and Off thalamic relay cells with receptive fields aligned in visual space. In turn, complex receptive fields originate from the convergence of simple cell inputs with similar orientation preferences but different spatial phases (Fig. 2). Thus, simple receptive fields emerge as the most direct approach to build orientation detectors from geniculate cells with circularly symmetric receptive fields. Complex cells, on the other, originate from the need to build orientation detectors that are independent of the contrast polarity and position of the stimulus within the receptive field. The initial success of this hierarchical (or feed forward) model was that it provided a mechanistic account of the generation of cortical receptive fields while it served to explain the functional roles of the main cellular components of the cortical circuit (Martinez and Alonso, 2003).

Over the years, the feed-forward model has received substantial experimental support and

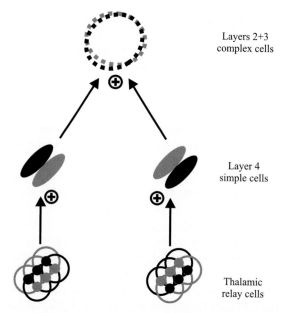

Fig. 2. Hierarchical model of receptive-field construction in primary visual cortex. Simple and complex cells represent two successive stages of cortical processing. Simple receptive fields are generated in layer 4 from the convergent input of geniculate neurons with receptive fields properly aligned in visual space. Complex receptive fields, in turn, pool the input from layer 4 simple cells with similar orientation preferences but different spatial phases. Black and gray code Off and On subfields respectively.

refinement. Concerning the first part of the model, some studies have reported that the majority of cells in layer 4 have simple receptive fields (Hubel and Wiesel, 1962; Gilbert, 1977; Gilbert and Wiesel, 1979; Hirsch et al., 1998a, b; Martinez et al., 2002, 2005; see also, Orban, 1984; Jacob et al., 2003; Ringach, 2004). Cross-correlation analysis demonstrated that monosynaptic connections between relay cells in the LGN and cortical cells are largely restricted to cases for which relay cells share the same sign and spatial position with the cortical targets (Tanaka, 1983; Reid and Alonso, 1995; Alonso et al., 2001). In addition, other studies have also provided compelling evidence that the generation of orientation selectivity in cat V1 relays primarily on thalamocortical mechanisms (see Ferster and Miller, 2000; Hirsch and Martinez, 2006, for review). For example, recordings from the axonal arbors of LGN cells in the cortex showed that thalamic receptive fields line up along the axis of orientation of local cortical cells (Chapman et al., 1991). Intracellular recordings from V1 neurons made when cortical firing was greatly suppressed by cooling (Ferster et al., 1996) or inhibition (Chung and Ferster, 1998), showed that the remaining excitatory (presumably thalamic) input is tuned for orientation.

In relation to the second part of the model, earlier studies combining anatomical reconstruction and functional characterization had shown that cells in layer 4 with simple receptive fields send a strong axonal projection to the superficial layers (Gilbert and Wiesel, 1979; Martin and Whitteridge, 1984; Hirsch et al., 1998a; Martinez et al., 2002, 2005), where most cells are complex (Hubel and Wiesel., 1962; Gilbert, 1977; Gilbert and Wiesel, 1979; Hirsch et al., 2002; Martinez et al., 2002, 2005; cf. Orban, 1984; Jacob et al., 2003; Ringach 2004). Recent evidence based on cross-correlation analysis and pharmacological blockades suggests that complex receptive fields in the superficial layers are constructed by a mechanism that requires monosynaptic inputs from simple cells in layer 4 (Alonso and Martinez, 1998; Martinez and Alonso, 2001). In addition, Tony Movshon and colleagues demonstrated that complex cell's responses depend on nonlinear interactions between at least two positions in space and time within their receptive fields (Movshon et al., 1978b). Thus, when complex cells are tested with sets of two bars with different spatiotemporal configurations, their responses along the direction perpendicular to the cells' preferred orientation display On and Off linear subunits resembling simple-cells' subregions (Movshon et al., 1978b; Heggelund, 1981; Baker and Cynader, 1986; Szulborski and Palmer, 1990; Gaska et al., 1994; Touryan et al., 2005). Movshon et al.'s (1978b) results were extended by others to show that the functional properties of the underlying subunits could explain the emergence of direction-selective complex cells (Emerson et al., 1992) and binocular complex cells (Ohzawa et al., 1990, 1997; Ohzawa and Freeman, 1986; Anzai et al., 1999).

Alternatives to the hierarchical model

Alternative lines of evidence suggest that a simple feed-forward circuit may not explain the emergence of cortical receptive fields and their related functional response properties, like orientation selectivity. For example, earlier versions of the feed-forward model cannot account for various aspects of cortical responses such as the maintenance of orientation selectivity over a range of stimulus contrasts (Sclar and Freeman, 1982; Ohzawa et al., 1985; Geisler and Albrecht, 1992; Troyer et al., 1998; Lauritzen and Miller, 2003); or the sharpening overtime of orientation and spatial frequency tuning due to the presence of delayed suppression extending the classical receptive field (Ringach et al., 1997, 2002, 2003; cf. Gillespie et al., 2001). However, perhaps the most widely used argument against hierarchical models is the nature of the cortical circuit itself. Since the number of excitatory synapses provided by feed-forward connections is only a small fraction (less than 10%) of the total excitatory synapses made onto cortical cells (LeVay and Gilbert, 1976; Kisvarday et al., 1986; Peters and Payne, 1993; Ahmed et al., 1994; Braitenberg and Schuz, 1998; Binzegger et al., 2004), it is usually assumed that the influence of the feed-forward input on cortical-response properties is weak. Therefore, some authors have proposed that cortical responses, even those in thalamorecipient layers, should be determined mostly by cortical inputs and not by thalamic inputs (for review see references, Martin, 2002; Nelson, 2002; Douglas and Martin, 2004). Although it is true that cortical cells (simple or complex) receive most of their input from other cortical neurons both inhibitory and excitatory (Ahmed et al., 1994; Fitzpatrick, 1996; Braitenberg and Schuz, 1998; Callaway, 1998; Thomson et al., 2002; Thomson and Bannister, 2003), it is usually chancy to estimate the strength of a given pathway based solely on the number of synaptic contacts. If this type of assumption was correct, the thalamic receptive fields should resemble more cortical receptive fields than retinal receptive fields, since the number of corticothalamic synapses onto LGN cells ($\sim$40%), largely outnumber retinal inputs (about 4%; Van Horn et al., 2000). Moreover, thalamocortical synapses have many features that make it possible for thalamocortical excitatory postsynaptic potentials (EPSPs) to be larger than corticocortical EPSPs (Stratford et al., 1996; Gil et al., 1999); they are located more proximally in the dendrites, are bigger (Ahmed et al., 1994) and have more release sites (Gil et al., 1999).

Another important critique to the hierarchical model originated in the discovery that some complex cells, like simple cells, receive direct thalamic input (Hoffmann and Stone, 1971). Since then, an agreement on whether or not there is a laminar segregation of simple cells and complex cells has been difficult to attain (see for example, Jacob et al., 2003 and Martinez et al., 2005, in the cat; and Kagan et al., 2002 and Ringach et al., 2002, in the monkey; for review see Martinez and Alonso, 2003; Ringach, 2004; Hirsch and Martinez, 2006). Moreover, results from other studies blurred the distinction between simple and complex receptive fields. For example, Debanne et al. (1998) showed that, in few cases, the relative strength of cortical On and Off responses, and hence their spatial distribution, could be subtly modified by pairing visual stimuli with current injections. More pronounced changes could be generated when the precise balance between cortical excitation and inhibition was manipulated (Sillito, 1975; Nelson et al., 1994; Rivadulla et al., 2001; see also Martinez and Alonso, 2003, for an alternative explanation).

These diverse views about how simple and complex receptive fields are made in the cortex have been likely developed for two main reasons (Hirsch and Martinez, 2006). First, experiments with different species are often pooled in discussion, even though the organization of primary visual cortex changes with phylogeny (Lund et al., 1979; Gilbert, 1983; Martin and Whitteridge, 1984; Coogan and Burkhalter, 1990; Fitzpatrick, 1996; Callaway, 1998; Dantzker and Callaway, 2000; Binzegger et al., 2004; Douglas and Martin, 2004; Martinez et al., 2005). Second, Hubel and Wiesel's (1962) description did not provide a quantitative test to clearly distinguish between the two populations of cortical receptive fields. In fact, complex receptive fields were originally defined by exclusion and therefore, they comprise a heterogeneous population. The lack of such a quantitative test led to the proposal of many different classification criteria (Palmer and Rosenquist, 1974; Schiller et al., 1976; Henry, 1977; Toyama et al., 1981; see also Orban, 1984; Mechler and Ringach, 2002; Martinez and Alonso, 2003, for review). Simple

cells and complex cells have been classified based on the presence and degree of overlap of On and Off subregions, spontaneous activity level, response amplitude, length summation, responses to patterns of random dots, responses to moving light and dark bars, responses to moving edges, responses to drifting or contrast-reversal gratings, responses to flashed light bars and reverse correlation maps (Skottun et al., 1991; Mechler and Ringach, 2002; Martinez and Alonso, 2003). A quantitative test to classify simple and complex cells was introduced by De Valois et al. (1982) and further refined by other authors (Skottun et al., 1991). The new method was inspired by linear system analysis and spatial and temporal frequency methods that had been very successful in explaining the functional response properties of retinal and thalamic neurons (Shapley and Hochstein, 1975; Enroth-Cugell and Robson, 1984; Shapley and Lennie, 1985). This new classification criterion is based on the different response modulation of simple and complex cells to drifting sinusoidal gratings. While simple cells tend to modulate their firing rate in phase with the stimulus, complex cells elevate their firing rate with little or no modulation. Thus, the ratio between the amplitude of the first response harmonic (F1) and the mean spike rate (F0) can be used as a quantitative index of 'response linearity' or 'receptive field complexity'. Interestingly, the F1/F0 method rendered two clearly discrete cell populations that allegedly corresponded to the Hubel-and-Wiesel simple/complex cells (Skottun et al., 1991).

However it is highly controversial how response modulation correlates with the degree of overlap of receptive field subregions (Kagan et al., 2002; Priebe et al., 2004; Mata and Ringach, 2005), indicating that the spatial organization of receptive fields cannot reliably be predicted from response modulation values. Therefore, classifying simple and complex cells by the F1/F0 method rather than receptive field structure could not be equivalent, making it even more difficult to compare results from different studies to evaluate the classical model of cortical receptive fields.

The controversy about the generation of cortical receptive fields also reached the field of computational neuroscience (for a more comprehensive review see Martinez and Alonso, 2003; Ringach, 2004). Influenced by Hubel and Wiesel (1962) hierarchical model and Movshon et al's. (1978b) results, many authors have modeled complex receptive fields as the result of a square sum of the output of four simple cells with similar orientation and spatial frequencies but with phases that differ in steps of 90 degrees (a quadrature pair of subunits and their mirror image; Pollen et al., 1989; Ohzawa et al., 1990, 1997; Emerson et al., 1992; Fleet et al., 1996; Qian and Zhu, 1997; Sakai and Tanaka, 2000 Okajima and Imaoka, 2001). Mathematically, this can be described as a square sum of two linear operators, each characterized by a Gabor function of the same frequency but with phases 90° apart from each other. These models are collectively known as energy models (Adelson and Bergen, 1985) and recently, Okajima and Imaoka (2001) demonstrated that they render complex cells that are optimally designed from an information theory point of view. Other authors have looked for developmental and coding strategies that would generate a hierarchy of simple and complex cells (Olshausen and Field, 1996; Hyvarinen and Hoyer, 2001; Einhauser et al., 2002).

Alternative approaches moved the field's focus from feed-forward connections to networks of cortical neurons that are reciprocally connected. Most of these new models consider that thalamic inputs confer cortical cells with only a small bias in orientation and spatial-phase (or subfield segregation) preferences. Therefore, cortical receptive-field structures and functional response properties must emerge from the computations performed by local connections that are isotropic and nonspecific. This new framework has been widely used to generate theoretical models of the simple receptive field (Somers et al., 1995, 1998; Wielaard et al., 2001) and orientation and direction selectivity (Ben-Yishai et al., 1995; Douglas et al., 1995; Carandini and Ringach, 1997; Sompolinsky and Shapley, 1997). In recent years, it has been extended to obtain both simple and complex receptive fields following different strategies (Debanne et al., 1998; Chance et al., 1999; Tao et al., 2004). For example, Debanne et al. (1998) used recurrent cortical circuits to generate simple and complex cells. In their model, both cell types

receive input from LGN cells with overlapping On and Off receptive-field centers (and from other cortical cells). Simple-cell responses are generated when cortical inhibition is strong enough to impose a bias toward either On or Off responses. Complex-cell responses are obtained when inhibition is reduced to unmask On–Off responses. Chance et al.'s (1999) model, in contrast, uses local, recurrent excitatory connections to generate the spatial-phase invariance of complex-cell responses. In their model, a first layer of simple cells (equivalent to layer 4) feeds into a second layer of cells with similar orientation preferences (equivalent to layers 2 + 3). The model assumes that connections from layer 4 to layers 2 + 3 are weaker than connections within layers 2 + 3 and that vertical connections link cells with similar spatial phase while recurrent connections link cells with a broader range of spatial phases. Under these conditions, neurons in the second layer exhibit simple-cell responses when recurrent connections are weak and complex-cell responses when they are strong. Tao et al. (2004) presented a related model in which the balance between lateral connections and thalamic drive determines whether individual neurons in the recurrent cortical circuit behave like simple or complex cells.

Thus, this new breed of recurrent models suggests that simple and complex receptive fields might actually represent two functional states of a unique cortical circuit. Similarly, Mechler and Ringach (2002) have recently suggested that simple and complex cells, rather than two discrete categories, might represent the ends of a continuum distribution of receptive-field structures (see also reference 1). Their new proposal originated in the demonstration that the bimodality seen in the F1/F0 distribution (Skottun et al., 1991) could be the consequence of the output non-linearity imposed by the spike threshold acting upon what appears to be a continuous distribution in the spatial organization of synaptic inputs (Mechler and Ringach, 2002; Priebe et al., 2004).

A fresh look into receptive-field structure

Investigating the organization of receptive-field structure in primary visual cortex is therefore not a minor issue. Whether simple and complex cells are discrete classes or ends of a continuum has strong implications on our understanding of the precise nature of thalamocortical and cortico-cortical circuits, on how they interact to generate cortical response properties and, finally, on the specific contribution of each cell class to visual processing.

Since most previous quantitative studies of receptive-field structure have been made with extra-cellular recordings, we wanted to use a technique, whole-cell recordings in vivo, that would allow us to explore directly the patterns of synaptic inputs that build receptive fields in the cat (Hirsch and Martinez, 2006). The method also provided a means to label cells intracellularly so that we could determine their laminar location and morpholog-ical class. Our main result is that simple cells are restricted to regions where thalamic afferents ter-minate, layers 4 and upper 6 (Hirsch et al., 1998a; Martinez et al., 2002, 2005). By contrast, complex receptive fields are found throughout the cortical depth, though the precise synaptic structure of the complex receptive field changes with laminar lo-cation (Hirsch et al., 2002, 2003; Martinez et al., 2002, 2005). Further, we found two populations of inhibitory interneurons in layer 4 with receptive-field structures that resembled those of excitatory simple cells and complex cells, respectively. The orientation selectivity of these two classes of inhibitory cells helps us understand how orienta-tion sensitivity is built and preserved over a wide range of contrasts (Hirsch et al., 2003). All told, our most recent results are consistent with feed-forward, thalamocortical models (Hubel and Wiesel, 1962; Troyer et al., 1998; Ferster and Miller, 2000; Lauritzen and Miller, 2003; Hirsch and Martinez, 2005) for cortical receptive fields and with the general idea that each layer of the cortex is adapted to serve unique functional demands.

Receptive-field structure in the thalamus and cortex

To explore how the synaptic structure of cortical receptive fields is transformed at successive stages in the early visual pathway, we mapped the spatial distribution of excitation and inhibition in the re-ceptive fields of neurons at identified anatomical sites. An overview of the receptive fields we re-corded is provided in Fig. 1 and includes record-ings from the thalamus, the main thalamorecipient zone in the cortex, layer 4, and its main postsy-naptic target, layers 2 + 3. The stimulus was a sparse-noise protocol (Jones and Palmer, 1987), light and dark squares singly flashed on pseudo-random order. The detailed maps in Fig. 1 are arrays of trace pairs in which each spatial coordi-nate is represented by the averaged responses to bright and dark stimuli flashed there.

The receptive field of an Off center relay cell is representative of recordings made in the LGN, Figure 1a. In both the center (dashed gray circle) and the surround (dashed black circle), stimuli of the reverse contrast evoked responses of the op-posite sign, or push–pull. That is, dark squares at the center coordinates evoked an initial depo-larization (push) followed by a hyperpolarization that corresponded to termination of the stimu-lus. Bright squares flashed at the same positions evoked the opposite response, a hyperpolarization (pull) followed by a depolarization. The responses from the surround, though weak (flashed spots are suboptimal stimuli for the surround), revealed a push–pull pattern as well.

The large majority of cells in thalamorecipient zones of the cortex, layer 4 and upper layer 6, were also built of On and Off subregions with push–pull (Martinez et al., 2005; Hirsch and Martinez, 2006). Unlike thalamic subfields, however, cortical sub-regions are elongated and parallel to each other, as first noted in the extracellular studies of Hubel and Wiesel (1962) and subsequent quantitative studies (Movshon et al., 1978a; Palmer and Davis, 1981; Dean and Tolhurst, 1983; Jones and Palmer, 1987; DeAngelis et al., 1993b). Fig. 1b shows an example of a simple field with an upper On subregion and a lower Off subregion. At every position along the length of the Off subregion, dark squares evoked excitation where bright squares evoked inhibition. A complementary pattern was obtained in the On subregion. Most remaining cells in layer 4 had complex receptive fields built of superimposed On and Off subregions (Hirsch et al., 2002, 2003; Martinez et al., 2005). Because bright and dark excitation overlapped, these fields had a push–push rather than a push–pull infrastructure (Fig. 1c). Many cells at later stages of processing, layers 2 + 3, 5 and lower 6, failed to respond to sparse static stimuli, although they were strongly driven by richer stimuli such as those including motion (Hirsch et al., 2002; Martinez et al., 2002). Of the responsive group, there was often a strong, or virtually absolute preference, for stimuli of one polarity (Hirsch et al., 2002; Martinez et al., 2005); the map shown in Fig. 1d is from a cell that responded to dark but not bright squares.

Quantification of receptive-field structure

Our results show that simple and complex receptive fields have different synaptic signatures (Martinez et al., 2005). To clearly establish whether they represent two different functional classes that correlate with specific locations of the cortical microcircuit, it is necessary to quantify the visually evoked responses of the recorded cells. Thus, we chose two measures that capture the most salient spatial features of cortical receptive fields. First, we used an overlap index (Schiller et al., 1976) to measure the spatial configuration of On and Off subregions within the receptive fields. Values equal or smaller than 0 indicate segregated subregions and those approaching 1 denote symmetrically overlapped subfields. A graphical explanation of the index is given in Fig. 3a beneath the distribution of values we measured. The distribution divides into two statistically significant modes; the columns containing cells whose receptive fields had a simple arrangement of On and Off subregions are shaded gray in this and the following histograms. The distribution was equally bimodal when measures were based on spiking responses rather than on membrane potential responses (see Fig. 3d in Martinez et al., 2005). Second, we used a push–pull index (Martinez et al., 2005) to determine the balance of antagonistic responses to stimuli of the opposite contrast within individual subregions; the absolute values of the index range from 0 for push–pull to 2 for push–push or pull–pull. As for the overlap index, the resulting distribution of push–pull

index values was not unimodal, as seen in histogram above the graphical explanation of the index (Fig. 3b).

In addition, cells that had segregated subregions also had push–pull, as did two cells with only one subregion recorded in layer 4. By contrast, cells with overlapping On and Off subregions had high values of the push–pull index. When values for the overlap index were plotted against those for the push–pull index, the points divided into two statistically independent clusters; one composed of simple cells and the other of complex cells (note that only cells that responded to bright and dark stimuli are plotted here) (Fig. 3c). Similar distributions were found for plots of the correlation coefficient against push–pull or against overlap index, not shown. Thus simple receptive fields are easily defined by two features, segregated On and Off subregions and the presence of push–pull, that are the quantitative counterpart of the qualitative classification criteria originally used by Hubel and Wiesel in 1962.

Other authors (Priebe et al., 2004), using Pearson's correlation coefficient as an alternative to Schiller's overlap index, have found that cortical cells form two distinct populations when measures were based on spiking responses, whereas they formed a single, continuous distribution when measures were based on membrane potential responses (as proposed by Mechler and Ringach, 2002, see discussion above). Using the same method, however, we have obtained a bimodal distribution also when measuring membrane potential responses (Fig. 3d, bimodality was determined with Hartigan's dip test; the probability of rejection for a unimodal distribution was 0.99). Part of the controversy may arise because any measure of subfield overlap is susceptible to certain artifacts (Martinez et al., 2005). For example, stimuli that cross the borders between subregions can conflate boundaries by evoking various balances of push and pull simultaneously. As well, it is important to record from cells whose resting levels are well above the threshold for inhibition. Had we made recordings near the reversal potential for inhibition, then our values for the push–pull or the cross-correlation indices could have also formed unimodal distributions.

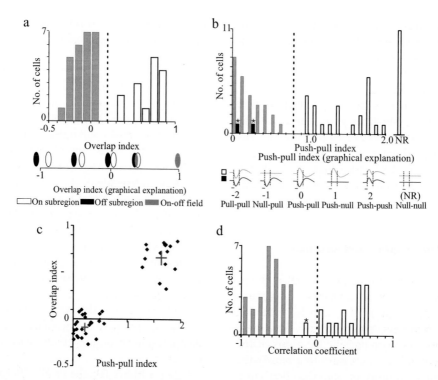

Fig. 3. Quantification of receptive-field structure in the visual cortex. (a) Histogram showing the distribution of values for the overlap index. Dashed line separates simple cells with segregated On and Off subregions (filled columns) from complex cells with overlapped subfields. The difference between the two populations is statistically significant. The overlap index is depicted graphically below the histogram and is defined as

$$\text{Overlap index} = \frac{0.5Wp + 0.5Wn - d}{0.5Wp + 0.5Wn + d}$$

where W_p and W_n are the widths of the On and Off subregions and d the distance between the peak positions of each subregion. The parameters W_p, W_n and d were determined by separately fitting each On and Off excitatory response region with an elliptical Gaussian (Schiller et al., 1976). (b) The distribution of the values for the push–pull index was also bimodal; cells that had simple scores on the overlap index are shown in filled columns; NR indicates that there was no response to the flashed stimulus. The push-pull index is depicted graphically below the histogram and is defined as

$$\text{push-pull} = |P + N|$$

where P and N represent synaptic responses to bright and dark stimuli, respectively. (c) A scatter plot of values for subregion overlap vs. push–pull forms two clusters; the leftmost defining simple cells. The intersection of the crosses in each cluster corresponds to the mean, and the length of each line to the 95% confidence intervals. (d) The distribution of values for Pearson's correlation coefficient also suggests that simple cells and complex cells form two distinct populations in cat primary visual cortex. Bimodality was determined with Hartigan's dip test; the probability of rejection for a unimodal distribution was 0.99 for all distributions. Adapted from Martinez et al. (2005) and Hirsch and Martinez (2006).

Receptive-field structure and laminar location

But even if simple and complex cells had represented the ends of a continuum rather than two distinct populations, that would not necessarily argue against a feed-forward model of receptive-field construction in primary visual cortex. The model would still essentially be correct if both cell types tended to appear at different laminar locations: simple cells in layer 4 and complex cells outside the reach of the thalamic afferents. Thus, we examined the relationship between receptive-field structure and location in the cortical microcircuit. We found that all cells with simple receptive fields had dendrites in regions where thalamic afferents terminate, layer 4, its borders or upper layer 6 (Martinez et al., 2005). Figure 4a presents a view of simple receptive fields plotted as a function of laminar depth of the soma, with the deepest cells in each layer at the left and the most superficial at the right. The plot reveals a trend for cells with more compact subregions to lie in lower aspects of layer 4 and those with elongated subregions to occupy the superficial half of the layer. This intralaminar distribution recalls the primate cortex, in which lower layer 4 is supplied by the parvocellular layers of the thalamus and the upper tiers by the magnocellular layers (Callaway, 1998).

Receptive-field structure does not appear to vary with general morphological class (Gilbert and Wiesel, 1979; Gilbert, 1983; Martin and Whitteridge, 1984; Hirsch et al., 2002; Martinez et al., 2002, 2005; Hirsch, 2003). Simple cells take various shapes including spiny stellate, pyramidal and interneuronal profiles (Fig. 4b). We have, however, found a correlation between receptive field and anatomical structure for patterns of interlaminar connectivity. Specifically, simple cells in layer 6 have robust dendritic and axonal arborizations in layer 4, where simple cells dominate, while complex cells in layer 6 target the superficial layers and hence prefer other complex cells (Hirsch et al., 1998b). Katz (1987) found that cells in layer 6 with dense arbors in layer 4 projected to the thalamus. Thus, it is possible that only simple cells provide geniculocortical feedback. To date, we find that all excitatory cells in layer 4 project heavily to the superficial layers, but do not have sufficient information to assay for intralaminar preferences nor for differences in projections to the deep layers or extrastriate regions.

Complex cells, on the other hand, populate all cortical layers and, like simple cells, they formed an anatomically diverse population (Fig. 4c). However, response patterns of complex cells change with laminar location (Hirsch et al., 2002; Martinez et al., 2002, 2005; Hirsch, 2003). Complex receptive fields in layer 4 had co-spatial On and Off subfields (large values of the push–pull and overlap indices; Hirsch et al., 2002, 2003;

Martinez et al., 2005). Like simple cells, complex cells in thalamorecipient layers responded robust and reliably to the flashed spots and the time course of their responses followed the temporal envelop of thalamic activity (Hirsch et al., 2002). Thus all cells in layer 4, whether simple or complex, seem to capture and relay ascending patterns of thalamic drive. Conversely, complex cells in positions further removed from the thalamus, layers 2 + 3, 5 and lower 6, responded poorly if at all to the static stimuli, even though they are easily driven by richer stimuli like moving bars (Movshon et al., 1978b; Hirsch et al., 2002; Martinez et al., 2002, 2005). Moreover, when they did respond to the flashed spots, they usually showed preference for just one stimulus polarity, either bright or dark (Hirsch et al., 2002; Martinez et al., 2005).

Remarkably, not only the structure of the complex receptive field changes with laminar location. Our results show that the relative orientation tuning of excitatory and inhibitory inputs in complex cells also varies with position in the cortical microcircuit (Martinez et al., 2002; Hirsch et al., 2003). In layer 4, complex cells, at least the inhibitory ones, are insensitive to stimulus angle (Bullier and Henry, 1979a, b; Hirsch et al., 2003; Usrey et al., 2003). In layers 2 + 3 and 5 both excitatory and inhibitory complex cells are orientation tuned (Hirsch et al., 2000; Martinez et al., 2002). However, while in the superficial layers tuning curves for excitation and inhibition have similar peaks and bandwidths (Fig. 5b), in layer 5 their preferred orientation diverges such that the peaks of the tuning curves for excitation and inhibition can be as far as 90° apart (Martinez et al., 2002) (Fig. 5c).

All told, at the thalamocortical stage, a lot of energy must be expended to translate temporal patterns of thalamic input into either excitation or inhibition to generate new functional response properties, like orientation selectivity. By contrast, the transmission of information is strongly filtered and reconfigured at later stages of processing. As a result, only stimuli that meet certain standards, including particular motion patterns, evoke activity (Hirsch et al., 2002; Martinez et al., 2002).

84

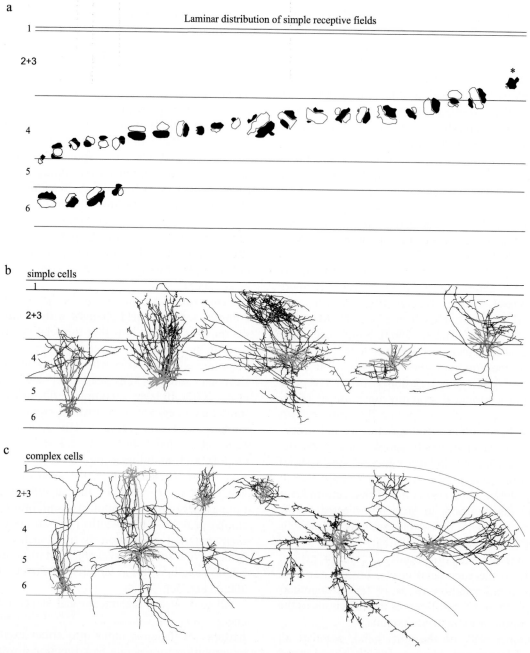

Fig. 4. Laminar distribution of receptive fields and morphology in cat primary visual cortex. (a) Cells with simple receptive fields were found exclusively in layer 4, its borders or in upper layer 6. The receptive fields are ordered from left to right according to the depth of the soma. Back and white code Off and On subregions, respectively. The asterisk marks the receptive field of a pyramidal cell recorded in layer 3 whose basal dendrites branched extensively into layer 4. (b,c) A sample of our three-dimensional reconstructions taken from the simple cell (a) and the complex cell (b) populations. The figure shows coronal views (from left to right, top) of a pyramid in upper layer 6, a pyramid at the 4–5 border, a spiny stellate cell in layer 4, a smooth cell in layer 4 and a pyramid at the 3–4 border; and (from left to right, bottom) of a pyramid in mid layer 6, a pyramid in layer 5, two pyramids in the superficial layers, a basket cell in layer 4 and a spiny stellate cell in the same layer. Cell bodies and dendrites are gray and axons are black. Adapted from Martinez et al. (2005) and Hirsch and Martinez (2006).

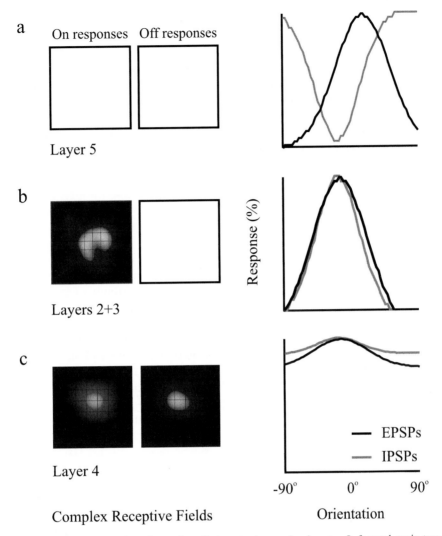

Fig. 5. Receptive fields and orientation tuning of complex cells in cat primary visual cortex. Left panel, main type of receptive field structure found in complex cells in layers 5, (a) 2 + 3 (b) and 4 (c). Right panel shows the cartoon versions of the profiles of orientation tuning found in complex cells. (a) In layer 5, the preferred orientation of excitation and inhibition diverges such that their peaks can be as far as 90° apart. (b) In layers 2 + 3, tuning curves for excitation and inhibition have similar peaks and bandwidths. (c) In layer 4 excitatory and inhibitory inputs are not tuned for orientation.

The extended feed-forward (push–pull) model of the simple receptive field

Hubel and Wiesel (1962) proposed that simple cell subregions were constructed from excitatory inputs provided by thalamic afferents properly aligned in visual space, a hypothesis that has received substantial experimental support (Tanaka, 1983, 1985; Chapman et al., 1991; Reid and Alonso, 1995; Ferster et al., 1996; Chung and Ferster, 1998; Hirsch et al., 1998a; Martinez et al., 2005). Since Hubel and Wiesel (1962) used extracellular recordings, the way in which inhibitory inputs contribute to the simple receptive field was left open in their model, and remains an issue of strong debate (Sillito, 1975; Pei et al., 1994; Allison et al., 1996; Borg-Graham et al., 1998; Crook et al., 1998; Hirsch et al., 1998a; Murthy and

Humphrey, 1999; Anderson et al., 2000; Martinez et al., 2002, 2005; Monier et al., 2003; Marino et al., 2005). We have recently recorded from a population of smooth cells in layer 4 whose receptive fields were indistinguishable from those of their neighboring, spiny neurons (Hirsch et al., 2003). Receptive fields of excitatory and inhibitory simple cells are similar in terms of geometry and number of component subregions and the layout of push–pull (Hirsch et al., 2003), which strongly suggests that they are built by a common mechanism (Fig. 6). In addition, a combination of excitatory and inhibitory simple receptive fields reciprocally connected explains why layer 4 simple cells have orientation tuning curves for excitation and inhibition with similar peaks and bandwidths (Anderson et al., 2000; Martinez et al., 2002). Such a feed-forward model of the simple receptive field is also consistent with the observation that the sharpness of orientation selectivity depends primarily on the degree of elongation of the On and Off subregions (Lampl et al., 2001; Martinez et al., 2002).

Inhibitory contributions to contrast invariant orientation tuning

What early feed-forward models (Hubel and Wiesel, 1962; Troyer et al., 1998; Ferster and Miller, 2000; Troyer et al., 2002; Lauritzen and Miller, 2003; Hirsch and Martinez, 2006) of orientation selectivity fail to explain is how cortical

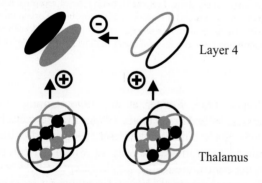

Fig. 6. Push–pull circuitry in layer 4. Wiring diagram for inputs that built simple excitatory neurons (left) and inhibitory neurons (right) in layer 4. See the text for further explanation. Black and gray code Off and On subfields, respectively.

neurons retain their orientation sensitivity over a wide range of stimulus contrasts (Sclar and Freeman, 1982; Ohzawa et al., 1985; Geisler and Albrecht, 1992; Priebe and Ferster, 2002). While cortical responses to stimuli at or near the preferred orientation grow stronger with increasing stimulus strength, responses to orthogonal stimuli remain small. Thus, stimulus contrast has little effect on the bandwidth of cortical orientation tuning curves (Sclar and Freeman, 1982; Ohzawa et al., 1985; Geisler and Albrecht, 1992). The situation is different for relay cells; as contrast grows stronger these neurons fire harder to stimuli of any orientation (Sclar and Freeman, 1982; Ohzawa et al., 1985; Geisler and Albrecht, 1992). Therefore, feed-forward models hold that a subset of the afferent input to each simple cell is 'untuned' as it is activated by stimuli of any orientation, including the orthogonal. Yet these models do not provide a means to counter the contrast-dependent increases in untuned thalamic firing that should elevate cortical tuning curves and increase bandwidth (Troyer et al., 1998, 2002). For at the orthogonal orientation, inhibitory simple cells are minimally active (Troyer et al., 1998, 2002).

The remaining inhibitory neurons we have recorded from in layer 4 had complex rather than simple receptive fields (Hirsch et al., 2003). These cells are insensitive to both stimulus polarity, they lack segregated subregions, and to stimulus angle, they respond equally well to stimuli of any orientation. In addition, their responses to static stimuli are as strong as those of simple cells, and they also follow temporal patterns of thalamic activity which indicates that they could be built by convergent input from On and Off relay cells with spatially overlapping receptive fields (Hirsch et al., 2002, 2003; Hirsch and Martinez, 2006). Untuned inhibitory cells are not an exclusive feature of the primary visual cortex, they are present in layer 4 of the somatosensory cortex as well (Simons, 1978; Swadlow and Gusev, 2002). In a recent report, Lauritzen and Miller (2003) demonstrated that a combination of simple and complex smooth cells in layer 4, tuned and untuned for orientation, respectively (Fig. 7), would supply the inhibition needed to explain simple-cell response properties. First, it would provide a component of feed-forward

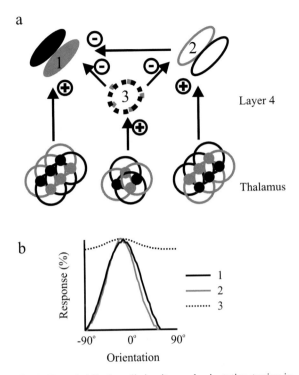

Layer 4

Thalamus

Fig. 7. Extended Push–pull circuitry and orientation tuning in layer 4. (a) Wiring diagram for inputs that built simple excitatory neurons, (1) simple inhibitory neurons (2) and complex inhibitory neurons (3) in layer 4. Black and gray code Off and On subfields respectively. (b) Orientation tuning curves (Gaussian fits) of one representative simple excitatory neurons, (1) one simple inhibitory cells (2) and one complex inhibitory neuron (3) in layer 4. Further details are given in the text.

inhibition that is untuned for orientation (complex smooth cells) to generate contrast invariant orientation tuning and regulate general levels of excitability. Second, orientation tuned push–pull inhibition (from simple smooth cells) would contribute to sharpen spatial frequency tuning, lower responses to low temporal frequency stimuli and control the excitability of the cortical network.

Summary

Anatomical evidence emphasizes the importance of laminar specialization. Evidence discussed here suggests that connections in different layers are specialized for different tasks. In particular, receptive field-structure changes and new response properties arise at successive stages of visual cortical processing. In layer 4, the computations performed by a synaptic network consisting of feed-forward inputs from the thalamus and two different sources of intracortical inhibition (push–pull tuned and push–push untuned) contribute to the generation of the simple receptive field and contrast invariant orientation tuning (Hirsch and Martinez, 2006).

This extended version of the feed forward (push–pull) circuit of layer 4 simple cells could be common to other parts of the visual pathway as well. For example, in the visual thalamus we also find push–pull and two different types of inhibitory inputs. The first one is analogous to that provided by layer 4 inhibitory neurons with simple receptive fields; it is sensitive to stimulus polarity (On or Off) and shares receptive field structure (center–surround) with neighboring excitatory relay cells. This type of inhibition is provided by local interneurons in the LGN. The second one resembles the inhibition provided by layer 4 inhibitory complex cells; it is insensitive to stimulus polarity (On–Off) and it is provided by neurons in the adjacent perigeniculate nucleus. It is thus conceivable that other stations along the visual pathway might also have similar circuit designs including a source of unselective inhibition. Such inhibitory circuits could be responsible for normalizing functional response properties and could also be involved in regulating excitability or mediating contextual effects of complex visual stimuli at a local level.

Finally, circuits outside layer 4 change the synaptic structure of the complex receptive field and orientation tuning with each step of cortical integration. As a result, new functional response properties, like the sensitivity to stimulus motion, emerge. Systematic changes in response properties with laminar location are observed in other sensory modalities as well. Our hope is that a better understanding of the structure and function of the visual cortical microcircuit will expose fundamental principles of neocortical function.

Acknowledgments

I would like to specially thank J.A. Hirsch for contributing to all aspects of this research and for her

88

support, encouragement and inspiration over the years. I am also deeply grateful to Jose-Manuel Alonso for many helpful discussions and for his contribution to the early phases of these projects. I also thank T.N. Wiesel for support and guidance, R.C. Reid for participation in early experiments, F.T. Sommer and Q. Wang for contributions to analysis and C. Gallagher Marshall, K. Desai Naik, C. Pillai and J.M. Provost for skillful anatomical reconstructions. Funding the experiments reported here was provided by NIH EY09593 to J.A. Hirsch.

References

Abbott, L.F. and Chance, F.S. (2002) Rethinking the taxonomy of visual neurons. Nat. Neurosci., 5: 391–392.

Adelson, E.H. and Bergen, J.R. (1985) Spatiotemporal energy models for the perception of motion. J. Opt. Soc. Am. A, 2: 284–299.

Ahmed, B., Anderson, J.C., Douglas, R.J., Martin, K.A. and Nelson, J.C. (1994) Polyneuronal innervation of spiny stellate neurons in cat visual cortex. J. Comp. Neurol., 341: 39–49.

Allison, J.D., Kabara, J.F., Snider, R.K., Casagrande, V.A. and Bonds, A.B. (1996) GABA B-receptor-mediated inhibition reduces the orientation selectivity of the sustained response of striate cortical neurons in cats. Visual Neurosci., 13: 559–566.

Alonso, J.M. and Martinez, L.M. (1998) Functional connectivity between simple cells and complex cells in cat striate cortex. Nat. Neurosci., 1: 395–403.

Alonso, J.M., Usrey, W.M. and Reid, R.C. (2001) Rules of connectivity between geniculate cells and simple cells in cat primary visual cortex. J. Neurosci., 21: 4002–4015.

Allman, J., Miezin, F. and McGuinness, E. (1985) Stimulus specific responses from beyond the classical receptive field: neurophysiological mechanisms for local–global comparisons in visual neurons. Ann. Rev. Neurosci., 8: 407–429.

Anderson, J.S., Carandini, M. and Ferster, D. (2000) Orientation tuning of input conductance, excitation, and inhibition in cat primary visual cortex. J. Neurophysiol., 84: 909–926.

Angelucci, A., Levitt, J.B. and Lund, J.S. (2002) Anatomical origins of the classical receptive field and modulatory surround field of single neurons in macaque visual cortical area V1. Prog. Brain Res., 136: 373–388.

Anzai, A., Ohzawa, I. and Freeman, R.D. (1999) Neural mechanisms for processing binocular information II. Complex cells. J. Neurophysiol., 82: 909–924.

Bair, W. (2005) Visual receptive field organization. Curr. Opin. Neurobiol., 15: 459–464.

Baker Jr., C.L. and Cynader, M.S. (1986) Spatial receptive-field properties of direction-selective neurons in cat striate cortex. J. Neurophysiol., 55: 1136–1152.

Barlow, H.B. (1953) Summation and inhibition in the frog's retina. J. Physiol., 119: 69–88.

Barlow, H.B. and Levick, W.R. (1976) Threshold setting by the surround of cat retinal ganglion cells. J. Physiol. (Lond.), 259: 737–757.

Ben-Yishai, R., Bar-Or, R.L. and Sompolinsky, H. (1995) Theory of orientation tuning in visual cortex. Proc. Natl. Acad. Sci. USA, 92: 3844–3848.

Binzegger, T., Douglas, R.J. and Martin, K.A. (2004) A quantitative map of the circuit of cat primary visual cortex. J. Neurosci., 24: 8441–8453.

Borg-Graham, L.J., Monier, C. and Fregnac, Y. (1998) Visual input evokes transient and strong shunting inhibition in visual cortical neurons. Nature, 393: 369–373.

Bosking, W.H., Zhang, Y., Schofield, B. and Fitzpatrick, D. (1997) Orientation selectivity and the arrangement of horizontal connections in tree shrew striate cortex. J. Neurosci., 17: 2112–2127.

Braitenberg, V. and Schuz, A. (1998) Cortex: Statistics and Geometry of Neuronal Connectivity (Second Edition). Springer, Berlin.

Bullier, J. and Henry, G.H. (1979a) Ordinal position of neurons in cat striate cortex. J. Neurophysiol., 42: 1251–1263.

Bullier, J. and Henry, G.H. (1979b) Laminar distribution of first-order neurons and afferent terminals in cat striate cortex. J. Neurophysiol., 42: 1271–1281.

Cai, D., DeAngelis, G.C. and Freeman, R.D. (1997) Spatiotemporal receptive field organization in the lateral geniculate nucleus of cats and kittens. J. Neurophysiol., 78: 1045–1061.

Callaway, E.M. (1998) Local circuits in primary visual cortex of the macaque monkey. Ann. Rev. Neurosci., 21: 47–74.

Carandini, M. and Ringach, D.L. (1997) Predictions of a recurrent model of orientation selectivity. Vision Res., 37: 3061–3071.

Cavanaugh, J.R., Bair, W. and Movshon, J.A. (2002) Selectivity and spatial distribution of signals from the receptive field surround in macaque V1 neurons. J. Neurophysiol., 88: 2547–2556.

Chance, F.S., Nelson, S.B. and Abbott, L.F. (1999) Complex cells as cortically amplified simple cells. Nat. Neurosci., 2: 277–282.

Chapman, B., Zahs, K.R. and Stryker, M.P. (1991) Relation of cortical cell orientation selectivity to alignment of receptive fields of the geniculocortical afferents that arborize within a single orientation column in ferret visual cortex. J. Neurosci., 11: 1347–1358.

Chisum, H.J., Mooser, F. and Fitzpatrick, D. (2003) Emergent properties of layer 2/3 neurons reflect the collinear arrangement of horizontal connections in tree shrew visual cortex. J. Neurosci., 23: 2947–2960.

Chung, S. and Ferster, D. (1998) Strength and orientation tuning of the thalamic input to simple cells revealed by electrically evoked cortical suppression. Neuron, 20: 1177–1189.

Cleland, B.G., Dubin, M.W. and Levick, W.R. (1971) Sustained and transient neurones in the cat's retina and lateral geniculate nucleus. J. Physiol. (Lond.), 217: 473–496.

Cleland, B.G. and Lee, B.B. (1985) A comparison of visual responses of cat lateral geniculate nucleus neurones with those of ganglion cells afferent to them. J. Physiol. (Lond.), 369: 249–268.

Coogan, T.A. and Burkhalter, A. (1990) Conserved patterns of cortico-cortical connections define areal hierarchy in rat visual cortex. Exp. Brain Res., 80: 49–53.

Crook, J.M., Kisvarday, Z.F. and Eysel, U.T. (1998) Evidence for a contribution of lateral inhibition to orientation tuning and direction selectivity in cat visual cortex: reversible inactivation of functionally characterized sites combined with neuroanatomical tracing techniques. Eur. J. Neurosci., 10: 2056–2075.

Dantzker, J.L. and Callaway, E.M. (2000) Laminar sources of synaptic input to cortical inhibitory interneurons and pyramidal neurons. Nat. Neurosci., 3: 701–707.

De Valois, R.L., Albrecht, D.G. and Thorell, L.G. (1982) Spatial frequency selectivity of cells in macaque visual cortex. Vision Res., 22: 545–559.

Dean, A.F. and Tolhurst, D.J. (1983) On the distinctness of simple and complex cells in the visual cortex of the cat. J. Physiol., 344: 305–325.

DeAngelis, G.C., Ohzawa, I. and Freeman, R.D. (1993a) Spatiotemporal organization of simple-cell receptive fields in the cat's striate cortex I. General characteristics and postnatal development. J. Neurophysiol., 69: 1091–1117.

DeAngelis, G.C., Ohzawa, I. and Freeman, R.D. (1993b) Spatiotemporal organization of simple-cell receptive fields in the cat's striate cortex II. Linearity of temporal and spatial summation. J. Neurophysiol., 69: 1118–1135.

DeAngelis, G.C., Ohzawa, I. and Freeman, R.D. (1995) Receptive field dynamics in central visual pathways. Trends Neurosci., 18: 451–458.

Debanne, D., Shultz, D.E. and Fregnac, Y. (1998) Activity-dependent regulation of 'on' and 'off' responses in cat visual cortical receptive fields. J. Physiol., 508: 523–548.

Douglas, R.J., Koch, C., Mahowald, M., Martin, K.A. and Suarez, H.H. (1995) Recurrent excitation in neocortical circuits. Science, 269: 981–985.

Douglas, R.J. and Martin, K.A. (2004) Neuronal circuits of the neocortex. Ann. Rev. Neurosci., 27: 419–451.

Einhauser, W., Kayser, C., Koning, P. and Kording, K.P. (2002) Learning the invariance properties of complex cells from their responses to natural stimuli. Eur. J. Neurosci., 15: 475–486.

Emerson, R.C., Bergen, J.R. and Adelson, E.H. (1992) Directionally selective complex cells and the computation of motion energy in cat visual cortex. Vision Res., 32: 203–218.

Enroth-Cugell, C. and Lennie, P. (1975) The control of retinal ganglion cell discharge by receptive field surrounds. J. Physiol., 247: 551–578.

Enroth-Cugell, C. and Robson, J.G. (1984) Functional characteristics and diversity of cat retinal ganglion cells. Basic characteristics and quantitative description. Invest. Ophthalmol. Vis. Sci., 25: 250–267.

Ferster, D. (1988) Spatially opponent excitation and inhibition in simple cells of the cat visual cortex. J. Neurosci., 8: 1172–1180.

Ferster, D., Chung, S. and Wheat, H. (1996) Orientation selectivity of thalamic input to simple cells of cat visual cortex. Nature, 380: 249–252.

Ferster, D. and Miller, K.D. (2000) Neural mechanisms of orientation selectivity in the visual cortex. Ann. Rev. Neurosci., 23: 441–471.

Fitzpatrick, D. (1996) The functional organization of local circuits in visual cortex: insights from the study of tree shrew striate cortex. Cereb. Cortex, 6: 329–341.

Fitzpatrick, D. (2000) Seeing beyond the receptive field in primary visual cortex. Curr. Opin. Neurobiol., 10: 438–442.

Fleet, D.J., Wagner, H. and Heeger, D.J. (1996) Neural encoding of binocular disparity: energy models, position shifts and phase shifts. Vision Res., 36: 1839–1857.

Gaska, J.P., Jacobson, L.D., Chen, H.W. and Pollen, D.A. (1994) Space–time spectra of complex cell filters in the macaque monkey: a comparison of results obtained with pseudowhite noise and grating stimuli. Visual Neurosci., 11: 805–821.

Geisler, W.S. and Albrecht, D.G. (1992) Cortical neurons: isolation of contrast gain control. Vision Res., 32: 1409–1410.

Gil, Z., Connors, B.W. and Amitai, Y. (1999) Efficacy of thalamocortical and intracortical synaptic connections: quanta, innervation, and reliability. Neuron, 23: 385–397.

Gilbert, C.D. (1977) Laminar differences in receptive field properties of cells in cat primary visual cortex. J. Physiol., 268: 391–421.

Gilbert, C.D. (1983) Microcircuitry of the visual cortex. Ann. Rev. Neurosci., 6: 217–247.

Gilbert, C.D. and Wiesel, T.N. (1979) Morphology and intracortical projections of functionally characterised neurones in the cat visual cortex. Nature, 280: 120–125.

Gillespie, D.C., Lampl, I., Anderson, J.S. and Ferster, D. (2001) Dynamics of the orientation-tuned membrane potential response in cat primary visual cortex. Nat. Neurosci., 4: 1014–1019.

Hartline, H.K. (1938) The response of single optic nerve fibers of the vertebrate eye to illumination of the retina. Am. J. Physiol., 121: 400–415.

Heggelund, P. (1981) Receptive field organization of complex cells in cat striate cortex. Exp. Brain Res., 42: 90–107.

Heggelund, P. (1986) Quantitative studies of enhancement and suppression zones in the receptive field of simple cells in cat striate cortex. J. Physiol., 373: 293–310.

Henry, G.H. (1977) Receptive field classes of cells in the striate cortex of the cat. Brain Res., 133: 1–28.

Hirsch, J.A. (2003) Synaptic physiology and receptive field structure in the early visual pathway of the cat. Cereb. Cortex, 13: 63–69.

Hirsch, J.A., Alonso, J.M., Reid, R.C. and Martinez, L.M. (1998a) Synaptic integration in striate cortical simple cells. J. Neurosci., 18: 9517–9528.

Hirsch, J.A., Gallagher, C.A., Alonso, J.M. and Martinez, L.M. (1998) Ascending projections of simple and complex cells in layer 6 of the cat striate cortex. J. Neurosci., 18: 8086–8094.

Hirsch, J.A. and Martinez, L.M. (2006) Circuits that build visual cortical receptive fields. Trends Neurosci., 29: 30–39, doi:10.1016/j.tins.2005.11.001.

Hirsch, J.A., Martinez, L.M., Alonso, J.M., Desai, K., Pillai, C. and Pierre, C. (2002) Synaptic physiology of the flow of information in the cat's visual cortex in vivo. J. Physiol., 540: 335–350.

Hirsch, J.A., Martinez, L.M., Alonso, J.M., Pillai, C. and Pierre, C. (2000) Simple and complex inhibitory cells in layer 4 of cat visual cortex. Soc. Neurosci. Abstr., 26: 108.

Hirsch, J.A., Martinez, L.M., Pillai, C., Alonso, J.M., Wang, Q. and Sommers, F.T. (2003) Functionally distinct inhibitory neurons at the first stage of visual cortical processing. Nat. Neurosci., 6: 1300–1308.

Hoffmann, K.P. and Stone, J. (1971) Conduction velocity of afferents to cat visual cortex: a correlation with cortical receptive field properties. Brain Res., 32: 460–466.

Hubel, D.H. and Wiesel, T.N. (1959) Receptive fields of single neurons in the cat's striate cortex. J. Physiol., 148: 574–591.

Hubel, D.H. and Wiesel, T.N. (1961) Integrative action in the cat's lateral geniculate body. J. Physiol., 155: 385–398.

Hubel, D.H. and Wiesel, T.N. (1962) Receptive fields, binocular interaction and functional architecture in the cat's visual cortex. J. Physiol., 160: 106–154.

Hubel, D.H. and Wiesel, T.N. (1968) Receptive fields and functional architecture of monkey striate cortex. J. Physiol. (Lond.), 195: 215–243.

Hyvarinen, A. and Hoyer, P.O. (2001) A two-layer sparse coding model learns simple and complex cell receptive fields and topography from natural images. Vision Res., 41: 2413–2423.

Jacob, M.S., Peterson, M.R., Wu. A. and Freeman, R.D. (2003) Laminar differences in response characteristics of cells in the primary visual cortex. Soc. Neurosci. Abstr., Program No. 910.913.

Jones, J.P. and Palmer, L.A. (1987) The two-dimensional spatial structure of simple receptive fields in cat striate cortex. J. Neurophysiol., 58: 1187–1211.

Kagan, I., Gur, M. and Snodderly, D.M. (2002) Spatial organization of receptive fields of V1 neurons of alert monkeys: comparison with responses to gratings. J. Neurophysiol., 88: 2557–2574.

Katz, L.C. (1987) Local circuitry of identified projection neurons in cat visual cortex brain slices. J. Neurosci., 7: 1223–1249.

Kisvarday, Z.F., Martin, K.A., Freund, T.F., Magloczky, Z., Whitteridge, D. and Somogyi, P. (1986) Synaptic targets of HRP-filled layer III pyramidal cells in the cat striate cortex. Exp. Brain Res., 64: 541–552.

Kuffler, S. (1953) Discharge patterns and functional organization of the mammalian retina. J. Neurophysiol., 16: 37–68.

Lampl, I., Anderson, J.S., Gillespie, D.C. and Ferster, D. (2001) Prediction of orientation selectivity from receptive field architecture in simple cells of cat visual cortex. Neuron, 30: 263–274.

Lauritzen, T.Z. and Miller, K.D. (2003) Different roles for simple- and complex-cell inhibiton in V1. J. Neurosci., 23: 10201–10213.

LeVay, S. and Gilbert, C.D. (1976) Laminar patterns of geniculocortical projection in the cat. Brain Res., 113: 1–19.

Levick, W.R., Cleland, B.G. and Dubin, M.W. (1972) Lateral geniculate neurons of cat: retinal inputs and physiology. Invest. Ophthalmol., 11: 302–311.

Lund, J.S., Henry, G.H., MacQueen, C.L. and Harvey, A.R. (1979) Anatomical organization of the primary visual cortex (area 17) of the cat. A comparison with area 17 of the macaque monkey. J. Comp. Neurol., 184: 599–618.

Marino, J., Schummers, J., Lyon, D.C., Schwabe, L., Beck, O., Wiesing, P., Obermayer, K. and Sur, M. (2005) Invariant computations in local cortical networks with balanced excitation and inhibition. Nat. Neurosci., 8: 194–201.

Martin, K.A. (2002) Microcircuits in visual cortex. Curr. Opin. Neurobiol., 12: 418–425.

Martin, K.A. and Whitteridge, D. (1984) Form, function and intracortical projections of spiny neurones in the striate visual cortex of the cat. J. Physiol., 353: 463–504.

Martinez, L.M. and Alonso, J.M. (2001) Construction of complex receptive fields in primary visual cortex. Neuron, 32: 515–525.

Martinez, L.M. and Alonso, J.M. (2003) Complex receptive fields in primary visual cortex. Neuroscientist, 8: 317–331.

Martinez, L.M., Alonso, J.M., Reid, R.C. and Hirsch, J.A. (2002) Laminar processing of stimulus orientation in cat visual cortex. J. Physiol., 540: 321–333.

Martinez, L.M., Wang, Q., Reid, R.C., Pillai, C., Alonso, J.M., Sommer, F.T. and Hirsch, J.A. (2005) Receptive field structure field varies with layer in the primary visual cortex. Nat. Neurosci., 8: 372–379.

Mata, M.L. and Ringach, D.L. (2005) Spatial Overlap of 'on' and 'off' subregions and its relation to response modulation ratio in macaque primary visual cortex. J. Neurophysiol., 93: 919–928.

McLean, J. and Palmer, L.A. (1994) Organization of simple cell responses in the three-dimensional (3-D) frequency domain. Visual Neurosci., 11: 295–306.

Mechler, F. and Ringach, D.L. (2002) On the classification of simple and complex cells. Vision Res., 42: 1017–1033.

Monier, C., Chavane, F., Baudot, P., Graham, L.J. and Fregnac, Y. (2003) Orientation and direction selectivity of synaptic inputs in visual cortical neurons: a diversity of combinations produces spike tuning. Neuron, 37: 663–680.

Mooser, F., Bosking, W.H. and Fitzpatrick, D. (2004) A morphological basis for orientation tuning in primary visual cortex. Nat. Neurosci., 8: 872–879.

Movshon, J.A., Thompson, I.D. and Tolhurst, D.J. (1978a) Spatial summation in the receptive fields of simple cells in the cat's striate cortex. J. Physiol., 283: 53–77.

Movshon, J.A., Thompson, I.D. and Tolhurst, D.J. (1978b) Receptive field organization of complex cells in the cat's striate cortex. J. Physiol., 283: 79–99.

Movshon, J.A., Thompson, I.D. and Tolhurst, D.J. (1978c) Spatial and temporal contrast sensitivity of neurones in areas 17 and 18 of the cat's visual cortex. J. Physiol., 283: 101–120.

Murthy, A. and Humphrey, A.L. (1999) Inhibitory contributions to spatiotemporal receptive-field structure and direction selectivity in simple cells of cat area 17. J. Neurophysiol., 81: 1212–1224.

Nelson, S. (2002) Cortical microcircuits. Diverse or canonical? Neuron, 36: 19–27.

Nelson, S., Toth, L., Sheth, B. and Sur, M. (1994) Orientation selectivity of cortical neurons during intracellular blockade of inhibition. Science, 265: 774–777.

Ohzawa, I., DeAngelis, G.C. and Freeman, R.D. (1990) Stereoscopic depth discrimination in the visual cortex: neurons ideally suited as disparity detectors. Science, 249: 1037–1041.

Ohzawa, I., DeAngelis, G.C. and Freeman, R.D. (1997) Encoding of binocular disparity by complex cells in the cat's visual cortex. J. Neurophysiol., 77: 2879–2909.

Ohzawa, I. and Freeman, R D. (1986) The binocular organization of complex cells in the cat's visual cortex. J. Neurophysiol., 56: 243–259.

Ohzawa, I. and Freeman, R.D. (1997) Spatial pooling of subunits in complex cell receptive fields. Soc. Neurosci. Abstr., 23: 1669.

Ohzawa, I., Sclar, G. and Freeman, R.D. (1985) Contrast gain control in the cat's visual system. J. Neurophysiol., 54: 651–667.

Okajima, K. and Imaoka, H. (2001) A complex cell-like receptive field obtained by information maximization. Neural Comput., 13: 547–562.

Olshausen, B.A. and Field, D.J. (1996) Emergence of simple-cell receptive field properties by learning a sparse code for natural images. Nature, 381: 607–609.

Orban, G.A. (1884) Neuronal Operations in the Visual Cortex. Springer, Berlin.

Palmer, L.A. and Davis, T.L. (1981) Receptive-field structure in cat striate cortex. J. Neurophysiol., 46: 260–276.

Palmer, L.A. and Rosenquist, A.C. (1974) Visual receptive fields of single striate cortical units projecting to the superior colliculus in the cat. Brain Res., 67: 27–42.

Peters, A. and Payne, B.R. (1993) Numerical relationships between geniculocortical afferents and pyramidal cell modules in cat primary visual cortex. Cereb. Cortex, 3: 69–78.

Pei, X., Vidyasagar, T.R., Volgushev, M. and Creutzfeldt, O.D. (1994) Receptive field analysis and orientation selectivity of postsynaptic potentials of simple cells in cat visual cortex. J. Neurosci., 14: 7130–7140.

Pollen, D.A., Gaska, J.P. and Jacobson, L.D. (1989) Physiological constrains on models of visual cortical function. In: Cotterill, R.M.J. (Ed.), Models of Brain Function. Cambridge University Press, Cambridge, NY, pp. 115–135.

Priebe, N.J. and Ferster, D. (2002) A new mechanism for neuronal gain control (or how the gain in brains has mainly been explained). Neuron, 35: 602–604.

Priebe, N.J., Mechler, F., Carandini, M. and Ferster, D. (2004) The contribution of spike threshold to the dichotomy of cortical simple and complex cells. Nat. Neurosci., 7: 1113–1122.

Qian, N. and Zhu, Y. (1997) Physiological computation of binocular disparity. Vision Res., 37: 1811–1827.

Reid, R.C. and Alonso, J.M. (1995) Specificity of monosynaptic connections from thalamus to visual cortex. Nature, 378: 281–284.

Ringach, D.L. (2004) Mapping receptive fields in primary visual cortex. J. Physiol. (Lond.), 558.3: 717–728.

Ringach, D.L., Hawken, M.J. and Shapley, R. (1997) Dynamics of orientation tuning in macaque primary visual cortex. Nature, 387: 281–284.

Ringach, D.L., Hawken, M.J. and Shapley, R. (2003) Dynamics of orientation tuning in macaque V1: the role of global and tuned supression. J. Neurophysiol., 90: 342–352.

Ringach, D.L., Shapley, R.M. and Hawken, M.J. (2002) Orientation selectivity in macaque V1: diversity and laminar dependence. J. Neurosci., 22: 5639–5651.

Rivadulla, C., Sharma, J. and Sur, M. (2001) Specific roles of NMDA and AMPA receptors in direction-selective and spatial phase-selective responses in visual cortex. J. Neurosci., 21: 1710–1719.

Sakai, K. and Tanaka, S. (2000) Spatial pooling in the second-order spatial structure of cortical complex cells. Vision Res., 40: 855–871.

Saul, A.B. and Humphrey, A.L. (1990) Spatial and temporal response properties of lagged and nonlagged cells in cat lateral geniculate nucleus. J. Neurophysiol., 64: 206–224.

Schiller, P.H., Finlay, B.L. and Volman, S.F. (1976) Quantitative studies of single-cell properties in monkey striate cortex. I. Spatiotemporal organization of receptive fields. J. Neurophysiol., 39: 1288–1319.

Sclar, G. and Freeman, R.D. (1982) Orientation selectivity in the cat's striate cortex is invariant with stimulus contrast. Exp. Brain Res., 46: 457–461.

Shapley, R. and Hochstein, S. (1975) Visual spatial summation in two classes of geniculate cells. Nature, 256: 411–413.

Shapley, R. and Lennie, P. (1985) Spatial frequency analysis in the visual system. Ann. Rev. Neurosci., 8: 547–583.

Sillito, A.M. (1975) The contribution of inhibitory mechanisms to the receptive field properties of neurones in the striate cortex of the cat. J. Physiol. (Lond.), 250: 305–329.

Simons, D.J. (1978) Response properties of vibrissa units in rat somatosensory neocortex. J. Neurophysiol., 50: 798–820.

Skottun, B.C., De Valois, R.L., Grosof, D.H., Movshon, J.A., Albrecht, D.G. and Bonds, A.B. (1991) Classifying simple and complex cells on the basis of response modulation. Vision Res., 31: 1079–1086.

Somers, D.C., Nelson, S.B. and Sur, M. (1995) An emergent model of orientation selectivity in cat visual cortical simple cells. J. Neurosci., 15: 5448–5465.

Somers, D.C., Todorov, .E.V., Siapas, A.G., Toth, L.J., Kim, D.S. and Sur, M. (1998) A local circuit approach to understanding integration of long-range inputs in primary visual cortex. Cereb. Cortex, 8: 204–217.

Sompolinsky, H. and Shapley, R. (1997) New perspectives on the mechanisms for orientation selectivity. Curr. Opin. Neurobiol., 7: 514–522.

Stratford, K.J., Tarczy-Hornoch, K., Martin, K.A., Bannister, N.J. and Jack, J.J. (1996) Excitatory synaptic inputs to spiny stellate cells in cat visual cortex. Nature, 382: 258–261.

Swadlow, H.A. and Gusev, A.G. (2002) Receptive-field construction in cortical inhibitory interneurons. Nat. Neurosci., 5: 403–404.

Szulborski, R.G. and Palmer, L.A. (1990) The two-dimensional spatial structure of nonlinear subunits in the receptive fields of complex cells. Vision Res., 30: 249–254.

Tanaka, K. (1983) Cross-correlation analysis of geniculostriate neuronal relationships in cats. J. Neurophysiol., 49: 1303–1318.

Tanaka, K. (1985) Organization of geniculate inputs to visual cortical cells in the cat. Vision Res., 25: 357–364.

Tao, L., Shelley, M., McLaughlin, D. and Shapley, R. (2004) An egalitarian network model for the emergence of simple and complex cells in visual cortex. Proc. Natl. Acad. Sci. USA, 101: 366–371.

Thomson, A.M. and Bannister, A.P. (2003) Interlaminar connections in the neocortex. Cereb. Cortex, 13: 5–14.

Thomson, A.M., West, D.C., Wang, Y. and Bannister, A.P. (2002) Synaptic connections and small circuits involving excitatory and inhibitory neurons in layers 2–5 of adult rat and cat neocortex: triple intracellular recordings and biocytin labelling in vitro. Cereb. Cortex, 12: 936–953.

Touryan, J., Felsen, G. and Dan, Y. (2005) Spatial structure of complex cell receptive fields measured with natural images. Neuron, 45: 781–791.

Toyama, K., Kimura, M. and Tanaka, K. (1981) Organization of cat visual cortex as investigated by cross-correlation analysis. J. Neurophysiol., 46: 202–214.

Troyer, T.W., Krukowski, A.E. and Miller, K.D. (2002) LGN input to simple cells and contrast-invariant orientation tuning: an analysis. J. Neurophysiol., 87: 2741–2752.

Troyer, T.W., Krukowski, A.E., Priebe, N.J. and Miller, K.D. (1998) Contrast-invariant orientation tuning in cat visual cortex: thalamocortical input tuning and correlation-based intracortical connectivity. J. Neurosci., 18: 5908–5927.

Usrey, W.M., Reppas, J.B. and Reid, R.C. (1999) Specificity and strength of retinogeniculate connections. J. Neurophysiol., 82: 3527–3540.

Usrey, W.M., Sceniak, M.P. and Chapman, B. (2003) Receptive fields and response properties of neurons in layer 4 of ferret visual cortex. J. Neurophysiol., 89: 1003–1015.

Van Hooser, S.D., Heimel, J.A. and Nelson, S.B. (2003) Receptive field properties and laminar organization of lateral geniculate nucleus in the gray squirrel (*Sciurus carolinensis*). J. Neurophysiol., 90: 3398–3418.

Van Horn, S.C., Erisir, A. and Sherman, S.M. (2000) Relative distribution of synapses in the A-laminae of the lateral geniculate nucleus of the cat. J. Comp. Neurol., 416: 509–520.

Victor, J.D. and Shapley, R.M. (1979) Receptive field mechanisms of cat X and Y retinal ganglion cells. J. Gen. Physiol., 74: 275–298.

Volgushev, M., Pei, X., Vidyasagar, T.R. and Creutzfeldt, O.D. (1993) Excitation and inhibition in orientation selectivity of cat visual cortex neurons revealed by whole-cell recordings in vivo. Visual Neurosci., 10: 1151–1155.

Wielaard, D.J., Shelley, M., Mclaughlin, D. and Shapley, R. (2001) How simple cells are made in a nonlinear network model of the visual cortex. J. Neurosci., 21: 5203–5211.

Wiesel, T.N. (1959) Recording inhibition and excitation in the cat's retinal ganglion cells with intracellular electrodes. Nature, 183: 264–265.

Wiesel, T.N. and Gilbert, C.D. (1983) The Sharpey–Schafer lecture. morphological basis of visual cortical function. Quart. J. Exp. Physiol., 68: 525–543.

Martinez-Conde, Macknik, Martinez, Alonso & Tse (Eds.)
Progress in Brain Research, Vol. 154
ISSN 0079-6123

CHAPTER 5

Contribution of feedforward, lateral and feedback connections to the classical receptive field center and extra-classical receptive field surround of primate V1 neurons

Alessandra Angelucci[1,*] and Paul C. Bressloff[2]

[1]*Department of Ophthalmology and Visual Science, Moran Eye Center, University of Utah, 50 North Medical Drive, Salt Lake City, UT 84132, USA*
[2]*Department of Mathematics, University of Utah, 155 South 1400 East, Salt Lake City, UT 84112, USA*

Abstract: A central question in visual neuroscience is what circuits generate the responses of neurons in the primary visual cortex (V1). V1 neurons respond best to oriented stimuli of optimal size within their receptive field (RF) center. This size tuning is contrast dependent, i.e. a neuron's optimal stimulus size measured at high contrast (the high-contrast summation RF, or hsRF) is smaller than when measured using low-contrast stimuli (the low-contrast summation RF, or lsRF). Responses to stimuli in the RF center are usually suppressed by iso-oriented stimuli in the extra-classical RF surround. Iso-orientation surround suppression is fast and long range, extending well beyond the size of V1 cells' lsRF.

Geniculocortical feedforward (FF), V1 lateral and extrastriate feedback (FB) connections to V1 could all contribute to generating the RF center and surround of V1 neurons. Studies on the spatio-temporal properties and functional organization of these connections can help disclose their specific contributions to the responses of V1 cells. These studies, reviewed in this chapter, have shown that FF afferents to V1 integrate signals within the hsRF of V1 cells; V1 lateral connections are commensurate with the size of the lsRF and may, thus, underlie contrast-dependent changes in spatial summation, and modulatory effects arising from the surround region closer to the RF center (the "near" surround). The spatial and temporal properties of lateral connections cannot account for the dimensions and onset latency of modulation arising from more distant regions of the surround (the "far" surround). Inter-areal FB connections to V1, instead, are commensurate with the full spatial range of center and surround responses, and show fast conduction velocity consistent with the short onset latency of modulation arising from the "far" surround.

We review data showing that a subset of FB connections terminate in a patchy fashion in V1, and show modular and orientation specificity, consistent with their proposed role in orientation-specific center–surround interactions.

We propose specific mechanisms by which each connection type contributes to the RF center and surround of V1 neurons, and implement these hypotheses into a recurrent network model. We show physiological data in support of the model's predictions, revealing that modulation from the "far" surround is not always suppressive, but can be facilitatory under specific stimulus conditions.

Keywords: extrastriate; geniculocortical; surround modulation; horizontal connections; striate cortex; macaque.

---

*Corresponding author. Tel.: + 1-801-585-7489; Fax: + 1-801-585-1295; E-mail: alessandra.angelucci@hsc.utah.edu

DOI: 10.1016/S0079-6123(06)54005-1

Introduction

The perception of a visual "figure" often relies upon the overall spatial arrangement of its local elements. The global attributes of a visual stimulus can affect the response of visual cortical neurons to the local attributes of the stimulus. For example, in early visual cortex, the response of neurons to stimuli within their receptive field (RF) center can be modulated by contextual stimuli outside their RF, in the RF surround. The long-term goal of our research is to identify the neural circuits that generate responses within and outside the RF of visual cortical neurons. The anatomical substrates and mechanisms involved are likely to be the cornerstone of contour integration, and figure-ground segregation. Thus, these studies ultimately will help our understanding of the neural substrates for higher visual cortical processing and perception.

Here we review anatomical and physiological data suggesting potential circuits involved in center–surround interactions in primate V1, and propose specific mechanisms by which these circuits could generate such interactions. We present a neural network model implementing these ideas, whose architecture is constrained to fit the anatomical and physiological data presented in this review. Finally, we discuss specific predictions of the model, and report our recent physiological data that support such predictions.

Spatial properties of the receptive field center and surround of V1 neurons in the primate

A basic feature of neurons at the early stages of visual cortical processing is that they respond to presentation of specific visual stimuli within a localized region of space, i.e. the neuron's RF. Presentation of similar stimuli outside the RF typically does not evoke a response from the neuron, but can modulate (facilitate or suppress) the neuron's response to stimulation of its RF center (Blakemore and Tobin, 1972; Nelson and Frost, 1978; Allman et al., 1985; Gilbert and Wiesel, 1990; Li et al., 2001).

The RF center of a V1 cell consists of a low-threshold spiking region (the minimum response field, mRF; or classical RF, cRF) that can be mapped using small drifting gratings of optimal parameters for the neuron and delimiting the area of visual space that elicits spikes (Hubel and Wiesel, 1962; Barlow et al., 1967). The mRF is surrounded by a higher threshold depolarizing field (the high-contrast summation RF, or hsRF) that facilitates the cell's response when stimulated together with the mRF. The hsRF is typically measured by centering a high-contrast drifting grating of optimal parameters for the cell on the cell's mRF, and systematically increasing its diameter (DeAngelis et al., 1994; Sceniak et al., 2001; Cavanaugh et al., 2002; Levitt and Lund, 2002). A typical V1 cell increases its response with increasing grating diameter, up to a peak, and it is then suppressed ("surround suppression") for further increases in stimulus diameter (Fig. 1a). The size of the hsRF is the stimulus diameter at peak response (gray arrow in Fig. 1a). In macaques, the hsRF size of V1 neurons at 2–8° eccentricity averages $1° \pm 0.1$ (Fig. 1b), and is about 2.2-fold larger than the mean mRF size of the same cells (Angelucci et al., 2002b; Levitt and Lund, 2002). The summation RF (sRF) has been shown to be on average 2.3-fold larger when measured using low-contrast gratings (Sceniak et al., 1999). We refer to the summation RF measured using low-contrast gratings as the "low-contrast sRF", or lsRF (Fig. 1d).

Many studies have examined the influence of stimuli outside the RF on the response of V1 cells to stimuli inside their RF. A distinctive property of center–surround interactions is their orientation specificity. However, there is less agreement among studies on the specific "sign" (i.e. facilitation or suppression) of the orientation-specific interactions. We (Angelucci and Bullier, 2003) have previously argued that this discrepancy may be due to the different definitions of RF center and surround adopted in the different studies. Specifically, in some studies, the RF center size was defined as the neuron's sRF (hsRF or lsRF), and the surround as the region outside it. In the reminder of this chapter, we will refer to this region outside the lsRF as the "far" surround (Fig. 2a, left). In these studies, large gratings of similar orientation to the center stimulus placed in the far surround most often suppressed the RF center's

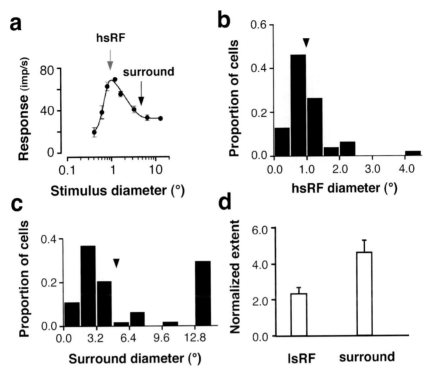

Fig. 1. Spatial extent of the RF center and surround of macaque V1 neurons. (a) Size-tuning curve of a typical V1 cell showing surround suppression, measured using high-contrast gratings of increasing diameter. Gray arrow: stimulus diameter at peak response, or hsRF size; black arrow: stimulus diameter at asymptotic response, or surround size. (b) Distribution of hsRF diameters and (c) of surround field diameters for the same population of macaque V1 neurons ($n = 59$). Arrowheads: means. (d) Average extent ($\pm$ s.e.m.) of the lsRF and surround field normalized to the size of the hsRF for each individual cell. (Modified from Angelucci et al., 2002b.)

response, while less suppression or sometimes facilitation was observed for cross-oriented center and surround stimuli (DeAngelis et al., 1994; Sillito et al., 1995; Levitt and Lund, 1997, 2002; Sengpiel et al., 1998; Sceniak et al., 2001; Cavanaugh et al., 2002). Center–surround interactions in these studies were well fit with a difference of Gaussians (DOG) model (Fig. 2b), i.e. they could be modeled as the summation of a center excitatory Gaussian, corresponding to the lsRF, overlapping a spatially more extensive surround inhibitory Gaussian, corresponding to the suppressive surround (for a review see Shapley, 2004). The DOG model predicts that large gratings placed in the far surround suppress the cell's response to optimal gratings in the center; this model is inconsistent with facilitation arising from the far surround. While most studies in primates have observed suppression for iso-oriented gratings in the RF center and in the far

surround, later in this chapter we demonstrate that in primate V1, under appropriate stimulus conditions, stimuli in the far surround of the same orientation as the center stimulus can also facilitate responses to optimally oriented stimuli in the RF center.

In other studies, the RF center size was defined as the neuron's mRF, and the surround as the region outside it. In this chapter we refer to the region outside the mRF, but within the lsRF, as the "near" surround (Fig. 2a, right). In these studies, iso-oriented spatially discrete stimuli, such as bars or Gabor patches, placed in the center and in the near surround often facilitated the RF center's response (Kapadia et al., 1995; Polat et al., 1998; Mizobe et al., 2001; Chisum et al., 2003). A specific case of such short-range modulation is "collinear facilitation", an enhancement of the mRF response to an optimally oriented bar (or

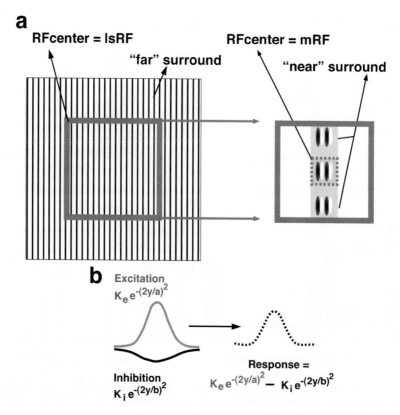

Fig. 2. "Near" and "far" surround, and the "DOG" model. (a) Different definitions of RF center and surround, and different types of stimuli used in different studies of center–surround interactions. Left panel, surround suppression: large gratings of the same orientation as the center grating placed in the "far" surround, i.e. outside the lsRF, suppress the RF center response. Right panel, collinear facilitation: iso-oriented and co-aligned spatially discrete stimuli in the mRF and in the "near" surround (i.e. the region between the mRF and the lsRF) result in facilitation of the mRF response. Note that the stimuli in the mRF and in the "near" surround occupy a region of space (delineated by the gray box) roughly co-extensive with the size of the same cell's lsRF (horizontal arrows). (b) The difference of Gaussians (DOG) model of center–surround interactions. The neuron's response (dashed Gaussian) is described as the sum of a center excitatory mechanism (gray Gaussian) overlapping a surround inhibitory mechanism (black Gaussian), both having a Gaussian spatial sensitivity profile. (Modified from Shapley, 2004.)

Gabor patch) stimulus by flanking co-oriented and co-axial high-contrast bar (or Gabor) stimuli (Fig. 2a, right). This phenomenon is thought to underlie perceptual "grouping" of contour elements (Kapadia et al., 1995; Hess and Field, 2000). In many instances, this short range modulation has been shown to be contrast dependent (Polat et al., 1998; Mizobe et al., 2001), with surround facilitation of low-contrast center stimuli turning to suppression for high-contrast center stimuli. We (Angelucci and Bullier, 2003) have previously suggested that in most of these studies the stimuli outside the mRF most likely involved the lsRF of the cell (or near surround), i.e. the center excitatory mechanism of the DOG model (Fig. 2b). In other words, near surround facilitation could reflect the same mechanism underlying the expansion of the sRF at low stimulus contrast. This interpretation is consistent with the DOG model. However, it appears that not all short-range surround modulation effects are consistent with this interpretation, as several other types of short-range interactions, both facilitatory and suppressive have been reported for different center stimulus contrasts and different center and surround stimulus configurations (Chen et al., 2001; Mizobe et al., 2001). Furthermore, later in this chapter we demonstrate that iso-orientation surround facilitation can also be induced by stimuli in the far surround.

Using single unit extracellular recordings, we (Angelucci et al., 2002b) and others (Sceniak et al., 2001; Cavanaugh et al., 2002; Levitt and Lund, 2002) have measured the extent of the far surround of macaque V1 neurons by presenting high-contrast (75%) gratings of increasing diameter to the cell, and defining the surround size as the stimulus diameter at asymptotic response (black arrow in Fig. 1a). Between $2°$ and $8°$ eccentricity, V1 cells' surround sizes averaged $5.1° \pm 0.6$, ranging up to $13°$ (largest stimulus size we tested; Fig. 1c), and were on average 4.6 times larger than the same cells' hsRF size (Fig. 1d). More recently, using larger surround gratings, we have found neurons in macaque V1 with even larger surround diameter, i.e. up to $28°$ (Ichida et al., 2005).

Candidate anatomical substrates for the receptive field center and surround of V1 neurons

Feedforward (FF) connections to V1 from the lateral geniculate nucleus (LGN), long-range V1 lateral (or horizontal) connections, and feedback (FB) connections to V1 from extrastriate cortex, could all contribute to generating the RF center and surround of V1 neurons. In the following sections, we review previous and more recent studies on the basic properties of these systems of connections and, on the basis of such properties, provide a hypothesis on the relative contributions of each connection type to the RF center and surround of V1 cells.

Area V1 receives its main driving FF inputs from the LGN of the thalamus. The geniculocortical pathway in the macaque consists of at least three channels, magnocellular (Magno), parvocellular (Parvo) and koniocellular (K) or interlaminar, originated in the retina from different ganglion cell populations (for reviews see: Casagrande, 1994; Hendry and Reid, 2000; Casagrande and Xu, 2004; Kaplan, 2004). LGN axons belonging to these different channels terminate in different V1 layers (Fig. 3), with the relays from each eye segregated into interleaved terminal stripe-like territories,

or ocular dominance bands (for reviews see: Casagrande and Kaas, 1994; Levitt et al., 1996). The principal termination zones of Magno thalamic axons are layers $4C\alpha$ and 6; Parvo thalamic afferents terminate in layers 4A, $4C\beta$ and 6, while K afferents terminate in layer 1 and in the cytochrome oxidase (CO)-rich regions (the CO blobs) of layer 3 (Hubel and Wiesel, 1972; Hendrickson et al., 1978; Lund, 1988; Hendry and Yoshioka, 1994; Yazar et al., 2004).

Geniculocortical connections are spatially restricted, i.e. they connect discrete regions of the LGN and V1 (Perkel et al., 1986). Such precise organization represents the anatomical basis for the orderly retinotopic map seen in the responses of postsynaptic neurons in input layer 4C (Blasdel and Fitzpatrick, 1984). Unlike in cats, where most cells in layer 4 are orientation selective (Hubel and Wiesel, 1962), in primates many cells in layer 4C have circularly symmetric LGN-like RFs that are not tuned, or are broadly tuned for orientation (Poggio et al., 1977; Bauer et al., 1980; Bullier and Henry, 1980; Blasdel and Fitzpatrick, 1984; Hawken and Parker, 1984; Leventhal et al., 1995). Orientation-selective neurons in macaque, however, first appear within layer 4C (Ringach et al., 2002; Gur et al., 2005). The generation of orientation specificity is generally thought to partly involve the convergence to single neurons of retinotopic information distributed along the axis of the neurons' preferred orientation (Hubel and Wiesel, 1962). In cats, this convergence is believed to be from LGN axons to single layer 4 neurons (for a review see Reid and Usrey, 2004); however, in macaque it is more likely to occur within layer 4C itself (Lund et al., 1995; Adorjan et al., 1999; Lund et al., 2003). In the macaque, FF afferents from the LGN could instead contribute to generating the non-oriented, circularly symmetric RFs of layer 4C.

Neurons in layer 4C show a gradient of decreasing mRF size and contrast sensitivity from the top to the bottom of the layer (Blasdel and Fitzpatrick, 1984; Hawken and Parker, 1984). Individual thalamic axons belonging to the different channels show very different terminal arbor size, with the Magno axons terminating at the top of layer $4C\alpha$ showing the largest intracortical

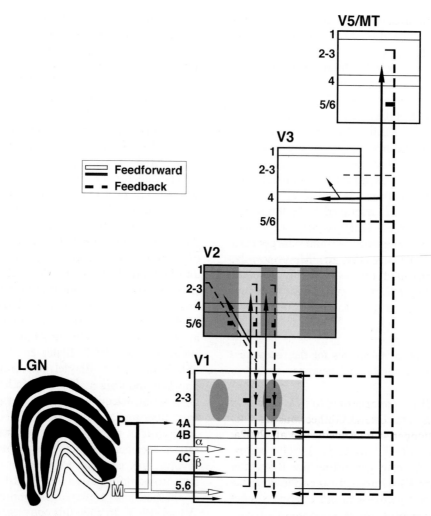

Fig. 3. Feedforward pathways to and from V1, and feedback pathways to V1. Schematic diagram illustrating the main feedforward pathways (solid arrows) to and from V1 and the extrastriate feedback pathways (dashed arrows) to V1. Thinner arrows indicate weaker projections. The six-layered macaque LGN is shown to the left. Afferent feedforward pathways to V1 arise from the LGN. For simplicity only the Parvo (P) and Magno (M) geniculocortical channels are illustrated, the K pathway and the feedback connections from V1 to the LGN are omitted. Schematic drawings of transverse, pia-to-white matter, sections of areas V1, V2, V3 and MT are shown to the right (boxes), with the six cortical layers indicated to the left of each box (numbers). In layers 2/3 of V1 and in V2, cytochrome oxidase (CO) compartments are indicated by different shades of gray (dark gray: V1 CO blobs and V2 thick and thin stripes; light gray: V1 interblobs and V2 interstripes). Feedforward and feedback connections between V1 and V2 connect specific CO compartments in these two areas (see text). Weak feedforward and feedback connections between V1 layers 2/3 and area V3 have been previously described, but are omitted in the diagram, for simplicity of illustration.

spread, and the Parvo axons terminating in layer 4Cβ showing the smallest spread (Blasdel and Lund, 1983; Freund et al., 1989). Modeling studies have suggested that feedforward convergence of Parvo and Magno inputs onto layer 4C spiny stellate neurons is sufficient to explain the observed

gradual change in mRF size and contrast sensitivity seen along the depth of this layer (Bauer et al., 1999).

In summary, because of all the properties described above, and because they drive their target V1 cells, geniculate FF afferents have traditionally been thought to underlie the spatial and tuning properties of the V1 cell RF center. More recently, based on the observation that blockade of intra-cortical inhibition (by iontophoresis of bicuculline, a GABA$_A$ receptor antagonist) does not abolish iso-orientation surround suppression in cat V1, it has been suggested that FF thalamic afferents could also underlie the RF surround of V1 neurons (Ozeki et al., 2004). More specifically, it has been proposed that surround suppression in V1 could result from withdrawal of FF excitation, due to extra-classical surround suppression of LGN afferents (see below). Later in this chapter we will report evidence arguing against this hypothesis.

In macaque V1, excitatory neurons in layers 2/3, 4B/upper 4Cα and 5/6 form long-range, reciprocal, intralaminar projections (Rockland and Lund, 1983). In layers 2/3, these connections contact predominantly excitatory (~80%), but also inhibitory (~20%) neurons (McGuire et al., 1991), show a "patchy" pattern of origin and termination, and link preferentially cortical domains of similar functional properties, such as orientation preference, ocular dominance and CO compartment (Malach et al., 1993; Yoshioka et al., 1996). It is noteworthy that lateral connections in macaque layers 4B/upper 4Cα (Angelucci et al., 2002a; Lund et al., 2003) and 6 (Li et al., 2003) show different patterns of termination from those in layers 2/3, and do not link cortical domains of like functional properties. In tree shrew (Bosking et al., 1997), cat (Schmidt et al., 1997) and new world primates (Sincich and Blasdel, 2001), lateral connections in layers 2/3 are anisotropic, and their axis of anisotropy has been shown to be collinear in space with the orientation preference of their neurons of origin. While such collinear organization has not been examined for lateral connections in macaque, we previously showed (Angelucci et al., 2002b) that, in this primate species, lateral connections are not anisotropic in visual space. This is due to the fact that their cortical anisotropy can be accounted for by the anisotropy in cortical magnification factor across ocular dominance columns (Blasdel and Campbell, 2001). In contrast to FF thalamic axons, horizontal axons do not drive their target neurons, but only elicit subthreshold responses (Hirsch and Gilbert, 1991; Yoshimura

et al., 2000), thus having a modulatory influence. Because of all these properties, layers 2/3 lateral connections have been thought to represent the anatomical substrate for orientation-specific surround modulation of V1 neurons' responses (Gilbert et al., 1996; Dragoi and Sur, 2000; Fitzpatrick, 2000; Somers et al., 2002; Stettler et al., 2002), and perceptual grouping of line segments (Kapadia et al., 1995; Hess and Field, 2000; Rajeev et al., 2003). Later in this chapter we argue that lateral connections, instead, cannot fully account for all center–surround interactions in V1.

On the basis of laminar origin and termination, inter-areal connections in visual cortex have been classified as FF and FB, and a hierarchical organization of cortical areas has been proposed (Rockland and Pandya, 1979; Felleman and Van Essen, 1991). V1, at the bottom of this hierarchy, sends partially segregated FF projections to several extrastriate cortical areas, including areas V2, V3 and V5/middle temporal (MT), which in turn send FB projections to V1 (Fig. 3). In the hierarchy of cortical areas, FF connections are broadly defined as terminating mainly in layer 4 and deep layer 3, and FB connections as arising from superficial and/or deep layers and terminating outside layer 4, although many exceptions to this rule have been reported. More specifically, FF projections from V1 to V2 arise predominantly from layers 2/3 and, less so, from 4B; projections to V3 arise predominantly from layer 4B and to a lesser extent from 2/3 (not shown in Fig. 3), while MT receives FF projections mainly from layer 4B. All three extrastriate areas also receive a small projection from V1 layer 6 (Fig. 3) (for reviews see Rockland, 1994; Salin and Bullier, 1995). Inter-areal FB projections arise from layers 2/3A and/or 5/6 of extrastriate cortex and, with the exception of layer 1, terminate in the same V1 layers that give rise to FF projections (i.e. 2/3, 4B and 6; Fig. 3). Within the V1 layers of FB termination, the cells of origin of striate FF projections and the terminal territories of extrastriate FB projections are spatially overlapped (Angelucci et al., 2002b). This precise alignment of FB terminals and FF efferent cells has also been shown in cat V1 (Salin et al., 1995) and in rat V1 (Johnson and Burkhalter, 1997), and is well suited to fast transfer of specific functional information to and from V1.

Functionally, while FF connections drive their target cells, FB connections modulate striate responses (Haenny et al., 1988; Crick and Koch, 1998; Ito and Gilbert, 1999; Bullier et al., 2001). In particular, the modulatory influence of FB connections on the RF center response of their target neurons has long been known. Many studies have shown that inactivation of areas V2 and MT reduces the response of neurons in lower-order areas to visual stimulation of their RF center (Sandell and Schiller, 1982; Mignard and Malpeli, 1991; Hupé et al., 1998; Hupé et al., 2001a). This suggests that FB input normally sums with FF inputs to increase the activity of their target neurons.

Like FF connections, FB connections arise from excitatory neurons. However, unlike FF connections (both geniculocortical and corticocortical), which in several mammalian species (cat, rat and monkey) target both excitatory (80–90%) and inhibitory (10–20%) neurons (Garey and Powell, 1971; Peters and Feldman, 1976; Freund et al., 1989; Johnson and Burkhalter, 1996; Anderson et al., 1998), FB connections, at least in rat V1, target almost exclusively excitatory cells (97–98%); only 2–3% of their postsynaptic targets are inhibitory interneurons (Johnson and Burkhalter, 1996). It is presently unknown whether a similar synaptic arrangement of FB axons exists also in the primate, since data on postsynaptic targets of FB connections in this species are still lacking. Consistent with the specific synaptic organization of FF and FB pathways, intracellular recordings in slices of rat visual cortex indicate that, irrespective of stimulus intensity and membrane potential, stimulation of FB pathways evokes purely depolarizing monosynaptic responses in most (80%) V1 cells; weak inhibitory effects are observed only in response to higher stimulus intensities, as a slight acceleration in the decay of the EPSPs. In contrast, stimulation of V1 FF connections to extrastriate cortex evokes EPSPs followed by strong IPSPs in most (90%) cells (Domenici et al., 1995; Shao and Burkhalter, 1996). The latter responses resemble those seen in 90% of area 17 cells after stimulation of LGN afferents (Douglas and Martin, 1991; Shao and Burkhalter, 1996) and after stimulation of V1 horizontal connections (Hirsch and Gilbert, 1991; Shao and Burkhalter, 1996). Thus, while stimulation of

FF, FB and V1 horizontal pathways has both depolarizing and hyperpolarizing effects on the postsynaptic V1 cell, the relative balance of excitation and inhibition induced by the three different circuits is somewhat different, with the FB pathway modulating V1 responses via synaptic mechanisms in which weak inhibition is strongly dominated by excitation. This is due to the fact that, while for FF and horizontal circuits the balance between excitation and inhibition is mainly dependent on the level of activation, in FB circuits additional anatomical constraints favor excitation.

FF and FB pathways have also long been thought to differ in their spatial and functional organization. Specifically, FF projections are retinotopically organized (Perkel et al., 1986), and both their cells of origin and terminal axon arbors are typically clustered into "patches". At least for the FF projections from V1 to V2, it has been demonstrated that these "patches" relate to specific CO compartments in the two areas (Fig. 3) (Sincich and Horton, 2002, 2005). FB connections, instead, have been shown to have a less precise retinotopic organization than FF projections (Perkel et al., 1986; Salin and Bullier, 1995). Furthermore, they have traditionally been described as terminating in a "diffuse" fashion in V1, and to be unspecific with respect to the CO, or other functional compartments contacted by their axon terminals (reviewed in Salin and Bullier, 1995; see also: Stettler et al., 2002; Rockland, 2003). Later in this chapter we will review recent studies demonstrating that at least a subset of FB connections is instead highly patterned and functionally specific.

FB connections have traditionally been thought to provide a rather unspecific contribution to the response of V1 cells, namely a mere enhancement of the RF center response. This modulatory effect of FB onto V1 cell's responses has been thought to serve as the neural mechanism underlying top–down effects in area V1, such as attention (Ito and Gilbert, 1999; Stettler et al., 2002; Treue, 2003), "reverse hierarchy" processing of vision (Ahissar and Hochstein, 2000) and visual awareness (Tong, 2003; Silvanto et al., 2005). More recently, we have proposed a more fundamental role for inter-areal FB connections to V1; namely, that they directly contribute to the response of V1 neurons to simple visual patterns (Angelucci et al., 2002a, b; Angelucci and Bullier, 2003; Schwabe et al., 2006). This hypothesis is discussed extensively later in this chapter.

To summarize, based on the studies reviewed above, the traditional view on the roles of FF, lateral and FB connections has been that FF connections underlie the spatial and tuning properties of V1 cells' RF center, lateral connections underlie surround modulation of RF center responses and FB connections to V1 provide unspecific enhancement of RF center responses that could underlie attention or similar top–down effects. However, recent data on the spatio-temporal properties and functional organization of these three systems of connections have challenged these traditional views of the roles of these connections in the responses of V1 neurons. In the following sections, we review these data.

Spatial extent of feedforward, lateral and feedback connections to V1 and its relation to the size of V1 neurons' receptive field center and surround

In previous studies in macaque, we quantitatively compared the spatial dimensions of V1 neurons' RF center and surround with the visuotopic extent of FF LGN afferents to V1, V1 lateral connections and inter-areal FB connections to V1 (Angelucci et al., 2002a, b; Angelucci and Sainsbury, 2006). The rationale for these studies is that, at least in retinotopically organized visual cortical areas, the spatial scale of a connectional system must be commensurate with the spatial scale of the specific neuronal response that it generates, in the cortical layers where this response first appears. Although RF responses and surround modulation of such responses most likely reflect the weight of influence from multiple connectional sources, these kinds of studies can help us identify anatomical and physiological constrains that will allow us to rule out possible mechanisms involved in these responses, while also pointing at candidate circuits.

Using a combination of anatomical-tract tracing and physiological recording methods, we delivered injections of bidirectional tracers (cholera toxin B, CTB; biotinylated dextran amine, BDA; dextran

tetramethylrhodamine or fluororuby) in electro-physiologically characterized V1 sites of macaque monkeys. V1 tracer injections labeled intra-V1 horizontal connections as well as the cells of origin in the LGN and extrastriate cortex of FF and FB connections to V1, respectively. This allowed us to directly compare the extent of three different systems of connections (FF, FB and lateral) to the same column of V1 cells. Specifically, we measured the extent of the retrogradely labeled fields of cells in the LGN, V1 and extrastriate cortex (areas V2, V3 and V5/MT), and converted this measurement into visual field coordinates. Such conversion was achieved by overlaying anatomical reconstructions of label to electrophysiologically recorded retino-topic maps from the same cortical region (using electrolytic lesions for alignment), and/or using published equations relating magnification factor and RF scatter to eccentricity in the LGN, V1 and extrastriate cortex.

Using this methodological approach, in a first set of studies (Angelucci and Sainsbury, 2006) we found that the visuotopic extent of LGN afferents to V1 layer 4C is commensurate with the hsRF size of their target V1 cells (Fig. 4). More specifically, we found the aggregate classical RF center of LGN neurons projecting to a small tracer-injected site in the V1 input layers to match the aggregate mRF size of their target V1 neurons (Fig. 4b; the ratio LGN RFc/V1 mRF = 1.1 ± 0.091 for Parvo and 1 ± 0.077 for Magno). This suggests that the classical RF centers of geniculate FF afferents converging to a given V1 neuron could provide the substrate for the size of the V1 neuron's mRF. On the other hand, we found a spatial match between the size of the aggregate classical RF center plus surround of LGN afferents and the aggregate hsRF size of their target V1 neurons (Fig. 4b; the ratio LGN RFc + s/V1 hsRF = 0.9 ± 0.071 for Parvo and 0.9 ± 0.03 for Magno). This indicates that FF connections from the LGN could, in fact, influence the response of V1 cells over a region of space larger than the V1 cells' mRF, and commensurate with their hsRF size. However, it is not clear if, and how, the classical RF surround of LGN cells contributes to the response of V1 cells. One possibility is that in primates it contributes to generating the antagonistic surround of

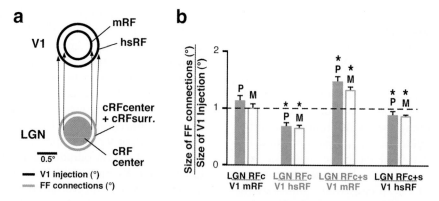

Fig. 4. Spatial extent of feedforward geniculocortical connections. (a) Visual field map of an example CTB injection site in V1 layer 4Cα (black circles), and of resulting retrogradely labeled feedforward fields of neurons in Magno layer 1 of the LGN (gray circles). Small and large black circle: aggregate mRF (0.67° diameter) and aggregate hsRF (1.02° diameter) size, respectively, of neurons at the CTB injection site; filled and empty gray circles: aggregate classical RF center (0.67° diameter) and aggregate classical RF center + classical surround (0.9° diameter) size, respectively, of FF-projecting neurons in the LGN labeled by the CTB injection. Upward pointing arrows: FF projections. (b) Population means of the "relative visuotopic extent" of retrogradely labeled geniculate FF connections. The visuotopic extent is expressed as the ratio of the FF connections' aggregate RF size to the V1 injection's aggregate RF size. The different measures of aggregate RF size used for the LGN labeled fields and for the V1 injection site are indicated on the abscissa. These are: RFc (aggregate classical RF center) and RFc + s (aggregate classical RF center plus classical surround) for the LGN neurons, and aggregate mRF and aggregate hsRF for the V1 neurons. Filled gray bars: mean visuotopic ratios for LGN label in LGN Parvo (P) layers; empty bars: mean visuotopic ratios for retrograde label in LGN Magno (M) layers. Error bars are s.e.m. The dashed horizontal line marks a ratio of 1. All two-way comparisons between the means of the different ratios on the abscissa were statistically significant (asterisks). (Modified from Angelucci and Sainsbury, 2006.)

non-oriented layer 4C neurons' RFs having a concentric, LGN-like center–surround organization. Recently, studies in new world primates and cats have shown that LGN neurons, much like V1 cells, are tuned to the size of the visual stimulus, with stimulus sizes larger or smaller than optimal evoking smaller responses (Felisberti and Derrington, 1999, 2001; Solomon et al., 2002; Nolt et al., 2004; Bonin et al., 2005). Interestingly, the stimulus size that evokes the largest response from the LGN cell (i.e. the LGN cell's hsRF size) is significantly larger than the cell's classical RF center, and significantly smaller than the hsRF size of single V1 cells (Kremers and Weiss, 1997; Jones et al., 2000; Solomon et al., 2002; Ozeki et al., 2004; Bonin et al., 2005), approximating the V1 cell's mRF size. This implies that at stimulus sizes coextensive with the V1 cell's mRF, the response of an LGN cell lying at the center of the stimulus is at the peak of its size-tuning curve. For larger stimuli, while the V1 cell is still increasing its response, the LGN cell's response begins to decline and reaches its maximum suppression for stimuli the

size of the V1 cell's hsRF. As a result, when a stimulus extends beyond the mRF of a V1 cell, the cell continues to summate inputs from attenuated (i.e. surround suppressed) geniculate cell responses. We propose that the mRF of V1 cells, i.e. the low-threshold spiking region of their RF, is generated by summation of inputs from the RF centers of LGN cells, whose response is at their peak spatial summation. The higher threshold, hsRF region of V1 neurons instead would result from integration of excitatory inputs from the classical RF center and surround of partially suppressed LGN cells.

It has recently been proposed that the suppressive surround of V1 neurons could be accounted for by surround suppression of LGN cells relayed to V1 via geniculate FF connections (Ozeki et al., 2004). As mentioned above, LGN cells are tuned to the size of a visual stimulus, and are suppressed by large stimuli. This suppression is thought to originate from the extra-classical surround of LGN neurons, as it occurs for gratings of optimal spatial frequency for the LGN cells, known not to activate the classical RF surround. While a FF component is likely to contribute to surround suppression in V1 (see below), our studies on the spatial scales of surround modulation in V1 and of FF LGN afferents suggest that surround suppression in the LGN cannot fully account for the large size of the modulatory surround in V1 (Angelucci et al., 2002b; Angelucci and Sainsbury, 2006). Specifically, the extra-classical surround of LGN neurons has been shown to be of similar size as these neuron's classical surround (Solomon et al., 2002; Bonin et al., 2005). As discussed above, the far surround of V1 neurons extends well beyond (on average 4.6 times) their hsRF (Fig. 1d), and the classical (and thus the extra-classical) surround of LGN neurons can only account for the hsRF size of their target V1 cells (Fig. 4b). Thus, LGN extra-classical surrounds cannot account for the average size of V1 surrounds, and certainly not for the largest surrounds measured in V1. There is additional evidence suggesting that FF afferents are unlikely to underlie surround suppression in V1. First, in primates and cats surround suppression in the LGN is not tuned for orientation, spatial and temporal frequency (Solomon et al., 2002; Bonin et al., 2005), unlike surround suppression in cat and primate V1, which is orientation tuned and broadly tuned for other stimulus dimensions. Second, differences in the hsRF size of LGN and V1 cells (see above), in both cats (Jones et al., 2000; Ozeki et al., 2004) and primates (Levitt et al., 1998; Solomon et al., 2002), indicate that a stimulus of optimal size for V1 cells is instead suppressive for LGN cells (see above). Thus, surround suppression in the LGN does not necessarily lead to suppression of V1 cells' responses. To summarize, the spatial scale of geniculate FF afferents to V1 suggests that these connections underlie the hsRF size of their target V1 neurons and cannot account for the full spatial scale of suppressive surrounds in V1.

What then underlies the lsRF and the modulatory surrounds of V1 neurons? Using a similar methodological approach as described above for FF connections, in a second set of experiments (Angelucci et al., 2002b), we found that the monosynaptic spread of V1 lateral connections is commensurate with the lsRF size of V1 cells

(Fig. 5a, b). The spatial scale of these connections led us to hypothesize that they may mediate near surround modulation of RF center responses. This modulation includes facilitation from the near surround, such as the expansion of the summation RF at low stimulus contrast (Kapadia et al., 1999; Sceniak et al., 1999) and collinear facilitation (Fig. 2a, right) (Kapadia et al., 1995; Polat et al., 1998; Mizobe et al., 2001; Chisum et al., 2003), as well as suppression from the near surround region located outside the hsRF but within the lsRF (Chen et al., 2001; Sceniak et al., 2001; Cavanaugh et al., 2002; Levitt and Lund, 2002; Chisum et al., 2003). A role for lateral connections in collinear facilitation is also consistent with the finding that the latter is reduced by GABA inactivation of laterally displaced V1 sites (Crook et al., 2002). Recently it has been proposed that the lsRF of V1 cells could be partially accounted for by the contrast-dependent increase in peak spatial summation seen also in LGN cells (Kremers et al., 2001; Solomon et al., 2002; Nolt et al., 2004; Bonin et al., 2005). However, in primates such increase for LGN cells is much smaller (lsRF = ~1.3xhsRF) than that seen in V1 cells (lsRF = ~2.3xhsRF), and thus cannot fully account for the lsRF of V1 neurons; additional integration from V1 lateral and extrastriate FB inputs needs to be invoked.

Because the far surround extends well beyond the lsRF (Fig. 1c–d), and thus beyond the monosynaptic spread of lateral connections, contrary to previous suggestions (see above), these connections cannot monosynaptically underlie far surround modulation of RF center responses. Polysynaptic chains of lateral connections are also unlikely to underlie influences from the far surround, due to the slow conduction velocity of their axons (see below).

In contrast, we showed that FB connections to V1 from extrastriate areas V2, V3 and MT have the appropriate spatial scale to underlie the large-scale modulatory effects arising from the far surround of V1 neurons (Fig. 5c–d) (Angelucci et al., 2002b). Consistent with our hypothesis that FB connections play a role in center–surround interactions is evidence that inactivation of area MT, by cooling, reduces the suppressive effect, sometimes even turning it to facilitation, of surround

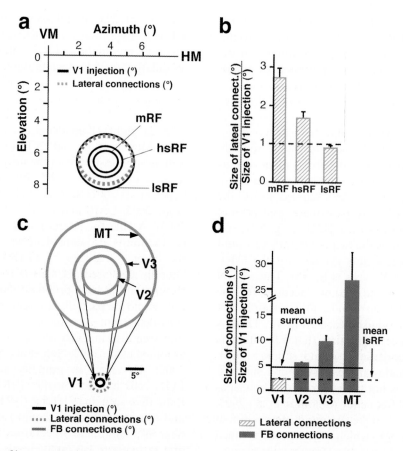

Fig. 5. Spatial extent of intra-areal lateral and inter-areal feedback connections. (a) Visual field map of an example CTB injection site (black circles) and of resulting labeled lateral connections (dashed gray circle). The different-sized black circles indicate the aggregate mRF, hsRF and lsRF sizes of neurons at the injection site, as indicated. (b) Population means ($n = 21$) of the "relative visuotopic extent" (normalized to the aggregate RF size of neurons at the V1 injection site) of labeled lateral connections, shown for each of three different methods of measuring the RF size of cells at the injection site. Dashed horizontal line marks a ratio of 1. Error bars are s.e.m. Lateral connections are commensurate with the lsRF size of their V1 cells of origin. (c) Visual field maps of the same CTB injection site as in (a) (black circle), and of resulting retrogradely labeled neuronal fields of FB connections (solid gray circles) from the lower layers of V2, V3, MT and of V1 lateral connections (dashed gray circle). All circles are aggregate hsRF sizes. Note that FB connections convey information to V1 from regions of space much larger than that conveyed by lateral connections to the same V1 column. The size of this convergent FB region increases with cortical distance from V1, being larger in MT than in V2. (d) Population means and s.e.m. of the "relative visuotopic extent" (aggregate hsRF size of FB connections normalized to the aggregate hsRF size of neurons at the V1 injection site) of FB connections to V1 from the lower layers of V2 ($n = 6$), V3 ($n = 5$) and MT ($n = 3$), and of V1 lateral connections ($n = 21$). Horizontal lines: population means of the "relative extents" of V1 cells lsRF (dashed line: lsRF/hsRF = 2.3) and surround field (solid line: surround diameter/hsRF diameter = 4.6), obtained from the histogram in Fig. 1d. The mean surround size of V1 cells is commensurate with the mean visuotopic extent of FB fields from V2 to V1 (both ~5xhsRF). The larger FB fields from V3 and MT are coextensive with the larger surround sizes measured physiologically in V1. Thus, the spatial scale of extrastriate FB connections is commensurate with the full range of RF center and surround sizes in V1. (Modified from Angelucci et al., 2002b.)

motion stimulation in V3, V2 and V1 cells (Hupé et al., 1998). The strongest effect of inactivating area MT occurs for low salience stimuli (saliency is defined as the ratio of the contrasts of the center and surround stimuli). These results indicate that

FB connections normally act to enhance the suppression arising from the extra-classical RF surround. However, inactivation of area V2 loci by GABA injections does not seem to affect the modulation of V1 responses generated by static texture patterns in the surround (Hupé et al., 2001a). The latter result does not rule out an involvement of FB connections from other extrastriate areas in surround modulation of V1 responses generated by static texture patterns. Furthermore, the stimulus used by Hupé et al. (2001a) may be a bit V2 specific, and thus may not have isolated the contribution from area V2; it is therefore possible that FB connections from V2 play a role in center–surround interactions generated by different types of center–surround stimuli. Most importantly, the stimulus used in the inactivation study by Hupé et al. (2001a) was confined to the near surround; thus, its suppressive effect was most likely mediated by both lateral and FB axons. Inactivation of FB from V2 would leave activity in the lateral connections unperturbed, with no resulting effect on surround suppression. While these experiments emphasize the role of FB connections in surround suppression, it has long been known that these connections also facilitate the responses of their target neurons to stimuli in their RF center (see above). Although we are suggesting that horizontal and FB connections play a crucial role in shaping the extra-classical surround of V1 neurons, the extra-classical surround of LGN afferents is also likely to contribute to the V1 cell's surround. A recent study (Webb et al., 2005) suggested that surround suppression in primate V1 is generated by two concurrently operating mechanisms, an intracortical mechanism that is stimulus tuned and binocularly driven, and an untuned and monocularly driven mechanism that may result from the non-orientation specific extra-classical surround suppression of LGN afferents (Solomon et al., 2002; Bonin et al., 2005).

Temporal properties of feedforward, lateral and feedback connections to V1 and their relation to the timing of surround modulation

As discussed above, monosynaptic lateral connections are too short to mediate the modulation arising from the far surround of V1 neurons. Could a cascade of lateral connections mediate these long-range contextual effects? Experimental evidence suggests that horizontal axons are too slow to mediate the fast onset of suppression from the far surround, but could mediate the suppression arising from the near surround. Several studies have shown that the onset of orientation- or pattern-specific surround suppression is delayed relative to the response of the RF center. Such delay ranges between 15 and 60 ms, depending on the stimulus configuration and methodological approach used in the different studies to measure the latency of surround modulation (Knierim and Van Essen, 1992; Lamme, 1995; Lamme et al., 1999; Nothdurft et al., 1999, 2000; Li et al., 2000; Hupé et al., 2001a). Recently, Bair et al. (2003) measured the onset latency of surround suppression using stimulus configurations similar to those previously used by us and others to measure the spatial extent of the RF surround (see above). Specifically, these authors presented a high-contrast grating fitted to the hsRF size of the recorded V1 neuron, together with an annular grating in the surround. By varying the inner diameter of the surround grating, the surround stimulus could effectively be moved farther from the RF center. Using this stimulus configuration, on average surround suppression was delayed by 9 ms (SD 15) relative to the onset of the RF center response, and the delay was on average 30 ms longer for weakly suppressed cells than for strongly suppressed cells. The average delay found in this study is significantly shorter than the 15–20 ms delay reported in most previous studies. The latter, however, used sparse field stimuli or bars, which are "weaker" stimuli, and/or used averages across populations. In contrast, Bair et al. (2003) used high-contrast gratings and performed a cell-by-cell analysis of the delay. Importantly, these authors found that, for a substantial fraction of cells, the latency of suppression induced by stimuli in the far surround was almost as short as the latency of suppression induced by stimuli in the near surround. If chains of lateral connections were to mediate the suppression from the far surround, one would expect the latency of suppression to increase with distance from the RF center. This was not observed. Additionally, in the same study, the propagation speed of suppression was estimated to be > 1 m/s for 40% of cells. This propagation speed is significantly faster than that of signals traveling along long-range lateral connections, but is consistent with the fast conduction velocity of FB axons (see below). In summary, the results from Bair et al. (2003) are not consistent with polysynaptic relays of lateral connections mediating far surround suppression, but support a role for FB connections in the generation of these long-distance suppressive effects.

Using a variety of methods to measure the speed of signal propagation within area V1, different studies have concluded that lateral propagation of signals is slow. Using real-time optical imaging based on voltage sensitive dyes, the wave of activity within V1 was estimated to travel at a speed of 0.1–0.25 m/s (Grinvald et al., 1994; Slovin et al., 2002). A similar propagation speed (0.1–0.2 m/s) of depolarizing potentials traveling across the RF of neurons was measured using intracellular recordings in cat area V1 (Bringuier et al., 1999). The conduction velocity of most horizontal axons in macaque V1, measured directly by electrical stimulation (Girard et al., 2001), was found to be ~0.1 m/s (median 0.3 m/s). As discussed above surround sizes in V1 can reach up to 16–28° in diameter, i.e. the surround signal can arise from V1 regions displaced by 8–14° from the RF center. In parafoveal V1 (between 2 and 8° eccentricity), this corresponds to a cortical distance of approximately 1.8–6.4 cm [using a magnification factor of 2.3 mm/deg at 5° eccentricity (Van Essen et al., 1984)]. Signals traveling along lateral axons with conduction velocities of 0.1 m/s would take 180–320 ms to span such cortical distances, while those traveling along lateral axons with faster conduction velocities of 0.3 m/s would take approximately 60–100 ms. Furthermore, since lateral connections in macaque V1 are limited in extent to 4.5 mm radius (Angelucci et al., 2002b), polysynaptic chains of lateral connections would add integration times of about 5–20 ms (Nowak and Bullier, 1997; Azouz and Gray, 1999) at each synaptic relay. Thus, clearly lateral connections are too slow to account for the relatively fast onset of suppression (9–60 ms, see above) arising from the far surround of V1 neurons, but could mediate modulation from the near surround. The near surround of V1 neurons, as defined in this chapter, has an average extent of 1.2° (radius), equivalent

to ~2.8 mm of V1 at 5° eccentricity; it takes 9–28 ms for monosynaptic lateral connections to carry signals across such a distance, a latency that is consistent with the onset delay of near surround suppression.

Contrary to the widespread belief that inter-areal FB connections are slower than lateral connections (e.g. Lamme et al., 1998), electrical stimulation studies on the conduction velocities of FF and FB axons between macaque areas V1 and V2 have shown both types of connections to conduct at 2–6 m/s, i.e. about 10 times faster than V1 lateral connections (Girard et al., 2001). This is consistent with the lack of delay observed in the effect of inactivating area MT on the responses of neurons in areas V1, V2 and V3 (Hupé et al., 2001b), or of inactivating area V2 on the responses of V1 neurons (Hupé et al., 2001a). Fast and highly divergent FB connections can also explain how the onset latency of surround suppression can be almost independent of distance (Bair et al., 2003).

In summary, on the basis of propagation speed and spatial extent, we propose that FB connections underlie influences from the far surround of V1 neurons, while modulatory effects from the near surround are mediated by both lateral and FB connections. Although polysynaptic chains of lateral connections cannot account for the fast onset of suppression from the far surround, they could in principle contribute to the late phase of the suppression. In V1 not all neurons show surround suppression; the size tuning curves of these V1 neurons show a plateau at the peak response, or an increase in response for increasing stimulus sizes (e.g. Levitt and Lund, 2002). There is experimental evidence suggesting that at least some of these neurons are inhibitory (Hirsch and Gilbert, 1991), but excitatory neurons could also exhibit these response profiles. The latter could in principle propagate horizontal signals from the far surround to the RF center, thus contributing to the late phase of the suppression.

Patterning and functional specificity of feedback connections

The stimulus specificity of center–surround interactions in V1 requires their underlying anatomical substrates to be stimulus specific. Thus, a central question that needs to be addressed is whether FB connections to V1 are patterned or diffuse, and specific relative to the functional compartments and orientation preference domains contacted by their terminals.

There appear to be conflicting data on the patterning and functional specificity of FB connections to V1. Studies in the 1970s and 1980s (Rockland and Pandya, 1979; Maunsell and Van Essen, 1983; Ungerleider and Desimone, 1986), hampered by the unavailability of sensitive anterograde tracers, concluded that FB connections to V1 terminate in a diffuse and unspecific manner (reviewed in Salin and Bullier, 1995; Rockland, 2003). However, using more sensitive tracers (CTB, BDA and Fluororuby), we (Angelucci et al., 2002a, b, 2003; Jeffs et al., 2003) and others (Shmuel et al., 2005) have recently reported that FB connections from V2 and V3 in old and/or new world primates are highly patterned (Fig. 6) and specific with respect to CO compartments (Fig. 7). Specifically, injections of BDA in the thick or pale CO stripes of area V2 in owl (Shmuel et al., 2005) and macaque (Angelucci et al., 2002b) monkeys (Fig. 6a), and injections of CTB in similar CO stripes of marmoset V2 (Angelucci et al., 2003) (Fig. 6d) result in patches of FB terminations in V1 layers 2/3 (Fig. 6c, f), 4B (mainly in macaque) and 5/6 (Fig. 6g) that are in vertical register preferentially with the CO interblobs. Similarly, CTB injections in macaque area V3 (Angelucci and Levitt, 2002) result in clusters of terminals mainly in V1 layers 4B and 6 that either preferentially align with the CO blobs (Fig. 7a–f) or with the interblobs (Fig. 7g–j). In addition to demonstrating CO-specificity of FB connections from V3, the latter result also suggests the existence of anatomically distinct compartments in area V3.

FB axons from V2 in new world primates (marmoset and owl monkeys) have also been shown to link regions of broadly similar orientation preference in V2 and V1. Furthermore, their terminal fields in V1 are anisotropic (Angelucci et al., 2002b), and their axis of major anisotropy has been shown to be parallel to a retinotopic axis in V1 matching the preferred orientation of the FB cells of origin in V2 (Angelucci et al., 2003; Shmuel et al., 2005).

Other recent studies in macaque monkey are not consistent with the above findings. Specifically, Stettler et al. (2002), using an adenovirus bearing the gene for the enhanced green fluorescent protein (EGFP), reported that FB projections from V2 terminate in V1 in a diffuse manner, and are not specific with respect to the CO compartments or orientation maps in V1. Diffuse FB projections from V2 to V1 have also been emphasized in other studies where PHA-L, BDA or fluororuby were used as tracers (reviewed in Rockland, 2003). The discrepancy between the different studies could be attributed to methodological differences. Specifically, the anatomical tracers used in our studies and in the study by Shmuel et al. (2005) can be delivered in small quantities, are transported bidirectionally even from very small injection sites, yield very good filling of fibers and terminals, and are not transported transneuronally. While bidirectional tracers can be desirable, one possible confound is that they could produce retrograde labeling of local axon collaterals; thus, the patterning of FB terminals labeled with these tracers may in fact reflect the patterning of the reciprocal FF pathway, which is known to be patterned and CO-specific (e.g. Sincich and Horton, 2002). However, while this is possible for BDA and fluororuby labeling, we have previously demonstrated that CTB labels retrogradely cell bodies but not the axon fibers arising from them (Angelucci et al., 1996; Angelucci and Bullier, 2003; Angelucci and Sainsbury, 2006). Furthermore, patterned FB terminations also occur in V1 layers that do not send FF projections, such as layer 6 (Figs. 6g, 7c–d) and the lower half of layer 1 (Fig. 6h). The viral-EGFP method used in the study by Stettler et al. (2002), instead, is a new method that is awaiting proper validation as an effective axonal tracer. In particular, it remains to be demonstrated whether the EGFP is effectively transported anterogradely, i.e. whether it fills the entire axon arbor of the neurons expressing it, or only parts of it. Incomplete filling of axon arbors, especially of inter-areal axons traveling longer distances, could yield the false impression of diffuse, unclustered FB projections, as in Stettler et al. (2002). In this respect, it is noteworthy that EGFP labeling of FB axons appears significantly less dense than that obtained

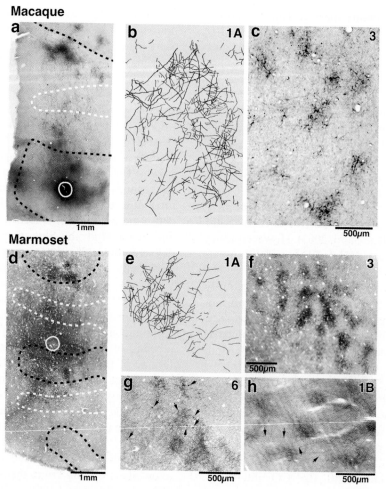

Macaque

a

b 1A

c 3

1mm

500µm

Marmoset

d

e 1A

f 3

g 6

h 1B

1mm

500µm

500µm

500µm

Fig. 6. Patchy and diffuse systems of feedback connections to macaque and marmoset V1. (a) Micrograph of a tangential section through layer 3 of macaque area V2, showing a BDA injection site (white circle) centered on a thick CO stripe, and resulting labeled V2 horizontal connections to one side of the injection. Dashed black and white contours: outlines of the thick and thin CO stripes, respectively, drawn on adjacent CO-stained sections and overlaid to the CTB section in (a) using the blood vessel pattern for alignment. The V1/V2 border is to the left of the figure. (b) Serial section reconstruction of FB axons in layer 1A of area V1 labeled by the BDA injection in (a). Note the diffuse termination pattern of FB connections in this layer. Number at top right corner in (b), (c), (e–h) indicates the V1 layer. (c) Micrograph of a tangential section through layer 3 of macaque V1, showing BDA-labeled patches of feedforward and feedback connections, resulting from the tracer injection in (a). Sections in (b) and (c) are in vertical register; note different patterning of FB terminations in these two sections. Scale bar in (c) valid also for (b). (d) Micrograph of a tangential section through layer 3 of marmoset area V2, showing a CTB injection site in a pale CO stripe, and resulting labeled V2 horizontal connections. Conventions are as in (a). V1/V2 border is to the left. (e) Serial section reconstruction of FB axons in layer 1A of area V1 labeled by the CTB injection in (d). As in macaque (b), in marmoset FB connections in this V1 layer form diffuse terminations. (f) Micrograph of a tangential section through layer 3 of marmoset V1, showing CTB-labeled patches of feedforward and feedback connections, resulting from the tracer injection in (d). Sections in (e) and (f) are in vertical register; note different patterning of label in these two sections. Scale bar in (f) valid also for (e). (g) Micrograph of a tangential section through layer 6 of marmoset V1, showing CTB-labeled patches of feedback axons, resulting from the injection in (d). This section is located directly under those shown in (e–f). In layer 6, the label is almost exclusively anterograde; the few labeled cell bodies are indicated by arrows. (h) Micrograph of a tangential section through layer 1B of marmoset V1, showing patches of anterogradely labeled FB axons resulting from a different injection of CTB in V1 (injection site not shown). No retrograde label is present in this layer. In addition to patchy axon terminals in layer 1B, long unbranched axons (arrows) can be seen traveling in the focal plane above that of the ''patchy'' terminations, i.e. in layer 1A.

with similar-sized CTB injections in the same area of the same species (compare EGFP-labeled corticocortical connections in Stettler et al., 2002, with CTB-labeled corticocortical connections in Angelucci et al. 2002b). We conclude that, at present, CTB is a superior axonal tracing method, and is better suited to labeling long-range cortico-cortical connections than the adenovirus-EGFP method. The studies by Rockland' s group (Rockland, 2003), instead, may have failed to observe patterned FB terminations because these authors sectioned V1 in the pia-to-white matter plane. This was motivated by their need to recon-struct individual axons, which can be best followed in this plane of sectioning. Tangential sectioning of flattened and unfolded cortex can reveal patterns that are otherwise difficult to observe in the pia-to-white matter plane (Sincich et al., 2003).

While differences in tracing methods may ex-plain the discrepancy between different studies on the patterning of FB connections, an alternative or additional explanation is that differences in the patterning and specificity of FB connections may reflect placement of tracer injections in different cortical layers or different compartments of extra-striate cortex, or it may reflect the V1 layers of FB termination analyzed in the different studies. Our unpublished data strongly suggests that clustering of FB terminals in area V1 is independent of the extrastriate area, CO compartment and cortical layer of origin of the FB connections. It is also independent of primate species. Specifically, tracer injections (BDA or CTB) in the thick or pale CO stripes of marmoset or macaque area V2 (Fig. 6) and in macaque area V3 (Fig. 7) produce patchy FB terminations in layers 2/3, 4B and 5/6. Fol-lowing V3 injections, the patches in V1 align either with the CO blobs or interblobs, suggesting place-ment of tracer injections in different compartments of V3. In layers 2/3 and 4B, the patches of FB terminals are spatially coincident with patches of retrogradely labeled cells sending ascending FF projections (Figs. 6c, f, 7b, h). The FB patches in layer 6, instead, consist predominantly of antero-grade label with only occasional retrogradely labe-led cell bodies (Figs. 6g, 7c–d). This FB pathway to V1 layers 6 is only labeled when the tracer in-jections involve layer 6 of extrastriate cortex, but not if the injections involve only the upper layers. Thus, there appears to exist a layer-6-to-layer-6 FB pathway that is not reciprocated by a signifi-cant FF pathway from V1 layer 6. We conclude that FB connections to V1 layers 2/3, 4B and 6, much like the FF connections from these V1 layers to areas V2 and V3, terminate in a patchy fashion, and that this patterning is independent of the extrastriate area or CO compartment of origin or primate species. Patchy FB connections to V1 lay-ers 2/3, 4B and 5/6 are also observed regardless of whether the tracer injection involves only layers 2/3 (Fig. 7i) or all layers of extrastriate cortex (Fig. 7e).

In agreement with previous reports of diffuse FB connections to V1 layer 1 (Rockland and Pandya, 1979; Ungerleider and Desimone, 1986; Shmuel et al., 2005), our data show that the patterning of FB terminations to V1 is instead *dependent* on the target layer, as FB connections to layer 1 terminate in a diffuse fashion. The same BDA or CTB injec-tions in V2 of macaque and marmoset monkeys that produced patchy FB connections in the deep V1 layers, labeled sparse and diffuse fibers in layer 1 (Fig. 6b, e). What is novel about our data on FB connections to layer 1, is that we found these con-nections to form different terminal patterns in the upper and lower halves of this layer. Specifically, while FB connections to V1 layer 1A terminate in a diffuse fashion (Fig. 6b, e), they form terminal clusters in layer 1B (Fig. 6h). As there are no FF projections arising from layer 1 of V1, the clusters of anterograde label in layer 1B occur in the ab-sence of any retrograde label in the same layer. That layer 1 can be subdivided into an upper (1A) and lower (1B) layer was previously proposed on the basis of distinct patterns of anatomical mar-kers and anatomical connections observed in these sub-layers (Ichinohe et al., 2003; Ichinohe and Rockland, 2004). These studies have also demon-strated a modular organization of specific markers in lower layer 1 (Ichinohe et al., 2003). On the basis of these results, it is possible that Stettler et al. (2002) failed to observe patterned FB terminations in V1, because they may have collapsed onto a single plane their anatomical reconstructions of FB label in layers 1–3, thus intermingling clustered and unclustered FB terminations in these different

110

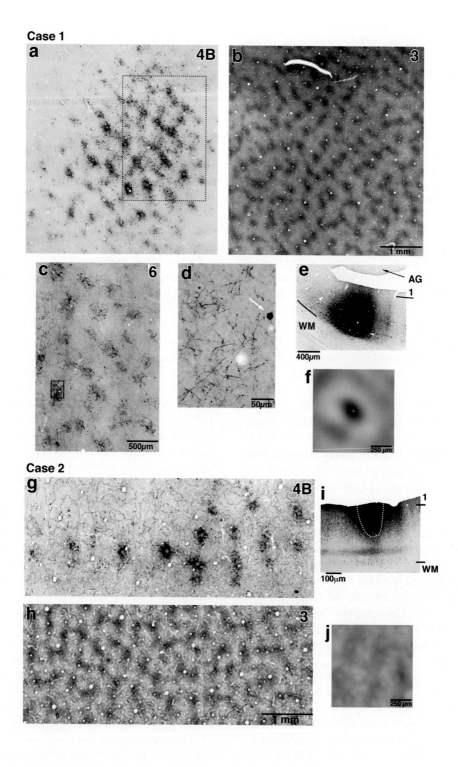

Case 1

a   4B

b   3

1 mm

c   6

d

50μm

e   AG   1

WM

400μm

f

250 μm

Case 2

g   4B

i   1

WM

100μm

h   3

1 mm

j

250 μm

layers. Indeed, it appears that these authors did not always have interleaved CO-stained sections in V1, and thus their identification of V1 layers may not have always been accurate.

Thus, our results suggest the existence of at least two differently organized systems of FB connections to V1, one patchy and specific, terminating in layers 1B, 2/3, 4B and 5/6, the other diffuse and unspecific, terminating in layer 1A. The clustering and functional specificity of the patchy FB system are consistent with their proposed role in mediating orientation-specific and attribute-specific influences from the far extra-classical surround of V1 neurons. A role for the diffuse FB systems remains to be determined.

The contribution of feedforward, lateral and feedback connections to the receptive field center and surround of V1 neurons: a neural network model and its experimental validation

The data reviewed above on the spatio-temporal properties and functional organization of FF, lateral and FB connections have led us to suggest a specific hypothesis on the relative contribution of these connections to the RF center and surround of V1 neurons. We suggest that geniculate FF connections generate responses within the hsRF

size of V1 neurons. V1 lateral and extrastriate FB connections enhance such responses. Modulation of RF center responses from the near surround region outside the hsRF is mediated by lateral and FB connections, while far surround modulation is generated by FB connections. This hypothesis, and the main data underlying it, is schematically depicted in Fig. 8a.

Data on the spatio-temporal properties and functional organization of FF, lateral and FB connections were incorporated in a simplified recurrent network model of center–surround interactions, specifically designed to test our hypothesis and to make specific predictions (Schwabe et al., 2006). The basic model architecture, constrained by these data, is depicted in Fig. 8b. The model consists of two interconnected visual areas, one corresponding to V1 and the other to an extrastriate area such as MT. In the absence of feedback, our model essentially reduces to a class of recurrent network models previously used to explore the role of lateral connections in center–surround interactions (Somers et al., 1998; Dragoi and Sur, 2000). These models assume that stimulation of the near surround modulates the response to a center stimulus via lateral connections targeting both excitatory and inhibitory neurons in the center. If the interneurons are taken to have higher threshold and gain than the local excitatory

Fig. 7. Modular specificity of feedback connections to V1. The figure shows patterns of FB projections to V1, and their relations to the CO blobs, resulting from two different CTB injections (case 1 and 2, respectively) in macaque area V3d. (a, g) Micrographs of tangential sections through layer 4B of macaque V1 showing patches of spatially overlapped anterogradely labeled feedback projections and retrogradely labeled feedforward-projecting neurons resulting from the CTB injections in area V3d shown in (e) and (i), respectively. Red dots mark the centers of the CTB patches. Dashed box in (a): area directly overlaying the section shown in (c). (b, h) Micrographs of tangential sections through layer 3 of macaque V1 stained for CO, showing the pattern of CO blobs. The centers of layer 4B CTB patches (red dots) lay preferentially within the CO blobs in (b) and outside the CO blobs in (h). Red box in (b): 1×1 mm window used to compute average CO intensity for the spatial cross-correlation analyses shown in (f, j). Scale bars in (b) and (h) valid also for (a) and (g), respectively. (c) Micrograph of tangential section through layer 6 of macaque V1 in case 1, showing patches of CTB-labeled terminals of feedback axons. Centers of CTB-labeled patches in layer 4B (red dots) align well with patches of FB axons in layer 6. The label in layer 6 is almost exclusively anterograde. The patch in the black box is shown at high power in (d). Note that only one neuron is labeled in this patch (white arrow). (e, i) CTB injection sites in area V3d in case 1 and 2, respectively. Note that the injection involves all cortical layers in (e) (layer 1 is spared by the injection in this specific section, but not in adjacent sections), but only layers 1–3 in (i). AG, annectant gyrus; 1, layer 1; WM, white matter. (f, j) Two-dimensional (2D) CO–CTB spatial cross-correlation analyses used to quantify the relationship of the CTB patches with the CO blobs, in case 1 and 2, respectively. The cross-correlation was performed within a 1×1 mm window (red box in b) centered on the corresponding location on the CO map of each CTB patch center. Within this window the average CO intensity was calculated pixel by pixel, and all the windows were then averaged pixel by pixel to create an average density image (Boyd and Casagrande, 1999; Sincich and Horton, 2003). Note that 2D cross-correlation analysis confirms the correlation of the CTB patches with the CO blobs for case 1, and no correlation with the CO blobs for case 2. Red cross marks the center of the window.

112

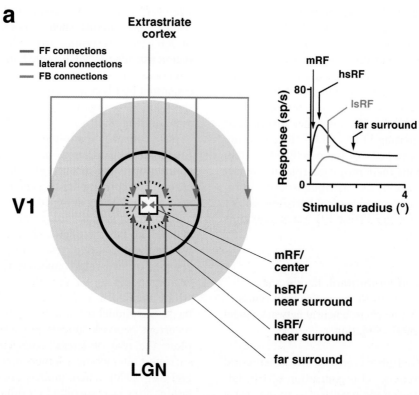

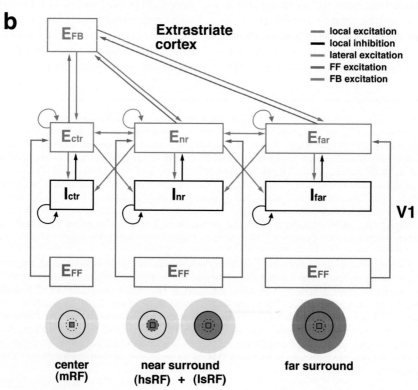

neurons whose output they control (Lund et al., 1995), then they only generate suppression under sufficiently high levels of excitation, for example, when the contrast and/or size of a grating stimulus in the RF center is sufficiently large. At low levels of excitation, such as for small or low-contrast stimuli, the inhibitors are inactive and stimulation of the near surround facilitates the center response, thus providing an explanation for the expansion of the sRF at low stimulus contrast.

In order to account for the fast onset and large spatial extent of surround suppression, we have extended these models by assuming that (i) FB from extrastriate cortex provides an additional source of excitation to excitatory neurons in the center and near surround; (ii) fast suppression from the far surround occurs via FB, rather than via a cascade of lateral connections. It is noteworthy that our proposed role for FB connections in far surround suppression may appear to be inconsistent with the experimental finding (see above) that FB axons arising from excitatory neurons in extrastriate cortex target almost exclusively (97–98%)

excitatory neurons in V1 (Johnson and Burkhalter, 1996). Our model, instead, provides an explanation for how exclusively excitatory inter-areal FB, targeting predominantly excitatory neurons, can mediate far surround suppression of the center neurons; namely by targeting excitatory neurons in the near surround, which in turn, via lateral connections, excite the local inhibitory neurons in the center. We have recently shown how our model can account for a wide range of physiological data regarding the static and dynamic effects of surround suppression (Schwabe et al., 2006). These include: (1) the expansion of the sRF at low stimulus contrast (Kapadia et al., 1999; Sceniak et al., 1999); (2) the size-tuning curves of V1 neurons (Sceniak et al., 2001; Cavanaugh et al., 2002; Levitt and Lund, 2002); (3) far surround suppression, as shown in experiments in which the afferent drive to near surround neurons sending lateral connections to the center was partially withdrawn by interposing a blank stimulus between a grating in the RF center and a grating in the far surround (Levitt and Lund, 2002); (4) FB-mediated facilitation of

Fig. 8. Hypothetical circuits underlying V1 cells' RF center and surround, and the basic architecture of the recurrent network model. (a) Schematic diagram showing the spatial scale of the different components of the RF center (white area) and surround (gray area) of V1 neurons and that of their underlying anatomical substrates. The spatial extent of each RF component is also indicated on the size-tuning curves (top right) of a representative V1 neuron, measured at high (black curve) and low (gray curve) stimulus contrast. White square area: minimum response field (mRF) or RF center; this is the RF region over which presentation of small high contrast optimally oriented stimuli evokes spikes from the cell. Dashed ring: high contrast summation RF (hsRF); the region between the mRF and the hsRF is the region over which presentation of high contrast gratings at the same orientation as the center grating facilitates the cell's response to optimally oriented gratings in the center. Continuous ring: low contrast summation RF (lsRF); the region between the hsRF and the lsRF is the region over which presentation of gratings at the same orientation as the center grating suppresses or facilitates the cell's response to optimally oriented gratings in the center, depending on the grating's contrast. Note the shift to the right of the peak response at low contrast (Sceniak et al., 1999). Gray area: RF surround. We consider two separate regions of the surround depending on their proximity to the RF center: (1) the near surround is the region between the mRF and the lsRF, (2) the far surround is the region outside the lsRF over which presentation of stimuli at the same orientation as the center stimulus usually suppresses the cell's response to optimally oriented high contrast gratings in the center. FF LGN afferents to V1 (green) are commensurate with the size of the hsRF of their target V1 neurons (Angelucci and Sainsbury, 2006). V1 lateral connections (red) are commensurate with the lsRF size of their V1 neurons of origin, while extrastriate FB (blue) connections to V1 are commensurate with the full spatial scale of the center and surround field of V1 neurons (Angelucci et al., 2002b). (b) Schematic diagram of the connections used in the network model, based on the anatomical and physiological data summarized in (a). Different connection types are indicated as color-coded arrows. Purple and black boxes represent populations of excitatory (E) or inhibitory (I) V1 neurons, respectively, labeled according to the position of their RF center relative to that of the center neurons, i.e. ctr, neurons in the RF center or mRF; nr, neurons in the near surround including the hsRF and the lsRF; far, neurons in the far surround; E_{FF}, excitatory neurons in other V1 layers sending feedforward afferents to the E neurons in V1 layers 2/3; E_{FB}, excitatory neurons in extrastriate cortex sending feedback projections to the E neurons in V1. FB connections are spatially highly divergent and convergent, and have larger spread than lateral and FF connections. There are no direct FB inputs to I neurons. The latter receive monosynaptic inputs only from V1 lateral connections (red arrows) and from local E neurons via local recurrent connections (purple arrows). Icons at the bottom: different regions of the RF center and surround (same conventions as in (a)), with red areas indicating the RF regions that are activated when each respective submodule is consecutively (from left to right) activated by a stimulus of increasing radius.

responses to stimuli in the RF center, and FB-mediated suppression of responses to stimuli in the RF center by stimuli in the surround, as demonstrated by experiments in which FB connections were inactivated (Hupé et al., 1998; Bullier et al., 2001); (5) the latency and dynamics of surround suppression (Bair et al., 2003).

Our model makes the novel prediction that stimulation of the far surround can suppress or facilitate the RF center response, depending on the total amount of excitatory drive to the local inhibitors. In Fig. 9a we show the results of a model simulation, in which a high contrast annular grating was presented in the far surround together with a central high-contrast grating fitted to the size of the neuron's hsRF. The inner radius of the annulus was systematically decreased from the far surround to a size no smaller than that of the lsRF, so that the near surround neurons in the lsRF but beyond the hsRF never received afferent stimulation. This minimized the stimulation of neurons

sending monosynaptic lateral connections to the center neuron and allowed us to unmask FB-mediated influences from the far surround. As the inner radius of the annulus was decreased, more neurons in the far surround receive afferent stimulation. This led to suppression of the excitatory neurons in the center (solid black curve in Fig. 9a), as previously observed experimentally (Levitt and Lund, 2002). On the other hand, when a low-contrast central stimulus was presented together with a high-contrast annulus in the far surround, decreasing the inner radius of the annulus could lead to initial facilitation of the center excitatory neurons followed by suppression (solid gray curve in Fig. 9a). This is due to the fact that the low-contrast central stimulus is too weak to activate the center interneurons by itself; thus, stimulation of the far surround initially facilitates the response of the center neurons, because FB inputs to the center sum with afferent and lateral inputs, until a critical annulus size is reached, beyond which the

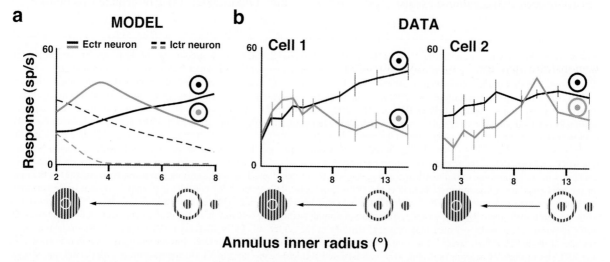

Fig. 9. Contrast-dependent suppression and facilitation from the far surround of V1 neurons: model and experiments. (a) Computer simulations: response of the center excitatory (solid curves) and inhibitory (dashed curves) neurons to a high contrast (85%; black curves) or a low contrast (15%; gray curves) central stimulus fit to the size of the neuron's hsRF plotted against the inner radius of a high contrast (85%) annular stimulus of 8° outer radius presented together with the central stimulus. The stimulus configuration is shown under the x axis; the rightmost data point on the x axis is the response to the central stimulus alone without the surround stimulus. Icons above each curve in (a) and in (b) indicate the contrasts of the center and surround stimulus used. (b) Experimental data: response of two example V1 cells to the same stimulus configuration used in (a), as shown under the x axis. For cell 1, we measured the response to a high contrast (70%; black curve) or low contrast (25%; gray curve) central stimulus fit to the cell's hsRF size as a function of the inner radius of a high contrast (70%) annular stimulus of 14° outer radius presented together with the central stimulus. For cell 2, the black curve is the response to high contrast (75%) center and surround stimuli, while the gray curve is the response to low contrast (25%) center and surround stimuli.

interneurons are activated (dashed gray curve in Fig. 9a) and suppression occurs. This prediction is consistent with our recent physiological data (Ichida et al., 2005); two sample cells are shown in Fig. 9b. As in the model, and as previously demonstrated (Levitt and Lund, 2002), presentation of a high-contrast annular grating in the far surround suppresseseded the RF center response to a high-contrast central grating (black curves in Fig. 9b). However, as predicted by the model, as the inner radius of a high-or low-contrast annular grating in the surround was reduced, we observed first facilitation and then suppression of the RF center response to a low-contrast central grating (gray curves in Fig. 9b). We have observed facilitation from the far surround in >50% of cells. Furthermore, for any given cell facilitation peaked at smaller inner radii when the stimulus contrast in the surround was also lowered (not shown). Note that cascading lateral connections are unlikely to contribute to the modulatory effects of the far surround in these experiments, because excitatory neurons whose RFs lie in the visual field location of the blank stimulus do not receive afferent drive, and thus cannot effectively relay signals to their postsynaptic V1 neurons. These findings demonstrate that the "suppressive surround" of V1 neurons is not always suppressive, and are thus inconsistent with the DOG model of center–surround interactions (Fig. 2b). More generally, our model provides a general mechanism of how top–down signals can shape the extra-classical RF of cortical neurons. The same model can also be easily extended to account for the effects of spatial attention in V1.

Conclusions

Studies on the spatio-temporal properties, patterning and functional specificity of FF, lateral and FB connections are beginning to shed some light on the relative roles of each system of connections in the generation of responses within and outside the classical RF of V1 neurons. We have now gathered enough anatomical and physiological data to constrain our neural network models, and to use such models to test specific hypotheses on the mechanisms by which these connections can generate V1 neuron responses.

Our data and modeling results so far have led us to think that the RF center and surround of V1 neurons result from integration of signals from all three sets of connections, FF, lateral and FB. These connections operate at different spatial scales, and all send excitatory inputs to neurons in the RF center. Lateral connections, but possibly not FF or FB connections, also contact local inhibitory neurons, which in turn control the output of the center excitatory neurons. FB connections can exert suppression via lateral excitation of local inhibitors. As these inhibitors have higher threshold and gain than the center excitatory neurons, they are only engaged when a significant amount of excitation leads them to fire. Thus, weak stimuli (such as bars, low-contrast or small high-contrast gratings) most often facilitate, whereas strong stimuli (such as high-contrast large gratings) most often suppress the center neurons.

The RF center of V1 neurons is initially generated by driving FF inputs arising from the LGN. Specifically, the mRF, i.e. the spiking region of the RF, is generated by converging inputs from the classical RF center of geniculate cells whose response is at the peak of their size tuning curve. The hsRF of V1 cells, i.e. the subthreshold depolarizing region surrounding the mRF, results from converging inputs from the classical RF center plus surround of geniculate neurons whose response is partially attenuated, due to surround suppression of LGN cells. V1 lateral connections and extrastriate FB connections also contribute to the mRF and hsRF of V1 neurons, by enhancing the neuron response. Stimuli outside the hsRF can influence the center neuron's response only via lateral and FB connections, and can evoke facilitation or suppression depending on the total amount of excitatory drive reaching the local inhibitors. Thus, low-contrast gratings fitted to the lsRF or smaller are facilitatory, but high-contrast gratings fitted to the lsRF or larger are suppressive. Similarly, gratings engaging the far, but not the near, surround can facilitate the response to low contrast gratings fitted to the hsRF, but are suppressive when the contrast of the central grating is increased.

What has emerged is the concept that the response of V1 neurons even to simple spatial patterns, such as gratings or bars, is the result of integration of inputs from FF, lateral and FB connections.

Abbreviations

BDA	biotinylated dextran amine
CO	cytochrome oxidase
cRF	classical receptive field
CTB	cholera toxin B
DOG	difference of Gaussians
EGFP	enhanced green fluorescent protein
FB	feedback
FF	feedforward
hsRF	high-contrast summation receptive field
K	koniocellular or interlaminar
lsRF	low contrast summation receptive field
LGN	lateral geniculate nucleus
mRF	minimum response field
Magno (or M)	magnocellular
Parvo (or P)	parvocellular
RF	receptive field
sRF	summation receptive field

Acknowledgments

This work was supported by grants from the National Science Foundation (IBN 0344569, and DMS 0515725), the National Eye Institute (EY 015262 and EY 015609), the Wellcome Trust (061113), the University of Utah Research Foundation and by a grant from Research to Prevent Blindness, Inc., New York, NY, to the Department of Ophthalmology, University of Utah, UT.

References

Adorjan, P., Levitt, J.B., Lund, J.S. and Obermayer, K. (1999) A model of the intracortical origin of orientation preference and tuning in macaque striate cortex. Visual Neurosci., 16: 303–318.

Ahissar, M. and Hochstein, S. (2000) The spread of attention and learning in feature search: effects of target distribution and task difficulty. Vision Res., 40: 1349–1364.

Allman, J., Miezin, F. and Mc, G.E. (1985) Stimulus-specific responses from beyond the classical receptive field: neurophysiological mechanisms for local–global comparisons in visual neurons. Ann. Rev. Neurosci., 8: 407–430.

Anderson, J.C., Binzegger, T., Martin, K.A.C. and Rockland, K.S. (1998) The connection from cortical area V1 to V5: a light and electron microscopic study. J. Neurosci., 18: 10525–10540.

Angelucci, A. and Bullier, J. (2003) Reaching beyond the classical receptive field of V1 neurons: horizontal or feedback axons? J. Physiol. (Paris), 97: 141–154.

Angelucci, A., Clasca, F. and Sur, M. (1996) Anterograde axonal tracing with the subunit B of cholera toxin: a highly sensitive immunohistochemical protocol for revealing fine axonal morphology in adult and neonatal brains. J. Neurosci. Methods, 65: 101–112.

Angelucci, A. and Levitt, J.B. (2002) Convergence of color, motion and form pathways in macaque V3. Soc. Neurosci. Abstr. Online, Program No. 658.2.

Angelucci, A., Levitt, J.B. and Lund, J.S. (2002a) Anatomical origins of the classical receptive field and modulatory surround field of single neurons in macaque visual cortical area V1. Prog. Brain Res., 136: 373–388.

Angelucci, A., Levitt, J.B., Walton, E., Hupé, J.M., Bullier, J. and Lund, J.S. (2002b) Circuits for local and global signal integration in primary visual cortex. J. Neurosci., 22: 8633–8646.

Angelucci, A. and Sainsbury, K. (2006) Contribution of feedforward thalamic afferents and corticogeniculate feedback to the spatial summation area of macaque V1 and LGN neurons. J. Comp. Neurol., in press.

Angelucci, A., Schiessl, I., Nowak, L. and McLoughlin, N. (2003) Functional specificity of feedforward and feedback connections between primate V1 and V2. Soc. Neurosci. Abstr. Online, Program No. 911.2.

Azouz, R. and Gray, C.M. (1999) Cellular mechanisms contributing to response variability of cortical neurons in vivo. J. Neurosci., 19: 2209–2223.

Bair, W., Cavanaugh, J.R. and Movshon, J.A. (2003) Time course and time–distance relationships for surround suppression in macaque V1 neurons. J. Neurosci., 23: 7690–7701.

Barlow, H.B., Blakemore, C. and Pettigrew, J.D. (1967) The neural mechanisms of binocular depth discrimination. J. Physiol. (Lond.), 193: 327–342.

Bauer, R., Dow, B.M. and Vautin, R.G. (1980) Laminar distribution of preferred orientations in foveal striate cortex of the monkey. Exp. Brain Res., 41: 54–60.

Bauer, U., Scholz, M., Levitt, J.B., Lund, J.S. and Obermayer, K. (1999) A model for the depth dependence of receptive field size and contrast sensitivity of cells in layer 4C of macaque striate cortex. Vision Res., 39: 613–629.

Blakemore, C. and Tobin, E.A. (1972) Lateral inhibition between orientation detectors in the cat's visual cortex. Exp. Brain Res., 15: 439–440.

Blasdel, G.G. and Campbell, D. (2001) Functional retinotopy of monkey visual cortex. J. Neurosci., 21: 8286–8301.

Blasdel, G.G. and Fitzpatrick, D. (1984) Physiological organization of layer 4 in macaque striate cortex. J. Neurosci., 4: 880–895.

Blasdel, G.G. and Lund, J.S. (1983) Terminations of afferent axons in macaque striate cortex. J. Neurosci., 3: 1389–1413.

Bonin, V., Mante, V. and Carandini, M. (2005) The suppressive field of neurons in the lateral geniculate neurons. J. Neurosci., 25: 10844–10856.

Bosking, W.H., Zhang, Y., Schofield, B. and Fitzpatrick, D. (1997) Orientation selectivity and the arrangement of horizontal connections in tree shrew striate cortex. J. Neurosci., 17: 2112–2127.

Boyd, J.D. and Casagrande, V.A. (1999) Relationships between cytochrome oxidase (CO) blobs in primate primary visual cortex (V1) and the distribution of neurons projecting to the middle temporal area (MT). J. Comp. Neurol., 409: 573–591.

Bringuier, V., Chavane, F., Glaeser, L. and Frégnac, Y. (1999) Horizontal propagation of visual activity in the synaptic integration field of area 17 neurons. Science, 283: 695–699.

Bullier, J. and Henry, G.H. (1980) Ordinal position and afferent input of neurons in monkey striate cortex. J. Comp. Neurol., 193: 913–935.

Bullier, J., Hupe, J.M., James, A.C. and Girard, P. (2001) The role of feedback connections in shaping the responses of visual cortical neurons. Prog. Brain Res., 134: 193–204.

Casagrande, V.A. (1994) A third parallel visual pathway to primate area V1. Trends Neurosci., 17: 305–310.

Casagrande, V.A. and Kaas, J.H. (1994) The afferent, intrinsic, and efferent connections of primary visual cortex. In: Peters, A. and Rockland, K.S. (Eds.) Primary Visual Cortex of Primates, Vol. 10. Plenum Press, New York, NY, pp. 201–259.

Casagrande, V.A. and Xu, X. (2004) Parallel visual pathways: a comparative perspective. In: Chalupa, L.M. and Werner, J.S. (Eds.) The Visual Neurosciences, Vol. 1. MIT Press, Cambridge, MA, pp. 494–506.

Cavanaugh, J.R., Bair, W. and Movshon, J.A. (2002) Nature and interaction of signals from the receptive field center and surround in macaque V1 neurons. J. Neurophysiol., 88: 2530–2546.

Chen, C., Kasamatsu, T., Polat, U. and Norcia, A.M. (2001) Contrast response characteristics of long-range lateral interactions in cat striate cortex. Neuroreport, 12: 655–661.

Chisum, H.J., Mooser, F. and Fitzpatrick, D. (2003) Emergent properties of layer 2/3 neurons reflect the collinear arrangement of horizontal connections in tree shrew visual cortex. J. Neurosci., 23: 2947–2960.

Crick, F. and Koch, C. (1998) Constrains on cortical and thalamic projections: the no-strong-loops hypothesis. Nature, 391: 245–250.

Crook, J.M., Engelmann, R. and Löwell, S. (2002) GABA-inactivation attenuates colinear facilitation in cat primary visual cortex. Exp. Brain Res., 143: 295–302.

DeAngelis, G.C., Freeman, R.D. and Ohzawa, I. (1994) Length and width tuning of neurons in the cat's primary visual cortex. J. Neurophysiol., 71: 347–374.

Domenici, L., Harding, G.W. and Burkhalter, A. (1995) Patterns of synaptic activity in forward and feedback pathways within rat visual cortex. J. Neurophysiol., 74: 2649–2664.

Douglas, R.J. and Martin, K.A. (1991) A functional microcircuit for cat visual cortex. J. Physiol. (Lond.), 440: 735–769.

Dragoi, V. and Sur, M. (2000) Dynamic properties of recurrent inhibition in primary visual cortex: contrast and orientation dependence of contextual effects. J. Neurophysiol., 83: 1019–1030.

Felisberti, F. and Derrington, A.M. (1999) Long-range interactions modulate the contrast gain in the lateral geniculate nucleus of cats. Visual Neurosci., 16: 943–956.

Felisberti, F. and Derrington, A.M. (2001) Long-range interactions in the lateral geniculate nucleus of the New-World monkey, Callithrix jacchus. Vis. Neurosci., 18: 209–218.

Felleman, D.J. and Van Essen, D.C. (1991) Distributed hierarchical processing in the primate cerebral cortex. Cereb. Cortex, 1: 1–47.

Fitzpatrick, D. (2000) Seeing beyond the receptive field in primary visual cortex. Curr. Opin. Neurobiol., 10: 438–443.

Freund, T.F., Martin, K.A., Soltesz, I., Somogyi, P. and Whitteridge, D. (1989) Arborisation pattern and postsynaptic targets of physiologically identified thalamocortical afferents in striate cortex of the macaque monkey. J. Comp. Neurol., 289: 315–336.

Garey, L.J. and Powell, T.P.S. (1971) An experimental study of the termination of the lateral geniculo-cortical pathway in the cat and monkey. Proc. R. Soc. (Biol.), 179: 21–40.

Gilbert, C., Das, A., Ito, M., Kapadia, M. and Westheimer, G. (1996) Spatial integration and cortical dynamics. Proc. Natl. Acad. Sci. USA, 93: 615–622.

Gilbert, C.D. and Wiesel, T.N. (1990) The influence of contextual stimuli on the orientation selectivity of cells in primary visual cortex of the cat. Vision Res., 30: 1689–1701.

Girard, P., Hupé, J.M. and Bullier, J. (2001) Feedforward and feedback connections between areas V1 and V2 of the monkey have similar rapid conduction velocities. J. Neurophysiol., 85: 1328–1331.

Grinvald, A., Lieke, E.E., Frostig, R.D. and Hildesheim, R. (1994) Cortical point-spread function and long-range lateral interactions revealed by real-time optical imaging of macaque monkey primary visual cortex. J. Neurosci., 14: 2545–2568.

Gur, M., Kagan, I. and Snodderly, D.M. (2005) Orientation and direction selectivity of neurons in V1 of alert monkeys: functional relationships and laminar distributions. Cereb. Cortex, 15: 1207–1221.

Haenny, P.E., Maunsell, J.H. and Schiller, P.H. (1988) State dependent activity in monkey visual cortex. Exp. Brain Res., 69: 245–259.

Hawken, M.J. and Parker, A. (1984) Contrast sensitivity and orientation selectivity in laminar IV of the striate cortex of old world monkeys. Exp. Brain Res., 54: 367–372.

Hendrickson, A.E., Wilson, J.R. and Ogren, M.P. (1978) The neuroanatomical organization of pathways between the dorsal lateral geniculate nucleus and visual cortex in Old World and New World primates. J. Comp. Neurol., 182: 123–136.

118

Hendry, S.H. and Reid, R.C. (2000) The koniocellular pathway in primate vision. Ann. Rev. Neurosci., 23: 127–153.

Hendry, S.H. and Yoshioka, T. (1994) A neurochemically distinct third channel in the macaque dorsal lateral geniculate nucleus. Science, 264: 575–577.

Hess, R. and Field, D. (2000) Integration of contours: new insights. Trends Cogn. Sci., 3: 480–486.

Hirsch, J.A. and Gilbert, C.D. (1991) Synaptic physiology of horizontal connections in the cat's visual cortex. J. Neurosci., 11: 1800–1809.

Hubel, D.H. and Wiesel, T.N. (1962) Receptive fields, binocular interaction and functional architecture in the cat's visual cortex. J. Physiol. (Lond.), 160: 106–154.

Hubel, D.H. and Wiesel, T.N. (1972) Laminar and columnar distribution of geniculo-cortical fibers in the macaque monkey. J. Comp. Neurol., 146: 421–450.

Hupé, J.M., James, A.C., Girard, P. and Bullier, J. (2001a) Response modulations by static texture surround in area V1 of the macaque monkey do not depend on feedback connections from V2. J. Neurophysiol., 85: 146–163.

Hupé, J.M., James, A.C., Girard, P., Lomber, S.G., Payne, B.R. and Bullier, J. (2001b) Feedback connections act on the early part of the responses in monkey visual cortex. J. Neurophysiol., 85: 134–145.

Hupé, J.M., James, A.C., Payne, B.R., Lomber, S.G., Girard, P. and Bullier, J. (1998) Cortical feedback improves discrimination between figure and background by V1, V2 and V3 neurons. Nature, 394: 784–787.

Ichida, J.M., Schwabe, L., Bressloff, P.C. and Angelucci, A. (2005) Feedback-mediated facilitation and suppression from the receptive field surround of macaque V1 neurons. Soc. Neurosci. Abstr. Online, Program No. 820.4.

Ichinohe, N., Fujiyama, F., Kaneko, T. and Rockland, K.S. (2003) Honeycomb-like mosaic at the border of layers 1 and 2 in the cerebral cortex. J. Neurosci., 23: 1372–1382.

Ichinohe, N. and Rockland, K.S. (2004) Region-specific micromodularity in the uppermost layers in the primate cerebral cortex. Cereb. Cortex, 14: 1173–1184.

Ito, M. and Gilbert, C.D. (1999) Attention modulates contextual influences in the primary visual cortex of alert monkeys. Neuron, 22: 593–604.

Jeffs, J., Ichida, J., Lund, J.S. and Angelucci, A. (2003) Modular specificity of feedforward and feedback pathways to and from marmoset V3. Soc. Neurosci. Abstr. Online, Program No. 911.11.

Johnson, R.R. and Burkhalter, A. (1996) Microcircuitry of forward and feedback connections within rat visual cortex. J. Comp. Neurol., 368: 383–398.

Johnson, R.R. and Burkhalter, A. (1997) A polysynaptic feedback circuit in rat visual cortex. J. Neurosci., 17: 7129–7140.

Jones, H.E., Andolina, I.M., Oakely, N.M., Murphy, P.C. and Sillito, A.M. (2000) Spatial summation in lateral geniculate nucleus and visual cortex. Exp. Brain Res., 135: 279–284.

Kapadia, M.K., Ito, M., Gilbert, C.D. and Westheimer, G. (1995) Improvement in visual sensitivity by changes in local context: parallel studies in human observers and in V1 of alert monkeys. Neuron, 15: 843–856.

Kapadia, M.K., Westheimer, G. and Gilbert, C.D. (1999) Dynamics of spatial summation in primary visual cortex of alert monkeys. Proc. Natl. Acad. Sci. USA, 96: 12073–12078.

Kaplan, E. (2004) The M, P, and K pathways of the primate visual system. In: Chalupa, L.M. and Werner, J.S. (Eds.) The Visual Neurosciences, Vol. 1. MIT Press, Cambridge, MA, pp. 481–493.

Knierim, J.J. and Van Essen, D. (1992) Neuronal responses to static texture patterns in area V1 of the alert macaque monkey. J. Neurophysiol., 67: 961–980.

Kremers, J., Silveira, L.C.L. and Kilavik, B.E. (2001) Influence of contrast on the responses of marmoset lateral geniculate cells to drifting gratings. J Neurophysiol., 85: 235–246.

Kremers, J. and Weiss, S. (1997) Receptive field dimensions of lateral geniculate cells in the common marmoset (Callithrix jacchus). Vision Res., 37: 2171–2181.

Lamme, V.A., Super, H. and Spekreijse, H. (1998) Feedforward, horizontal, and feedback processing in the visual cortex. Curr. Opin. Neurobiol., 8: 529–535.

Lamme, V.A.F. (1995) The neurophysiology of figure-ground segregation in primary visual cortex. J. Neurosci., 15: 1605–1615.

Lamme, V.A.F., Rodriguez-Rodriguez, V. and Spekreijse, H. (1999) Separate processing dynamics for texture elements, boundaries and surfaces in primary visual cortex of the macaque monkey. Cereb. Cortex, 9: 406–413.

Leventhal, A.G., Thompson, K.G., Liu, D., Zhou, Y. and Ault, S.J. (1995) Concomitant sensitivity to orientation, direction, and color of cells in layers 2, 3, and 4 of monkey striate cortex. J. Neurosci., 15: 1808–1818.

Levitt, J.B. and Lund, J.S. (1997) Contrast dependence of contextual effects in primate visual cortex. Nature, 387: 73–76.

Levitt, J.B. and Lund, J.S. (2002) The spatial extent over which neurons in macaque striate cortex pool visual signals. Vis. Neurosci., 19: 439–452.

Levitt, J.B., Lund, J.S. and Yoshioka, T. (1996) Anatomical substrates for early stages in cortical processing of visual information in the macaque monkey. Behav. Brain Res., 76: 5–19.

Levitt, J.B., Tyler, C.J. and Lund, J.S. (1998) Receptive field properties of neurons in marmoset striate cortex. Soc. Neurosci. Abstr., 24: 645.

Li, H., Fukuda, M., Tanifuji, M. and Rockland, K.S. (2003) Intrinsic collaterals of layer 6 Meynert cells and functional columns in primate V1. Neuroscience, 120: 1061–1069.

Li, W., Their, P. and Wehrhahn, C. (2000) Contextual influence on orientation discrimination of humans and responses of neurons in V1 of alert monkeys. J. Neurophysiol., 83: 941–954.

Li, W., Thier, P. and Wehrhahn, C. (2001) Neuronal responses from beyond the classical receptive field in V1 of alert monkeys. Exp. Brain Res., 139: 359–371.

Lund, J.S. (1988) Anatomical organization of macaque monkey striate visual cortex. Ann. Rev. Neurosci., 11: 253–288.

Lund, J.S., Angelucci, A. and Bressloff, P. (2003) Anatomical substrates for functional columns in macaque monkey primary visual cortex. Cereb. Cortex, 12: 15–24.

Lund, J.S., Wu, Q., Hadingham, P.T. and Levitt, J.B. (1995) Cells and circuits contributing to functional properties in area V1 of macaque monkey cerebral cortex: bases for neuroanatomically realistic models. J. Anat., 187: 563–581.

Malach, R., Amir, Y., Harel, M. and Grinvald, A. (1993) Relationship between intrinsic connections and functional architecture revealed by optical imaging and in vivo-targeted biocytin injections in primate striate cortex. Proc. Natl. Acad. Sci. USA, 90: 10469–10473.

Maunsell, J.H. and Van Essen, D.C. (1983) The connections of the middle temporal visual area (MT) and their relationship to a cortical hierarchy in the macaque monkey. J. Neurosci., 3: 2563–2586.

McGuire, B.A., Gilbert, C.D., Rivlin, P.K. and Wiesel, T.N. (1991) Targets of horizontal connections in macaque primary visual cortex. J. Comp. Neurol., 305: 370–392.

Mignard, M. and Malpeli, J.G. (1991) Paths of information flow through visual cortex. Science, 251: 1249–1251.

Mizobe, K., Polat, U., Pettet, M.W. and Kasamatsu, T. (2001) Facilitation and suppression of single striate-cell activity by spatially discrete pattern stimuli presented beyond the receptive field. Vis. Neurosci., 18: 377–391.

Nelson, J.I. and Frost, B. (1978) Orientation selective inhibition from beyond the classical receptive field. Brain Res., 139: 359–365.

Nolt, M.J., Kumbhani, R.D. and Palmer, L.A. (2004) Contrast-dependent spatial summation in the lateral geniculate nucleus and retina of the cat. J. Neurophysiol., 92: 1708–1717.

Nothdurft, H.C., Gallant, J.L. and Van Essen, D.C. (1999) Response modulation by texture surround in primate area V1: correlates of "popout" under anesthesia. Vis. Neurosci., 16: 15–34.

Nothdurft, H.C., Gallant, J.L. and Van Essen, D.C. (2000) Response profiles to texture border patterns in area V1. Vis. Neurosci., 17: 421–436.

Nowak, L.G. and Bullier, J. (1997) The timing of information transfer in the visual system. In: Rockland, K.S., Kaas, J.H. and Peters, A. (Eds.) Extrastriate Cortex in Primates, Vol. 12. Plenum Press, New York, pp. 205–241.

Ozeki, H., Sadakane, O., Akasaki, T., Naito, T., Shimegi, S. and Sato, H. (2004) Relationship between excitation and inhibition underlying size tuning and contextual response modulation in the cat primary visual cortex. J. Neurosci., 24: 1428–1438.

Perkel, D.J., Bullier, J. and Kennedy, H. (1986) Topography of the afferent connectivity of area 17 in the macaque monkey: a double-labelling study. J. Comp. Neurol., 253: 374–402.

Peters, A. and Feldman, M.L. (1976) The projection of the lateral geniculate nucleus to area 17 of the rat cerebral cortex. I. General description. J. Neurocytol., 5: 63–84.

Poggio, G.F., Doty, R.W.J. and Talbot, W.H. (1977) Foveal striate cortex of behaving monkey: single-neuron responses to square-wave gratings during fixation of gaze. J. Neurophysiol., 40: 1369–1391.

Polat, U., Mizobe, K., Pettet, M.W., Kasamatsu, T. and Norcia, A.M. (1998) Collinear stimuli regulate visual responses depending on cell's contrast threshold. Nature, 391: 580–584.

Rajeev, D., Raizada, S. and Grossberg, S. (2003) Towards a theory of the laminar architecture of cerebral cortex: computational cues from the visual system. Cereb. Cortex, 13: 100–113.

Reid, R.C. and Usrey, W.M. (2004) Functional connectivity in the pathway from retina to striate cortex. In: Chalupa, L.M. and Werner, J.S. (Eds.) The Visual Neurosciences, Vol. 1. MIT Press, Cambridge, MA, pp. 673–679.

Ringach, D.L., Shapley, R.M. and Hawken, M.J. (2002) Orientation selectivity in macaque V1: diversity and laminar dependence. J. Neurosci., 22: 5639–5651.

Rockland, K.S. (1994) The organization of feedback connections from area V2 (18) to V1 (17). In: Peters, A. and Rockland, K.S. (Eds.) Primary Visual Cortex in Primates, Vol. 10. Plenum Press, New York, pp. 261–299.

Rockland, K.S. (2003) Feedback connections: splitting the arrow. In: Kaas, J.H. and Collins, C.E. (Eds.), The Primate Visual System. CRC Press, pp. 387–405.

Rockland, K.S. and Lund, J.S. (1983) Intrinsic laminar lattice connections in primate visual cortex. J. Comp. Neurol., 216: 303–318.

Rockland, K.S. and Pandya, D.N. (1979) Laminar origins and terminations of cortical connections to the occipital lobe in the rhesus monkey. Brain Res., 179: 3–20.

Salin, P.A. and Bullier, J. (1995) Corticocortical connections in the visual system: structure and function. Physiol. Rev., 75: 107–154.

Salin, P.A., Kennedy, H. and Bullier, J. (1995) Spatial reciprocity of connections between areas 17 and 18 in the cat. Can. J. Physiol. Pharmacol., 73: 1339–1347.

Sandell, J.H. and Schiller, P.H. (1982) Effect of cooling area 18 on striate cortex cells in the squirrel monkey. J. Neurophysiol., 48: 38–48.

Sceniak, M.P., Hawken, M.J. and Shapley, R.M. (2001) Visual spatial characterization of macaque V1 neurons. J. Neurophysiol., 85: 1873–1887.

Sceniak, M.P., Ringach, D.L., Hawken, M.J. and Shapley, R. (1999) Contrast's effect on spatial summation by macaque V1 neurons. Nat. Neurosci., 2: 733–739.

Schmidt, K.E., Goebel, R., Löwell, S. and Singer, W. (1997) The perceptual grouping criterion of colinearity is reflected by anisotropies of connections in the primary visual cortex. Eur. J. Neurosci., 9: 1083–1089.

Schwabe, L., Obermayer, K., Angelucci, A. and Bressloff, P.C. (2006) The role of feedback in shaping the extra-classical receptive field of cortical neurons: a recurrent network model. J. Neurosci., in press.

Sengpiel, F., Baddley, R.J., Freeman, T.C.B., Harrad, R. and Blakemore, C. (1998) Different mechanisms underlie three inhibitory phenomena in cat area 17. Vision Res., 38: 2067–2080.

Shao, Z. and Burkhalter, A. (1996) Different balance of excitation and inhibition in forward and feedback circuits of rat visual cortex. J. Neurosci., 16: 7353–7365.

Shapley, R. (2004) A new view of the primary visual cortex. Neural Networks, 17: 615–623.

Shmuel, A., Korman, M., Sterkin, A., Harel, M., Ullman, S., Malach, R. and Grinvald, A. (2005) Retinotopic axis

120

specificity and selective clustering of feedback projections from V2 to V1 in the owl monkey. J. Neurosci., 25: 2117–2131.

Sillito, A.M., Grieve, K.L., Jones, H.E., Cudeiro, J. and Davis, J. (1995) Visual cortical mechanisms detecting focal orientation discontinuities. Nature, 378: 492–496.

Silvanto, J., Lavie, N. and Walsh, V. (2005) Double dissociation of V1 and V5/MT activity in visual awareness. Cereb. Cortex, 15: 1736–1741.

Sincich, L.C., Adams, D.L. and Horton, J.C. (2003) Complete flatmounting of the macaque cerebral cortex. Vis. Neurosci., 20: 663–686.

Sincich, L.C. and Blasdel, G.G. (2001) Oriented axon projections in primary visual cortex of the monkey. J. Neurosci., 21: 4416–4426.

Sincich, L.C. and Horton, J.C. (2002) Divided by cytochrome oxidase: a map of the projections from V1 to V2 in macaques. Science, 295: 1734–1737.

Sincich, L.C. and Horton, J.C. (2003) Independent projection streams from macaque striate cortex to the second visual area and middle temporal area. J. Neurosci., 23: 5684–5692.

Sincich, L.C. and Horton, J.C. (2005) The circuitry of V1 and V2: integration of color, form and motion. Ann. Rev. Neurosci., 28: 303–326.

Slovin, H., Arieli, A., Hildesheim, R. and Grinvald, A. (2002) Long-term voltage-sensitive dye imaging reveals cortical dynamics in behaving monkeys. J. Neurophysiol., 88: 3421–3438.

Solomon, S.G., White, A.J.R. and Martin, P.R. (2002) Extra-classical receptive field properties of parvocellular, magnocellular, and koniocellular cells in the primate lateral geniculate nucleus. J. Neurosci., 22: 338–349.

Somers, D., Dragoi, V. and Sur, M. (2002) Orientation selectivity and its modulation by local and long-range connections in visual cortex. In: Payne, B.R. and Peters, A. (Eds.), The Cat Primary Visual Cortex. Academic Press, San Diego, CA, pp. 471–520.

Somers, D.C., Todorov, E.V., Siapas, A.G., Toth, L.J., Kim, D.S. and Sur, M. (1998) A local circuit approach to understanding integration of long-range inputs in primary visual cortex. Cereb. Cortex, 8: 204–217.

Stettler, D.D., Das, A., Bennett, J. and Gilbert, C.D. (2002) Lateral connectivity and contextual interactions in macaque primary visual cortex. Neuron, 36: 739–750.

Tong, F. (2003) Primary visual cortex and visual awareness. Nat. Rev. Neurosci., 4: 219–229.

Treue, S. (2003) Visual attention: the where, what, how and why of saliency. Curr. Opin. Neurobiol., 13: 428–432.

Ungerleider, L.G. and Desimone, R. (1986) Cortical connections of visual area MT in the macaque. J. Comp. Neurol., 248: 190–222.

Van Essen, D.C., Newsome, W.T. and Maunsell, J.H. (1984) The visual field representation in striate cortex of the macaque monkey: asymmetries, anisotropies, and individual variability. Vision Res., 24: 429–448.

Webb, B.S., Dhruv, N.T., Solomon, S.G., Taliby, C. and Lennie, P. (2005) Early and late mechanisms of surround suppression in striate cortex of macaque. J. Neurosci., 25: 11666–11675.

Yazar, F., Mavity-Hudson, J.A., Ding, Y., Oztas, E. and Casagrande, V.A. (2004) Layer IIIB (IVA) of primary visual cortex (V1) and its relationship to the koniocellular (K) pathway in macaque monkeys. Soc. Neurosci. Abstr.Viewer/Itinerary Planner Online, Program No. 300.17.

Yoshimura, Y., Sato, H., Imamura, K. and Watanabe, Y. (2000) Properties of horizontal and vertical inputs to pyramidal cells in the superficial layers of the cat visual cortex. J. Neurosci., 20: 1931–1940.

Yoshioka, T., Blasdel, G.G., Levitt, J.B. and Lund, J.S. (1996) Relation between patterns of intrinsic lateral connectivity, ocular dominance and cytochrome oxidase reactive regions in macaque monkey striate cortex. Cereb. Cortex, 6: 297–310.

Martinez-Conde, Macknik, Martinez, Alonso & Tse (Eds.)
Progress in Brain Research, Vol. 154
ISSN 0079-6123

CHAPTER 6

Cortical cartography revisited: a frequency perspective on the functional architecture of visual cortex

Amit Basole[1], Vincenzo Kreft-Kerekes[1], Leonard E. White[1,2] and David Fitzpatrick[1,*]

[1]*Department of Neurobiology, Box 3209, Duke University Medical Center, Durham, NC 27710, USA*
[2]*Department of Community and Family Medicine, Duke University Medical Center, Durham, NC 27710, USA*

Abstract: Viewed in the plane of the cortical surface, the visual cortex is composed of overlapping functional maps that represent stimulus features such as edge orientation, direction of motion, and spatial frequency. Spatial relationships between these maps are thought to ensure that all combinations of stimulus features are represented uniformly across the visual field. Implicit in this view is the assumption that feature combinations are represented in the form of a place code such that a given pattern of activity uniquely signifies a specific combination of stimulus features. Here we review results of experiments that challenge the place code model for the representation of feature combinations. Rather than overlapping maps of stimulus features, we suggest that patterns of activity evoked by complex stimuli are best understood in the context of a single map of spatiotemporal energy.

Keywords: optical imaging; population activity; spatiotemporal energy; orientation columns; spatial frequency; Gabor filters

Introduction

The pioneering studies of functional architecture by Hubel and Wiesel provided the foundation for understanding the radial and tangential organization of primary visual cortex (Hubel and Wiesel, 1962, 1968, 1977). Building on the work of Mountcastle (1957), they noted that neurons displaced along the radial axis (perpendicular to the cortical surface) formed columns sharing similar response properties such as receptive field location, preference for edge orientation, and relative effectiveness of the input from each eye (ocular dominance). In the tangential dimension (in the plane of the cortical surface), adjacent columns were found to exhibit slightly different stimulus properties, forming orderly two-dimensional (2-D) representations that have come to be known as cortical maps.

The existence of distinct columnar maps for different properties raises two fundamental and inter-related questions. First, what are the spatial relationships among these maps? And second, what is the spatial pattern of neural activity evoked by a stimulus with a particular combination of features? Hubel and Wiesel's experiments suggested a simple model that addressed both of these issues: maps for ocular dominance and orientation each iterate at a scale sufficient to achieve coverage — i.e., to adequately represent all combinations of ocular dominance and orientation for each point in visual space. As a result of this orderly arrangement, a simple line stimulus presented to one eye was expected to produce a

*Corresponding author. Tel.: + 1-919-684-8510;
Fax: + 1-919-684-4431.

DOI: 10.1016/S0079-6123(06)54006-3

patchy distribution of activity that could be well predicted by the constraints imposed by each map (Hubel and Wiesel, 1977). Subsequent studies using 2-deoxyglucose and optical imaging methods confirmed these predictions.

Later studies provided evidence for additional columnar maps; notably, maps for direction of motion and spatial frequency (Shmuel and Grinvald, 1996; Weliky et al., 1996; Shoham et al., 1997; Everson et al., 1998; Issa et al., 2000). Given the iterative nature of these maps, and the orderly relationships between them, a simple extension of the original Hubel–Wiesel framework was all that seemed necessary to accommodate them (Weliky et al., 1996; Hubener et al., 1997; Issa et al., 2000; Swindale et al., 2000). Although the combinatorial problem obviously limits the number of features that can be represented in this way, computational studies estimate that up to six or seven separate cortical maps could coexist and be overlaid in a fashion that maintains uniform coverage for all feature combinations (Swindale, 2000). As with the original Hubel–Wiesel model, specific combinations of stimulus features (e.g., a vertical stimulus of high spatial frequency moving to the right) are assumed to be represented in the form of a place code, where the most active neurons are located at cortical sites whose positions satisfy the joint constraints of the individual maps, and this pattern of activity uniquely signifies the presence of a specific combination of stimulus features.

There are, however, reasons to question whether the place code model, which accounts so well for ocular dominance, orientation, and position, is appropriate for additional stimulus features such as direction of motion and spatial frequency. Most current views of multiple maps are based on the analysis of cortical responses to drifting gratings in which the range of motion and spatial frequency cues are limited to those that vary along an axis orthogonal to the grating's orientation (one-dimensional (1-D) stimuli) (Wallach, 1935; Adelson and Movshon, 1982; Wuerger et al., 1996). As a result, it has not been possible to examine a broader range of interactions between motion, orientation, and spatial frequency that are common in the visual environment. Moreover, studies of the responses of single neurons to the motion of two-dimensional (2-D) stimuli i.e. stimuli whose motion is not limited to axes orthogonal to their orientation, such as small bars or drifting texture patterns, indicate that the spatial properties of the stimulus (for example, its orientation and size) as well as its speed can significantly alter tuning for direction of motion (Hammond and Smith, 1983; Crook, 1990; Crook et al., 1994). The consequences of this interdependence for the population response and the implications for a place code model of feature representation in the primary visual cortex (V1) have not been fully appreciated.

We recently explored this issue using intrinsic signal optical-imaging techniques to examine the patterns of neural activity in V1 of the ferret evoked by the motion of texture stimuli composed of iso-oriented line segments (Fig. 1a). Because this stimulus is made up of short bar segments, its axis of motion can be varied independent of bar orientation. This allows us to test the population response (i.e., the pattern of columnar activity across V1) to many different combinations of stimulus features; something that is not possible with a full-field grating stimulus. By varying the orientation of the line segments, their length, axis of motion, and speed, we reached a different conclusion about the representation of stimulus features in V1 that is difficult to reconcile with the place code model (Basole et al., 2003). Here we review these observations and consider an alternative framework that we believe offers a more complete account of these phenomena and a more accurate view of population coding in V1.

Population response to texture stimuli

Moving texture stimuli evoke strong activation of cortical circuits, and this activation resembles the patterns of activity that are found with square-wave gratings. For motion of the texture stimulus along axes orthogonal to the orientation of the line segments, the patterns of activity are indistinguishable from those found with square-wave gratings of the same orientation; i.e., both kinds of oriented stimuli activate the very same set of cortical columns across that portion of the visual cortex that represents the location of the stimulus

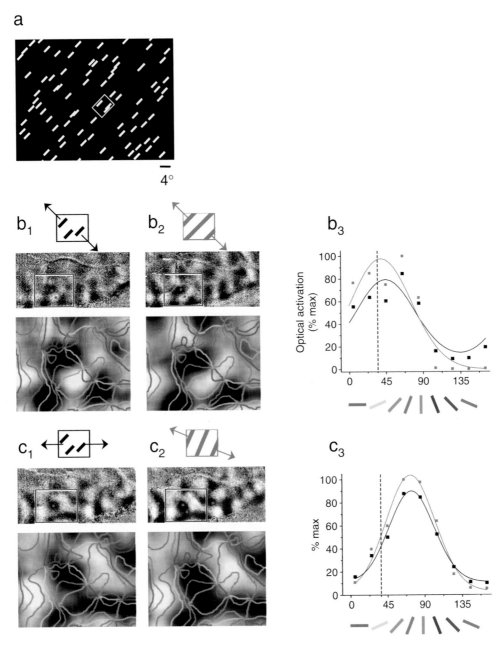

Fig. 1. Systematic shifts in the population response induced by changes in axis of motion. (a) The "texture" stimulus. (b) and (c) Each panel contains texture (b₁, c₁) or grating (b₂, c₂) difference images, with insets showing iso-orientation contours derived from the grating angle map overlaid on the boxed region of interest (see angle color key below graph, inset width = 1.7 mm). The red dots are placed over identical regions of each image to facilitate comparison. Graphs quantify the population tuning functions (b3, c3, and black curve: texture response; gray curve: grating response). The dashed vertical line is placed at the same position on the *x*-axis to compare the relative shift between the curves. (b) Orthogonally moving textures and gratings of the same orientation evoke similar responses. (c) A 45° anticlockwise shift in texture motion evokes a response similar to a 67.5° grating. (Reproduced with permission from Macmillan Magazines.)

monitor (see Fig. 1b). Motion along nonortho-gonal axes also evokes strong responses, but in this case, the pattern of activation departs significantly from that observed with a full-field grating of the same orientation. For example, the patterns of cortical activity evoked by a 45° texture stimulus (0° corresponds to the horizontal) composed of short bars (6° in length) moved along the horizontal axis and by a 45° square-wave grating are only weakly correlated ($R^2 = 0.16$). However, the population activity evoked by the 45° texture stimulus that moved along the horizontal axis does resemble the activity evoked by a grating stimulus in terms of its magnitude and the spacing of the active columns. In particular, we found that it strongly resembles the pattern of populat-ion activity evoked by a grating stimulus with a less acute orientation (R^2 for 67.5° grating pattern = 0.7). To quantify the patterns of popu-lation activity evoked by these stimuli, we devel-oped a population response tuning function that captures the relative activation of each pixel in the region of interest and expresses this value in terms of the pixel's preferred orientation, as assessed with grating stimuli. As illustrated in Figs. 1b$_3$ and c$_3$, changing the axis of motion of the stimulus results in striking shifts in the population tuning function such that the response resembles that found for grating stimuli whose orientation differs

substantially from the orientation of bars in the texture stimulus.

These results suggest that the iterated pattern of activity resulting from the presentation of a grating stimulus, which has been thought to represent the orientation of the grating can be elicited by a com-bination of a range of different orientation and axis of motion made possible by the use of texture stimuli. This fact is further demonstrated by the experiment whose result is illustrated in Fig. 2. Population responses were recorded for three dif-ferent combinations of texture orientation and axis of motion (shown as icons in the left column). The spatial distribution of population activity and the accompanying population tuning functions show that these very different orientations elicit similar responses in V1. Taken together, these observations seriously undermine the place code model, at least as it applies to the representation of orientation preference in V1. Rather than a given columnar pattern of population activity (i.e., a particular set of co-active cortical columns) invar-iably signaling the presence of a single orientation, these data show that any pattern of population activity in V1 may be evoked by a surprisingly broad range of stimulus orientations.

The dissociation of the activity patterns for tex-tures from patterns evoked by their orientation-matched gratings is also evident under conditions

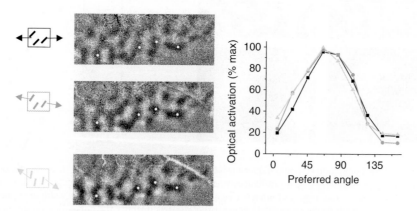

Fig. 2. Different combinations of stimulus features give similar responses. Optical difference maps (left) and corresponding population tuning functions (right) obtained with different combinations of texture orientation and axis of motion, shown as icons on the left. As before, dots are placed over identical regions of each image to facilitate comparison. Very similar patterns of activity are obtained for the different combinations of orientation and motion axis. Image width = 6.7 mm. (Reproduced with permission from Macmillan Magazines.)

where the length of the texture line segments is altered. Significant shifts in the distribution of population activity were found when the length of the line segments was altered without changing either the line orientation or the axis of motion. For example, for a 45° texture stimulus moving horizontally, changing the length of the line segments from 2° to 10° produced progressive shifts in the peak of the population tuning function activity from 87° for the short line segments to 56° for the longer ones (Fig. 3). This result complements those shown earlier by demonstrating that the same stimulus orientation does not invariably evoke the same columnar pattern of population activity in V1.

Frequency filtering and the population response

The results discussed above show that a given pattern of activity could be evoked by different combinations of line orientation, axis of motion, and length. This finding is difficult to reconcile with the place code model of feature representation in V1, which would predict that the orientation of line elements in a texture stimulus should always correspond to the columnar pattern of population activity in V1. Instead, these data are better interpreted by considering the tuning of individual neurons to a restricted range of spatial (along two dimensions) and temporal frequencies (Movshon et al., 1978a; De Valois et al., 1979; Skottun et al., 1994). In the following account we offer a qualitative explanation of the above results by considering both the stimulus and the receptive fields of cortical neurons in frequency space. We then test a specific prediction of this framework related to changes in stimulus speed. Finally, we review the results of our computational simulations, which confirm that a model based on Gabor filters exhibits shifts in population tuning that mimic the effects described in the preceding experiments.

The first step is to consider how the static texture stimuli that are used in these experiments are represented in 2-D spatial frequency space (Fig. 4). In this space, sine-wave gratings are represented as points whose positions reflect both the spatial

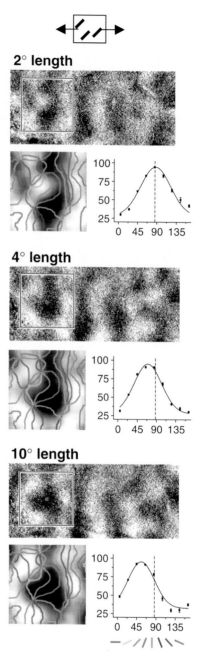

2° length

4° length

10° length

Fig. 3. Systematic shifts in the population response induced by changes in bar length. Difference images, contour overlays (inset width = 1.2 mm), and population tuning curves for texture stimuli of three lengths (2°, 4°, and 10°) moved nonorthogonally. Increasing bar length shifts the activity pattern and the peak of the response from 87° to 70° to 56° (from blue contours to green/yellow contours). (Reproduced with permission from Macmillan Magazines.)

126

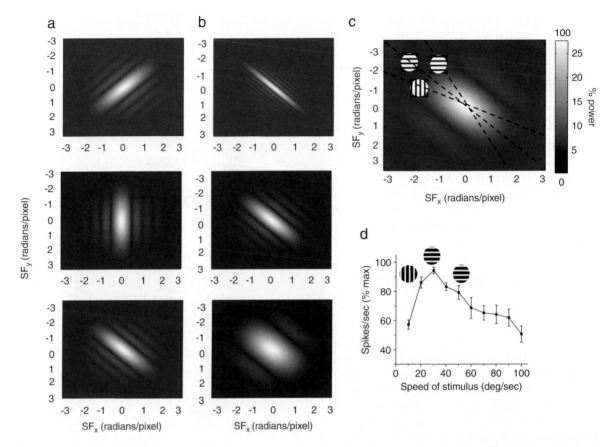

Fig. 4. Representing stimuli in frequency space. (a) 2-D power spectrum showing how stimulus orientation affects its representation in frequency space. The x-axis is the spatial frequency content along the x-axis in space and the y-axis is the spatial frequency content along the y-axis in space. The gray scale encodes power. The bar orientations are 135° (top), 0° (middle), and 45° (bottom). (b) 2-D power spectrums for a single bar of changing aspect ratio are shown. The axes and conventions are as described before. As the bar becomes shorter, the spectrum broadens and significant power is seen at orientations other than the bar orientation (45°). (c) A short bar can be thought of as consisting of several sine-wave components each of which modulates at a different speed. (d) A neural speed tuning curve showing that the effectiveness of a sine-wave component will depend on its speed.

frequency of the stimulus and its orientation. Vertical sine-wave gratings are represented as points along the x-axis (there is no modulation along the y-axis) and conversely, horizontal sine-wave gratings are represented as points along the y-axis. Texture stimuli are considerably broader in their power spectra than gratings. Fig. 4a shows the 2-D power spectrum for a short bar (an element of the texture stimulus) oriented at 135° (top), 0° (middle), and 45° (bottom). Figure 4b shows what happens to the power spectrum of a single bar oriented at 45° as its aspect ratio changes from 10:1 to 2:1. As before, the x- and y-axis correspond to spatial

frequency along the x and y-dimensions and the gray scale encodes magnitude of power at a given frequency. In addition to a concentration of power (or Fourier energy) expected based on the bar's orientation, the amount of power/energy present in the stimulus at other orientations in the frequency domain increases as bar length decreases (the representation in frequency space becomes broader) (Fig. 4c). Modulation of the neural response by Fourier components at these additional orientations could explain why the tuning of the population response does not always match that predicted from the orientation of the texture elements.

If the breadth of the stimulus in 2-D spatial frequency space was the sole factor influencing the population response, then the activity patterns should become broader with decreasing line length and remain centered on the orientation of the line segment. What explains the fact that the distribution of population activity shifts to a new center position consistent with a different stimulus orientation? When the stimulus moves, each Fourier component in the spatial domain moves at a different speed depending on its orientation relative to the direction of stimulus motion. This is because the speed of any given Fourier component is determined by its temporal frequency divided by its spatial frequency, and the temporal modulation of a drifting sine wave is proportional to the angular difference between the orientation of the grating and its direction of motion. Consequently, components that are oriented nearly orthogonal to the direction of motion are modulated at the highest temporal frequencies and therefore attain the highest speeds, while those that are oriented nearly parallel to the direction of motion are modulated at much lower temporal frequencies and therefore only move at the lowest speeds (Simoncelli and Heeger, 1998). Thus, the Fourier component whose speed is optimal, given the temporal tuning properties of V1 neurons, will determine the pattern of columnar activity in V1 and the peak of the active population (Fig. 4d). In the case of non-orthogonal motion of the texture stimulus, the optimal component can deviate significantly from the orientation of the line segments, and is a function of: (1) the length of the line segments, which determines the degree to which there is power in "off-orientation" components that can be modulated by motion; (2) the axis of motion, which determines the relative speed of the different frequency components; and (3) the speed of texture motion, which determines the absolute speed at which the various components modulate.

We have already demonstrated how alterations in the axis of motion and line length can produce significant shifts in the distribution of population activity. Another strong test of the ability of spatiotemporal frequency tuning to account for the shifts in population response is to determine whether the activity patterns evoked by texture stimuli are sensitive to the speed of texture motion, as has been described in single unit studies (Hammond and Smith, 1983; Crook, 1990; Skottun et al., 1994; Geisler et al., 2001). Changing the speed of texture motion without changing either the axis of motion or the orientation of the line segments indeed resulted in significant shifts in the pattern of population response (Fig. 5a). The peak of the population tuning function shifted by an average of 55° as the speed of the stimulus was changed from 10° to 100°/s (Fig. 5b). In one animal that was tested at 10 different speeds, the peak varied linearly over a 60° range as speed increased (Fig. 5c). Similarly, speed-dependent shifts were obtained for random dots, but not for full-field gratings, as is predicted from their respective power spectra (data not shown).

Quantitative modeling of the V1 population response to complex stimuli

The above account provides an intuitive sense of how frequency–space descriptions of the stimulus and cortical tuning properties can explain the empirical observations presented in Figs. 1–3 and 5. However, to formally test whether this model can account for our results, it is necessary to show that the magnitudes of the induced shifts in activity in a population of neurons with spatial and temporal tuning properties similar to those of ferret visual cortical neurons match those seen in our data. Two recent publications have addressed this question using frequency–domain models that compute the response of a bank of 3-D spatiotemporal frequency filters to texture stimuli that vary in line length, axis of motion, and speed (Baker and Issa, 2005; Mante and Carandini, 2005). Both analyses demonstrate shifts in the distribution of the active filters with changes in stimulus parameters, and the magnitude of these shifts is consistent with the experimental results described above.

Admittedly, one of the challenges of working in 3-D frequency space is the difficulty of visualizing relationships between stimuli and spatiotemporal receptive field envelopes, even when these relationships are depicted with great accuracy in static 2-D images (see Mante and Carandini, 2005).

128

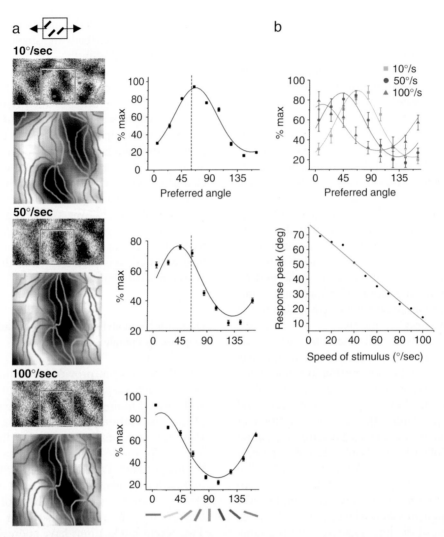

Fig. 5. Shifts in the population response induced by changes in stimulus speed. (a) Difference images, contour overlays (inset width = 1.3 mm), and population tuning functions obtained with 45° texture stimuli moving horizontally at three speeds: 10°/s (peak = 69°), 50°/s (peak = 42°), and 100°/s (peak = 14°). (b) Average population responses (±SEM) for three speeds (n = 7 animals); the peak shifts from 73° at low speed to 47° at intermediate speed to 17° at high speed. (c) The peak of best-fit Gaussian function to population tuning functions for textures drifting at 10 different speeds (10–100°/s at 10°/s intervals) are plotted against stimulus speed (R^2 of regression line = 0.99). (Reproduced with permission from Macmillan Magazines.)

In an attempt to acquire a better understanding of these relationships, we developed a program that calculates the 3-D power spectrum of a moving stimulus and the response of a set of realistic receptive field envelopes to that spectrum in real-time. This program (available at www.fitzpatricklab.net) can show interactively and in 3-D how changes to a stimulus' features such as line length, motion axis, or speed affect its power spectrum and how these changes in turn affect the activation of the receptive field envelopes.

The problem of understanding the population response to moving 2-D stimuli can also be approached in the space domain (as opposed to the frequency domain) using a Gabor function as a V1 receptive field model. (DeAngelis et al., 1993; Jones and Palmer, 1987). We modeled the population response using eight different Gabor functions with different orientations (as described in more detail below), each Gabor acting as a proxy for a cortical column with similar response properties. We computed the population response of our entire filter bank (consisting of eight filters) to a given stimulus and have reported the result as a "population tuning function" to be compared directly in its shape to the optically derived tuning functions shown earlier.

The stimulus was a single, drifting, white bar on a black background, constructed from MATLAB libraries written and made available by Prof. Eero Simoncelli of the Center for Neural Science, New York University. The bar orientation, direction of drift, length, and speed could be varied independently. The moving bar was represented as a 3-D matrix of luminance values (two spatial dimensions and one temporal dimension). The entire visual field was 128×128 pixels, while the bar length varied from 4 to 70 pixels. Bar widths were either 2 or 5 pixels. The length-to-width ratios (aspect ratios) were the same as those used in the physiological experiments (1:2, 1:4, and 1:10).

Receptive fields were constructed as follows. First, a family of sinusoidal functions (sine-wave gratings) of eight different orientations ($0°$–$180°$ at $22.5°$ intervals) was constructed. Next, each sinusoid was multiplied by a 2-D Gaussian function to produce Gabor filters of eight different orientations. Finally, the temporal aspect of the model receptive field was produced by multiplying the Gabor function with a temporal impulse function (adapted from Adelson and Bergen, 1985). The size of the receptive field was either 12 or 24 pixels (standard deviation of Gaussian envelope). The period of the sine wave was 30 (for 12 pixels SD) or 60 pixels (for 24 pixels SD). The receptive field size to bar size ratio was approximately the same as for the imaging experiments and the period of the sine wave was chosen to reproduce the relatively low spatial frequency-tuned responses in ferret V1. The orientation tuning bandwidth of these filters also agreed well with our experimentally observed bandwidths (see below).

The filter response was computed by multiplying the 2-D stimulus matrix with the 2D-Gabor receptive field matrix for each time step in the temporal impulse function. The response was rectified to mimic the purely positive-going spike output of a neuron's response to the drifting stimulus. The output for the entire duration of stimulus presentation was then summed (analogous to counting the number of spikes produced for the entire duration of the stimulus presentation) and normalized to the maximum response for a given length or speed; this output is referred to here as the "population tuning function".

We first tested the model response to a single bar moving in two different directions of motion. Figure 6a shows the model response for a $45°$ bar moving in two different directions of motion. The tuning functions were well fit by Gaussian functions. Moreover, the average tuning width (half-width of Gaussian function) for the model responses was $39.6°$, which is very close to the average physiological tuning width ($39.1°$) seen with texture stimuli. We quantified the shifts in the peak of the response by measuring the mean of the best-fit Gaussian. The $45°$ bar moving orthogonal to its orientation evoked the maximum response, as expected, in the $45°$-oriented Gabor (peak of the Gaussian = $40°$). When the bar direction of motion was changed from orthogonal to nonorthogonal (as shown in the stimulus icons), the model tuning changed as predicted, peaking at $76°$ for a $45°$ shift in direction anticlockwise ($0°$ motion axis), with the maximum responses being produced by the $90°$ Gabor filter. The magnitude of the peak shift ($36°$) is the same as that seen with imaging. This shows that shifts in population tuning obtained by simply changing the axis of motion of bar stimuli (without changing the orientation) can be reproduced by the rectified output of simple, linear filters.

We have seen earlier that the patterns of activity resulting from the presentation of an oriented grating can in fact be produced by a combination of a range of different orientation and axis of motion. Figure 6b shows the responses obtained from the model for the same combinations as were tested in Fig. 2. A comparison of the optically derived population tuning function with the model tuning function shows that the filter output is indeed similar for these different stimuli. Thus, the surprising behavior of the V1 population response becomes explicable if it is seen as resulting from

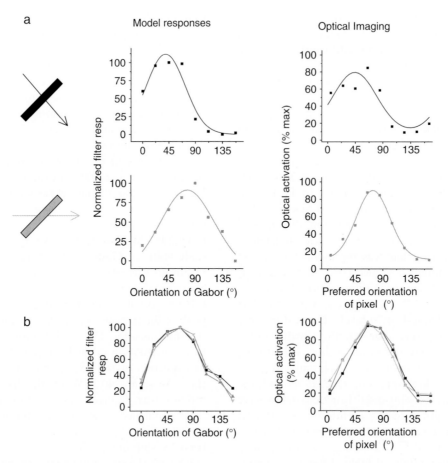

Fig. 6. Impact of varying axis of motion on filter response. (a) Filter responses to a single 45° bar moving along two different motion axes (as shown in icons). For orthogonal motion (top panel) the filter tuning curve peaks near 45° as expected from the bar orientation. Nonorthogonally moving bars elicit tuning shifts comparable to the shifts observed experimentally (reproduced from Fig. 1). The magnitudes of peak shifts for a 45° change in motion axis is 36°, which is the same as that seen with imaging (compare top and bottom panels). (b) Responses of the Gabor filter bank to the same three combinations of orientation and axis of motion shown in Fig. 2. The model was tested with single bars, while the optical data is for textures. Like the neural population (reproduced from Fig. 2), the filter bank responds indistinguishably to the three different combinations.

receptive fields possessing certain spatial and temporal tuning properties.

The model also succeeds in predicting the changes in the cortical patterns of activity that accompany changes in line length. The aspect ratios we tested were the same as the ones reported for the optical data (1:2, 1:4, and 1:10). Figure 7a shows the comparison between the optical data and the model response. Figure 7c plots the peaks of the best-fit Gaussian functions to optical tuning data against the same for the model tuning data. The regression line is a good fit to the data ($R^2 = 0.9$) and has a slope of 0.88 indicating that there is good agreement between the peak of the population response measured optically and the model output. However, there are also distinct departures between the experimental data and the model behavior. As is clear from the linear regression shown in Fig. 7c, the model exhibits a somewhat larger tuning shift than the optical data, and model tuning widths are usually larger for shorter stimuli (i.e., stimuli with broader spectra), for example, compare the 1:2 aspect ratio responses (black curves) in Fig. 7a.

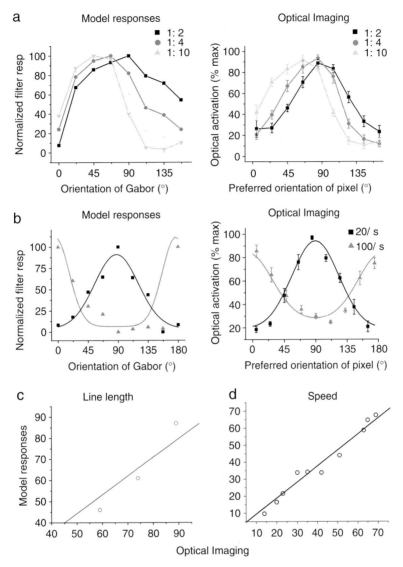

Fig. 7. Impact of change in stimulus length and speed on the filter response. (a) Change in filter tuning with change in length (from 1:2 to 1:10 aspect ratio). Compare this to the optically imaged change in tuning reproduced from Fig. 3. While there are differences in the tuning width between the model and the physiology (particularly at shorter lengths), the shift in the peak of response is comparable. (b) Responses of the model filter bank to a single dot moving at low and high speeds. Response inverts at high speed for the dot same as that observed for a dot field with imaging. (c) The regression between the peak of model response and the optically imaged response for the three different line lengths shown in (a) (R^2 of linear fit $= 0.9$ and slope $= 0.88$). (d) The regression between the peak of model response and the optically imaged response for the 10 different stimulus speeds shown in Fig. 5c (R^2 of linear fit $= 0.98$ and slope $= 0.94$).

These departures can be due to various cortical mechanisms such as recurrent excitation or inhibition that are not incorporated in the simple Gabor model.

Finally, we also verified that the speed-dependent changes in tuning could be reproduced by the Gabor filter bank. Figure 7b shows that tuning for dots changes with speed, as is predicted by the frequency–space framework outlined above. Figure 7d compares the shift in tuning for textures seen in the population response with the change in tuning of the filter bank to a single bar stimulus over the same range of speeds. The magnitude and direction of the shift are comparable to the population data ($\sim 60°$ for a 10-fold change in texture speed, R^2 of regression line $= 0.98$, and slope $= 0.94$).

An alternative framework: cortical maps in frequency space

Taken together, these observations provide a different perspective on the organization of functional maps in the primary visual cortex. The facts that multiple combinations of texture orientation and axis of motion can result in similar activation patterns, and that orientation-specific cortical patterns can be changed by changes in stimulus length and speed, are difficult to reconcile with the place code view that the intersection of the relevant feature maps can signal the presence of particular feature combinations. However, they can be easily accommodated within a spatiotemporal frequency framework, where the distribution of population activity satisfies the joint constraints imposed by the orderly mapping of receptive field preference for position in visual space, and the orderly mapping for receptive field preference for position in frequency space.

It has been argued that an independent mapping of spatial frequency preference is consistent with these results (Baker and Issa, 2005). It should be pointed out that the broadband stimuli employed in these experiments do not make it possible for us to explicitly address the existence of a separate columnar map of spatial frequency preference. However, additional experiments from our lab challenge this view, showing that activity patterns that have the appearance of a map of spatial frequency actually reflect a cardinal bias in the representation of high spatial frequencies: i.e., the patchy cortical activation patterns produced by high spatial frequencies coincide with regions of the cortex that respond preferentially to horizontal gratings (White et al., 2005). A full description of the map of preferred position in frequency space is currently under investigation.

As the modeling results emphasize, our observations of population response to 2-D stimuli should not have been a surprise given the extensive single unit analysis which supports the view that cortical neuron-receptive fields are best conceived as filters in frequency space rather than feature detectors (Movshon et al., 1978b; De Valois et al., 1979; Jones and Palmer, 1987; DeAngelis et al., 1993; Skottun et al., 1994; Carandini et al., 1999). Nevertheless, the majority of single unit studies that have explored the implications of the spatio-temporal frequency filter properties of V1 neurons for processing of motion information have focused on responses to 1-D stimuli, providing a clear description of the spatial and temporal tuning envelopes which predict 2-D tuning shifts, while not actually exploring the shifts themselves. Even among the studies that have explored tuning shifts with single unit recordings, there is considerable variation in the types of shifts that are reported, how prevalent these shifts are, and whether they are characteristic of all classes of cortical neurons (both simple and complex) (Hammond and MacKay, 1977; Hammond and Smith, 1983; Skottun et al., 1988; Crook et al., 1994).

But perhaps the failure to appreciate the implications of an energy perspective for patterns of population response lies less in the inconsistencies of the extracellular recording evidence gathered with 2-D stimuli than in the power and the simplicity of the prevailing view of how features are represented in cortical columns. It is intuitively satisfying to consider a specific pattern of columnar activity as the representation of a particular combination of visual features. A framework that specifically predicts similar patterns of activity for different visual stimuli, and offers little explanation for resolving such an ambiguity appears to pose more problems than it solves.

In this context, however, it is worth emphasizing that the interactions between speed, line length, and direction observed in the population response are consistent with studies showing similar interactions in perception. For example, the speed-dependent shifts in tuning described here could account for speed-dependent changes in a human observer's ability to detect the direction of a moving dot through oriented masks (Geisler, 1999).

At slow speeds, noise masks oriented parallel to the direction of motion of a fast-moving dot stimulus have no impact on dot detection, while masks oriented orthogonal to the direction of motion elevate detection thresholds. At higher speeds, the effects are reversed: parallel masks impair detection, while orthogonal masks do not. While these observations have been interpreted in the context of a "motion-streak" hypothesis (Geisler, 1999; Geisler et al., 2001; Burr and Ross, 2002), the effects are entirely consistent with the speed-dependent change in direction tuning for broadband stimuli that we see at the population level and that was first reported for single units by Hammond and Smith (1983).

Similarly, the observation that stimulus length is critical in determining the population response to moving stimuli has its counterpart in experiments by Lorenceau and colleagues showing that human observers systematically misjudge the direction of motion of a field of moving bars (similar to the texture stimuli used in our studies) as the bar length is altered (Lorenceau et al., 1993). Observers tend to perceive the veridical direction of motion of the pattern for shorter bar lengths, but their judgments are biased toward the direction orthogonal to the bar orientation for longer bar lengths. While these results have been explained in the context of a "contour-terminator" model of motion processing — a framework adopted by other studies that have probed responses to texture stimuli (Pack et al., 2001, 2003, 2004; Pack and Born, 2001) — it is equally well explained by the shifts in population response that accord with the frequency–space model described here.

In conclusion, our imaging results as well as the modeling and psychophysical evidence discussed here force us to revise our current notions of the functional architecture of visual cortex. While spatial coding schemes based on topological relationships between multiple feature maps are attractive, the actual behavior of a neural receptive field makes these schemes unlikely. Our results show that existing models of V1 which consider receptive fields as filters in spatiotemporal frequency space are better suited to explaining the patterns of population activity evoked by complex stimuli.

Acknowledgments

This work was supported by NEI Grant no. EY 11488.

References

Adelson, E.H. and Bergen, J.R. (1985) Spatiotemporal energy models for the perception of motion. J. Opt. Soc. Am. A, 2: 284–299.

Adelson, E.H. and Movshon, J.A. (1982) Phenomenal coherence of moving visual patterns. Nature, 300: 523–525.

Baker, T.I. and Issa, N.P. (2005) Cortical maps of separable tuning properties predict population responses to complex visual stimuli. J. Neurophysiol., 94: 775–787.

Basole, A., White, L.E. and Fitzpatrick, D. (2003) Mapping multiple features in the population response of visual cortex. Nature, 423: 986–990.

Burr, D.C. and Ross, J. (2002) Direct evidence that "speed-lines" influence motion mechanisms. J. Neurosci., 22: 8661–8664.

Carandini, M., Heeger, D.J. and Movshon, J.A. (1999) Linearity and gain control in V1 simple cells. In: Ulinski, P.S. (Ed.), Cerebral Cortex. New York, Kluwer Academic/Plenum, pp. 401–443.

Crook, J.M. (1990) Directional tuning of cells in area 18 of the feline visual cortex for visual noise, bar and spot stimuli: a comparison with area 17. Exp. Brain Res., 80: 545–561.

Crook, J.M., Worgotter, F. and Eysel, U.T. (1994) Velocity invariance of preferred axis of motion for single spot stimuli in simple cells of cat striate cortex. Exp. Brain Res., 102: 175–180.

DeAngelis, G.C., Ohzawa, I. and Freeman, R.D. (1993) Spatiotemporal organization of simple-cell receptive fields in the cat's striate cortex II. Linearity of temporal and spatial summation. J. Neurophysiol., 69: 1118–1135.

De Valois, K.K., De Valois, R.L. and Yund, E.W. (1979) Responses of striate cortex cells to grating and checkerboard patterns. J. Physiol., 291: 483–505.

Everson, R.M., Prashanth, A.K., Gabbay, M., Knight, B.W., Sirovich, L. and Kaplan, E. (1998) Representation of spatial frequency and orientation in the visual cortex. Proc. Natl. Acad. Sci. USA, 95: 8334–8338.

Geisler, W.S. (1999) Motion streaks provide a spatial code for motion direction. Nature, 400: 65–69.

Geisler, W.S., Albrecht, D.G., Crane, A.M. and Stern, L. (2001) Motion direction signals in the primary visual cortex of cat and monkey. Vis. Neurosci., 18: 501–516.

Hammond, P. and MacKay, D.M. (1977) Differential responsiveness of simple and complex cells in cat striate cortex to visual texture. Exp. Brain Res., 30: 275–296.

Hammond, P. and Smith, A.T. (1983) Directional tuning interactions between moving oriented and textured stimuli in complex cells of feline striate cortex. J. Physiol., 342: 35–49.

Hubel, D.H. and Wiesel, T.N. (1962) Receptive fields, binocular interaction and functional architecture in the cat's visual cortex. J. Physiol., 160: 106–154.

Hubel, D.H. and Wiesel, T.N. (1968) Receptive fields and functional architecture of monkey striate cortex. J. Physiol., 195: 215–243.

Hubel, D.H. and Wiesel, T.N. (1977) Ferrier lecture. Functional architecture of macaque monkey visual cortex. Proc. R. Soc. Lond. B Biol. Sci., 198: 1–59.

Hubener, M., Shoham, D., Grinvald, A. and Bonhoeffer, T. (1997) Spatial relationships among three columnar systems in cat area 17. J. Neurosci., 17: 9270–9284.

Issa, N.P., Trepel, C. and Stryker, M.P. (2000) Spatial frequency maps in cat visual cortex. J. Neurosci., 20: 8504–8514.

Jones, J.P. and Palmer, L.A. (1987) An evaluation of the two-dimensional Gabor filter model of simple receptive fields in cat striate cortex. J. Neurophysiol., 58: 1233–1258.

Lorenceau, J., Shiffrar, M., Wells, N. and Castet, E. (1993) Different motion sensitive units are involved in recovering the direction of moving lines. Vision Res., 33: 1207–1217.

Mante, V. and Carandini, M. (2005) Mapping of stimulus energy in primary visual cortex. J. Neurophysiol., 94: 788–798.

Mountcastle, V.B. (1957) Modality and topographic properties of single neurons of cat's somatic sensory cortex. J. Neurophysiol., 20: 408–434.

Movshon, J.A., Thompson, I.D. and Tolhurst, D.J. (1978a) Spatial and temporal contrast sensitivity of neurons in areas 17 and 18 of the cat's visual cortex. J. Physiol., 283: 101–120.

Movshon, J.A., Thompson, I.D. and Tolhurst, D.J. (1978b) Spatial summation in the receptive fields of simple cells in the cat's striate cortex. J. Physiol., 283: 53–77.

Pack, C.C., Berezovskii, V.K. and Born, R.T. (2001) Dynamic properties of neurons in cortical area MT in alert and anaesthetized macaque monkeys. Nature, 414: 905–908.

Pack, C.C. and Born, R.T. (2001) Temporal dynamics of a neural solution to the aperture problem in visual area MT of macaque brain. Nature, 409: 1040–1042.

Pack, C.C., Gartland, A.J. and Born, R.T. (2004) Integration of contour and terminator signals in visual area MT of alert macaque. J. Neurosci., 24: 3268–3280.

Pack, C.C., Livingstone, M.S., Duffy, K.R. and Born, R.T. (2003) End-stopping and the aperture problem: two-dimensional motion signals in macaque V1. Neuron, 39: 671–680.

Shmuel, A. and Grinvald, A. (1996) Functional organization for direction of motion and its relationship to orientation maps in cat area 18. J. Neurosci., 16: 6945–6964.

Shoham, D., Hubener, M., Schulze, S., Grinvald, A. and Bonhoeffer, T. (1997) Spatio-temporal frequency domains and their relation to cytochrome oxidase staining in cat visual cortex. Nature, 385: 529–533.

Simoncelli, E.P. and Heeger, D.J. (1998) A model of neuronal responses in visual area MT. Vision Res., 38: 743–761.

Skottun, B.C., Grosof, D.H. and De Valois, R.L. (1988) Responses of simple and complex cells to random dot patterns: a quantitative comparison. J. Neurophysiol., 59: 1719–1735.

Skottun, B.C., Zhang, J. and Grosof, D. (1994) On the directional selectivity of cells in the visual cortex to drifting dot patterns. Vis. Neurosci., 11: 885–897.

Swindale, N.V. (2000) How many maps are there in visual cortex? Cereb. Cortex, 10: 633–643.

Swindale, N.V., Shoham, D., Grinvald, A., Bonhoeffer, T. and Hubener, M. (2000) Visual cortex maps are optimized for uniform coverage. Nat. Neurosci., 3: 822–826.

Wallach, H. (1935) Uber visuell wahrgenommene Bewegungrichtung. Psycholog. Forsch., 20: 325–380.

Weliky, M., Bosking, W.H. and Fitzpatrick, D. (1996) A systematic map of direction preference in primary visual cortex. Nature, 379: 725–728.

White, L.E., Basole, A., Kreft-Kerekes, V. and Fitzpatrick, D. (2005) The mapping of spatial frequency in ferret visual cortex: relation to maps of visual space and orientation preference. SFN abstract, 508.14.

Wuerger, S., Shapley, R. and Rubin, N. (1996) On the visually perceived direction of motion by Hans Wallach, 60 years later. Perception, 25: 1317–1367.

Martinez-Conde, Macknik, Martinez, Alonso & Tse (Eds.)
Progress in Brain Research, Vol. 154
ISSN 0079-6123

CHAPTER 7

The sensitivity of primate STS neurons to walking sequences and to the degree of articulation in static images

Nick E. Barraclough[1,2], Dengke Xiao[1], Mike W. Oram[1] and David I. Perrett[1,*]

[1]*School of Psychology, St. Mary's College, University of St. Andrews, South Street, St. Andrews, Fife KY16 9JP, UK*
[2]*Department of Psychology, University of Hull, Hull HU6 7RX, UK*

Abstract: We readily use the form of human figures to determine if they are moving. Human figures that have arms and legs outstretched (articulated) appear to be moving more than figures where the arms and legs are near the body (standing). We tested whether neurons in the macaque monkey superior temporal sulcus (STS), a region known to be involved in processing social stimuli, were sensitive to the degree of articulation of a static human figure. Additionally, we tested sensitivity to the same stimuli within forward and backward walking sequences. We found that 57% of cells that responded to the static image of a human figure was also sensitive to the degree of articulation of the figure. Some cells displayed selective responses for articulated postures, while others (in equal numbers) displayed selective responses for standing postures. Cells selective for static images of articulated figures were more likely to respond to movies of walking forwards than walking backwards. Cells selective for static images of standing figures were more likely to respond to movies of walking backwards than forwards. An association between form sensitivity and walking sensitivity could be consistent with an interpretation that cell responses to articulated figures act as an implied motion signal.

Keywords: motion; implied motion; form; integration; temporal cortex; action

Introduction

Artists use many tricks to convey information about movement. One method commonly used is to illustrate a person with legs and arms outstretched or articulated as if the artist had captured a snapshot of the person mid-stride during walking or running. When we see such static images we commonly interpret the human as moving, walking or running forwards through the scene. Although no real movement occurs, the articulated human figure 'implies' movement forward by its

configuration or form. There is considerable evolutionary advantage in this ability to infer information about movement from the posture; we can interpret movement direction and speed from a momentary glimpse of a figure.

Traditionally, form and motion information have been thought to be processed along anatomically separate pathways; relatively little effort has been spent investigating how the pathways interact and how motion and form are integrated. Recently, however, three fMRI studies have shown that the brain structure that processes motion, hMT + /V5 (Zeki et al., 1991; Watson et al., 1993; Tootell et al., 1995), is more active to images implying motion when compared to similar images

*Corresponding author. Tel.: +44-1334-463044;
Fax: +44-1334-463042; E-mail: dp@st-andrews.ac.uk

DOI: 10.1016/S0079-6123(06)54007-5

where motion in not implied (Kourtzi and Kanwisher, 2000; Senior et al., 2000; Krekelberg et al., 2005). In each study very different images were used to imply motion; Kourtzi and Kanwisher used images of athletes and animals in action, Senior et al. used images of moving objects and Krekelberg et al. used 'glass patterns', i.e., arrangements of dots suggesting a path of motion. These papers all argue that information regarding the form of static images is made available to hMT + /V5 for coding motion.

Neurons in the monkey homologue of human hMT + /V5, the medial temporal (MT) and medial superior temporal (MST) areas, also respond to glass patterns, where motion is implied (Krekelberg et al., 2003). Areas MT and MST contain neurons that respond to motion (Dubner and Zeki, 1971; Desimone and Ungerleider, 1986) and respond in correlation with the monkey's perception of motion (Newsome et al., 1986; Newsome and Pare, 1988). Neurons in MT/MST area respond maximally to movement in one direction; Krekelberg et al. (2003) showed that they respond preferentially to both real dot motion and implied motion in the preferred direction. Presentation of contradictory implied motion and real motion results in a compromised MT/MST neural response and compromises the monkey's perception of coherent movement.

The blood-oxygen level-dependent (BOLD) activity seen in human hMT + /V5 to complex images implying motion (Kourtzi and Kanwisher, 2000; Senior et al., 2000) could be explained by input from other regions of the cortex. Measurement of event-related potentials (ERP) responses from a dipole pair in the occipital lobe, consistent with localization to hMT + /V5, showed that the responses to the real motion of a random-dot field were 100 ms earlier than responses to static images containing human figures implying motion (Lorteije et al., 2006). The delay in the implied motion response indicates that this information arrives via a different and longer pathway. Kourtzi and Kanwisher (2000) concluded that since inferring information about still images depends upon categorization and knowledge, this must be analysed elsewhere. The activation of hMT + /V5 by implied motion of body images could be due to

top-down influences. Senior et al. (2000) suggested that the activation they saw in hMT + /V5 is more likely due to processing of the form of the image in temporal cortex without the need for engagement of conceptual knowledge. At present, there is no evidence that cells in monkey MT are sensitive to articulated human figures implying motion despite active search (Jeanette Lorteije, personal communication).

Information about body posture and articulation in a human figure is likely to come from regions of the cortex that contain neurons sensitive to body form. The superior temporal sulcus (STS) in monkeys and the superior temporal gyrus (STG) and nearby cortex in humans is widely believed to be responsible for processing socially important information. Monkey STS contains neurons that respond to movement of human bodies (Bruce et al., 1981; Perrett et al., 1985), the form (view) of human bodies (Wachsmuth et al., 1994) and many appear to integrate motion and form to code walking direction (Oram and Perrett, 1996; Jellema et al., 2004). It is not known, however, if cells exist that are sensitive to the pattern of articulation that may differentiate postures associated with motion from those associated with standing still.

Giese and Poggio (2003) extended models of object recognition (Riesenhuber and Poggio, 1999, 2002) to generate a plausible feed-forward model of biological motion recognition. A critical postulate of Giese and Poggio's model is the existence of 'snapshot' neurons, neurons tuned to differing degrees of articulation of bodies. Giese and Poggio suggest that these neurons should be found in inferotemporal (IT) or STS cortex, and would feed-forward to neurons coding specific motion patterns, e.g., walking (Oram and Perrett, 1996; Jellema et al., 2004).

In this study we set out to investigate if neurons in temporal cortex can code the degree of articulation of a human figure. Video taping a person walking or running produces a series of stills capturing discrete moments in time. Some of these stills show the person in an articulated pose, others in less-articulated poses akin to standing still. We made use of such video footage in order to compare the responses of STS neurons to a human figure articulated and standing. Neurons in STS sensitive to non-walking articulated postures are also sensitive to actions leading to such postures (Jellema and Perrett, 2003). It is possible, however, to arrive at a posture from two different directions, by walking forwards, or by walking backwards, both movement directions are consistent with the same static form. We therefore used the video footage played forwards and backwards to investigate how form sensitivity was related to walking.

Following Giese and Poggio (2003) we hypothesized that STS neurons would discriminate articulated postures from standing postures. We also hypothesized that the ability to differentiate posture in static images would relate to sensitivity to motion type for the same neurons. To this end we explore the cells' sensitivity to images of static figures taken from video and movies containing the same images, played forward and in reverse. We also investigate the sensitivity to body view since cells sensitive to static and moving bodies exhibit viewpoint sensitivity (Perrett et al., 1991; Oram and Perrett, 1996).

Methods

Physiological subjects, recording and reconstruction techniques

One rhesus macaque, aged 9 years, was trained to sit in a primate chair with head restraint. Using standard techniques (Perrett et al., 1985), recording chambers were implanted over both hemispheres to enable electrode penetrations to reach the STS. Cells were recorded using tungsten microelectrodes inserted through the dura mater. The subject's eye position ($\pm 1°$) was monitored (IView, SMI, Germany). A Pentium IV PC with a Cambridge electronics CED 1401 interface running Spike 2 recorded eye position, spike arrival and stimulus on/offset times.

After each electrode penetration, X-ray photographs were taken coronally and para-sagitally. The positions of the tip of each electrode and its trajectory were measured with respect to the intra-aural plane and the skull's midline. Using the

distance of each recorded neuron along the penetration, a three-dimensional map of the position of the recorded cells was calculated. Coronal sections were taken at 1 mm intervals over the anterior–posterior extent of the recorded neurons. Alignment of sections with the X-ray co-ordinates of the recording sites was achieved using the location of microlesions and injection markers on the sections.

Stimuli and presentation

Stimuli consisted of four (16 bit colour) movies of a human walking and four images of the human in different poses. One movie (4326 ms duration) was made by filming (Panasonic, NV-DX110, 3CCD digital video camera) a human walking to the right across a room (walk right). Each individual frame of the movie was flipped horizontally to create a second movie of the human walking to the left (walk left). The frames of both of these movies were arranged in the reverse order to create two movies, one of the human walking to the right backwards (walk right backwards) and the second to a human walking to the left backwards (walk left backwards). There were thus two movies of compatible or forward walking (walk right, walk left) and two movies of incompatible or backward walking (walk right backwards, walk left backwards); two of these movies contained movement in the rightwards direction (walk right, walk right backwards) and two contained movement in the leftwards direction (walk left, walk left backwards).

Two frames from the walk right movie were selected, one when the human was in an articulated pose with legs and arms away from the body (articulated right) and one when the human appeared to be standing with legs and arms arranged vertically (standing right). In both frames the human was in the centre of the room and the time between the two poses was not more than 210 ms. Both frames were flipped horizontally to create two more images (articulated left and standing left). There were thus two images of an articulated human pose (articulated left, articulated right) and two images of a standing pose (standing left, standing right); two images contained a view of a human facing right (articulated right, standing right) and two images contained a view of a human facing left (articulated right, standing right).

Stimuli were stored on an Indigo2 Silicon Graphics workstation hard disk and presented centrally subtending $25° \times 20.5°$ on a black monitor screen (Sony GDM-20D11, resolution 25.7 pixels/deg, refresh rate 72 Hz), 57 cm from the subject. Movies were presented by rendering each frame of the movie on the screen in sequence, where each frame was presented for 42 ms. Occasionally, movies were presented in a shortened form (duration 1092 ms), where the earlier and later frames were removed from the sequence to show the human walking only across the centre of the room.

Testing procedure

Responses were isolated using standard techniques, and visualized using oscilloscopes. Responses were defined as arising from either single units or multiple units. Both are referred to hereafter as 'cells', 44% was multiple units. Pre-testing was performed with a search set of (on average 55) static images and movies of different objects, bodies and body parts previously shown to activate neurons in the STS (Foldiak et al., 2003; Barraclough et al., 2005). Within this search set were the four different movies of a human walking and four different static images of human forms. Initially, this screening set was used to test each cell with the images and movies presented in a pseudorandom sequence with a 500 ms inter-stimulus interval, where no stimulus was presented for the $n+1$ time until all had been presented n times. Presentation commenced when the subject fixated within $±3°$ of a yellow dot presented centrally on the screen for 500 ms. To allow for blinking, deviations outside the fixation window lasting < 100 ms were ignored. Fixation was rewarded with the delivery of fruit juice. Spikes were recorded during the period of fixation, if the subject looked away for longer than 100 ms, spike recording and presentation of stimuli stopped until the subject resumed fixation for > 500 ms. Responses to each stimulus in the screening set were displayed as online rastergrams and post-stimulus time histograms (PSTHs) aligned to stimulus onset. If after 4–6 trials the cell gave a substantial response to one of the four walking stimuli or four static human images as determined by observing the online PSTHs, the additional images and movies were removed and testing resumed. From this point, cell responses were saved to a hard disk for offline analysis.

Cell response analysis

Offline isolation of cells was performed using a template-matching procedure and principal components analysis (Spike2, CED, Cambridge, UK). Each cell's response to a stimulus in the experimental test set was calculated by aligning segments (duration > stimulus duration) in the continuous recording, on each occurrence of that particular stimulus (trials).

For each stimulus a PSTH was generated and a spike density function (SDF) calculated by summing across trials (bin size = 1 ms) and smoothing (Gaussian, $\sigma = 10$ ms). Background spontaneous activity (SA) was measured in the 250 ms period prior to stimulus onset. Response latencies to each stimulus were measured as the first 1 ms time bin, where the SDF exceeded 3 SD above the spontaneous activity for over 25 ms in the period following stimulus onset (Oram and Perrett, 1992; Edwards et al., 2003).

The response to each static image was measured within a 250 ms window starting at the stimulus response latency. The response to each walking movie was measured within a 500 ms window starting at the stimulus response latency. Subsequent analysis was performed if the cell's response to one of the stimuli was significantly (3 SD) above the spontaneous background activity.

For each cell showing a significant visual response, the responses to the static images were entered into a 2-way ANOVA [articulation (articulated, standing) by view (left, right) with trials as replicates]. Cells that showed a significant main effect of articulation ($p < 0.05$) or a significant interaction between articulation and view (PLSD post-hoc test, $p < 0.05$) were classified as sensitive to articulation. Cells that showed a significant main effect of view ($p<0.05$) or a significant interaction between articulation and view (PLSD post-hoc test, $p<0.05$) were classified as sensitive to view.

The responses to the walking stimuli were entered into a separate 2-way ANOVA [compatibility (forwards, backwards) by direction (left, right) with trials as replicates]. Cells that showed a significant main effect of compatibility ($p<0.05$) or a significant interaction between compatibility and direction (PLSD post-hoc test, $p<0.05$) were classified as sensitive to compatibility. Cells that showed a significant main effect of direction ($p<0.05$) or a significant interaction between compatibility and direction (PLSD post-hoc test, $p<0.05$) were classified as sensitive to direction.

We were also interested in the responses to the articulated human form as it occurred within the walking sequences; this was achieved by measuring responses within a 500 ms window centred around the point in time the articulated form frame occurred within each walking movie. The responses to the walking stimuli, measured in this fashion, were entered into a separate 2-way ANOVA [compatibility (forwards, backwards) by direction (left, right) with trials as replicates]. Cell responses were analysed and cells classified in an analogous fashion.

Results

We tested 55 cells that responded significantly (see methods) to either static images of a human figure or movies of a human walking.

Form-sensitive cells

Thirty-five of the 55 tested cells (64%) showed a significant response to at least one image of a human figure. The sensitivity to different images of a human figure was tested for each cell using ANOVA [articulation: (articulated or standing) view: (left or right)]. Twenty (57%) of the 35 cells responding to images of humans were sensitive to the degree of articulation of the human figure (see methods). Ten out of the twenty cells (50%)

responded significantly more to articulated human figures than standing human figures, while the remaining ten (50%) responded significantly more to standing human figures than articulated human figures. The mean response latency to the most effective stimulus for cells responding more to articulated figures was 83 ms (SEM 4.9 ms), significantly (t-test, $t_{[18]} = 2.11$, and $p = 0.049$) less than the mean response latency for cells responding more to standing figures, 111 ms (SEM 12.3 ms). For all other cells that did not differentiate between articulated and standing figures the mean response latency was 93 ms (SEM 13.7 ms). Figure 1 shows the responses of a multiple unit and a single unit sensitive to the degree of articulation of a human figure.

The middle row in Fig. 1 shows the responses of a multiple unit that responds significantly more to the human figure in an articulated pose than a standing pose. The mean response of the multiple unit to the articulated human facing right was 113.6 spikes/s, this decreased to 56.8 spikes/s when the human was standing facing right. The bottom row in Fig. 1 shows the response of a single unit that is also sensitive to the degree of articulation of a human figure, but prefers standing poses. The mean response to the articulated human facing right was 80.4 spikes/s, this increased to 107.6 spikes/s when the human was in a standing pose.

Of the 35 cells responding to a human form, 14 (40%) were sensitive to the direction the human was facing (view) (see methods). Seven out of 14 (50%) of these view-sensitive cells were also sensitive to the degree of articulation in the human figure. Figure 2 shows the responses of an example of a single unit that prefers articulated figures facing to the right (30.3 spikes/s). The responses to the articulated human facing left (19.8 spikes/s), standing human facing right (7.8 spikes/s) or standing human facing left (6.8 spikes/s) are significantly smaller.

Relationship between sensitivity to articulation and motion during walking

To test how the sensitivity to static images of human figures was related to sensitivity to movies of

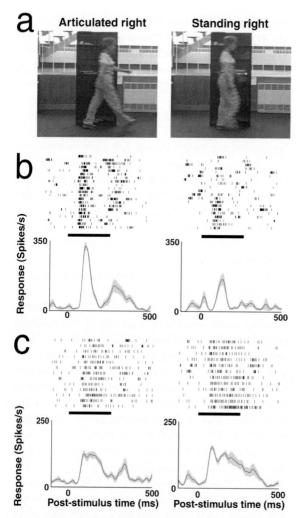

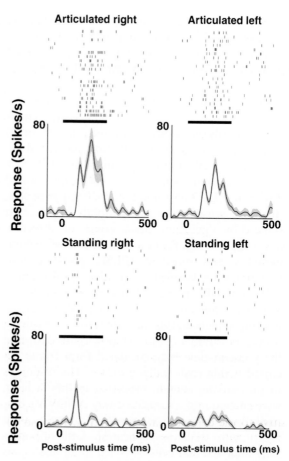

Fig. 2. Single cell responses to human figures articulated and standing, facing right and left. Rastergrams and SDFs plotted as in Fig. 1. The top row shows the responses to the articulated human figure facing right (trials = 21) and facing left (trials = 21). The bottom row shows the responses to the standing human figure facing right (trials = 21) and facing left (trials = 21). The responses to the articulated figures are greater than the responses to the standing figures, and the responses to the figure facing right are greater than the responses to the figure facing left (ANOVA: articulation $F_{[1,80]} = 40.214$, $p < 0.0001$, direction $F_{[1,80]} = 4.18$ and $p = 0.04$).

Fig. 1. Responses to human figures. (a) Grey-scale representations of the static visual images used to test responses. (b, c) Plots of responses of a multiple unit and a single unit to the images illustrated. The upper section of each plot shows individual trial responses as rastergrams, the lower section the spike density functions (SDFs) calculated from all trials (grey-SEM) and the black bar in between indicates the onset and duration of the stimulus. (b) Responses of a multiple unit to the image of an articulated body (trials = 21) and a standing body (trials = 20). The responses to articulated human forms were greater than the responses to standing human forms (ANOVA: articulation $F_{[1,77]} = 21.79$, $p < 0.0001$). (c) Responses of a single unit to the image of an articulated body (trials = 13) and a standing body (trials = 12). The response to the standing human form was greater than the response to the articulated human form (ANOVA: interaction articulation × view $F_{[1,45]} = 7.629$, $p = 0.0083$ and PLSD post-hoc test, $p < 0.05$).

walking, the sensitivity to different walking movies was calculated for all 35 cells showing a response to a static image of a human figure. We were interested in the responses to the key frames with articulated and non-articulated human figures as they occurred within a walking sequence; this was achieved by centring the 500 ms response measurement window around the point in time the articulated form frame occurred within each

Table 1. Test of sensitivity of cells to different movies of a human walking using ANOVA

Preferred static figure	Preferred walking			Total
	Compatible	Incompatible	Non-discriminative	
Articulated	5	1	4	10
Standing	0	6	4	10
Non-discriminative	2	3	10	15
Total	7	10	18	35

Association between cell sensitivity to body articulation in static images and sensitivity to type of walking movement. Compatible = body facing the direction of walking; incompatible = body facing away from the direction of walking. Cell tuning for the degree of articulation of static body images significantly predicted cell tuning for compatibility of walking.

movie. The sensitivity to different movies of a human walking [compatibility: (forwards or backwards) by direction: (left or right)] was tested for each cell using ANOVA (Table 1). Of the 20/35 cells sensitive to the degree of articulation in the static human figure, 12 (60%) were also sensitive to the compatibility of movement in the movies of the human walking (see methods). For the 10 cells that responded more to articulated human figures in static images, five responded significantly more to forward walking, one responded significantly more to backward walking and four were not sensitive to the compatibility of walking. For the 10 cells that responded more to standing human figures in static images, none responded significantly more to forward walking, six responded significantly more to backward walking and four were not sensitive to the compatibility of walking. For the 15/35 cells that were not sensitive to the degree of articulation in the human figures, two responded significantly more to forward walking, three responded significantly more to backward walking and 10 were not sensitive to the compatibility of walking. In the cells showing sensitivity to the degree of articulation in static images there was an association between the preferred degree of articulation for static images and the preferred compatibility of a walking human (Pearson $\chi^2 = 8.571$, df = 1, $p = 0.003$ and Fisher's exact test, $p = 0.015$).

Figure 3 illustrates the responses of a single unit to static images and walking movies. The cell responds significantly more to a static image of an articulated human figure facing to the right (25.6 spikes/s) than to a static image of a standing human figure facing right (16.6 spikes/s). The cell also responds significantly more to a movie of walking forwards to the right (26.8 spikes/s) than walking backwards to the right (14.0 spikes/s). The frame containing the articulated human form occurs within the walking forwards movie at 1092 ms and the walking backwards movie at 1218 ms. An increase in the response to the walking forwards movie can be seen both at the start of the movie and at the time of the occurrence of the articulated frame. No similar increase in the response to the walking backward movie can be seen at the start or at the time when the articulated frame occurs.

Figure 4 illustrates an equivalent relationship between responses to standing human figures and walking backwards movies. The cell responds more to a static image of standing human figure facing to the left (23.7 spikes/s) than a static image of an articulated human figure facing left (11.5 spikes/s). The cell also responds more to a movie of walking backwards to the right (22.2 spikes/s) than to a movie of walking forwards to the left (15.7 spikes/s). The frame containing the articulated human form (in this case, to the left) occurs within both the walking movies (walking to the left after 630 ms and walking backwards to the right after 714 ms).

Illustrated in Fig. 4b are the responses of the cell to the static images and walking movies in the other directions. The left-hand column shows the responses to a standing human form facing right (2.4 spikes/s) and to an articulated form facing right (5.0 spikes/s), the right column shows the responses to walking backwards to the left (5.6 spikes/s) and walking forwards to the right

142

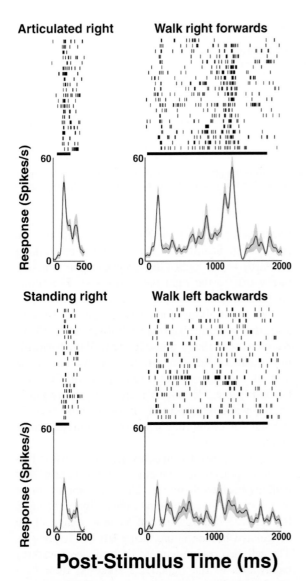

Post-Stimulus Time (ms)

Fig. 3. Responses of a single STS cell that prefers articulated human figures and walking forwards. Rastergrams and SDFs are plotted as in Fig. 1. The left column shows example of responses to images. Top: responses to an articulated human facing right (trials = 21) and bottom: responses to a standing human facing right (trials = 21). The cell responds more to articulated human figures than to standing human figures (ANOVA: articulation, $F_{[1,80]} = 7.472$ and $p = 0.0077$). The right column shows example of responses to movies of walking. Top: responses to walking right forwards (trials = 21) and bottom: responses to walking left backwards (trials = 20). The cell prefers compatible walking to incompatible walking (ANOVA: compatibility, $F_{[1,78]} = 4.365$, $p = 0.039$, interaction compatibility × direction, $F_{[1,78]} = 20.982$, $p < 0.0001$, PLSD post-hoc test, $p < 0.05$).

(3.2 spikes/s). The cell responds significantly more to the static image of a human standing facing left than any other static image and responds significantly more to the movie of a human walking backwards to the right than any other movie. The illustrated cell is thus additionally sensitive to the view of the human figure in the static images and the walking direction in the walking movies.

Cells sensitive to walking

Having established that a cell's sensitivity to the degree of articulation in a static image of a human form can predict the cell's sensitivity to the compatibility of walking stimuli, we wanted to know if the inverse was true. Could knowing the sensitivity to the compatibility of a walking human be used to predict sensitivity to static images of human figures? We measured the responses to the four walking stimuli in a 500 ms window starting at the stimulus response latencies.

Fifty-two of 55 tested cells (95%) showed a significant response to at least one movie of a human walking. The sensitivity to different walking directions [compatibility: (forwards or backwards), direction: (left or right)] was tested in each cell using ANOVA (Table 2). For 19 (36%) of the 52 cells the compatibility of walking was a significant factor (see methods). Seven out of the 19 cells (37%) responded more to walking forwards and 12 out of 19 cells (63%) responded more to walking backwards.

In the 19 cells sensitive to the compatibility of walking, 12 (63%) were also sensitive to the degree of articulation in the static image of the human figure. For the seven cells that responded more to walking forwards, four responded significantly more to articulated figures, none responded significantly more to standing figures in static images and three were insensitive to the degree of articulation of the static human figure. For the 12 cells that responded more to walking backwards, one responded significantly more to articulated figures, four responded significantly more to standing figures in static images and seven were insensitive to the degree of articulation of the

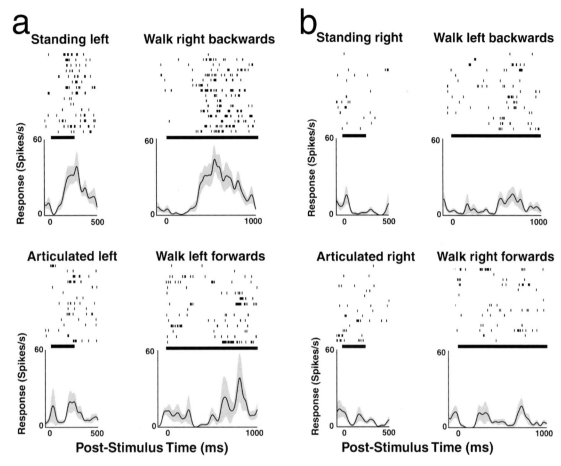

Fig. 4. Responses of a single STS cell to static images and movies of a human walking. Rastergrams and SDFs plotted as in Fig. 1. (a) The left column shows responses to static images, top: responses to a standing human facing left (trials = 15) and bottom: responses to articulated human facing left (trials = 15). The cell responds more to standing figures than articulated figures (ANOVA: articulation, $F_{[1,57]} = 5.030$, $p = 0.0288$, interaction articulation $\times$ view, $F_{[1,57]} = 10.950$, $p = 0.0016$, PLSD post-hoc test, $p < 0.05$). The right column shows the responses to movies of walking, top: responses to walking right backwards (trials = 16) and bottom: responses to walking left forwards (trials = 15). The cell responds more to incompatible walking to the right more than compatible walking to the left (ANOVA: interaction compatibility $\times$ direction, $F_{[1,57]} = 19.772$, $p < 0.00001$, PLSD post-hoc test, $p < 0.05$). (b) The left column shows responses to static images, top: responses to a standing human facing right (trials = 15) and bottom: responses to an articulated human facing right (trials = 16). The cell responds less to figures facing right than to the left (ANOVA: view, $F_{[1,57]} = 39.620$, $p < 0.001$), and responds more to a standing figure facing left than any other static image (ANOVA: interaction articulation $\times$ view, $F_{[1,57]} = 10.950$, $p = 0.0016$, PLSD post-hoc test, $p < 0.05$ each comparison). The right column shows the responses to movies of walking, top: responses to walking left backwards (trials = 15) and bottom: responses to walking right forwards (trials = 15). The cell responds more to incompatible walking to the right than any other walking stimulus (ANOVA: interaction compatibility $\times$ direction, $F_{[1,57]} = 19.772$, $p < 0.0001$, PLSD post-hoc test, $p < 0.05$ for each comparison).

static human figure. For the 33/52 cells that were not sensitive to the compatibility of the walking stimuli, 4 responded significantly more to articulated human figures, 5 responded significantly more to standing human figures and 24 were insensitive to the degree of articulation of the static human figure. In the cells showing sensitivity to the compatibility of walking, there was an association between the preferred compatibility and the preferred degree of articulation of a human figure (Pearson $\chi^2 = 5.76$, df = 1 and $p = 0.016$) (Fisher's exact test, $p = 0.048$).

Table 2. Test of sensitivity of cells to different walking directions using ANOVA

Preferred walking stimulus	Preferred static figure			Total
	Articulated	Standing	Non-discriminative	
Compatible	4	0	3	7
Incompatible	1	4	7	12
Non-discriminative	4	5	24	33
Total	9	9	34	52

Association between cell sensitivity to type of walking movement and degree of body articulation in static images. Compatible = body facing the direction of walking; incompatible = body facing away from the direction of walking. Cell tuning for compatibility of walking significantly predicted cell tuning in the degree of articulation of static body images.

Histological localization

Cells showing sensitivity to static images of human forms and movies of a human walking were found in the target area of the upper bank, lower bank and fundus of rostral STS and inferotemporal cortex. As defined in previous studies (Desimone and Gross, 1979; Bruce et al., 1981, 1986; Baylis et al., 1987; Hikosaka et al., 1988; Distler et al., 1993; Seltzer and Pandya, 1994; Saleem et al., 2000), rostral STS is the region of cortex in the upper bank (TAa, TPO), lower bank (TEa, TEm) and fundus (PGa, IPa) of the STS that lies rostral to the fundus of the superior temporal area (FST). The anterior–posterior extent of the recorded cells was from 6.9 to 10.5 mm posterior of the anterior commissure. We saw no apparent concentration of cells showing sensitivity to one figure or walking type within sub-regions of the STS. Figure 5 shows the position of all neurons that responded to at least one of the stimuli tested.

Discussion

The results of this study show two main findings: (1) Fifty-seven per cent of STS neurons that respond to static images of a human figure are sensitive to the degree of articulation of the figure itself. (2) There is an association between STS neuronal response sensitivity to the degree of articulation of a human figure and sensitivity to the compatibility between the direction of locomotion and view of the body of a human walking. For the cells that were sensitive to both the degree of articulation of a human figure in a static image and compatibility of walking: cells that 'preferred' articulated human figures 'preferred' compatible walking (walking forwards), cells that 'preferred' standing human figures 'preferred' incompatible walking (walking backwards).

STS neurons have been known to be sensitive to the form of faces (Bruce et al., 1981; Perrett et al., 1982), body parts (Perrett et al., 1989) and whole bodies (Wachsmuth et al., 1994) for a significant time. The cells sensitive to the articulation of a human figure described here are a novel subset of STS neurons that code the form of whole bodies. A previous study of STS cell responses to whole bodies described cells that were sensitive to the left or right view of the body (Wachsmuth et al., 1994). In this study many cells selectively responded to one articulation type irrespective of the view of the figure, others, however, were additionally sensitive to the view of the body.

The model of Giese and Poggio (2003) describing a feed-forward model of motion recognition was reliant upon the existence of a subset of neurons in IT or STS responding selectively to different 'snapshots' of a human walking. The neurons we describe here might potentially represent snapshots. Giese and Poggio's model, however, used 21 types of snapshot neuron, tuned to 21 different degrees of articulation of the human figure. We demonstrate only two types of snapshots, articulated and standing, we did not investigate intermediate poses. Further fine-grained analysis of the number of prototypical poses would be needed to reveal full supporting physiological evidence for Giese and Poggio's model. The number of

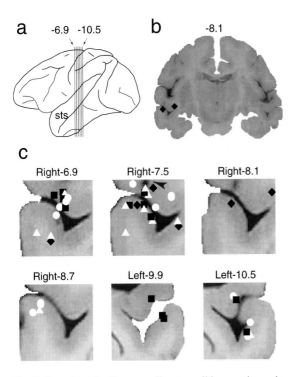

Fig. 5. Location of cells responding to walking movies or images of human poses. (a) Positions along the STS illustrated on a schematic representation of the brain of the six sections shown in (c). (b) Photograph of the section at −8.1 mm (posterior from the anterior commissure — ac). (c) Photographs of six sections (−6.9, −7.5, −8.1, −8.7, −9.9, −10.5) cropped and enlarged to illustrate the right and left STSs where cells were recorded. In order to illustrate the grey matter–white matter boundary, the contrasts of the photographs have been enhanced using Adobe Photoshop [The 8 bit contrast range from 91 to 177 was increased to the range from 0 to 255]. All 55 cells responding to at least one visual stimulus are plotted, where: cells sensitive to the articulation of a human figure in static images and compatibility during walking (white circles), cells sensitive to the articulation of a human figure (white triangles), cells sensitive to compatibility during walking (black squares), cells responsive to stimuli but not sensitive to articulation of a human figure or compatibility during walking (black diamonds). The black arrow indicates the position of a marker lesion used during the reconstruction.

prototypical poses could be <21, as acknowledged by Giese and Poggio, and depending upon the tuning to the degree of articulation of the human figure. It is worth noting that the STS neurons code all head views in the horizontal plane, but show a biased distribution for prototypical head views (namely face, left and right profile, and

back), and broad tuning — 60% bandwidth (Perrett et al., 1991). Several computational models (Ullman, 1989) use interpolation between key templates to 'recognize' intermediate views. A similar mechanism may interpolate between a small number of key postures.

Neurons sensitive to the degree of articulation of a human figure might be used in coding walking movements of other agents when visual information is degraded. 'Biological motion' stimuli, consisting of points of light attached to a human in the dark, result in a vivid perception of a walking figure where no form information is available to the visual system (Johansson, 1973). The conventional interpretation is that multiple local motion vectors for each point of light are first calculated within a motion processing system before integration to generate a form signal and the perception of a particular type of action. In a recent study, the position of each light dot on the human walking figure was moved to a different point on the figure between each frame of the biological motion movie (Beintema and Lappe, 2002). Perception of the form of the walking human was similar to the conventional biological motion stimulus even though local motion vectors could not be used to generate a global biological motion signal. Beintema and Lappe suggest that templates of different articulated human figures might be used to interpret these biological motion stimuli. Interestingly, some cells recorded in the STS that are sensitive to biological motion stimuli also respond entirely to static body views (unpublished observations). It remains to be seen if the cells recorded here are sensitive to biological motion stimuli and thus represent the templates proposed by Beintema and Lappe.

Similar to the finding of Jellema and Perrett (2003) we found an association between sensitivity to static forms and sensitivity to articulating actions. Jellema and Perrett's study, however, used actors moving and posing, thereby creating 3-D articulation between the head and trunk and between the upper and lower body. The association between sensitivity to static forms and articulated actions in this study were shown with video stimuli, in 2-D, and articulation was between the limbs and body in the cyclic action of walking.

This association shows how these cells code walking from images of human figures in the absence of motion information. Forward walking and backward walking, coded by STS neurons (Oram and Perrett, 1994, 1996), are two different meaningful actions. It is important for survival and social interaction to be able to interpret if a predator/prey/friend is approaching or backing away. You would gain a considerable advantage from detecting a predator approaching despite being afforded only a brief glimpse. During walking forwards and walking backwards, however, a figure cycles through the same repertoire of postures, only in a different sequence (the kinematics will be slightly different). It is possible that a glimpse of an articulated posture could arise from a figure walking either forwards or backwards. We however interpret articulated figures as walking forwards (Pavlova et al., 2002). Coding of forward walking in cells that respond significantly to articulated figures is consistent with this bias in our perception.

It is slightly more difficult to interpret the association between the sensitivity for standing postures and walking backwards. One approach is to regard the mutually exclusive postures and behaviours as opposite to each other. A standing figure cannot also be adopting an articulated posture, and we consistently interpret standing figures as not moving (unlike articulated figures). An agent backing away cannot also be approaching, since approaching and retreating behaviours involve opposite directions of motion. Increased responses to backward walking for cells that are selective for standing postures is consistent with this association and might represent an opponent population of cells to those that code forward walking and articulated postures.

Human cortex contains hMT+/V5 and posterior STS/superior temporal gyrus (STG), which are homologous regions to monkey MT and STS, respectively. Assuming the presence of similar cells in both species, the response sensitivity described here could explain activity to images implying motion in human hMT+/V5 (Kourtzi and Kanwisher, 2000), but less activity in STS. Images implying motion would activate STS cells that are tuned to articulated postures, and this information could be relayed to area hMT+/V5. The presence of an equal number of cells tuned to both articulated and standing postures in the STS result in a reduced net activation of this region with the contrast (implied motion — static) used by Kourtzi and Kanwisher to detect a response to implied motion.

One aspect we chose not to investigate here was the relationship between the direction the human figure faced in the static images (view) and the direction of movement during walking. This action was taken as our analysis consisted of repeatedly classifying responses and subdividing into separate cell populations. With an initial population of 55 cells, a further subdivision of cells into groups sensitive to each possible permutation of articulation and view would make each sub-group too small for meaningful statistics. Figure 4, however, illustrates the responses of a cell that could not only code motion (or in this cell's case, absence of motion), but also direction. The cell responded more to the static image of a human figure facing to the left; static images of humans facing to the right were ineffectual. The sensitivity to static images was consistent with the sensitivity to walking stimuli in only one direction. Thus, the responses of this cell can be interpreted as coding 'not moving', or 'not moving to the left'. More extensive studies of larger populations of neurons will enable analyses to determine if the neurons we describe here can signal motion or absence of motion in specific directions from static images.

Acknowledgments

This work was funded by grants from the EU and the Wellcome Trust.

References

Barraclough, N.E., Xiao, D.-K., Oram, M.W. and Perrett, D.I. (2005) Integration of visual and auditory information by STS neurons responsive to the sight of actions. J. Cogn. Neurosci., 17: 377–391.

Baylis, G.C., Rolls, E.T. and Leonard, C.M. (1987) Functional subdivisions of the temporal neocortex. J. Neurosci., 7: 330–342.

Beintema, J.A. and Lappe, M. (2002) Perception of biological motion without local image motion. Proc. Natl. Acad. Sci. USA, 99: 5661–5663.

Bruce, C.J., Desimone, R. and Gross, C.G. (1981) Visual properties of neurons in a polysensory area in superior temporal sulcus of the macaque. J. Neurophysiol., 46: 369–384.

Bruce, C.J., Desimone, R. and Gross, C.G. (1986) Both striate cortex and superior colliculus contribute to visual properties of neurons in superior temporal polysensory area of macaque monkey. J. Neurophysiol., 55: 1057–1075.

Desimone, R. and Gross, C.G. (1979) Visual areas in the temporal cortex of the macaque. Brain Res., 178: 363–380.

Desimone, R. and Ungerleider, L.G. (1986) Multiple visual areas in the caudal superior temporal sulcus of the macaque. J. Comp. Neurol., 248: 164–189.

Distler, C., Boussaoud, D., Desimone, R. and Ungerleider, L.G. (1993) Cortical connections of inferior temporal area TEO in macaque monkeys. J. Comp. Neurol., 334: 125–150.

Dubner, R. and Zeki, S.M. (1971) Response properties and receptive fields of cells in an anatomically defined region of the superior temporal sulcus in the monkey. Brain Res., 35: 528–532.

Edwards, R., Xiao, D.-K., Keysers, C., Foldiak, P. and Perrett, D.I. (2003) Color sensitivity of cells responsive to complex stimuli in the temporal cortex. J. Neurophysiol., 90: 1245–1256.

Foldiak, P., Xiao, D.-K., Keysers, C., Edwards, R. and Perrett, D.I. (2003) Rapid serial visual presentation for the determination of neural selectivity in area STSa. Prog. Brain Res., 144: 107–116.

Giese, M.A. and Poggio, T. (2003) Neural mechanisms for the recognition of biological movements. Nat. Rev. Neurosci., 4: 179–192.

Hikosaka, K., Iwai, E., Saito, H. and Tanaka, K. (1988) Polysensory properties of neurons in the anterior bank of the caudal superior temporal sulcus of the macaque monkey. J. Neurophysiol., 60: 1615–1637.

Jellema, T., Maassen, G. and Perrett, D.I. (2004) Single cell integration of animate form, motion and location in the superior temporal cortex of the macaque monkey. Cereb. Cortex, 14: 781–790.

Jellema, T. and Perrett, D.I. (2003) Cells in monkey STS responsive to articulated body motions and consequent static posture: a case of implied motion. Neuropsychologia, 41: 1728–1737.

Johansson, G. (1973) Visual perception of biological motion and a model for its analysis. Percept. Psychophys., 14: 201–211.

Kourtzi, Z. and Kanwisher, N. (2000) Activation in human MT/MST by static images with implied motion. J. Cogn. Neurosci., 12: 48–55.

Krekelberg, B., Dannenberg, S., Hoffmann, K.-P., Bremmer, F. and Ross, J. (2003) Neural correlates of implied motion. Nature, 424: 674–677.

Krekelberg, B., Vatakis, A. and Kourtzi, Z. (2005) Implied motion from form in human visual cortex. J. Neurophysiol., 94(6): 4373–4386.

Lorteije, J.A.M., Kenemans, J.L., Jellema, T., van der Lubbe, R.H.J., de Heer, F. and van Wezel, R.J.A. (2006) Delayed response to animate implied motion in human motion processing areas. J. Cogn. Neurosci., 18: 158–168.

Newsome, W.T., Mikami, A. and Wurtz, R.H. (1986) Motion selectivity in macaque visual cortex. III. Psychophysics and physiology of apparent motion. J. Neurophysiol., 55: 1340–1351.

Newsome, W.T. and Pare, E.B. (1988) A selective impairment of motion perception following lesions of the middle temporal visual area (MT). J. Neurosci., 8: 2201–2211.

Oram, M.W. and Perrett, D.I. (1992) Time course of neural responses discriminating views of the face and head. J. Neurophysiol., 68: 70–84.

Oram, M.W. and Perrett, D.I. (1994) Responses of anterior superior temporal polysensory (STPa) neurons to "biological motion" stimuli. J. Cogn. Neurosci., 6: 99–116.

Oram, M.W. and Perrett, D.I. (1996) Integration of form and motion in the anterior superior temporal polysensory area (STPa) of the macaque monkey. J. Neurophysiol., 76: 109–129.

Pavlova, M., Krageloh-Mann, I., Birbaumer, N. and Sokolov, A. (2002) Biological motion shown backwards: the apparent-facing effect. Perception, 31: 435–443.

Perrett, D.I., Harries, M.H., Bevan, R., Thomas, S., Benson, P.J., Mistlin, A.J., Chitty, A.J., Hietanen, J.K. and Ortega, J.E. (1989) Frameworks of analysis for the neural representation of animate objects and actions. J. Exp. Biol., 146: 87–113.

Perrett, D.I., Oram, M.W., Harries, M.H., Bevan, R., Hietanen, J.K., Benson, P.J. and Thomas, S. (1991) Viewer-centred and object-centred coding of heads in the macaque temporal cortex. Exp. Brain Res., 86: 159–173.

Perrett, D.I., Rolls, E.T. and Caan, W. (1982) Visual neurons sensitive to faces in the monkey temporal cortex. Exp. Brain Res., 47: 329–342.

Perrett, D.I., Smith, P.A.J., Mistlin, A.J., Chitty, A.J., Head, A.S., Potter, D.D., Broennimann, R., Milner, A.D. and Jeeves, M.A. (1985) Visual analysis of body movements by neurons in the temporal cortex of the macaque monkey: a preliminary report. Behav. Brain Res., 16: 153–170.

Riesenhuber, M. and Poggio, T. (1999) Hierarchical models of object recognition in cortex. Nat. Neurosci., 2: 1019–1025.

Riesenhuber, M. and Poggio, T. (2002) Neural mechanisms of object recognition. Curr. Opin. Neurobiol., 12: 162–168.

Saleem, K.S., Suzuki, W., Tanaka, K. and Hashikawa, T. (2000) Connections between anterior inferotemporal cortex and superior temporal sulcus regions in the macaque monkey. J. Neurosci., 20: 5083–5101.

Seltzer, B. and Pandya, D.N. (1994) Parietal, temporal and occipital projections to cortex of the superior temporal sulcus in the rhesus monkey: a retrograde tracer study. J. Comp. Neurol., 15: 445–463.

Senior, C., Barnes, J., Giampietro, V., Simmons, A., Bullmore, E.T., Brammer, M. and David, A.S. (2000) The functional neuroanatomy of implicit-motion perception 'representational momentum'. Curr. Biol., 10: 16–22.

Tootell, R.B., Reppas, J.B., Kwong, K.K., Malach, R., Born, R.T., Brady, T.J., Rosen, B.R. and Belliveau, J.W. (1995) Functional analysis of human MT and related visual cortical areas using magnetic resonance imaging. J. Neurosci., 15: 3215–3230.

Ullman, S. (1989) Aligning pictorial descriptions: An approach to object recognition. Cognition, 32: 193–254.

148

Wachsmuth, E., Oram, M.W. and Perrett, D.I. (1994) Recognition of objects and their component parts: Responses of single units in the temporal cortex of the macaque. Cereb. Cortex, 4: 509–522.

Watson, J.D., Myers, R., Frackowiak, R.S., Hajnal, J.V., Woods, R.P., Mazziotta, J.C., Shipp, S. and Zeki, S. (1993) Area V5 of the human brain: evidence from a combined study using positron emission tomography and magnetic resonance imaging. Cereb. Cortex, 3: 79–94.

Zeki, S., Watson, J.D., Lueck, C.J., Friston, K.J., Kennard, C. and Frackowiak, R.S. (1991) A direct demonstration of functional specialization in human visual cortex. J. Neurosci., 11: 641–649.

Eye Movements and Perception During Visual Fixation

Introduction

Our visual system contains a built-in contradiction: when we fixate our gaze on an object of interest, our eyes are never still. Instead we produce, several times each second, small eye movements of which we are unaware, called "microsaccades," "drifts" and "tremor". Microsaccades are miniature saccades produced during fixation, drifts are slow curvy motions that occur between microsaccades and tremor is a very fast, extremely small oscillation of the eye superimposed on drifts.

If we eliminate all these eye movements in the laboratory (using any number of retinal stabilization techniques), our visual perception of stationary objects fades, owing to neural adaptation. Since we fixate our gaze about 80% of the time during visual exploration, fixational eye movements are often responsible for driving most of our visual experience. When our eyes move across the image once again, after having stabilized the retinas, visual perception reappears. Owing to their role in counteracting adaptation, fixational eye movements are an important tool to understanding how the brain makes our environment visible. Moreover, because we are not aware of these eye movements, they can also help us understand the underpinnings of visual awareness.

Research in fixational eye movements bridges the gap between visual and oculomotor approaches and has gained a lot of interest over the last decade. The four chapters in this section discuss a wide spectrum of discoveries concerning perception, neural responses and oculomotor control during visual fixation.

Susana Martinez-Conde's chapter reviews the role of fixational eye movements in neural coding, in normal human perception and in visual and oculomotor disease. Ralf Engbert addresses the role of attention in modulating microsaccades as well as the importance of microsaccades for oculomotor control. Ikuya Murakami discusses visual motion illusions that arise when our visual system fails to compensate for fixational eye movements. Jose-María Delgado-García and colleagues present recent discoveries concerning the mechanisms underlying the generation of sustained activity, necessary for eye fixation, in neurons of the prepositus hypoglossi nucleus.

Susana Martinez-Conde

Martinez-Conde, Macknik, Martinez, Alonso & Tse (Eds.)
Progress in Brain Research, Vol. 154
ISSN 0079-6123

CHAPTER 8

Fixational eye movements in normal and pathological vision

Susana Martinez-Conde*

Department of Neurobiology, Barrow Neurological Institute, 350 W Thomas Road, Phoenix, AZ 85013, USA

Abstract: Most of our visual experience is driven by the eye movements we produce while we fixate our gaze. In a sense, our visual system thus has a built-in contradiction: when we direct our gaze at an object of interest, our eyes are never still. Therefore the perception, physiology, and computational modeling of fixational eye movements is critical to our understanding of vision in general, and also to the understanding of the neural computations that work to overcome neural adaptation in normal subjects as well as in clinical patients. Moreover, because we are not aware of our fixational eye movements, they can also help us understand the underpinnings of visual awareness. Research in the field of fixational eye movements faded in importance for several decades during the late 20th century. However, new electrophysiological and psychophysical data have now rejuvenated the field. The last decade has brought significant advances to our understanding of the neuronal and perceptual effects of fixational eye movements, with crucial implications for neural coding, visual awareness, and perception in normal and pathological vision. This chapter will review the type of neural activity generated by fixational eye movements at different levels in the visual system, as well as the importance of fixational eye movements for visual perception in normal vision and in visual disease. Special attention will be given to microsaccades, the fastest and largest type of fixational eye movement.

Fixational eye movements in normal vision

Eye movements during fixation are necessary to overcome loss of vision due to adaptive neural mechanisms that normalize responses across neurons in the face of unchanging or uniform visual stimulation. Thus, the goal of oculomotor fixational mechanisms may not be retinal stabilization, but rather controlled image motion adjusted so as to overcome adaptation in an optimal fashion for visual processing (Skavenski et al., 1979). In the early 1950s, it was shown that all eye movements could be eliminated in the laboratory, causing visual perception to fade to a homogeneous field (Ditchburn and Ginsborg, 1952; Riggs and Ratliff,

1952; Yarbus, 1967). Although this may seem counterintuitive at first, it is a common experience in all sensory modalities: we do not generally notice that our shoes are on for 16 hours a day. When the eyes were released from artificial stabilization, or if the stabilized image was changed, visual perception reappeared (Krauskopf, 1957; Ditchburn et al., 1959; Gerrits and Vendrik, 1970; Sharpe, 1972; Drysdale, 1975), just as, if we wiggle our toes, we once again notice that our shoes are on. Coppola and Purves (1996) found that images of entoptic vascular shadows (which are very stable) disappear in as little as 80 ms, suggesting that normal visual processing entails a rapid mechanism for image creation and erasure.

Even though retinal stabilization is most easily achieved under laboratory conditions, fading of objects in our visual periphery occurs quite often in normal vision: we are usually unaware of this.

---

*Corresponding author. Tel.: + 1-602-406-3484;
Fax: + 1-602-406-4172; E-mail: smart@neuralcorrelate.com

DOI: 10.1016/S0079-6123(06)54008-7

152

Peripheral fading of stationary objects was first noticed by Troxler (see Fig. 3A). Troxler (1804) reported that, under voluntary fixation, stationary objects in the periphery of vision tend to fade out and disappear. In the late 1950s, Clarke related Troxler fading to the fading of stabilized images in the laboratory (Ditchburn and Ginsborg, 1952; Riggs and Ratliff, 1952), and was the first to attribute both phenomena to neural adaptation (Clarke, 1957, 1960, 1961; Clarke and Belcher, 1962).

The three main types of fixational eye movements are tremor, drift and microsaccades. See Tables 1–4 of Martinez-Conde et al. (2004) for a review of fixational eye movement parameters in humans and primates.

Overall fixation stability does not appear to be affected significantly by age, although older observers show greater variability in their fixations along the horizontal meridian compared to the vertical meridian (Kosnik et al., 1986). Abadi and Gowen (2004) found that age is positively correlated with the amplitude, but not the frequency, of saccadic intrusions during fixation. The range of fixation for upward gaze may decrease somewhat with age (Ciuffreda and Tannen, 1995), whereas younger observers present more equivalent fixation variabilities along the two meridians (Kosnik et al., 1986).

Fixation instability is greater in the dark: microsaccades tend to become larger (Ditchburn and Ginsborg, 1953; Cornsweet, 1956; Snodderly, 1987), and drifts are both larger and more frequent (Ditchburn and Ginsborg, 1953; Nachmias, 1961). Deprivation of vision, whether congenital or acquired, can lead to severe fixation instability (Leigh and Zee, 1980; Leigh et al., 1989), reflecting an ocular motor system that has not been calibrated by experience. In the presence of severe foveal damage, a preferred retinal location or pseudo fovea may be developed (Barrett and Zwick, 2000). Fixation instability may also be greater in patients with attention-deficit/hyperactivity disorder than in normal subjects (Gould et al., 2001).

Tremor

Tremor is an aperiodic, wave-like motion of the eyes (Riggs et al., 1953), with a bandwidth of ~90 Hz (Carpenter, 1988) and an amplitude of ~8.5 s of arc (Eizenman et al., 1985). Tremor is the smallest of all eye movements (amplitudes are about the diameter of a cone in the fovea (Ratliff and Riggs, 1950; Yarbus, 1967; Carpenter, 1988)), making it difficult to record accurately: tremor amplitudes and frequencies are usually near the level of the recording system's noise. The contribution of tremor to the maintenance of vision is unclear. It has been argued that tremor frequencies are well over the flicker fusion frequencies in humans, and so the tremor of the visual image may be ineffective as a stimulus (Ditchburn, 1955; Gerrits and Vendrik, 1970; Sharpe, 1972). But some studies suggest that tremor frequencies can be below the flicker fusion limit (Spauschus et al., 1999). Greschner et al. (2002) have shown that low frequencies (5 Hz) of tremor-like motion generate strong synchronous firing in the turtle's retina. Furthermore, early visual neurons can follow high-frequency flickering that is above the perceptual threshold for flicker fusion (Martinez-Conde et al., 2002). Thus it is possible that even high-frequency tremor is adequate to maintain activity in the early visual system, which may then lead to visual perception. Hennig and colleagues have proposed that noise in the range of ocular tremor improves spatial resolution and may partly underlie the hyperacuity properties of the visual system (Hennig et al., 2002).

Tremor is generally thought to be independent in the two eyes. This imposes a physical limit to the ability of the visual system to match corresponding points in the retinas during stereovision (Riggs and Ratliff, 1951; Spauschus et al., 1999).

Patients with brainstem damage and alteration in their level of consciousness present tremor with lower frequencies than normal individuals (Shakhnovich and Thomas, 1977; Coakley, 1983; Ciuffreda and Tannen, 1995).

Drift

Drifts occur simultaneously with tremor and are slow motions of the eye that take place between microsaccades. During drifts, the image of the object being fixated moves across approximately a dozen photoreceptors (Ratliff and Riggs, 1950). Drifts appear to be random motions of the eye (Ditchburn and Ginsborg, 1953), generated by the instability of the oculomotor system (Cornsweet, 1956). Also, the orbital mechanics impose elastic restoring forces that pull the eye back to the center from eccentric positions. Drifts to the center at the end of saccades are actively avoided by the neural integrator through the sustained firing of ocular motoneurons (Leigh and Zee, 2006).

Drifts may have a compensatory role in maintaining accurate visual fixation when microsaccades are absent, or when compensation by microsaccades is poor (Nachmias, 1959, 1961; Steinman et al., 1967; St Cyr and Fender, 1969). Drifts have been reported to be both conjugate (Ditchburn and Ginsborg, 1953; Spauschus et al., 1999) and disconjugate (Krauskopf et al., 1960; Yarbus, 1967). As with tremor, drifts may result from the noise and variability of neural firing processes to the ocular muscles (Eizenman et al., 1985; Carpenter, 1988). However, if drifts and tremor are indeed conjugate in the two eyes, this may suggest a central origin (at least in part) for drifts and tremor. This agrees with observations of reduced or absent tremor in patients with brainstem lesions (Shakhnovich and Thomas, 1977).

Drifts have usually been characterized as the eye position change that occurs during the periods in between microsaccades. This categorization method has the potential complication that one may unintentionally attribute non-drift activity (such as activity produced by undetected tremor) to drifts. Gur et al. (1997) found drifts to cause less variability in neuronal responses in V1 than a combination of drifts and microsaccades.

Microsaccades

Microsaccades are involuntary jerk-like fixational eye movements that occur 3–4 times per second. They are the largest and fastest of the three fixational eye movements. They carry the retinal image across a range of several dozen (Ratliff and Riggs, 1950) to several hundred photoreceptor widths (Martinez-Conde et al., 2000, 2002, 2004; Hafed and Clark, 2002; Moller et al., 2002;

Engbert and Kliegl, 2003a, b, 2004) and are about 25 ms in duration (Ditchburn, 1980). Microsaccades cannot be defined according to magnitude alone, as the magnitude of voluntary saccades can be as small as that of fixational microsaccades. The one critical aspect that differentiates microsaccades from regular saccades is that microsaccades are produced *involuntarily while the subject is attempting to fixate*. Microsaccades in the macaque monkey are very similar to those in the human (Skavenski et al., 1975; Snodderly and Kurtz, 1985; Snodderly, 1987) (Fig. 1A) and they have been described in several other species (Carpenter, 1988), although they seem to be most important in species with foveal vision. Microsaccade velocities are parametrically related to microsaccade magnitudes, following the "main sequence" (Zuber and Stark, 1965; Martinez-Conde et al., 2000; Moller et al., 2002) (Figs. 3C–E). This is also true of large voluntary saccades, and therefore it has been proposed that microsaccades and voluntary saccades may be generated by the same oculomotor mechanisms (Zuber and Stark, 1965). Van Gisbergen and colleagues found that the activity of burst neurons is similar for saccades and microsaccades (Van Gisbergen and Robinson, 1977; Van Gisbergen et al., 1981). Microsaccades in the two eyes are generally conjugate (Lord, 1951; Ditchburn and Ginsborg, 1953; Yarbus, 1967; Moller et al., 2002). The fact that microsaccades are involuntary suggests a subcortical control mechanism for microsaccade production (Moller et al., 2002).

Recent studies suggest that microsaccades may increase the retinal refresh to counteract receptor adaptation on a short time-scale and help to correct fixation errors on a longer time-scale (Engbert and Kliegl, 2004). It must be noted that the concept of "refreshing" the retinal images is mainly metaphorical. It does not imply that the same region of a visual scene will stimulate the same set of photoreceptors over and over due to microsaccades (as if one were to flash a stationary stimulus on the retina). On the contrary, microsaccades are expected to produce retinal slippage. Consecutive microsaccades will generally shift any given set of photoreceptors to a slightly different region of the visual scene. However, sequential pairs of horizontal

154

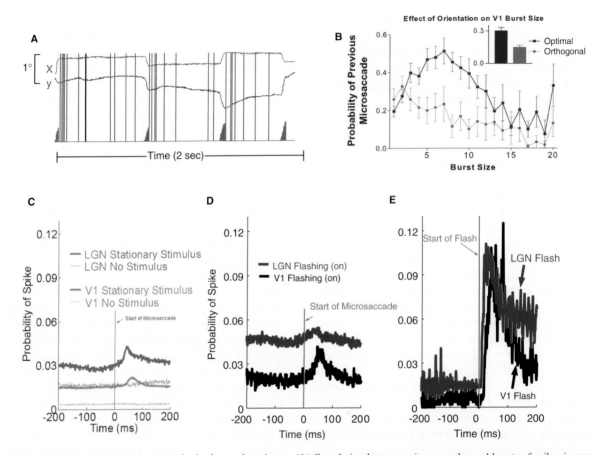

Fig. 1. Neural responses to microsaccades in the awake primate. (A) Correlation between microsaccades and bursts of spikes in area V1 during a 2-s period. The green and blue traces represent horizontal and vertical eye positions, respectively (tracked with a search coil). We identified microsaccades objectively with an automatic algorithm. The red triangles indicate where a microsaccade has occurred (the height of the triangles represents magnitudes of microsaccades). The vertical black lines represent the spikes of a single V1 neuron. Microsaccades tend to be followed by a rapid cluster, or burst, of spikes. (B) Average probability of microsaccades before V1 bursts of different sizes, for optimal (red) and orthogonal (blue) orientations ($n = 11$ neurons; each cell was tested in both the optimal and orthogonal conditions). Optimal latencies and interspike intervals were selected for each individual neuron. Inset: Average probability of microsaccades before all burst sizes. (C) Microsaccades increase spike probabilities in the lateral geniculate nucleus (LGN, $n = 57$ cells) and V1 ($n = 308$ cells). In the absence of visual stimulation, microsaccades do not generate spikes in the LGN ($n = 42$ cells) or in V1 ($n = 37$ cells). (D) Microsaccades increase spike probabilities in the LGN (purple trace, $n = 48$ neurons) and area V1 (black trace, $n = 6$ neurons) when a bar is flashing. Starts of all microsaccades are aligned at the vertical line. (E) The probability of a spike after a flashing bar turns on is about seven times higher than the probability of a spike after a microsaccade when that same flashing bar is on. The same data set from (D) (LGN and V1) has been replotted and realigned to the onset of the flashing bar (vertical line). Panel A: Reprinted from Martinez-Conde et al. (2000), with kind permission from Nature Publishing Group. Panels B, E: Reprinted from Martinez-Conde et al. (2002). Panels C, D: Modified from Martinez-Conde et al. (2002), with kind permission from the National Academy of Sciences of the United States of America.

microsaccades may sometimes be coupled in a square-wave pattern that moves the eye along one vector and then back along the reverse vector.

Several statistics of microsaccades are indicative of cognitive processes. During an attentional task,

microsaccade rates transiently decrease and then increase to a higher than baseline level. Moreover, the direction of microsaccades is biased toward the spatial locus of attention (Hafed and Clark, 2002; Engbert and Kliegl, 2003b).

Neural responses to microsaccades

The neural responses to microsaccades have been studied in the lateral geniculate nucleus (LGN) (Martinez-Conde et al., 2002; Reppas et al., 2002), area V1 (Leopold and Logothetis, 1998; Martinez-Conde et al., 2000; Snodderly et al., 2001; Martinez-Conde et al., 2002), and extrastriate cortex (Bair and O'Keefe, 1998; Leopold and Logothetis, 1998). Presumably, microsaccades first generate neural signals at the level of retinal photoreceptors by moving the receptive fields (RFs) of less adapted photoreceptors over otherwise stationary stimuli. This photoreceptor activity may then be transmitted to subsequent levels in the visual hierarchy.

In our experiments, macaque monkeys were trained to fixate their gaze on a small fixation spot while a stationary stimulus of optimal characteristics was placed over the RF of the recorded neuron (for instance, a bar with optimal dimensions and orientation when recording from area V1). Microsaccades were then correlated with subsequent neural activity. Because the visual stimulus did not move and the head was fixed, modulation of neural activity only occurred when fixational eye movements moved the visual RF over the stationary stimulus.

We found microsaccades to be predominantly excitatory in the LGN and area V1 (Martinez-Conde et al., 2000; Martinez-Conde et al., 2002) (Fig. 1A–D). Neuronal responses following microsaccades were purely visual in nature: microsaccades led to an increase in neural activity when a stationary bar of light was centered over the neuron's RF. However, when the bar was removed from the RF (and the monitor facing the monkey was blank except for the fixation spot) microsaccades did not lead to changes in neural activity (Fig. 1C). This demonstrated that microsaccade-induced activity in early visual neurons was visual (rather than motor) in nature. The neurons were excited only when their RFs swept across stimuli, and they were not excited during equivalent action by the motor system in the absence of a visual stimulus (Martinez-Conde et al., 2000, 2002).

Increases in firing rate after microsaccades were clustered in bursts of spikes. These bursts of spikes were better correlated with previous microsaccades than either single spikes or instantaneous firing rate (Fig. 2). Bursts that were highly correlated with previous microsaccades had large spike numbers and short inter-spike intervals (Martinez-Conde et al., 2000). Therefore long, tight bursts of spikes are the type of activity most effective in sustaining a visible image during fixation (Martinez-Conde et al., 2000, 2002). It is important to note that those bursts that were best indicators of previous microsaccades are not defined in terms of their biophysical properties, and may not share a common biophysical mechanism. Burst definitions that are solely based on specific biophysical parameters are unavoidably arbitrary, and they are not necessarily meaningful from a perceptual standpoint. On the contrary, when we correlate all possible burst parameters to previous microsaccades, we have the great advantage that we are letting perception tell us what an optimal burst is. The next section will establish that microsaccades are directly correlated with visibility; therefore bursts that are well correlated with previous microsaccades must encompass the neural code for visibility.

The optimality of the stationary visual stimulation had an effect on the size of bursts following microsaccades: when the stationary stimulus covering the neuron's RF had optimal characteristics (for instance, a bar of light with optimal orientation), microsaccades during fixation generated long bursts. When the stimulus centered on the RF had non-optimal characteristics, microsaccades produced shorter bursts. Thus long bursts were correlated with salient optimal stimuli, whereas short bursts were correlated with non-optimal visual stimulation (Martinez-Conde et al., 2002). Figure 1B plots the correlation between microsaccades and bursts for optimal vs. non-optimal stimuli in area V1.

To address the effectiveness of microsaccades in generating neural activity, we compared neural responses induced by microsaccades to neural responses induced by flashing bars. Onset responses to flashing bars in the LGN and area V1 were about 7 times larger than the responses to stationary bars moved across the neurons' RFs by microsaccades, perhaps because of the relative abruptness of flashes as stimuli (Martinez-Conde et al., 2002) (Fig. 1D, E). This experiment demonstrated that changes in retinal stimulation (which may or may not be due to retinal slippage)

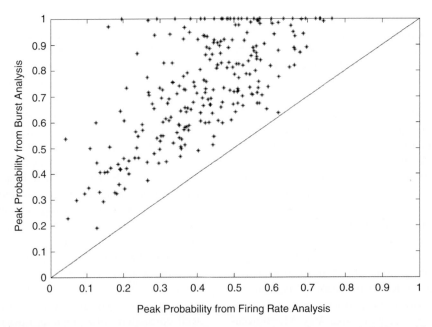

Fig. 2. Peak probability that a spike pattern is preceded by a microsaccade in V1 ($n = 246$ neurons). Each asterisk represents a single neuron. Peak probabilities using instantaneous firing rate are plotted against peak probabilities using bursts (for optimum burst sizes, across all latencies and all interspike intervals). In every neuron, preceding microsaccades were better indicated by the burst analysis than by the instantaneous firing rate analysis. Reprinted from Martinez-Conde et al. (2000), with kind permission from Nature Publishing Group.

are critical to generating neuronal responses in the visual system, thereby counteracting fading. Flashes (for which there is a low probability of slippage) are even more effective than microsaccades in generating neural responses in the LGN and V1.

Microsaccades could enhance spatial summation by synchronizing the activity of nearby neurons (Martinez-Conde et al., 2000). By generating bursts of spikes, microsaccades may also enhance temporal summation of responses from neurons with neighboring RFs (Martinez-Conde et al., 2000). Moreover, microsaccades may help disambiguate latency and brightness in visual perception, allowing us to use latency in our visual discriminations (Martinez-Conde et al., 2000). Changes in contrast can be encoded as changes in the latency of neuronal responses (Albrecht and Hamilton, 1982; Albrecht, 1995; Gawne et al., 1996). Since the brain knows when a microsaccade is generated, differential latencies in visual responses could be used by the brain to indicate differences in contrast and salience.

Suppression of perception and neural firing during large saccades is well known to exist (Wurtz, 1968, 1969; Macknik et al., 1991; Burr et al., 1994; Bridgeman and Macknik, 1995; Ross et al., 1996, 1997, 2001), but the existence of microsaccadic suppression is more controversial. Some studies have reported elevation of visual thresholds during microsaccades (Ditchburn, 1955; Beeler, 1967), but others have found little or no threshold elevation (Krauskopf, 1966; Sperling, 1990). In the early visual system (LGN and area V1), microsaccades generate increases in neural activity, but not suppression (Martinez-Conde et al., 2000, 2002). Murakami and colleagues have proposed that the extrastriate cortex, especially area MT (Murakami and Cavanagh, 2001; Sasaki et al., 2002) could be the locus for microsaccadic suppression. However, electrophysiological studies in macaque MT indicate that microsaccades induce strong excitatory responses in this area (Bair and O'Keefe, 1998). This seems to contradict a specific role of MT in microsaccadic suppression, although one cannot rule out that neural responses in MT may drive a microsaccadic suppression system later in the visual hierarchy: the question remains open.

Perceptual responses to microsaccades

The role of microsaccades during visual fixation was first discussed 50 years ago. Cornsweet (1956) originally proposed that microsaccades return the eyes to the fixation target, and thus serve to correct the intersaccadic drifts of the eye. It was also postulated that microsaccades may play an "important role in maintaining vision by counteracting retinal fatigue" (Ditchburn et al., 1959; Nachmias, 1961). Carpenter (1988) postulated that, of the three types of fixational eye movements, only microsaccades may contribute significantly to the maintenance of vision, as drift velocities are too low and the magnitude and frequency of tremor would make it more detrimental than otherwise. Not all studies agreed, however. Starting in the late 1960s, and through the 1970s, a lively discussion on the importance of microsaccades for the maintenance of vision took place. Its main representatives were Ditchburn (microsaccades play an essential part in normal vision) and Steinman (microsaccades serve no useful purpose). The strongest evidence against the role of microsaccades in preserving visual perception was as follows: (1) trained subjects can suppress their microsaccades for several seconds when asked to hold their gaze on a visible target (Fiorentini and Ercoles, 1966; Steinman et al., 1967, 1973), and (2) Microsaccades are naturally suppressed while subjects perform high-acuity tasks, such as when sighting a rifle or threading a needle (Winterson and Collewijn, 1976; Kowler and Steinman, 1977, 1979; Bridgeman and Palca, 1980).

In 1980, the controversy as to the perceptual effects of microsaccades came to an abrupt end. Kowler and Steinman famously concluded (as stated in the title of their *Vision Research* reply to Ditchburn) that "*Small saccades serve no useful purpose*" (Kowler and Steinman, 1980). This view largely dominated the field of fixational eye movements for the next several decades, as asserted by Malinov et al. (2000): "*By 1980, microsaccades were no longer interesting*". However,

the perceptual effect of microsaccades on visibility had not been tested directly (although several studies had specifically questioned the "usefulness [of microsaccades] for preserving vision by preventing fading" (Cunitz and Steinman, 1969; Steinman, 1975; Kowler and Steinman, 1979)).

My colleagues and I hypothesized that microsaccades would be sufficient for (and potentially causal to) visibility during fixation, based on two previous independent observations: (1) Microsaccades tend to be naturally suppressed during precise fixation (Winterson and Collewijn, 1976; Kowler and Steinman, 1979), and (2) Troxler fading tends to occur during precise fixation (Troxler, 1804). It followed that microsaccades may counteract Troxler fading.

To establish the correlation between microsaccades and visibility, we conducted a continuously sampled two-alternative forced choice (2-AFC) task in which the subject fixated a small spot, and simultaneously reported the visibility of a peripheral target (Fig. 3B), via button press (Martinez-Conde et al., 2006). Every millisecond of the experiment was coded as either visible (or intensifying) or invisible (or fading), according to the subject's report. A naïve subject later reported that she had thought the stimulus was modulating in brightness physically: she did not realize that her fixation behavior was in fact driving the perceptual alternations.

Eye position was simultaneously measured, and microsaccades automatically identified with an objective algorithm (Martinez-Conde et al., 2000, 2002, 2006). We found that microsaccade probabilities, rates, and magnitudes increased before transitions to a more visible state (Fig. 4A–C, black lines), and decreased before transitions into a period of invisibility (fading) (Fig. 4A–C, gray lines). Moreover, binocular microsaccades were more effective than monocular microsaccades (Fig. 4D–F). The results revealed, for the first time, a direct correlation between microsaccades and visibility during fixation, and suggested that microsaccades may cause the bi-stable dynamics seen during Troxler Fading.

The psychophysical link found between microsaccades and visibility matched the predictions from our previous primate studies, in which microsaccades generated visual responses in V1

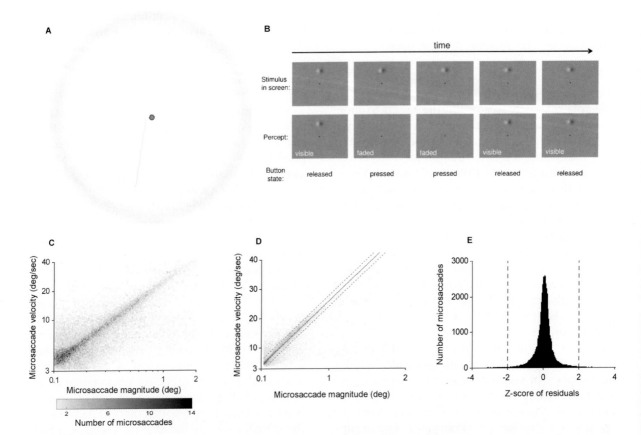

Fig. 3. The measurement of microsaccades during Troxler fading. (A) Demonstration of peripheral visual fading, or Troxler's effect. Fixate very precisely on the black spot, while paying attention to the gray annulus. After a few seconds of careful fixation, the annulus will fade, and the black spot will appear to be surrounded by a completely white field. Movements of the eyes will immediately bring the annulus back to perception. (B) An epoch from a trial during the experiment. The top row shows the stimulus, which does not change over time. The second row shows the percept of the stimulus, which is intermittently invisible due to Troxler fading. The third row shows the subjects' report of their perception. (C) Log–log main sequence of all microsaccades ($n = 106{,}692$). (D) Linear plot of main sequence in (C). Dotted lines indicate the 95% prediction intervals. The 95% confidence intervals are obscured by the regression line. Microsaccade binning as in (C). (E) Distribution of linear regression residuals for microsaccadic main sequence. Microsaccade distances from the regression line (D) follow a normal distribution (Kolmogorov–Smirnov test, $p > 0.1$), confirming that the microsaccades studied had an orderly relationship between magnitude and speed (Zuber and Stark, 1965). The vertical dotted lines contain 95% of all microsaccades, and correspond to the prediction intervals in (D). Panel A: Modified from Martinez-Conde et al. (2004). Panels B–E: Modified from Martinez-Conde et al. (2006), with kind permission from Elsevier.

and LGN neurons (Martinez-Conde et al., 2000, 2002). Our combined psychophysical and physiological data indicate that microsaccades drive the perception of visibility during Troxler fading, and that they increase the responses of neurons in the early visual system. We can therefore conclude that the neuronal responses produced by microsaccades in the LGN and V1 (and presumably also in the retina) are the neural correlates of the perception of visibility during Troxler fading. When

microsaccades are produced, early visual neurons fire and the stimulus appears visible. When microsaccades are suppressed, early visual neurons fall silent, and the stimulus becomes invisible, due to neural adaptation processes. Thus the neural adaptation necessary for the perception of Troxler fading takes place in the very first stages of the visual system.

Having shown that microsaccades contribute significantly to the maintenance of visibility, an

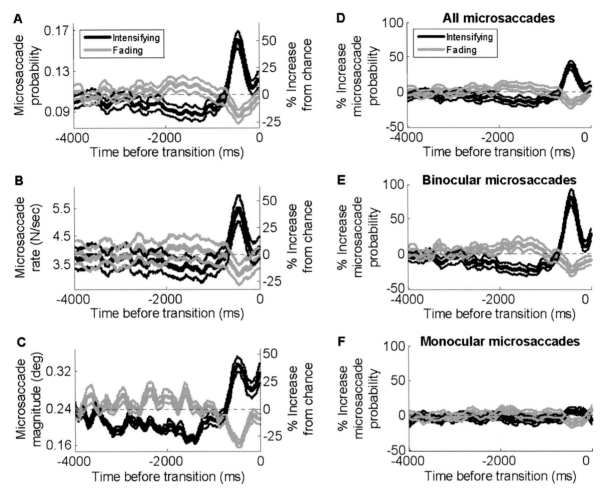

Fig. 4. Microsaccade dynamics before transitions toward perceptual intensification vs. fading. (A) Average probability of microsaccades before transitions toward perceptual intensification (black) vs. fading (gray). The horizontal dashed line indicates average probability of microsaccades during the recording session. (B) Average rate of microsaccades before perceptual transitions. (C) Average microsaccade magnitude before perceptual transitions. The combined results in (A–C), indicate that a reduction in microsaccade rates and magnitudes leads to perceptual fading, whereas increases in microsaccade rates and magnitudes lead to perceptual intensification, confirming our predictions. (D) Average percentage increase in microsaccade probabilities before transitions toward perceptual intensification vs. fading. All microsaccades (same data as in panel A). (E) Binocular microsaccades (all other details as in (D)). (F) Monocular microsaccades (all other details as in (D)). Thin lines indicate SEM between subjects ($n = 8$ subjects). Modified from Martinez-Conde et al. (2006), with kind permission from Elsevier.

open question remains: does the role of microsaccades differ from the role of drifts and tremor? Microsaccades may be more important for peripheral vision, whereas drifts/tremor may maintain foveal vision (Clowes, 1962; Gerrits and Vendrik, 1974) when microsaccades are suppressed during specific tasks. Foveal RFs may be so small that drifts and tremor could be sufficient to prevent visual fading in the absence of microsaccades.

RFs in the periphery may be so large that only microsaccades are large and fast enough (compared to drifts and tremor) to prevent visual fading, especially with low-contrast stimuli (Gerrits and Vendrik, 1974; Gerrits, 1978; Ditchburn, 1980; Martinez-Conde et al., 2000). But one should keep in mind that, if one could eliminate drifts and tremor altogether, while preserving microsaccades, microsaccades alone might then suffice to sustain both peripheral and foveal vision during fixation. In summary, all fixational eye movements may contribute to the maintenance of vision to some degree, and their relative contributions may depend on the specific task and stimulation conditions (Martinez-Conde et al., 2004).

Microsaccades may also drive perceptual flips in binocular rivalry (Sabrin and Kertesz, 1983) and other bi-stable percepts. Because microsaccades are correlated to visibility (Martinez-Conde et al., 2006) and they are suppressed during precise fixation, it follows that microsaccades must contribute (at least partially) to the perception of those classes of visual illusions that vary in strength depending on the accurateness of fixation. The experiments above demonstrate that Troxler fading is counteracted by microsaccades (Martinez-Conde et al., 2006). Second-order adaptation (such as the filling-in of artificial scotomas (Ramachandran and Gregory, 1991; Spillmann and Kurtenbach, 1992; Ramachandran et al., 1993)) is also facilitated by precise fixation. It therefore follows that microsaccades may similarly counteract filling-in of dynamic textures.

Many visual illusions are attenuated when the observer fixates his/her gaze carefully (thus suppressing microsaccades), suggesting that microsaccades drive (completely or partially) the generation of the illusory percept. Such illusions include the illusory motion of static patterns (Fermuller et al., 1997), such as Leviant's "Enigma" (Leviant, 1996), the Ouchi illusion (Ouchi, 1977), or Kitaoka's "Rotating Snakes" (Kitaoka and Ashida, 2003), and even some classical brightness illusions such as the Hermann grid (Hermann, 1870). A recent study has proposed that microsaccades can be ruled out as a contributor to the Enigma illusion (Kumar and Glaser, 2006). However, no eye movement measurements were carried out, and so the question remains open.

The fact that microsaccades can be transiently suppressed by carefully fixating our gaze provides us with a very useful tool to predict the potential involvement of microsaccades in a variety of percepts.

Head-unrestrained microsaccades
Steinman and Kowler proposed that microsaccades were a laboratory artifact: i.e. that microsaccades do not occur in normal viewing conditions, but only after prolonged fixation in the laboratory when the subject's head is restrained (for instance, with a bite bar) (Kowler and Steinman, 1980). They reasoned that during natural viewing conditions, normal head movements should suffice to maintain vision during fixation, and therefore very few or no microsaccades need be produced (Skavenski et al., 1979; Kowler and Steinman, 1980; Steinman and Collewijn, 1980). However, even if microsaccades have no significant effect on the visibility of moving stimuli (i.e. moving either on their own or due to head movements), they may be generated nevertheless. Indeed, it is possible that microsaccades could serve to enhance the visibility of a moving stimulus when the dynamics of head movements or the intrinsic stimulus motion do not have the ideal parameters to invoke visibility.

This is a critical issue to resolve: if microsaccades are a laboratory artifact, then their significance to normal and clinical vision is vastly diminished, even if they are correlated with visibility and neural activity. To address this possibility, we repeated the above experiment in head-unrestrained conditions (i.e. without the chinrest). We found that microsaccade characteristics and functional properties were equivalent with heads restrained and unrestrained (Fig. 5A–C). Thus microsaccades are a natural oculomotor behavior, and not a laboratory artifact of head fixation.

Moreover, evidence from a clinical subject (A.I.), who is unable to make eye movements, shows that in the absence of eye movements, *normal* head movements alone do not suffice to maintain vision. Although the authors of the A.I. studies did not address microsaccades per se, they established that A.I. learned to move her head in a "saccadic" fashion, in order to conduct visual tasks such as reading, and visuomotor tasks such as pouring tea (Gilchrist et al., 1997, 1998; Land et al., 2002). Therefore saccadic movements, either of the head or of the eye, may represent an optimal sampling strategy for the visual system. Microsaccades may thus provide us with a window into the visual sampling mechanisms used by the brain during fixation.

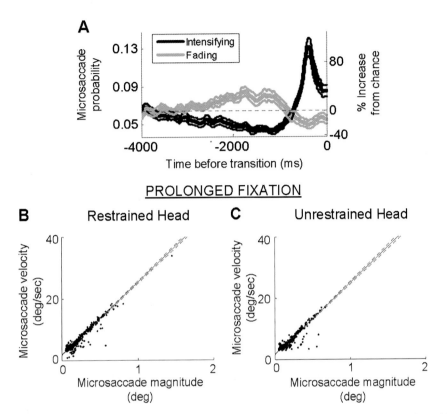

Fig. 5. Head-unrestrained microsaccade dynamics. (A) Average probability of microsaccades before perceptual transitions under head unrestrained conditions. Thin lines indicate SEM between subjects ($n = 7$ subjects). Modified from Martinez-Conde et al., 2006, with kind permission from Elsevier. (B, C) Microsaccade main sequences for visual fixation with restrained vs. unrestrained head. (B) Microsaccades ($n = 515$; rate $= 4.27$ Hz) from a human subject during 2 min of visual fixation, with head supported by a chinrest. (C) Microsaccades ($n = 444$; rate $= 3.7$ Hz) from the same subject during 2 min of visual fixation with head unrestrained. Dashed lines indicate the 95% confidence intervals.

Microsaccades during visual exploration

Steinman et al. (1973) and Kowler and Steinman (1979, 1980) reported that microsaccades are not helpful in tasks requiring complex visual information processing, and therefore they are much less common during brief fixations interposed between large saccades (in activities such as reading or counting) than during prolonged fixation. Winterson and Collewijn (1976) also reported that microsaccades are far less frequent during fine visuomotor tasks than during maintained fixation.

In the last decade, several laboratories have developed objective and automatic algorithms for microsaccade detection (Martinez-Conde et al., 2000; Engbert and Kliegl, 2003b). Current objective algorithms allow the identification of hundreds of thousands of microsaccades in a fast and

automatic manner. Thus, the results obtained in the 1970s with subjective microsaccade identification techniques must now be re-evaluated with modern and objective methods. Figure 6 plots the eye movements of a human subject during the visual exploration of a static image. We found that the periods of fixation accounted for approximately 80% of the time spent in free-viewing, with either restrained or unrestrained heads. During the other 20% of the time we are virtually blind, due to saccadic suppression mechanisms. Since fixational eye movements sustain visibility during fixation, it follows that fixational eye movements may drive up to 80% of our visual experience. Microsaccades were prominent during the fixation periods that naturally occur during visual exploration. Moreover, the dynamics of microsaccades

FREE-VIEWING

A
Restrained Head

B
Restrained Head

C
Unrestrained Head

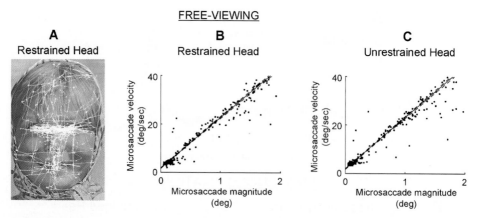

Fig. 6. Saccadic and microsaccadic eye movements during free-viewing. (A) Record of human eye movements during 2-min of visual exploration. Large saccades connect fixation periods. Modified from Yarbus (1967), with kind permission from Plenum Press. (B, C) The microsaccades produced in the "fixation periods" during free-viewing follow the main sequence, both with head restrained and unrestrained. No substantial differences can be observed for microsaccades produced during free-viewing with restrained ((B), $n = 278$; rate = 2.27 Hz) or unrestrained head ((C), $n = 265$; rate = 2.16 Hz). Dashed lines indicate the 95% confidence intervals.

produced during visual exploration were similar to microsaccades produced during prolonged fixation. Since microsaccades counteract peripheral fading during fixation (Martinez-Conde et al., 2006), it may be that microsaccades drive a large fraction of our visual experience. Future research will quantify the fraction of visual experience driven by microsaccades vs. the other fixational eye movements during various visual tasks.

We have also observed that microsaccades occur in untrained monkeys during spontaneous free-viewing, and in trained monkeys during guided-viewing (i.e. with the fixation point appearing at random locations over the image; Figs. 7 and 8). Livingstone et al. (1996) also reported, anecdotally, that microsaccades occur in fixations during free- and guided-viewing in monkeys.

Abnormal eye movements during fixation

Impaired fixational eye movements are observed in patients with a variety of central and peripheral pathologies (Shakhnovich and Thomas, 1974, 1977; Ciuffreda and Tannen, 1995). Although we spend about 80% percent of our waking lives fixating our gaze, the contribution of impaired fixational eye movements to vision loss is generally

overlooked. This gap in knowledge has prevented the field from developing new treatments and early diagnostic tools to ameliorate those visual deficits that are due to impaired fixational eye movements. The evaluation of fixational eye movements may prove useful in the differential diagnosis of disorders of the oculomotor system (especially at early stages) and their quantitative measurement (Yamazaki, 1968; Filin and Okhotsimskaya, 1977; Okhotsimskaia, 1977; Hotson, 1982). The clinical evaluation of fixational eye movements may also help to determine their potential role in therapies for visual deficits such as amblyopia, and in establishing the optimal duration of a given treatment (von Noorden and Burian, 1958; Ciuffreda et al., 1979a).

A non-exhaustive classification of abnormal eye movements during fixation follows below (Table 1). I have addressed the perceptual consequences of these abnormal eye movements when possible. See Leigh and Zee (2006) for detailed discussion on the pathophysiology of the various disorders of fixation. As the next several pages illustrate, the fixational eye movement system must achieve a very delicate balance: insufficient fixational eye movements lead to adaptation and visual fading, whereas excessive motion of the eyes produces blurring and unstable vision during fixation.

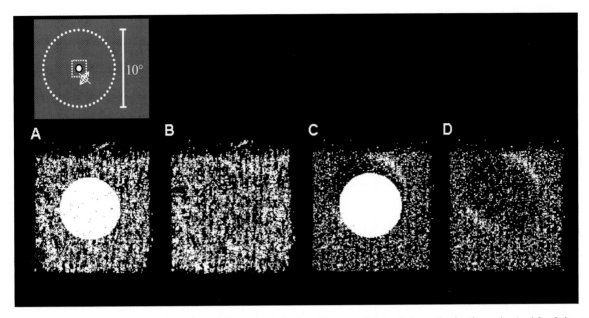

Fig. 7. Responses of a V1 cell to a white circle during guided-viewing. The upper left inset shows the fixation point (and 2 × 2 degree fixation window) in relationship to the cell's RF and the stimulus (large circle). In (A, B) the fixation spot jumped randomly to each location in a square grid around the stimulus and each dot represents the foveal position at the time of each spike (accounting for a 35 ms response latency); (C, D) same data filtered so that each dot represents a burst of 4 or more spikes within a 20 ms interval. (B, D) are the same as (A, C), with stimulus removed. The large responses in (C) and (D) are to microsaccades and show preference at the cell's orientation. Reprinted from Martinez-Conde et al. (2000), with kind permission from Nature Publishing Group.

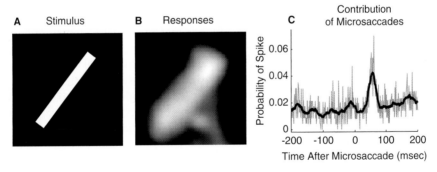

Fig. 8. Correlation of microsaccades to neural activity during a guided-viewing task in an awake monkey. (A) Visual stimulus scanned by a V1 RF during the guided-viewing task. (B) Responses (spike densities) of the V1 cell as a function of eye position. (C) Correlation of microsaccades to spikes from the data in (B).

Increased drift and paucity of microsaccades in amblyopia

Amblyopia is defined as a visual acuity loss that is not attributable to detectable pathology or uncorrected refractive error. It is generally associated with strabismus, anisometropia, or both (Bedell et al., 1990). It has a known relationship with unsteady fixation and impaired fixational eye movements. Amblyopes exhibit decreased microsaccades and increased drifts in the amblyopic (non-dominant) eye 75% of the time spent during fixation (Ciuffreda and Tannen, 1995). However, fixation in the dark-adapted state is normal or close to normal (Wald and Burian, 1944; von Noorden and Burian, 1958). The dominant eye

Table 1. Abnormal eye movements during fixation

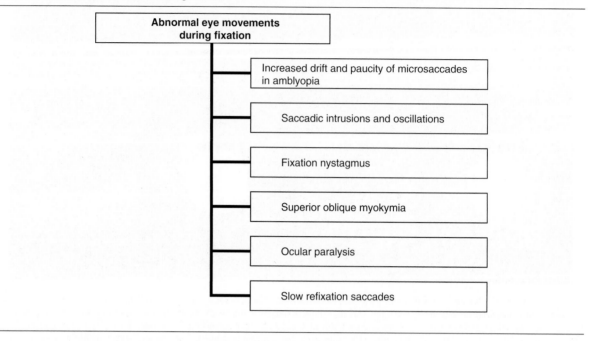

presents normal fixational eye movements, and binocular fixation is also normal (Ciuffreda et al., 1980). Fixation errors produced in amblyopes by excessive drift of gaze position are usually corrected by subsequent drifts in the opposite direction, rather than by microsaccades. Although strabismus often results in amblyopia, the abnormal fixational eye movement pattern described here is correlated to amblyopia irrespective of whether strabismus is present (Ciuffreda et al., 1979b, d, 1980; Srebro, 1983).

Increased drift and suppression of microsaccades in patients with severe amblyopia are often associated with rapid fading of the "small fixation spot, small and large acuity targets, and even portions of the laboratory" during monocular fixation with the amblyopic eye (Lawwill, 1968; Hess et al., 1978; Ciuffreda et al., 1979b, c). According to one patient's report, he "made saccades to revive the faded or blanked-out portions" of the image during fixation with the amblyopic eye (Ciuffreda et al., 1979b). Visual fading in amblyopic eyes (in which drifts, but not microsaccades, are common) is likely related to Troxler fading during suppression

of microsaccades in normal eyes (Martinez-Conde et al., 2006) (Fig. 4). The prevalence of perceptual fading in amblyopia also lends support to the theory that microsaccades may provide more optimal visual sampling dynamics than drifts.

Increased fixational drift can produce perceptual shifting of visual targets (Srebro, 1983). Increased drift amplitudes may also reduce visual acuity in amblyopia by moving the retinal images onto more eccentric positions, as well as contribute to increased variability in visual acuity measurements (Ciuffreda et al., 1979a, 1980). However, not all studies agree that increased drift leads to visual acuity loss (von Noorden and Burian, 1958; Srebro, 1983).

Increased drift velocities and amplitudes in amblyopia may be due to at least three factors: ineffectiveness of the microsaccadic system, ineffectiveness of the smooth pursuit velocity-correcting system, and/or normal drift characteristics for nonfoveal fixation, as amblyopic patients often fixate on eccentric locations (Ciuffreda et al., 1979b). Patients with macular scotomas, and other pathologies that lead to prolonged deprivation of vision from one eye, may also exhibit increased drift, with similar characteristics to drift in amblyopes (Leigh et al., 1989; Ciuffreda and Tannen, 1995).

Fixational eye movements in amblyopia tend to normalize during the course of successful orthoptics therapy (von Noorden and Burian, 1958; Ciuffreda et al., 1979a). However, all visual and oculomotor functions in the amblyopic eye may not improve concurrently (Ciuffreda et al., 1979a), bringing into question the relative importance of the sensory vs. motor deficit in amblyopia (von Noorden and Burian, 1958; Ciuffreda et al., 1978). It follows that amblyopic therapies should not be discontinued upon normalization of visual acuity and centralization of fixation, but should be extended until fixational eye movements are normalized or reach a stable state (Ciuffreda et al., 1979a). Ciuffreda et al. have suggested that the "critical period" for certain aspects of oculomotor plasticity in amblyopia may extend into adulthood. The lack of normalization of fixational eye movements in amblyopia may be responsible for some of the patients reverting to their former condition after the termination of the therapy (Ciuffreda et al., 1979a).

Saccadic intrusions and oscillations

Saccadic intrusions are abnormal horizontal saccades that "intrude" or interrupt accurate fixation (Ciuffreda and Tannen, 1995). Saccadic intrusions are biphasic: although one phase may be a smooth eye movement, the phase that takes the fovea away from its intended target is always a saccade (Sharpe and Fletcher, 1986). Saccadic intrusions tend to be 3–4 times larger than physiological microsaccades (Abadi and Gowen, 2004) and they may cause perceptual instability during fixation (Feldon and Langston, 1977).

Although saccadic intrusions are found in a variety of neurological disorders (Leigh and Zee, 2006), it is important to note that they also occur in normal subjects, with no adverse effect (although their frequency and amplitude are generally smaller than in patients). Abadi and Gowen (2004) found that, in a population of 50 healthy subjects, all 50 individuals presented saccadic intrusions during fixation.

Strabismus without amblyopia is often characterized by normal drift accompanied by saccadic intrusions during fixation (Ciuffreda et al., 1979c, 1980). These saccadic intrusions consist of an error-producing saccade followed (150–500 ms later) by an error-correcting saccade. Intrusion amplitudes usually range from 0.5 to 3.0°, and occur 1–4 times per second. Saccadic intrusions in strabismus without amblyopia do not appear to affect visual acuity (Ciuffreda et al., 1979b). Moreover, the production of saccadic intrusions in strabismus may result from local adaptation, and their occurrence may prevent and/or counteract visual fading (Ciuffreda et al., 1979c).

Saccadic oscillations are bursts of disruptions of fixation, and they may be thought of as salvos of saccadic intrusions (Sharpe and Fletcher, 1986). They may occur in normal subjects during blinks, vergence movements, or large vertical saccades, and moreover they may be produced voluntarily by some individuals (Ramat et al., 2005). Sustained high-frequency oscillations may give rise to oscillopsia (i.e. oscillating vision), due to the excessive motion of the retinal images (Leigh et al., 1994).

A key aspect to the identification of the various types of saccadic intrusions and oscillations is whether they present an intersaccadic interval, or not. Saccadic oscillations *with* an intersaccadic interval include square-wave jerks, macro square-wave jerks, square-wave oscillations, saccadic pulses, double saccadic pulses, and macro saccadic oscillations. Saccadic oscillations *without* an intersaccadic interval include microsaccadic oscillations, opsoclonus, microsaccadic opsoclonus, ocular flutter, microsaccadic flutter, and voluntary nystagmus (also called voluntary flutter). See Table 2 for a classification of saccadic intrusions and oscillations. Figure 9 illustrates several types of saccadic intrusions and oscillations.

Saccadic intrusions and oscillations with an intersaccadic interval

Saccadic intrusions and oscillations with saccadic intervals may share a common pathogenesis, usually involving dysfunction of saccade control due

Table 2. Saccadic intrusions and oscillations

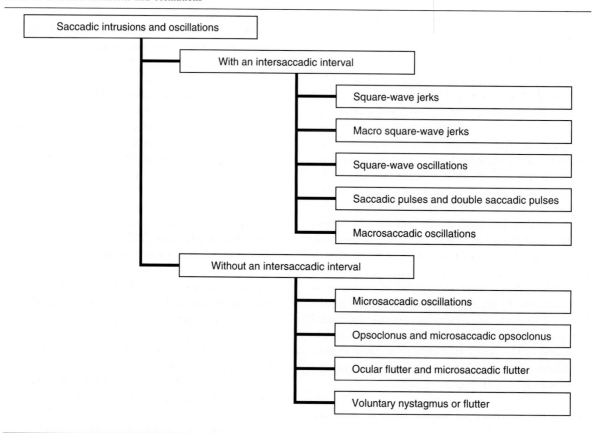

to lesions of the frontal eye field, the superior colliculus, or disruption of the inputs to the superior colliculus (Leigh and Zee, 2006).

Square-wave jerks. Square-wave jerks are the most common type of saccadic intrusion (Sharpe and Fletcher, 1986; Abadi and Gowen, 2004). Monophasic square-wave jerks are couplets of unconscious, involuntary, conjugate microsaccades that occur in opposite horizontal directions (a microsaccade moving away from the fixation target, and a corrective microsaccade about 200 ms later) (Ciuffreda and Tannen, 1995). Their name comes from their "square-wave" appearance on the electro-oculogram (Feldon and Langston, 1977). Biphasic square-wave jerks are microsaccade triplets: the first microsaccade moves the eye away from the fixation target; the second microsaccade is twice as large as the first one and travels toward the fixation target, but then takes the eye beyond its original position; the third and final microsaccade has an amplitude equivalent to the initial microsaccade, and it returns the eye to its original position (Abadi and Gowen, 2004).

Square-wave jerks usually range in amplitude from 0.5 to 5° (Ciuffreda and Tannen, 1995), follow the main sequence, and their frequency is equivalent to that of microsaccades (Feldon and Langston, 1977). Because square-wave jerk dynamics are so similar to microsaccade dynamics, and because normal microsaccades are absent in some square-wave jerk patients, it has been proposed that square-wave jerks are abnormal microsaccades (Feldon and Langston, 1977; Ohtsuka et al., 1986). It is known that even a minimal enlargement of horizontal microsaccades tends to

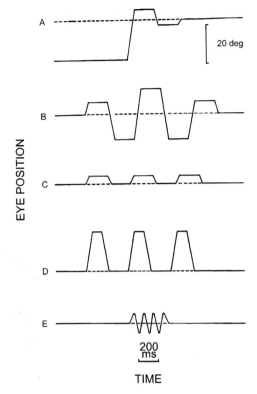

Fig. 9. Saccadic intrusions and oscillations. (A) Dysmetria (inaccurate saccades). (B) Macrosaccadic oscillations (hypermetric saccades that overshoot the fixation target in both directions). (C) Square-wave jerks. (D) Macro square-wave jerks. (E) Ocular flutter. Reprinted from Leigh and Zee (2006). Courtesy of R. John Leigh.

produce square-wave coupling in normal subjects (Ditchburn and Ginsborg, 1953; Feldon and Langston, 1977), and so it follows that square-wave jerks may be abnormally enlarged microsaccades (Feldon and Langston, 1977).

Although square-wave jerks are common in certain diseases, they also occur in most normal subjects, without adverse effect (Shallo-Hoffmann et al., 1990; Ciuffreda and Tannen, 1995; Abadi and Gowen, 2004). Square-wave jerks are produced in all conditions of illumination, and also with closed eyes (Shallo-Hoffmann et al., 1989; Abadi et al., 2000). Just as with microsaccades, mild occurrences of square-wave jerks may be voluntarily and transiently suppressed during strict fixation (Ciuffreda et al., 1979d; Herishanu and Sharpe, 1981).

Shallo-Hoffmann and colleagues proposed that more than 16 square-wave jerks per minute during fixation, and over 20 square-wave jerks per minute in the dark or with closed eyes should be considered abnormal (Shallo-Hoffmann et al., 1989). However, Abadi and Gowen have recently reported that normal subjects may present up to 42.5 square-wave jerks per minute during fixation in mesopic conditions (Abadi and Gowen, 2004).

Frequent square-wave jerks are common in patients with functional strabismus, and they may precede the postnatal appearance of congenital nystagmus (Hertle et al., 1988; Ciuffreda and Tannen, 1995).

Square-wave jerks may occur almost continuously in progressive supranuclear palsy, cerebellar, and local cerebral lesions (Sharpe and Fletcher, 1986; Leigh and Zee, 2006). *Cerebral* lesions lead to square-wave jerks that are usually smaller than square-wave jerks of *cerebellar* origin, but their rates are equivalent in both conditions (Sharpe and Fletcher, 1986; Leigh and Zee, 2006). Frequent square-wave jerks may reflect a disorder of the saccadic pause cells, or a dysfunctional cerebellar-related saccadic gain control system (Ciuffreda and Tannen, 1995).

Macro square-wave jerks. Macro square-wave jerks (also called *square-wave pulses*) are rare (Leigh and Zee, 2006). They are larger than square-wave jerks (5–15° or more) (Ciuffreda and Tannen, 1995). However, they may not be simply enlarged square-wave jerks, as they have shorter intersaccadic intervals (50–150 ms) between sequential saccades and they take place on just one side of the fixation target (right or left) (Sharpe and Fletcher, 1986). Macro square-wave jerks are produced in bursts and they vary in amplitude. They occur in light or darkness, but they may be suppressed during monocular fixation (Leigh and Zee, 2006). Macro square-wave jerks are associated with cerebellar disease and multiple sclerosis (Leigh and Zee, 2006).

Square-wave oscillations. Square-wave oscillations are characterized by horizontal oscillations, in which each half cycle is indistinguishable from a sporadic square-wave jerk. The eyes typically oscillate to just one side of the fixation position (Sharpe and Fletcher, 1986). This disorder can be found in Parkinson's disease combined with alcoholic cerebellar degeneration (Sharpe and Fletcher, 1986) and in progressive supranuclear palsy (Abel et al., 1984).

Saccadic pulses and double saccadic pulses. Saccadic pulses are high-frequency saccades that take the eyes away from the intended position. After each saccadic pulse, negative exponential smooth eye movement returns the eyes to the position previous to the saccade. Saccadic pulses are the least frequently observed type of saccadic intrusion in normal human subjects, although they can occur about once per minute (Abadi and Gowen, 2004).

Double saccadic pulses are intermittent and closely spaced saccadic couplets. They may occur in normal subjects sporadically (once every 2.5 min, Abadi and Gowen, 2004), especially in miniature form (Sharpe and Fletcher, 1986). Double saccadic pulses are the second most prevalent saccadic intrusion in normal subjects (after square-wave jerks) (Abadi and Gowen, 2004). Frequent saccadic pulses occur in patients with internuclear ophthalmoplegia (Leigh and Zee, 2006).

Macrosaccadic oscillations. Macrosaccadic oscillations look like bursts of conjugate, horizontal saccades, separated by intersaccadic intervals of about 200 ms (Sharpe and Fletcher, 1986). The saccades are so hypermetric that they overshoot the intended fixation target in both directions (Zee and Robinson, 1979; Sharpe and Fletcher, 1986; Leigh and Zee, 2006). The oscillations occur with increasing, then decreasing amplitudes, in a crescendo–decrescendo pattern (Selhorst et al., 1976). Saccade amplitudes may reach 40° or more (Selhorst et al., 1976; Sharpe and Fletcher, 1986). The bursts of oscillations last for several seconds and they are usually evoked by attempts to shift visual fixation (Selhorst et al., 1976). However, they may also occur during attempted fixation, or in the dark. Macrosaccadic oscillations may interfere with visual perception by changing the direction of gaze (i.e. losing one's place during reading; R. John Leigh, personal communication).

Macrosaccadic oscillations are associated with lesions affecting the fastigial nucleus and its output in the superior cerebellar peduncles, and they may also occur in some forms of spinocerebellar ataxia (Leigh and Zee, 2006).

Saccadic intrusions and oscillations without an intersaccadic interval
The pathogenesis of oscillations without saccadic intervals (such as flutter and opsoclonus) remains controversial, as no animal model exists. They probably reflect an inappropriate, repetitive, and alternating discharge pattern of different groups of burst neurons (Leigh and Zee, 2006). Ramat and colleagues have recently proposed a theoretical model for saccadic oscillations based on: (a) the coupling of excitatory and inhibitory burst neurons in the brainstem, and (b) the hypothesis that burst neurons show postinhibitory rebound discharge (Ramat et al., 1999, 2005).

Normal subjects present transient horizontal conjugate oscillations without an intersaccadic interval during vergence movements, combined saccade–vergence movements, vertical saccades, pure vergence, and blinks (Ramat et al., 2005).

Microsaccadic oscillations. Microsaccadic oscillations appear as bursts or spindles of horizontal microsaccades without an intersaccadic interval, in a crescendo–decrescendo pattern and with amplitudes under 1°. They may occur in normal subjects about twice per minute. They are also found in cerebellar patients with similar characteristics but higher rates (about eight times a minute) (Sharpe and Fletcher, 1986).

Opsoclonus and microsaccadic opsoclonus. Opsoclonus is characterized by multi-directional saccades of varying amplitudes, without an intersaccadic interval. The saccades are usually conjugate (Sharpe and Fletcher, 1986) and they occur in all three planes (horizontal, vertical and torsional) (Foroozan and Brodsky, 2004). Opsoclonus may result from a combination of uncontrolled saccades and microsaccades (Ellenberger et al., 1972). It may occur during smooth pursuit, convergence or blinks, and it often persists during eyelid closure or sleep (Leigh and Zee, 2006). Opsoclonus is typically associated with brainstem encephalitis (Ashe et al., 1991), diencephalic lesions (Ashe et al., 1991), and cerebellar lesions (Ellenberger et al., 1972; Sharpe and Fletcher, 1986; Ashe et al., 1991). Opsoclonus may also occur without evident cause (Leigh and Zee, 2006).

Microsaccadic opsoclonus can be described as a 1°- saw-tooth pattern of microsaccades in all three planes, without intersaccadic intervals (Leigh et al., 1994; Foroozan and Brodsky, 2004). Microsaccadic opsoclonus can be associated with blurred vision and oscillopsia. The etiology is currently unknown (Foroozan and Brodsky, 2004).

Ocular flutter and microsaccadic flutter. Unlike opsoclonus, ocular flutter is limited to one plane (typically the horizontal plane) and it consists of 1–5° saccades without intersaccadic intervals (Sharpe and Fletcher, 1986; Foroozan and Brodsky, 2004). On rare occasions, ocular flutter may be observed on the vertical plane (Hotson, 1982; Sharpe and Fletcher, 1986). Ocular flutter may be intermittent (Leigh and Zee, 2006) and it is often precipitated by a change in gaze (Ashe et al., 1991). It occurs in the dark as well as in the light, and it is typically conjugate (Cogan et al., 1982).

Opsoclonus and flutter appear closely related, and they may be seen in the same patient at different stages of an illness (Ellenberger et al., 1972; Sharpe and Fletcher, 1986; Ashe et al., 1991). However, opsoclonus is usually observed in the sickest patients (Ellenberger et al., 1972). Patients with multiple sclerosis and with signs of cerebellar and brainstem dysfunction often have flutter (Ellenberger et al., 1972; Sharpe and Fletcher, 1986). Opsoclonus and flutter are also associated with neuroblastoma and tumors of the lung, breast, and uterus (Ashe et al., 1991).

Opsoclonus and flutter may produce blurred vision (Zee and Robinson, 1979) and they frequently generate oscillopsia, due to the high frequency of the oscillations (which generates large retinal slip velocities), even when the oscillations themselves are of small amplitude (Leigh et al., 1994). Opsoclonus and flutter cannot be suppressed by voluntary effort (Leigh and Zee, 2006), but they diminish with eyelid closure (Ellenberger et al., 1972).

Microsaccadic flutter is characterized by abnormal microsaccadic oscillation, comprised of back-to-back horizontal microsaccades in a saw-tooth pattern, ranging from 15 to 50 Hz in frequency and with amplitudes of 0.1–0.5° (Carlow, 1986; Sharpe and Fletcher, 1986; Ashe et al., 1991). It usually causes disruptive oscillopsia (Sharpe and Fletcher, 1986; Ashe et al., 1991). The etiology is unknown (Sharpe and Fletcher, 1986; Ashe et al., 1991).

Voluntary nystagmus or flutter. About 8% of the population has the ability to generate (usually by making a vergence effort) bursts of high-frequency horizontal oscillations of back-to-back saccades, about 2–5° in amplitude (Sharpe and Fletcher, 1986; Ashe et al., 1991). Although this pattern (often used as a party trick) is usually referred to as voluntary nystagmus, it is not truly nystagmus, since slow eye movements are absent (Sharpe and Fletcher, 1986). Voluntary nystagmus causes oscillopsia; however it is not a clinical condition as it is generated voluntarily (Ashe et al., 1991). Voluntary nystagmus has equivalent dynamics and characteristics to pathological involuntary flutter, and it can be mistaken with it (Sharpe and Fletcher, 1986; Ashe et al., 1991). Thus by studying voluntary nystagmus we may gain insight into the nature of some disorders, such as involuntary flutter. Voluntary nystagmus may be an intrinsic and normally undeveloped capability that can be learned by most people (Hotson, 1984), potentially creating an untapped resource for the study of visual disorders.

Nystagmus during attempted fixation

This is a type of pathological oscillation that increases in size when the patient attempts to fixate (Sharpe and Fletcher, 1986). It is unaffected by illumination conditions and/or eyelid closure (Dell'osso and Daroff, 1975). It is characterized by a repetitive, to-and-fro motion of the eyes, initiated by a slow phase (Leigh et al., 1994), and it is often accompanied by impaired vision and oscillopsia (Sharpe and Fletcher, 1986; Leigh and Zee, 1991). However, the magnitude of the oscillopsia is usually smaller than the magnitude of the nystagmus (Leigh et al., 1994).

Nystagmus during attempted fixation commonly arises due to disturbance of the three main gaze-holding mechanisms: vestibular, neural integrator, and visual fixation. With nystagmus that is due to disturbance of the vestibular mechanism, the imbalance of the vestibular drives often causes constant velocity drifts. With nystagmus that is due to disturbance of the neural integrator, the eyes cannot be held in an eccentric position and thus drift back to the center of the orbit, giving rise to gaze-evoked nystagmus. With nystagmus that is due to disturbance of the visual fixation mechanism, the ability to suppress drifts (for example of vestibular origin) during attempted fixation may be deteriorated (Leigh and Zee, 2006).

Other forms of nystagmus are less well understood. Acquired pendular nystagmus (due, e.g., , to multiple sclerosis) has a quasi-sinusoidal waveform (Sharpe and Fletcher, 1986; Leigh and Zee, 2006) and it is most visually distressing, impairing clear vision, and causing oscillopsia (Sharpe and Fletcher, 1986; Leigh and Zee, 1991; Leigh et al., 1994; Ciuffreda and Tannen, 1995; Leigh and Zee, 2006).

Congenital nystagmus may also disrupt steady fixation. One type is due to visual disorders, such as congenital retinal disorders and albinism. Another type of congenital nystagmus is not associated with visual disorders: these individuals often show brief foveation periods when the eyes are relatively still and on target. Such individuals usually have near-normal vision and no illusion of oscillopsia (Leigh and Zee, 2006).

Congenital fixation nystagmus can occur in a variety of jerk and pendular waveforms. In pendular (i.e. sinusoidal) nystagmus, both phases of the oscillation are smooth movements. In jerk nystagmus, the second phase of the oscillation is a correcting saccade (Sharpe and Fletcher, 1986). However, some patients have complex waveforms that are not easy to characterize as either jerk or pendular (Dell'osso and Daroff, 1975; Yee et al., 1976), and many sub-varieties have been described (Dell'osso and Daroff, 1975). Also, the nystagmus may change from pendular to jerk for different gaze directions (Ciuffreda and Tannen, 1995). Psychological factors such as fatigue, stress, and especially attention can exert a strong influence on the intensity and waveform shape of the nystagmus

(Abadi and Dickinson, 1986). Figure 10 illustrates four common nystagmus waveforms.

About 50% of strabismic patients have nystagmus, and 15% of patients with congenital nystagmus have strabismus (Ciuffreda and Tannen, 1995). See Leigh and Zee (2006) for further details and discussion on the various nystagmus types.

Superior oblique myokymia

Superior oblique myokymia is a rapid, small amplitude, non-saccadic, rotatory ocular oscillation limited to one eye (Susac et al., 1973). Attacks usually last less than 10 s, but may occur many times a day (Leigh and Zee, 2006). Superior oblique myokymia can be distinguished from microsaccadic flutter because it is always monocular and has a strong torsional component (Ashe et al., 1991). Superior oblique myokimia is accompanied by monocular blurring (Leigh and Zee, 2006), monocular oscillopsia, and sometimes torsional diplopia (Susac et al., 1973). Although the causes are obscure, the clinical course is benign (Susac et al., 1973). An abnormality of the trochlear motor units may underlie this disorder (Leigh et al., 1994).

Ocular paralysis

Filin and Okhotsimskaya examined the dynamics of fixational eye movements in a large population of patients with orbital paralysis (including myasthenia, myopathy, and malignant exophthalmia), basal paralysis (including patients with isolated paresis and paralysis of the III and VI nerves and patients with total or incomplete ophthalmoplegia), and nuclear paralysis (Filin and Okhotsimskaya, 1977; Okhotsimskaia and Filin, 1977; Okhotsimskaia, 1977). In myasthenia and myopathy the rate and speed of saccades decreased, and in a number of cases drift also decreased in frequency and amplitude, resulting in considerable stabilization of the eyes. In patients with incomplete basal paralyses of the III and IV nerves, the frequency of microsaccades decreased, whereas drift increased in frequency and amplitude. In cases of mild paresis, fixational eye movements of the affected eye were comparable to the normal eye. The most

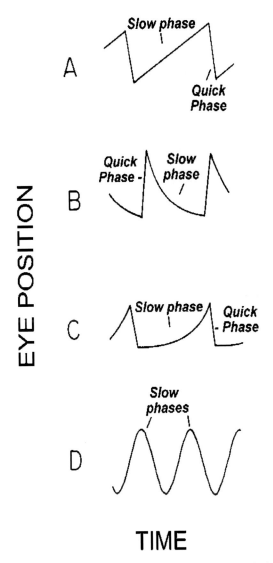

Fig. 10. Common slow-phase waveforms of nystagmus during attempted fixation. (A) Constant velocity drift of the eyes ("saw-tooth" nystagmus), due to peripheral or central vestibular disease, or to cerebral hemisphere lesions. (B) Drift of the eyes from an eccentric orbital position back to the center of the orbit (gaze-evoked nystagmus), with decreasing velocity. The unsustained eye position is caused by an impaired neural integrator. (C) Drift of the eyes away from the central position, with increasing velocity. The unsustained eye position suggests an unstable neural integrator. This disorder is found in the horizontal plane in congenital nystagmus and in the vertical plane in cerebellar disease. (D) Pendular nystagmus (either congenital or acquired). Reprinted from Leigh and Zee (2006). Courtesy of R. John Leigh.

pronounced changes in fixational eye movements occurred in patients with complete paralysis of the oculomotor nerves. None of these patients produced microsaccades and the drift was of small amplitude or completely absent: the eye was stabilized.

Stabilization of the eyes due to complete or incomplete paralysis presumably leads to decrease in visibility and ultimately visual fading, as it is the case with the fading of stabilized images in the laboratory (Ditchburn and Ginsborg, 1952; Riggs and Ratliff, 1952).

Slow refixation saccades

Slow saccades of restricted amplitude may reflect abnormalities in the oculomotor periphery, whereas slow saccades of normal amplitude are usually caused by central neurological disorders (Leigh and Zee, 2006). Refixation saccades can be pathologically slow in progressive nuclear palsy (Garbutt et al., 2003, 2004), especially when the disease is severe (Boghen et al., 1974; Troost et al., 1976).

The velocity–amplitude relationship of microsaccades in progressive nuclear palsy may be similarly affected (R. John Leigh, personal communication). If so, several questions need to be addressed: are slow microsaccades as effective in preventing adaptation and counteracting fading as normal microsaccades? If not, how effective are they? In our previous experiments, we found that both microsaccade amplitudes and microsaccade velocities were positively correlated with visibility during fixation. However, as microsaccade amplitude and microsaccade velocity covary in the main sequence, it is difficult to determine which of these two variables, amplitude or speed, is more critical to visibility. Future experiments that examine the effects on visibility of microsaccades with an abnormal velocity–amplitude slope should help answer these questions.

Conclusions

Approximately 80% of our visual experience happens during fixation. During the other 20% of the time we are virtually blind, due to saccadic suppression mechanisms. Therefore, understanding

172

the neural and perceptual effects of fixational eye movements is crucial to understanding vision.

Fixational eye movements were first measured in the 1950s, but sometime during the 1970s, the field arrived at an impasse due to difficulties in data collection, discrepancies in results by different laboratories, and disagreements over the interpretation of the available data. A revival in interest in the late 1990s was ushered in by the development of very accurate non-invasive eye movement measurement techniques, in addition to the advent of single-unit recording techniques in alert monkeys, and new computational approaches for eye-movement characterization and modeling. Research in fixational eye movements is one of the newest and fastest-moving fields in visual and oculomotor neuroscience today.

The evaluation of fixational eye movements may be critical to the early and differential diagnosis of oculomotor disease, to the assessment of ongoing treatments, and to develop therapies to restore visual function in patients who cannot produce normal eye movements during fixation. Many visual and oculomotor diseases include fixational eye movement defects that have gone untreated. Correcting these fixational eye movement deficiencies may provide a novel way to ameliorate some of the debilitating effects of these pathologies.

Finally, a large amount of psychophysical and physiological visual research has been carried out while subjects were engaged in visual fixation. Therefore, understanding the precise physiological and perceptual contributions of fixational eye movements may moreover be critical to the interpretation of previous and future vision research.

Abbreviations

LGN   lateral geniculate nucleus of the thalamus
RF   receptive field
V1   primary visual cortex

Acknowledgments

Drs. R. John Leigh, Stephen Macknik, and Xoana Troncoso read the manuscript and made helpful

comments. I am very grateful to Dr. R. John Leigh for graciously providing Figs. 9 and 10, and for his insights and discussion of slow saccades, square-wave jerks, and many other concepts addressed here. Thanks also to Dr. David Sparks for helpful discussion and for pointing me to the Van Gisbergen and colleagues' studies on the generation of microsaccades. Thomas Dyar and Dr. Xoana Troncoso helped with data analysis and figure formatting, and Dr. Xoana Troncoso acquired the data for Figs. 5b, c and 6. This study was supported by the Barrow Neurological Foundation.

References

Abadi, R.V. and Dickinson, C.M. (1986) Waveform characteristics in congenital nystagmus. Doc. Ophthalmol., 64: 153–167.

Abadi, R.V. and Gowen, E. (2004) Characteristics of saccadic intrusions. Vision Res., 44: 2675–2690.

Abadi, R.V., Scallan, C.J. and Clement, R.A. (2000) The characteristics of dynamic overshoots in square-wave jerks, and in congenital and manifest latent nystagmus. Vision Res., 40: 2813–2829.

Abel, L.A., Traccis, S., Dell'Osso, L.F., Daroff, R.B. and Troost, B.T. (1984) Square-wave oscillation: the relationship of saccadic intrusions and oscillations. Neuro-ophthalmology (Amsterdam), 4: 21–25.

Albrecht, D.G. (1995) Visual cortex neurons in monkey and cat: effect of contrast on the spatial and temporal phase transfer functions. Visual Neurosci., 12: 1191–1210.

Albrecht, D.G. and Hamilton, D.B. (1982) Striate cortex of monkey and cat: contrast response function. J. Neurophysiol., 48: 217–237.

Ashe, J., Hain, T.C., Zee, D.S. and Schatz, N.J. (1991) Microsaccadic flutter. Brain, 114(Pt 1B): 461–472.

Bair, W. and O'Keefe, L.P. (1998) The influence of fixational eye movements on the response of neurons in area MT of the macaque. Visual Neurosci., 15: 779–786.

Barrett, S.F. and Zwick, H. (2000) An eye movement technique for correlating fixational target eye movements with location on the retinal image. Biomed. Sci. Instrum., 36: 183–188.

Bedell, H.E., Yap, Y.L. and Flom, M.C. (1990) Fixational drift and nasal-temporal pursuit asymmetries in strabismic amblyopes. Invest. Ophthalmol. Vis. Sci., 31: 968–976.

Beeler, G.W. (1967) Visual threshold changes resulting from spontaneous saccadic eye movements. Vision Res., 7: 769–775.

Boghen, D., Troost, B.T., Daroff, R.B., Dell'Osso, L.F. and Birkett, J.E. (1974) Velocity characteristics of normal human saccades. Invest. Ophthalmol., 13: 619–623.

Bridgeman, B. and Palca, J. (1980) The role of microsaccades in high acuity observational tasks. Vision Res., 20: 813–817.

Bridgeman, B.B. and Macknik, S.L. (1995) Saccadic suppression relies on luminance information. Psycholo. Res., 58: 163–168.

Burr, D.C., Morrone, M.C. and Ross, J. (1994) Selective suppresion of the magnocellular visual pathway during saccadic eye movements. Nature, 371: 511–513.

Carlow, T.J. (1986) Medical treatment of nystagmus and ocular motor disorders. Int. Ophthalmol. Clin., 26: 251–264.

Carpenter, R.H.S. (1988) Movements of the Eyes. Pion Ltd., London, UK.

Ciuffreda, K.J., Kenyon, R.V. and Stark, L. (1978) Increased saccadic latencies in amblyopic eyes. Invest. Ophthalmol. Vis. Sci., 17: 697–702.

Ciuffreda, K.J., Kenyon, R.V. and Stark, L. (1979a) Different rates of functional recovery of eye movements during orthoptics treatment in an adult amblyope. Invest. Ophthalmol. Vis. Sci., 18: 213–219.

Ciuffreda, K.J., Kenyon, R.V. and Stark, L. (1979b) Fixational eye movements in amblyopia and strabismus. J. Am. Optom. Assoc., 50: 1251–1258.

Ciuffreda, K.J., Kenyon, R.V. and Stark, L. (1979c) Saccadic intrusions in strabismus. Arch. Ophthalmol., 97: 1673–1679.

Ciuffreda, K.J., Kenyon, R.V. and Stark, L. (1979d) Suppression of fixational saccades in strabismic and anisometropic amblyopia. Ophthalmol. Res., 11: 31–39.

Ciuffreda, K.J., Kenyon, R.V. and Stark, L. (1980) Increased drift in amblyopic eyes. Br. J. Ophthalmol., 64: 7–14.

Ciuffreda, K.L. and Tannen, B. (1995) Eye Movement Basics for the Clinician. Mosby-Year book, Inc., St. Louis, MO.

Clarke, F.J.J. (1957) Rapid light adaptation of localised areas of the extra-foveal retina. Opt. Acta, 4: 69–77.

Clarke, F.J.J. (1960) A study of Troxler's effect. Opt. Acta, 7: 219–236.

Clarke, F.J.J. (1961) Visual recovery following local adaptation of the perpheral retina (Troxler's effect). Opt. Acta, 8: 121–135.

Clarke, F.J.J. and Belcher, S.J. (1962) On the localization of Troxler's effect in the visual pathway. Vision Res., 2: 53–68.

Clowes, M.B. (1962) A note on colour discrimination under conditions of retinal image constraint. Opt. Acta, 9: 65–68.

Coakley, D. (1983) Miniature Eye Movement and Brain Stem Function. CRC Press, Boca Raton, FL.

Cogan, D.G., Chu, F.C. and Reingold, D.B. (1982) Ocular signs of cerebellar disease. Arch. Ophthalmol., 100: 755–760.

Coppola, D. and Purves, D. (1996) The extraordinarily rapid disappearance of entoptic images. Proc. Natl. Acad. Sci. USA, 93: 8001–8004.

Cornsweet, T.N. (1956) Determination of the stimuli for involuntary drifts and saccadic eye movements. J. Opt. Soc. Am., 46: 987–993.

Cunitz, R.J. and Steinman, R.M. (1969) Comparison of saccadic eye movements during fixation and reading. Vision Res., 9: 683–693.

Dell'osso, L.F. and Daroff, R.B. (1975) Congenital nystagmus waveforms and foveation strategy. Doc. Ophthalmol., 39: 155–182.

Ditchburn, R.W. (1955) Eye-movements in relation to retinal action. Opt. Acta (Lond.), 1: 171–176.

Ditchburn, R.W. (1980) The function of small saccades. Vision Res., 20: 271–272.

Ditchburn, R.W., Fender, D.H. and Mayne, S. (1959) Vision with controlled movements of the retinal image. J. Physiol., 145: 98–107.

Ditchburn, R.W. and Ginsborg, B.L. (1952) Vision with a stabilized retinal image. Nature, 170: 36–37.

Ditchburn, R.W. and Ginsborg, B.L. (1953) Involuntary eye movements during fixation. J. Physiol., 119: 1–17.

Drysdale, A.E. (1975) The visibility of retinal blood vessels. Vision Res., 15: 813–818.

Eizenman, M., Hallett, P.E. and Frecker, R.C. (1985) Power spectra for ocular drift and tremor. Vision Res., 25: 1635–1640.

Ellenberger Jr., C., Keltner, J.L. and Stroud, M.H. (1972) Ocular dyskinesia in cerebellar disease. Evidence for the similarity of opsoclonus, ocular dysmetria and flutter-like oscillations. Brain, 95: 685–692.

Engbert, R. and Kliegl, R. (2003a) Binocular coordination in microsaccades. In: Hyona, R.R.J. and Deubel, H. (Eds.), The Mind's Eyes: Cognitive and Applied Aspects of Eye Movements. Elsevier, Oxford, pp. 103–117.

Engbert, R. and Kliegl, R. (2003b) Microsaccades uncover the orientation of covert attention. Vision Res., 43: 1035–1045.

Engbert, R. and Kliegl, R. (2004) Microsaccades keep the eyes' balance during fixation. Psychol. Sci., 15: 431–436.

Feldon, S.E. and Langston, J.W. (1977) Square-wave jerks: a disorder of microsaccades. Neurology, 27: 278–281.

Fermuller, C., Pless, R. and Aloimonos, Y. (1997) Families of stationary patterns producing illusory movement: insights into the visual system. Proc. Biol. Sci., 264: 795–806.

Filin, V.A. and Okhotsimskaya. (1977) Use of a photoelectronic apparatus to record saccadic eye movements in neurological practice. Zh. Nevropat. Psikhiat. Korsakov, 77: 221–225.

Fiorentini, A. and Ercoles, A.M. (1966) Involuntary eye movements during attempted monocular fixation. Atti. Fond. Giorgio Ronchi, 21: 199–217.

Foroozan, R. and Brodsky, M.C. (2004) Microsaccadic opsoclonus: an idiopathic cause of oscillopsia and episodic blurred vision. Am. J. Ophthalmol., 138: 1053–1054.

Garbutt, S., Harwood, M.R., Kumar, A.N., Han, Y.H. and Leigh, R.J. (2003) Evaluating small eye movements in patients with saccadic palsies. Ann. NY Acad. Sci., 1004: 337–346.

Garbutt, S., Riley, D.E., Kumar, A.N., Han, Y., Harwood, M.R. and Leigh, R.J. (2004) Abnormalities of optokinetic nystagmus in progressive supranuclear palsy. J. Neurol. Neurosurg. Psychiat., 75: 1386–1394.

Gawne, T.J., Kjaer, T.W. and Richmond, B.J. (1996) Latency: another potential code for feature binding in striate cortex. J. Neurophysiol., 76: 1356–1360.

Gerrits, H.J. (1978) Differences in peripheral and foveal effects observed in stabilized vision. Exp. Brain Res., 32: 225–244.

Gerrits, H.J. and Vendrik, A.J. (1970) Artificial movements of a stabilized image. Vision Res., 10: 1443–1456.

Gerrits, H.J. and Vendrik, A.J. (1974) The influence of stimulus movements on perception in parafoveal stabilized vision. Vision Res., 14: 175–180.

Gilchrist, I.D., Brown, V. and Findlay, J.M. (1997) Saccades without eye movements. Nature, 390: 130–131.

Gilchrist, I.D., Brown, V., Findlay, J.M. and Clarke, M.P. (1998) Using the eye-movement system to control the head. Proc. R. Soc. Lond. B Biol. Sci., 265: 1831–1836.

Gould, T.D., Bastain, T.M., Israel, M.E., Hommer, D.W. and Castellanos, F.X. (2001) Altered performance on an ocular fixation task in attention-deficit/hyperactivity disorder. Biol. Psychiat., 50: 633–635.

Greschner, M., Bongard, M., Rujan, P. and Ammermuller, J. (2002) Retinal ganglion cell synchronization by fixational eye movements improves feature stimation. Nature, 5: 341–347.

Gur, M., Beylin, A. and Snodderly, D.M. (1997) Response variability of neurons in primary visual cortex (V1) of alert monkeys. J. Neurosci., 17: 2914–2920.

Hafed, Z.M. and Clark, J.J. (2002) Microsaccades as an overt measure of covert attention shifts. Vision Res., 42: 2533–2545.

Hennig, M.A., Kerscher, N.J., Funke, K. and Worgotter, F. (2002) Stochastic resonance in visual cortical neurons: does the eye-tremor actually improve visual acuity. Neurocomputing, 44–46: 115–120.

Herishanu, Y.O. and Sharpe, J.A. (1981) Normal square-wave jerks. Invest. Ophthalmol. Vis. Sci., 20: 268–272.

Hermann, L. (1870) Eine Erscheinung des simultanen Contrastes. Pflügers Arch. Gesamte Physiol., 3: 13–15.

Hertle, R.W., Tabuchi, A., Dell'Osso, L.F., Abel, L.A. and Weisman, B.M. (1988) Saccadic oscillations and intrusions preceding the postnatal appearance of congenital nystagmud. Neuro-ophthalmology, 8: 37.

Hess, R.F., Campbell, F.W. and Greenhalgh, T. (1978) On the nature of the neural abnormality in human amblyopia; neural aberrations and neural sensitivity loss. Pflugers Arch., 377: 201–207.

Hotson, J.R. (1982) Cerebellar control of fixation eye movements. Neurology, 32: 31–36.

Hotson, J.R. (1984) Convergence-initiated voluntary flutter: a normal intrinsic capability in man. Brain Res., 294: 299–304.

Kitaoka, A. and Ashida, H. (2003) Phenomenal characteristics of the peripheral drift illusion. Vision, 15: 261–262.

Kosnik, W., Fikre, J. and Sekuler, R. (1986) Visual fixation stability in older adults. Invest. Ophthalmol. Vis. Sci., 27: 1720–1725.

Kowler, E. and Steinman, R.M. (1977) The role of small saccades in counting. Vision Res., 17: 141–146.

Kowler, E. and Steinman, R.M. (1979) Miniature saccades: eye movements that do not count. Vision Res., 19: 105–108.

Kowler, E. and Steinman, R.M. (1980) Small saccades serve no useful purpose: reply to a letter by R.W. Ditchburn. Vision Res., 20: 273–276.

Krauskopf, J. (1957) Effect of retinal image motion on contrast thresholds for maintained vision. J. Opt. Soc. Am., 47: 740–744.

Krauskopf, J. (1966) Lack of inhibition during involuntary saccades. Am. J. Psychol., 79: 73–81.

Krauskopf, J., Cornsweet, T.N. and Riggs, L.A. (1960) Analysis of eye movements during monocular and binocular fixation. J. Opt. Soc. Am., 50: 572–578.

Kumar, T. and Glaser, D.A. (2006) Illusory motion in enigma: a psychophysical investigation. Proc. Natl. Acad. Sci. USA, 103: 1947–1952.

Land, M.F., Furneaux, S.M. and Gilchrist, I.D. (2002) The organization of visually mediated actions in a subject without eye movements. Neurocase, 8: 80–87.

Lawwill, T. (1968) Local adaptation in functional amblyopia. Am. J. Ophthalmol., 65: 903–906.

Leigh, R.J., Averbuch-Heller, L., Tomsak, R.L., Remler, B.F., Yaniglos, S.S. and Dell'Osso, L.F. (1994) Treatment of abnormal eye movements that impair vision: strategies based on current concepts of physiology and pharmacology. Ann. Neurol., 36: 129–141.

Leigh, R.J., Thurston, S.E., Tomsak, R.L., Grossman, G.E. and Lanska, D.J. (1989) Effect of monocular visual loss upon stability of gaze. Invest. Ophthalmol. Vis. Sci., 30: 288–292.

Leigh, R.J. and Zee, D.S. (1980) Eye movements of the blind. Invest. Ophthalmol. Vis. Sci., 19: 328–331.

Leigh, R.J. and Zee, D.S. (1991) Oculomotor disorders. In: Carpenter, R.H.S. (Ed.), Eye Movements. Macmillan, Boca Raton, FL.

Leigh, R.J. and Zee, D.S. (2006) The Neurology of Eye Mmovements. Oxford University Press, Oxford.

Leopold, D.A. and Logothetis, N.K. (1998) Microsaccades differentially modulate neural activity in the striate and extrastriate visual cortex. Exp. Brain Res., 123: 341–345.

Leviant, I. (1996) Does 'brain power' make Enigma spin? Proc. R. Soc. Lond. B, 263: 997–1001.

Livingstone, M.S., Freeman, D.C. and Hubel, D.H. (1996) Visual responses in V1 of freely viewing monkeys. Cold Spring Harb. Sym. Quant. Biol., 765: 27–37.

Lord, M.P. (1951) Measurement of binocular eye movements of subjects in the sitting position. Br. J. Ophthalmol., 35: 21–30.

Macknik, S.L., Fisher, B.D. and Bridgeman, B. (1991) Flicker distorts visual space constancy. Vision Res., 31: 2057–2064.

Malinov, I.V., Epelboim, J., Herst, A.N. and Steinman, R.M. (2000) Characteristics of saccades and vergence in two types of sequential looking tasks. Vision Res., 40: 2083–2090.

Martinez-Conde, S., Macknik, S.L. and Hubel, D.H. (2000) Microsaccadic eye movements and firing of single cells in the striate cortex of macaque monkeys. Nat. Neurosci., 3: 251–258.

Martinez-Conde, S., Macknik, S.L. and Hubel, D.H. (2002) The function of bursts of spikes during visual fixation in the awake primate lateral geniculate nucleus and primary visual cortex. Proc. Natl. Acad. Sci. USA, 99: 13920–13925.

Martinez-Conde, S., Macknik, S.L. and Hubel, D.H. (2004) The role of fixational eye movements in visual perception. Nat. Rev. Neurosci., 5: 229–240.

Martinez-Conde, S., Macknik, S.L., Troncoso, X. and Dyar, T.A. (2006) Microsaccades counteract visual fading during fixation. Neuron, 49: 297–305.

Moller, F., Laursen, M.L., Tygesen, J. and Sjolie, A.K. (2002) Binocular quantification and characterization of microsaccades. Graefes Arch. Clin. Exp. Ophthalmol., 240: 765–770.

Murakami, I. and Cavanagh, P. (2001) Visual jitter: evidence for visual-motion-based compensation of retinal slip due to small eye movements. Vision Res., 41: 173–186.

Nachmias, J. (1959) Two-dimensional motion of the retinal image during monocular fixation. J. Opt. Soc. Am., 49: 901–908.

Nachmias, J. (1961) Determiners of the drift of the eye during monocular fixation. J. Opt. Soc. Am., 51: 761–766.

Ohtsuka, K., Mukuno, K., Ukai, K. and Ishikawa, S. (1986) The origin of square wave jerks: conditions of fixation and microsaccades. Jpn. J. Ophthalmol., 30: 209–215.

Okhotsimskaia, D.A. and Filin, V.A. (1977) Micromovements of the eyes in isolated pareses and paralyses of the abducens and oculomotor nerves. Zh. Vopr. Neirokhir. Im. N. N. Burdenko, 1: 47–54.

Okhotsimskaia, S.A. (1977) Eye micromovements in patients with oculomotor myopathies. Oftalmol. Zh., 32: 123–128.

Ouchi, H. (1977) Japanese Optical and Geometrical Art. Dover, New York.

Ramachandran, V.S. and Gregory, R.L. (1991) Perceptual filling in of artificially induced scotomas in human vision (see comments). Nature, 350: 699–702.

Ramachandran, V.S., Gregory, R.L. and Aiken, W. (1993) Perceptual fading of visual texture borders. Vision Res., 33: 717–721.

Ramat, S., Leigh, R.J., Zee, D.S. and Optican, L.M. (2005) Ocular oscillations generated by coupling of brainstem excitatory and inhibitory saccadic burst neurons. Exp. Brain Res., 160: 89–106.

Ramat, S., Somers, J.T., Vallabh, E.D. and Leigh, R.J. (1999) Conjugate ocular oscillations during shifts of the direction and depth of visual fixation. Invest. Ophthalmol. Vis. Sci., 40: 1681–1686.

Ratliff, F. and Riggs, L.A. (1950) Involuntary motions of the eye during monocular fixation. J. Exp. Psychol., 40: 687–701.

Reppas, J.B., Usrey, W.M. and Reid, R.C. (2002) Saccadic eye movements modulate visual responses in the lateral geniculate nucleus. Neuron, 35: 961–974.

Riggs, L.A. and Ratliff, F. (1951) Visual acuity and the normal tremor of the eyes. Science, 114: 17–18.

Riggs, L.A. and Ratliff, F. (1952) The effects of counteracting the normal movements of the eye. J. Opt. Soc. Am., 42: 872–873.

Riggs, L.A., Ratliff, F., Cornsweet, J.C. and Cornsweet, T.N. (1953) The disappearance of steadily fixated visual test objects. J. Opt. Soc. Am., 43: 495–501.

Ross, J., Burr, D. and Morrone, C. (1996) Suppression of the magnocellular pathway during saccades. Behav. Brain Res., 80: 1–8.

Ross, J., Morrone, M.C. and Burr, D.C. (1997) Compression of visual space before saccades (see comments). Nature, 386: 598–601.

Ross, J., Morrone, M.C., Goldberg, M.E. and Burr, D.C. (2001) Changes in visual perception at the time of saccades (see comments). Trends Neurosci., 24: 113–121.

Sabrin, H.W. and Kertesz, A.E. (1983) The effect of imposed fixational eye movements on binocular rivalry. Percept. Psychophys., 34: 155–157.

Sasaki, Y., Murakami, I., Cavanagh, P. and Tootell, R.H. (2002) Human brain activity during illusory visual jitter as revealed by functional magnetic resonance imaging. Neuron, 35: 1147–1156.

Selhorst, J.B., Stark, L., Ochs, A.L. and Hoyt, W.F. (1976) Disorders in cerebellar ocular motor control. II. Macrosaccadic oscillation. An oculographic, control system and clinico-anatomical analysis. Brain, 99: 509–522.

Shakhnovich, A.R. and Thomas, J.G. (1974) Proceedings: micro-tremor of the eyes as an index of motor unit activity, and of the functional state of the brain stem. J. Physiol., 238: 36P.

Shakhnovich, A.R. and Thomas, J.G. (1977) Micro-tremor of the eyes of comatose patients. Electroencephalogr. Clin. Neurophysiol., 42: 117–119.

Shallo-Hoffmann, J., Petersen, J. and Muhlendyck, H. (1989) How normal are "normal" square wave jerks. Invest. Ophthalmol. Vis. Sci., 30: 1009–1011.

Shallo-Hoffmann, J., Sendler, B. and Muhlendyck, H. (1990) Normal square wave jerks in differing age groups. Invest. Ophthalmol. Vis. Sci., 31: 1649–1652.

Sharpe, C.R. (1972) The visibility and fading of thin lines visualized by their controlled movement across the retina. J. Physiol. (Lond.), 222: 113–134.

Sharpe, J.A. and Fletcher, W.A. (1986) Disorders of visual fixation. In: Lawton Smith, J. (Ed.), Neuro-Ophthalmology Now! Field, Rich and Assoc., Chicago, pp. 267–284.

Skavenski, A.A., Hansen, R.M., Steinman, R.M. and Winterson, B.J. (1979) Quality of retinal image stabilization during small natural and artificial body rotations in man. Vision Res., 19: 675–683.

Skavenski, A.A., Robinson, D.A., Steinman, R.M. and Timberlake, G.T. (1975) Miniature eye movements of fixation in rhesus monkey. Vision Res., 15: 1269–1273.

Snodderly, D.M. (1987) Effects of light and dark environments on macaque and human fixational eye movements. Vision Res., 27: 401–415.

Snodderly, D.M., Kagan, I. and Gur, M. (2001) Selective activation of visual cortex neurons by fixational eye movements: implications for neural coding. Visual Neurosci., 18: 259–277.

Snodderly, D.M. and Kurtz, D. (1985) Eye position during fixation tasks: comparison of macaque and human. Vision Res., 25: 83–98.

Spauschus, A., Marsden, J., Halliday, D.M., Rosenberg, J.R. and Brown, P. (1999) The origin of ocular microtremor in man. Exp. Brain Res., 126: 556–562.

Sperling, G. (1990) Comparison of perception in the moving and stationary eye. In: Kowler, E. (Ed.), Eye Movements and Their Role in Visual and Cognitive Processes. Elsevier, Amsterdam, The Netherlands, pp. 307–351.

Spillmann, L. and Kurtenbach, A. (1992) Dynamic noise backgrounds facilitate target fading. Vision Res., 32: 1941–1946.

Srebro, R. (1983) Fixation of normal and amblyopic eyes. Arch. Ophthalmol., 101: 214–217.

St Cyr, G.J. and Fender, D.H. (1969) The interplay of drifts and flicks in binocular fixation. Vision Res., 9: 245–265.

Steinman, R.M. (1975) Oculomotor effects on vision. In: Lennerstrand, G. and Bach-Y-Rita, P. (Eds.), Basic Mechanisms of Ocular Motility and Their Clinical Implications. Pergamon Press, New York.

Steinman, R.M. and Collewijn, H. (1980) Binocular retinal image motion during active head rotation. Vision Res., 20: 415–429.

Steinman, R.M., Cunitz, R.J., Timberlake, G.T. and Herman, M. (1967) Voluntary control of microsaccades during maintained monocular fixation. Science, 155: 1577–1579.

Steinman, R.M., Haddad, G.M., Skavenski, A.A. and Wyman, D. (1973) Miniature eye movement. Science, 181: 810–819.

Susac, J.O., Smith, J.L. and Schatz, N.J. (1973) Superior oblique myokymia. Arch. Neurol., 29: 432–434.

Troost, B.T., Daroff, R.B. and Dell'Osso, L.F. (1976) Quantitative analysis of the ocular motor deficit in progressive supranuclear palsy (PSP). Trans. Am. Neurol. Assoc., 101: 60–64.

Troxler, D. (1804). Ueber das Verschwinden gegebener Gegenstande innerhalb unseres Gesichtskreises. In: Himley, K. and Schmidt, J.A. (Eds.), Ophthalmologische Bibliothek, Springer, Jena, pp. 1–53.

Van Gisbergen, J.A., Robinson, D.A. and Gielen, S. (1981) A quantitative analysis of generation of saccadic eye movements by burst neurons. J. Neurophysiol., 45: 417–442.

Van Gisbergen, J.A.M. and Robinson, D.A. (1977) Generation of micro and macrosaccades by burst neurons in the monkey. In: Baker, R. and Berthoz, A. (Eds.), Control of Gaze by Brain Stem Neurons. Elsevier, New York.

von Noorden, G.K. and Burian, H.M. (1958) An electro-oculographic study of the behavior of the fixation in amblyopic eyes in light- and dark-adpated state: a preliminary report. Am. J. Ophthalmol., 46: 68.

Wald, G. and Burian, H.M. (1944) The dissociation of form and light perception in strabismus amblyopia. Am. J. Ophthalmol., 27: 950–963.

Winterson, B.J. and Collewijn, H. (1976) Microsaccades during finely guided visuomotor tasks. Vision Res., 16: 1387–1390.

Wurtz, R.H. (1968) Visual cortex neurons: response to stimuli during rapid eye movements. Science, 162: 1148–1150.

Wurtz, R.H. (1969) Comparison of effects of eye movements and stimulus movements on striate cortex neurons of the monkey. J. Neurophysiol., 32: 987–994.

Yamazaki, A. (1968) Electrophysiological study on "flick" eye movements during fixation. Nippon Ganka Gakkai Zasshi, 72: 2446–2459.

Yarbus, A.L. (1967) Eye Movements and Vision. Plenum Press, New York.

Yee, R.D., Wong, E.K., Baloh, R.W. and Honrubia, V. (1976) A study of congenital nystagmus: waveforms. Neurology, 26: 326–333.

Zee, D.S. and Robinson, D.A. (1979) A hypothetical explanation of saccadic oscillations. Ann. Neurol., 5: 405–414.

Zuber, B.L. and Stark, L. (1965) Microsaccades and the velocity–amplitude relationship for saccadic eye movements. Science, 150: 1459–1460.

Martinez-Conde, Macknik, Martinez, Alonso & Tse (Eds.)
Progress in Brain Research, Vol. 154
ISSN 0079-6123

CHAPTER 9

Microsaccades: a microcosm for research on oculomotor control, attention, and visual perception

Ralf Engbert*

Computational Neuroscience, Department of Psychology, University of Potsdam, PO Box 601553, 14415 Potsdam, Germany

Abstract: Miniature eye movements occur involuntarily during visual fixation. The most prominent contribution to these fixational eye movements is generated by microsaccades, which are rapid small-amplitude saccades with a rate of about one per second. Recent work demonstrates that microsaccades are optimized to counteract perceptual fading during perception of a stationary scene. Furthermore, microsaccades are modulated by visual attention and turned out to generate rich spatio-temporal dynamics. We conclude that the investigation of microsaccades will evolve into a new research field contributing to many facets of oculomotor control, visual perception, and the allocation of attention.

Keywords: fixational eye movements; microsaccades; random walk; visual perception; covert attention

Introduction

Visual perception is based on motion. This fact is obvious for some nervous systems in animals: a resting fly is invisible to a frog. The situation rapidly changes as soon as the fly starts to move. Thus, motion is an essential prerequisite for sensation in frogs (Lettvin et al., 1959). Because motion information is critical in predator–prey relationships, the detection of motion might have been a key advantage for the evolution of visual systems.

The human visual system shows a rapid adaptation to stationary objects. The adaptation causes *perceptual fading* when the retinal image is artificially stabilized in the experimental paradigm of *retinal stabilization* (Ditchburn and Ginsborg, 1952; Riggs et al., 1953). This is a potential fingerprint of evolutionary history in the human visual system. Equipped with such a visual system optimized for the detection of motion and change,

we were unable to process fine details of a completely stationary scene without active refresh of the retinal image, because retinal adaptation would induce a bleaching of constant input. To counteract retinal adaptation our oculomotor system generates miniature eye movements (Ratliff and Riggs, 1950). Thus, ironically, the term "visual fixation" is a misnomer, since there is rich dynamical behavior during each fixation. To capture this built-in paradox, the term *fixational eye movements* is used most often.

It has been well documented long before the advent of research into modern eye movement in the 1950s that our eyes are never motionless. For example, Helmholtz (1866), one of the pioneers of eye-movement research, noticed the difficulty of producing perfect fixation. Interestingly, Helmholtz had already suggested the prevention of retinal fatigue as a perceptual function for fixational eye movements. Fixational eye movements are rather erratic (Fig. 1) — the eye's trajectory represents a *random walk*. Therefore, adequate mathematical methods for the analysis of fixational eye movements

*Corresponding author. Tel.: +49-331-9772869;
Fax: +49-331-9772793; E-mail: engbert@rz.uni-potsdam.de

DOI: 10.1016/S0079-6123(06)54009-9

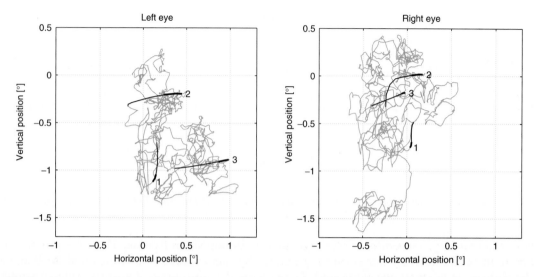

Fig. 1. Fixational eye movements during fixation. A human observer fixated a small stimulus for a duration of 3 s (left and right panels refer to data from the left and right eye, respectively). The trajectory is rather irregular, but contains three microsaccades, which can be characterized by a linear movement episode. The numbers refer to the sequences of microsaccades and indicate the endpoints of the microsaccades.

come from statistical physics, a discipline which can be traced back to Einstein's work on Brownian motion (Einstein, 1905).

Fixational eye movements in human observers fall into three different physiological categories: microsaccades, tremor, and drift. All three types of fixational eye movements occur involuntarily and unconsciously. Microsaccades are rapid small-amplitude movements of the eyes with an average rate of $1–2 s^{-1}$. As a consequence, the eyes move more linearly during a microsaccade than during the remaining part of the trajectory (Fig. 1).[1] The linear movement during a microsaccade might be an effect of the eye's inertia, i.e., microsaccades are ballistic, which is compatible with the fact that the kinematic properties of microsaccades are very similar to those of voluntary saccades (Zuber et al., 1965). Drift is a low-velocity movement with a peak velocity below $30 min arc s^{-1}$. Tremor is an oscillatory component of fixational eye movements within a frequency range from 30 to 100 Hz superimposed to drift.

It has been an open research problem over the last 30 years to find a specific function for microsaccades. Cornsweet (1956) originally suggested that microsaccades might correct the errors produced by the drift component of fixational eye movements. There are, however, two main lines of evidence against such a straightforward functional interpretation of microsaccades. First, microsaccades can be suppressed voluntarily for several seconds without perceptual bleaching (Steinman et al., 1967). This result lends support to the hypothesis that the contribution of microsaccades to the prevention of retinal adaptation can simply be replaced by slow movements. Second, microsaccades are naturally suppressed in laboratory analogs of high-acuity tasks like threading a needle or shooting a rifle (Winterson and Collewijn, 1976; Bridgeman and Palca, 1980). Therefore, a specific function for microsaccades was rejected. While Ditchburn (1980) argued that the fact that humans can learn to prevent microsaccades does not contradict a specific role of microsaccades for normal vision, Kowler and Steinman (1980) concluded that microsaccades serve no useful purpose. The debate was unresolved and presumably generated a decrease in the interest in microsaccades and fixational eye movements during the 1980s until the mid-1990s, which is evident in the number of publications on or relevant to microsaccades (Fig. 2).

[1]The data used to illustrate properties of microsaccades were obtained from a simple fixation task. For details of the methods see Engbert and Kliegl (2004).

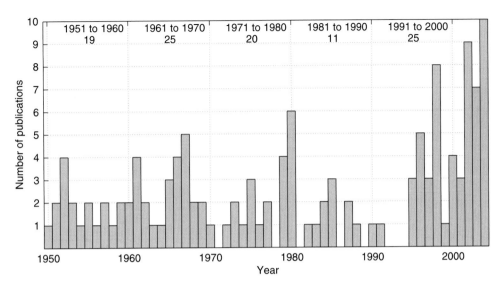

Fig. 2. Number of publications on microsaccades and fixational eye movements per year based on the references cited in the review article by Martinez-Conde et al. (2004).

A renaissance of research on microsaccades has been triggered by (i) new techniques in eye-movement recording, which facilitate the measurement of microsaccades in laboratory situations and the computational analysis of large datasets and (ii) neurophysiological findings, which demonstrate the impact of microsaccades on visual information processing. As an example for the latter line of research, it has been discovered that microsaccades are correlated with bursts of spikes in the primary visual cortex (Martinez-Conde et al., 2000, 2002). Since this and related findings were summarized and discussed in a recent review article by Martinez-Conde et al. (2004), we focus on the behavioral aspects of microsaccades here. The article is organized as follows: we discuss the problem of microsaccade detection, summarize the main kinematic and dynamic properties, and address the modulation of microsaccade rate by visual attention.

Detection of microsaccades

Microsaccades can be detected in eye-movement data because of their higher velocity compared to drift. We developed an algorithm based on a two-dimensional (2D) velocity space for the eyes'

trajectories (Engbert and Kliegl, 2003a).[2] First, the time series of eye positions is transformed to velocities by

$$\vec{v}_n = \frac{\vec{x}_{n+2} + \vec{x}_{n+1} - \vec{x}_{n-1} - \vec{x}_{n-2}}{6\Delta t} \tag{1}$$

which represents a weighted moving average over five data samples to suppress noise (for details see Engbert and Kliegl, 2003b). As can be seen in Fig. 3, the transformation of the trajectory into 2D velocity space results in a scattering of data samples around the origin of the coordinate system. In this representation, microsaccades are "outliers" with much higher velocity than what one would expect from assuming a normal distribution.

Second, we compute the standard deviation $\sigma_{x,y}$ to estimate the noise level of an individual trial and take a fixed multiple λ of this value as a threshold $\eta_{x,y}$ for the detection of a microsaccade:

$$\sigma_{x,y} = \sqrt{\left\langle v_{x,y}^2 \right\rangle - \left\langle v_{x,y} \right\rangle^2} \tag{2}$$

$$\eta_{x,y} = \lambda \, \sigma_{x,y} \tag{3}$$

where $\langle \cdot \rangle$ denotes the median estimator to suppress the influence of the high-velocity samples arising

---

[2]Our algorithm is available on the internet at: http://www.agnld.uni-potsdam.de/~ralf/micro/.

180

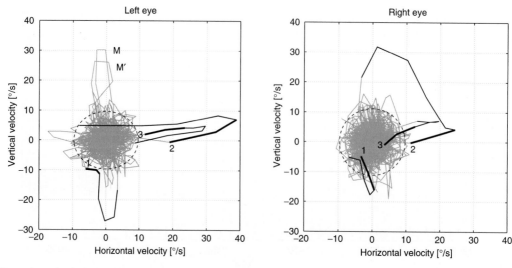

Fig. 3. Detection of microsaccades. After transformation of the trajectory into 2D velocity space, microsaccades can easily be identified by their high velocity (data samples outside ellipse, where $\lambda = 5$ compared to the slower drift component (inside ellipse). The raw data correspond to the same trials as the data plotted in Fig. 1.

from the microsaccades. Because the analysis is performed separately for horizontal and vertical components, the corresponding thresholds η_x and η_y define an ellipse in the velocity space (Fig. 3). As a necessary condition for a microsaccade, we require that all samples $\vec{v}_k = (v_{k,x}, v_{k,y})$ fulfill the criterion

$$\left(\frac{v_{k,x}}{\eta_x}\right)^2 + \left(\frac{v_{k,y}}{\eta_y}\right)^2 > 1 \tag{4}$$

Third, we apply a lower cutoff of 6 ms (or three data samples) for the microsaccade durations to reduce noise.

Fourth, microsaccades are traditionally defined as binocular eye movements (Lord, 1951; Ditchburn and Ginsborg, 1953; Krauskopf, 1960; see also Ciuffreda and Tannen, 1995). With modern video-based eye-tracking technology, binocular recording is easily available. Because (i) both eye traces can be analyzed independently and (ii) microsaccades are relatively rare events (roughly less than every 100th data sample belongs to a microsaccade), the amount of noise can be reduced dramatically, if we require binocularity. So far, we detected candidate microsaccades based on monocular data streams according to Eqs. (2)–(4). Operationally, we define binocular

microsaccade by a temporal overlap criterion. If we observe a microsaccade in the right eye with onset time r_1 and endpoint at r_2, we look through the set of microsaccade candidates in the left eye with corresponding onset time l_1 and offset time l_2. A temporal overlap of (at least) one data sample is equivalent to

$$r_2 > l_1 \quad \text{and} \quad r_1 < l_2 \tag{5}$$

It is straightforward to check the symmetry of the criterion by exchanging the roles of the left and right eyes. Figure 3 shows an example with three binocular microsaccades (numbered as 1–3). The two monocular microsaccades occurring in the left eye (labeled M and M') are unlikely to represent outliers or noise, given the high peak velocity. Thus, we speculate that monocular microsaccades might also exist (Engbert and Kliegl, 2003b). We also identified differences in average orientations for monocular microsaccades (see Fig. 5). To implement a conservative detection algorithm, however, we generally use the temporal overlap criterion (Eq. (5)). More importantly, most researchers record the movements of one eye only, which precludes the use of the more-conservative binocularity criterion.

Additionally, we would like to remark that, if a temporal overlap is verified, the binocular

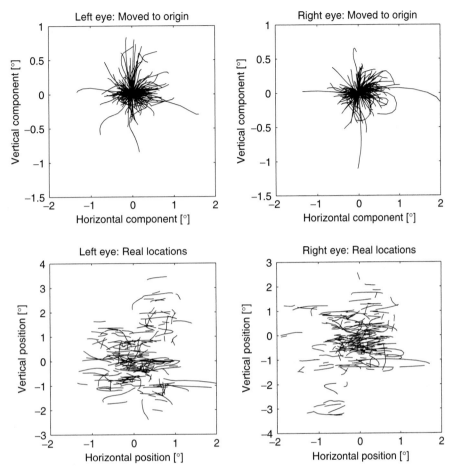

Fig. 4. Patterns of 220 microsaccades generated by a participant during 100 trials of a simple fixation task, each with duration of 3 s. Top panels: microsaccades translated so that the starting point is the origin of the coordinate system. Bottom panels: same data as in the top panels with microsaccades at their real locations relative to the fixation stimulus.

microsaccade is defined to start at time $t_1 = \min(r_1, l_1)$ and to end at time $t_2 = \max(r_2, l_2)$. As a consequence of this definition, microsaccades 1 and 3 in Fig. 3 in the right eye appear to start within the ellipse, i.e., with sub-threshold velocity. In the next section, we perform a quantitative analysis of the kinematics of microsaccades, which were detected by the algorithm discussed here.

Kinematic properties

Some of the kinematic properties of microsaccades can already be seen from visual inspection, if we plot all microsaccades generated by one participant

during 100 trials, each with a duration of 3 s, in a simple fixation task (Fig. 4). The top panels in the figure show all microsaccades generated by the participant with the starting point translated to the origin. The corresponding plots for the left and right eyes clearly indicate the preference for horizontal and vertical microsaccades, with oblique microsaccades only in rare exceptions.

The bottom panels in Fig. 4 show the same data with all microsaccades at their real locations. A glance at these plots verifies that microsaccades cover almost all parts of the central 2° of the visual field. Obviously, the majority of microsaccades is horizontally oriented. These descriptive results already demonstrate that microsaccades generate

182

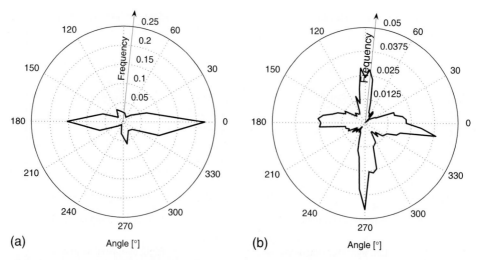

Fig. 5. Polar plot of angular orientations. (a) Binocular microsaccades, plotted in Fig. 4, show a clear preference for horizontal orientations. (b) Monocular microsaccades, generated by the same participant consist of comparable contribution from horizontal and vertical orientations.

rich statistical patterns, which might reflect important requirements of the oculomotor and/or perceptual systems. Thus, patterns of microsaccades may be exploited to help to understand the oculomotor system and even aspects of visual perception and the dynamics of allocation of visual attention.

A polar plot of microsaccade orientations (Fig. 5) shows that the majority of binocular microsaccades is horizontally oriented. For those microsaccades that were detected monocularly, horizontal and vertical orientations are comparable in frequency (Fig. 5b). To understand these properties of microsaccades from the neural foundations, it is important to note that different nuclei in the brain stem circuitry for saccade generation are responsible for the control of horizontal and vertical saccades (Sparks, 2002). While neural mechanisms for the control of oblique movement vectors exist for voluntary saccades, we speculate that for microsaccades, which are generated involuntarily, such mechanisms are questionable. As a consequence, the distribution of angular orientations for microsaccades is dominated by the main contributions, i.e., purely horizontal or vertical orientations. Since binocular movements are related to the control of disparity, binocular microsaccades might contribute to changes of binocular disparity. Therefore, a preference of binocular microsaccades

for horizontal orientations seems compatible with oculomotor needs. From this perspective, the investigation of binocular coordination in microsaccades (Engbert and Kliegl, 2003b) might contribute to the general problem of monocular vs. binocular control of eye movements (Zhou and King, 1998).

A key property of microsaccades is that they share the fixed relation between peak velocity and movement amplitude with voluntary saccades (Zuber et al., 1965). This finding is a consequence of the ballistic nature of microsaccades. To underline the importance of the result, the fixed relation of peak velocity and amplitude is often referred to as the "main sequence" (Bahill et al., 1975).[3] Given the validity of the main sequence, one useful application of this property of microsaccades is to check detection algorithms. Deviations from the main sequence might reflect noise in the detection algorithm.

In a larger study of fixational eye movements, 37 participants were required to fixate a small stimulus (black square on a white background, 3×3 pixels on a computer display with a spatial extension of 0.12°). Each participant performed 100 trials with duration of 3 s. Eye movements were

---

[3]The astronomical term "main sequence" refers to the relationship between the brightness of a star and its temperature.

recorded using an EyeLink-II system (SR Research, Osgoode, Ont., Canada) with a sampling rate of 500 Hz and an instrument whose spatial resolution was better than 0.01° (for details see the methods section in Engbert and Kliegl, 2004). Figure 6a shows the main sequence for about 20.000 microsaccades generated by all participants in the simple fixation task. Corresponding distributions of amplitude and peak velocity are shown in Figs. 6b and 6c, respectively. The analysis of the large number of microsaccades indicates that our detection algorithm reproduces the main sequence well. The smallest microsaccade detected in the set of about 20.000 microsaccades had an amplitude of 0.0361° or about 2 min arc. We conclude that

video-based recording techniques can be applied to detect even small-amplitude microsaccades. The distribution of microsaccade durations (Fig. 6d) is less smooth with a lower cutoff at 6 ms (or three samples) owing to the detection criteria.

A straightforward hypothesis on their function is that microsaccades help to scan fine details of an object during fixation. This hypothesis would imply that fixational eye movements represent a search process. According to this analogy, the statistics of microsaccades can be compared to other types of random searches. An important class of random-search processes are Lévy flights, which have been found in foraging animals (Viswanathan et al., 1996). Moreover, it has been shown that this

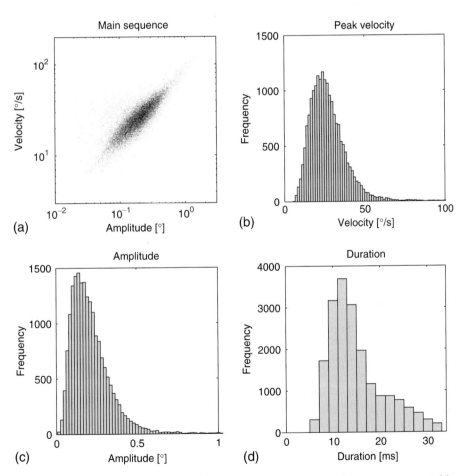

Fig. 6. Kinematic properties of microsaccades. The figure contains data from 19.163 microsaccades generated by 37 participants during a simple fixation task (100 trials, 3 s fixation duration). (a) The main sequence (see text). Histograms for: (b) peak velocity, (c) amplitude, and (d) duration.

184

class of random searches is optimal, if target sites are sparsely distributed (Viswanathan et al., 1999). With respect to research on eye movements, Brockmann and Geisel (2000) suggested that inspection of saccades during free picture viewing represent an example of a Lévy flight. Given these results, we check whether the amplitude distribution of microsaccades follows a similar law.

Lévy flights are characterized by a certain distribution function of flight lengths l_j. The flight lengths correspond to microsaccade amplitudes in fixational eye movements. For a Lévy flight, the distribution function of flight lengths has the functional form

$$P(l_j) \propto l_j^{-\mu} \tag{6}$$

where the exponent μ is limited to the range $1 < \mu \leq 3$. If the distribution decays with $\mu \geq 3$, we obtain a normal distribution for the search process, owing to the central limit theorem (Metzler and Klafter, 2000).

To investigate the distribution of microsaccade amplitude in our dataset of 20.000 microsaccades, we plot the tail of the distribution in Fig. 7 on a double-logarithmic scale. This plot clearly indicates a power-law decay of the tail of the distribution for

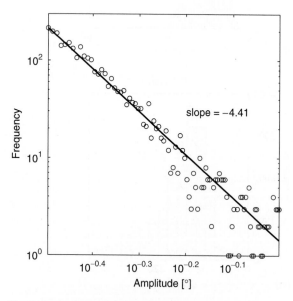

Fig. 7. Power-law distribution for amplitudes. The tail of the amplitude distribution obeys a power-law with an exponent given by the slope in the log–log plot.

amplitudes ranging from 0.3° to 1° of visual angle. The exponent, however, has a value of $\mu = 4.41$, which rejects the hypothesis of a Lévy flight. Thus, we conclude that even if the statistics of microsaccades during free viewing of a stationary scene might represent a Lévy flight (Brockmann and Geisel, 2000), this law does not transfer to the small scale in fixational eye movements.

Temporal correlations

While kinematic properties of microsaccades have long been investigated from the beginnings of eye-movement research (e.g., Zuber et al., 1965), the question of temporal correlations did not receive a similar amount of attention. The main reason is obviously that a typical fixation duration in free viewing or reading is roughly between 200 and 500 ms. Therefore, the probability of observing more than one microsaccade in a single fixation will be rather small. Temporal correlations in the series of microsaccades, however, might also be indicative of specific functions and/or neural mechanisms underlying microsaccades. In the simplest case, the probability that a microsaccade is generated is time-independent. More precisely, the probability of observing a microsaccade in an arbitrary time interval $(t, t + \Delta t)$ is $\rho \Delta t$, where ρ is the rate constant and Δt a small time interval, i.e., $\Delta t \to 0$, so that only one event can happen in the time interval of length Δt. This assumption constitutes a Poisson process. It is important to note that the number of events found in time interval $(tt + \Delta t)$ is completely independent of the number of occurrences in $(0, t)$. Experimentally, a simple observable parameter is the *waiting time* between two events of the process. For the Poisson process, the corresponding probability density of waiting time τ is an exponential function (Cox and Miller, 1965),

$$p(\tau) = \rho \, e^{-\rho\tau} \tag{7}$$

as in the general case of Markov processes. The waiting-time distribution is normalized in the sense that with certainty an event will occur, if we wait infinitely long, i.e., $\int_0^\infty p(\tau)\, d\tau = 1$.

To check whether the temporal patterns of microsaccades represent a Poisson process, we

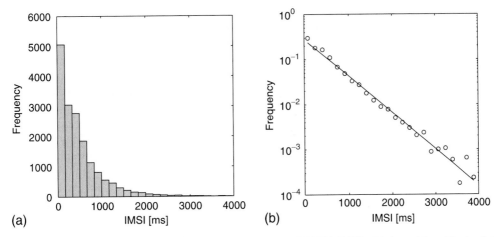

Fig. 8. Statistics of inter-microsaccadic intervals (IMSIs). (a) Histogram of 16.931 IMSIs. (b) A semi-logarithmic plot verifies an exponential distribution, compatible with the assumption of a Poisson process.

computed *inter-microsaccade intervals* (IMSI) for the microsaccade data shown in Fig. 8. This procedure is only possible if at least two microsaccades occur during a single trial. Obviously, this method is only an approximation to the problem of estimating the probability density, Eq. (7), which would ideally require recordings for several minutes to extract many IMSIs from a single time series. The rate constant ρ can be read off from the slope of the plot and has a numerical value of $1.86\,s^{-1}$, which is equivalent to a mean IMSI interval of $538\,ms$. Thus, we conclude that the temporal pattern of microsaccades represents a Poisson process, i.e., there are no temporal correlations of the numbers of microsaccades during two different time intervals. In a non-stationary situation with changing visual input and/or attentional shifts, however, the microsaccade rate will no longer be constant (see the section on modulation of microsaccades below). In this situation the temporal pattern of microsaccades can be described by an inhomogeneous Poisson process.

Dynamic properties: time-scale separation

A typical trajectory generated by the eyes during fixational movements shows the clear features of a random walk (Fig. 1). Different classes of random walks can be distinguished by their statistical correlations between subsequent increments (Metzler and Klafter, 2000). Such correlations can be investigated, if we plot the mean square displacement $\langle \Delta x^2 \rangle$ of the process as a function of the travel time Δt. A key finding related to Brownian motion is that the mean square displacement $\langle \Delta x^2 \rangle$ increases linearly with the time interval Δt (Einstein, 1905). This result is equivalent to the property of Brownian random walks in which the increments of the process are uncorrelated.

A generalization of classical Brownian motion was introduced by Mandelbrot and Van Ness (1968) to account for processes showing a power-law of the functional form

$$\langle \Delta x^2 \rangle \propto \Delta t^H \qquad (8)$$

where the scaling exponent H can be any real number between 0 and 2. In classical Brownian motion, we find $H = 1$, which is a direct consequence of the fact that the increments of the random walk are uncorrelated. When $H > 1$, increments are positively correlated, i.e., the random walk shows the tendency to continue to move in the current direction. This behavior is called *persistence*. In the case $H < 1$, the random walk generates negatively correlated increments and is *anti-persistent*.

Motivated by the study of Collins and De Luca (1993) on human postural data, we applied a random-walk analysis to fixational eye movements

(Engbert and Kliegl, 2004). To characterize the behavior in the eye's random walk during fixation, we introduce a displacement estimator (Collins and De Luca, 1993),

$$D^2(m) = \frac{1}{N-m} \sum_{i=1}^{N-m} \left\| \vec{x}_{i+m} - \vec{x}_i \right\|^2 \qquad (9)$$

which is based on the two-dimensional time series $\{\vec{x}_1, \vec{x}_2, \vec{x}_3, \ldots, \vec{x}_N\}$ of eye positions. The scaling exponent H can be obtained by calculating the slope in a log–log plot of D^2 vs. lag $\Delta t = m \cdot T_0$, where $T_0 = 2\,\mathrm{ms}$ is the sampling time interval. The analysis of 100 trials generated by one subject indicates two different power-laws by linear regions in the log–log plot (Fig. 9). On a short time scale (2–20 ms), the random walk is persistent with $H = 1.28$, whereas on a long time scale (100–800 ms), we find anti-persistent behavior. Thus, we observe a time-scale separation with two qualitatively different types of motion.

The fact that microsaccades are embedded in the drift and tremor components of fixational eye movements poses a difficult problem for the investigation of potential behavioral functions of

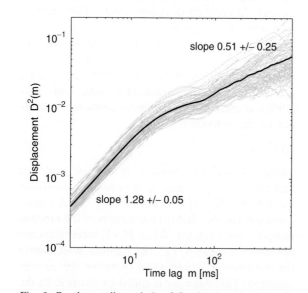

Fig. 9. Random-walk analysis of fixational eye movements. Linear regions in this log–log plot of the mean square displacement as a function of time indicate a power-law. On a short time scale, the random walk is persistent with a slope $H > 1$, while on a long time scale, we observe anti-persistence, i.e., slope $H < 1$.

microsaccades, because microsaccades will probably interact with these two other sources of randomness. To this aim, we cleaned the raw time series from microsaccades and applied the random-walk analysis (Engbert and Kliegl, 2004). First, we found that microsaccades contribute to persistence on the short time scale. Even after cleaning from microsaccades the drift still produced persistence. This tendency is increased by the presence of microsaccades. Second, on the long time scale, the anti-persistent behavior is specifically created by microsaccades. Since the persistent behavior on the short time scale helps to prevent perceptual fading and the anti-persistent behavior on the long time scale is error correcting and prevents loss of fixation, we can conclude that microsaccades are optimal motor acts to contribute to visual perception.

The random-walk analysis discussed here was motivated by the successful application of the framework to data from human postural control, where the center-of-pressure trajectory is statistically very similar to fixational eye movements, the dynamics unfold on a longer timescale. In a landmark study, Collins and De Luca (1993) observed the time-scale separation with the transition from anti-persistence to persistence. This discovery generated numerous publications since then. Combined with the observation that input noise can enhance sensory and motor functions (e.g., Collins et al., 1996; Douglass et al., 1993; Wiesenfeld and Moss, 1995), this research even inspired a technical application with the construction of vibrating insoles to reduce postural sway in elderly people (Priplata et al., 2002, 2003).

Modulation of microsaccade statistics by visual attention

The dynamic properties discussed in the last section support the view that microsaccades enhance visual perception and, therefore, represent a fundamental motor process with a specific purpose for visual fixation. Recent work demonstrated, however, that microsaccades are strongly modulated by display changes and visual attention in spatial cuing paradigms (Engbert and Kliegl, 2003a; see

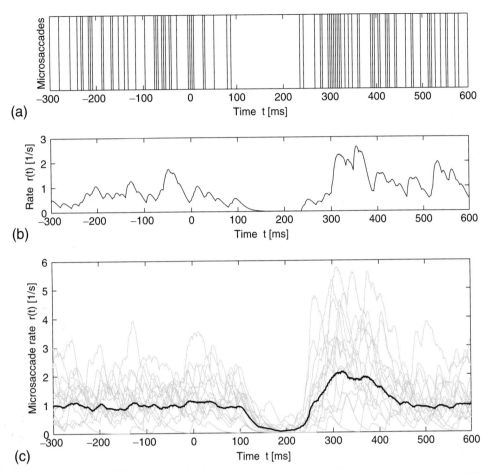

Fig. 10. Modulation of microsaccade rate by a display change. (a) Series of microsaccades generated during 100 trials by one participant. All microsaccades are plotted relative to cue onset at $t = 0$ ms. (b) Microsaccade rate computed from the series of events shown in (a). The rate was estimated by a causal window (see text). (c) Average modulation (bold line) induced in all 30 participants (thin lines). An early microsaccade inhibition is followed by a later enhancement.

also Hafed and Clark, 2002). Effects were related to microsaccade rate (*rate effect*) and to the angular orientation of microsaccades (*orientation effect*). Thus, while microsaccades might be essential for visual perception, they are — nevertheless — highly dynamic and underlie top-down modulation by high-level attentional influences.

In a variant of a classical spatial cuing paradigm (Posner, 1980), we instructed participants to prepare a saccade or manual reaction in response to a cue, but to wait for a target stimulus before the reaction was executed (Engbert and Kliegl, 2003a, b). As the rate effect related to cue onset, we reported a rapid decrease of the rate of microsaccades

from a baseline level[4] to a very low level, followed by an enhancement or supra-baseline level, until the rate resettled at baseline level.

The generation of microsaccades is a point process, i.e., discrete events are generated in continuous time (Fig. 10a).[5] Therefore, the problem of computing a continuously changing rate over time from many discrete events is equivalent to the

---

[4]The baseline of the microsaccade rate is generally about 1 microsaccade per second, but will depend on the specifics of the paradigm used.

[5]Figure 10 represents raw data and re-analyses of Experiment 1 by Engbert and Kliegl (2003a).

estimation of neuronal discharge rates from a series of action potentials in single-cell research. Following Dayan and Abbott (2001), we can formally describe the series of $i = 1, 2, 3, \ldots, N$ microsaccadic "spikes" at times t_i by

$$\rho(t) = \sum_{i=1}^{N} \delta(t - t_i) \tag{10}$$

where $\delta(t)$ denotes Dirac's δ function. For the computation of a continuous microsaccade rate $r(t)$, a window function $w(\tau)$ must be chosen, since any rate estimate must be based on temporal averaging, i.e.,

$$r(t) = \int_{-\infty}^{\infty} d\tau \, w(\tau)\rho(t - \tau) \tag{11}$$

Because the approximate firing rate at time t should depend only on spikes fired before t, a causal window of the form

$$w(\tau) = \alpha^2 \tau \exp(-\alpha\tau) \tag{12}$$

which is defined for $\tau \geq 0$ only and vanishes for $\tau < 0$. Figure 10b shows the estimated microsaccade rate computed from the raw data in Fig. 10a using Eqs. (10)–(12). The corresponding rate estimates from 30 participants are plotted in Fig. 10c (thin lines), which were averaged to obtain a stable estimate of the microsaccade rate (bold line). The reliability of the estimate can be seen from the relatively constant baseline over the pre-cue interval (-300–0 ms). Microsaccadic inhibition starts 100 ms after cue onset and lasts about 150 ms. Microsaccadic enhancement starts roughly 250 ms after cue onset, until the rate resettles at baseline level approximately 450 ms after cue onset.

As the orientation effect, Engbert and Kliegl (2003a) found a bias of microsaccade orientations toward the cue direction. While in Experiment 1 of their study, arrows used as endogenous (central) cues induced the orientation effect during the enhancement interval, the data from Experiment 2 with color cues showed a later (and weaker) orientation effect. Thus, the results of Engbert and Kliegl (2003a) already indicated that the rate and orientation effects are not mandatorily coupled. Finally, in Experiment 3, it was shown that a relatively small display change, i.e., the same stimuli applied in a simple fixation task without instructions for attentional cuing, was sufficient to replicate the rate effect without producing an orientation effect. Therefore, the rate effect alone could also be interpreted as a response to a display change.

In a series of experiments, the relation between rate modulation, covert attention, and microsaccade orientation was further investigated (Laubrock et al., 2005; Rolfs et al., 2005). Depending on the details of the task, these studies replicated the rate effect with both inhibition of microsaccades about 150 ms after the first display change related to the appearance of the cue (Table 1) and an enhancement of microsaccade rate, which was found around 400 ms after cue onset. More importantly, the rate effect was reproduced in a simple display change condition (Engbert and Kliegl, 2003a) and in purely auditory cuing, i.e., without any visual display change (Rolfs et al., 2005). Therefore, the rate effect might be interpreted as a stereotyped response to a sudden change in — possibly multisensory input. The patterns of results on the orientation effect, however, turned out to be more complex. The list of

Table 1. Rate and orientation effects for microsaccades

Reference	Cue type	Rate effect (ms)		Orientation effect (ms)	
		Inhibition	Enhancement	Cue-congruent	Cue-incongruent
Engbert and Kliegl (2003a)	Endogenous, arrows	150	350	300–400	—
	Endogenous, color	150	350	350–600	—
	Display change	150	350	—	—
Laubrock et al. (2005a)	Exogenous, flash	180–200	450	Early: 50–200, late: 600–800	250–550
	Endogenous, color	200	400–600	—	500–700
Rolfs et al. (2005)	Exogenous, visual	150–225	350–500	—	250–500
	Auditory	150	300–350	150–300[a]	—

[a] A significant effect was observed only for cues to the left.

effects for the orientation effect in Table 1 shows that there are combinations of early and late effects with both cue-congruent and cue-incongruent modulations of microsaccade orientations.

Can we explain the rate and orientation effects from our knowledge on the neural circuitry controlling saccadic eye movements? Numerous publications have contributed to the current view that multiple voluntary and reflexive pathways exist to generate saccades (Fig. 11). Most of these pathways are convergent to the superior colliculus (SC), a structure controlling brain stem saccade generation equipped with several sensory and motor maps (Robinson, 1972; Schiller and Stryker, 1972; see also Bergeron et al., 2003). Deeper layers of the SC are multisensory, which can explain the auditory effects reported by Rolfs et al. (2005). In the rostral pole of the SC, fixation neurons are activated to keep our gaze stable. According to Krauzlis et al. (2000), fixation neurons are better conceived of as rostral, i.e., small-amplitude, build-up neurons. Thus, the amount of activation of the fixation neurons in the SC will determine the rate of microsaccades.

A key dynamical feature of the spatio-temporal evolution of activation in the SC has inspired models of local excitation and global inhibition (e.g., Findlay and Walker, 1999; Trappenberg et al., 2001). As a consequence of the latter property, a global change in sensory input will induce a higher mean-field activation, which will lead to an increased inhibition of the fixation neurons in the rostral pole of the SC. Consequently, we can expect a decrease of the microsaccade rate after a display change (or a sudden auditory stimulus). Given the time course with the fast inhibitory part of the rate effect, we further suggest that the "direct" or retinotectal connection from sensory input to the SC is responsible for the inhibition effect. In addition to the inhibition, the subsequent enhancement of microsaccade rate could also be generated by the intrinsic dynamics of global inhibition and local excitation, since after the global inhibition of fixation neurons the global increase of activation will fade out with the consequence that the remaining activation of fixation neurons will locally rise. The enhancement, however, is more difficult to interpret, because a latency of 350–400 ms is sufficient for multiple pathways to contribute to this effect.

For the interpretation of the orientation effect, a primary difference between experiments is whether endogenous or exogenous cues were used (Table 1).

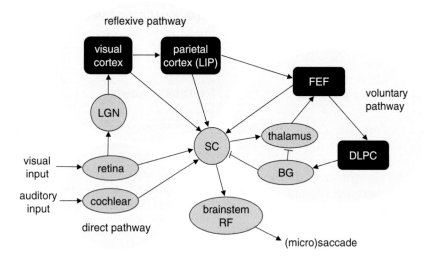

Fig. 11. Neural circuitry underlying saccade generation (modified after Munoz and Everling, 2004). Three separate pathways can be distinguished (BG = basal ganglia, DLPC = dorso-lateral prefrontal cortex, FEF = frontal eye fields, LGN = lateral geniculate nucleus, SC = superior colliculus, and RF = reticular formation). The direct pathway generates very fast, stereotyped responses. The reflexive pathway is related to computation of spatial orientation. The voluntary pathway can mediate excitatory as well as inhibitory influences.

We start with a discussion of cue-congruent effects. For endogenous cues, microsaccade orientations were biased toward the cue direction as early as 300 ms after cue onset (Engbert and Kliegl, 2003a). Such an orientation effect is likely to be controlled by the "reflexive pathway" with contributions from occipital/visual and parietal cortex (Fig. 11). Rolfs et al. (2005) found a similar cue-congruent effect induced by auditory cues with a somewhat shorter latency. For exogenous flash cues, Laubrock et al. (2005) reported an early cue-congruent effect starting <100 ms after cue onset. From the timescale alone, we can conclude that the direct pathway must generate this effect.

Cue-incongruent effects might represent inhibitory influences from cortex. From the neural circuitry in Fig. 11, we see that inhibitory control is mediated via frontal eye fields (FEF), dorso-lateral prefrontal cortex (DLPC), and basal ganglia (BG). As a consequence of this long-loop control circuit, we expect longer latencies for effects mediated via the "voluntary" pathway. Experimentally, we observed results highly compatible with this picture: Cue-incongruent effects have latencies roughly around 300 ms for exogenous cues (Laubrock et al., 2005; Rolfs et al., 2005) and 600 ms for endogenous color cues (Laubrock et al., 2005). Therefore, a detailed analysis of spatio-temporal response patterns of microsaccades in human behavioral testing seems a promising approach to investigate principles of the neural circuitry of saccade generation.

Summary: microsaccades as a new toolbox?

Visual fixation is a platform for almost all visual perception. As a consequence, we can expect new insights into the organization of perception, when we carefully investigate the dynamics of fixational eye movements, in particular, the dynamics of microsaccades. Converging evidence from research on basic oculomotor control (Engbert and Kliegl, 2004; Engbert and Mergenthaler, 2006), attentional cuing (Engbert and Kliegl, 2003a; Rolfs et al., 2004, 2005; Laubrock et al., 2005), and neurophysiology (Martinez-Conde et al., 2000, 2002) suggests that microsaccades represent highly optimized motor behavior essential to visual

perception. Furthermore, a recent study (Rolfs et al., 2006) demonstrates that microsaccades interact with ongoing saccade programming, which can produce both increased or decreased latencies for upcoming saccades.

On the one hand, research on microsaccades will benefit from our understanding of the visual system, on the other, new findings on the properties and on the functional role of microsaccades will clearly prove our current knowledge of many systems from oculomotor control to the allocation of visual attention. An important application of the orientation effect is straightforward: microsaccades can be exploited to map the time course of the allocation of visual attention. For example, Deaner and Platt (2003) observed reflexive effects of attention in monkeys and humans, which were also present in fixational eye movements. Rolfs et al. (2004) used microsaccade orientations to replicate the effect of attentional enhancement opposite to a peripheral cue (Tse et al., 2003). Galfano et al. (2004) reproduced the effect of inhibition of return from analyses of microsaccades (see Klein, 2000, for a recent overview). These results demonstrate that microsaccades might be used as an independent indicator of covert attention.

In summary, we can expect that the investigation of microsaccades will become a productive field of research with contributions to many aspects of perception, oculomotor control, and the control of visual attention.

Acknowledgments

The author wishes to thank Reinhold Kliegl, Jochen Laubrock, Konstantin Mergenthaler, Hannes Noack, Antje Nuthmann, Claudia Paladini, and Martin Rolfs for many discussions and comments on the manuscript. This work was supported by Deutsche Forschungsgemeinschaft (Grant no. KL 955/3).

References

Bahill, A.T., Clark, M.R. and Stark, L. (1975) The main sequence. A tool for studying human eye movements. Math. Biosci., 24: 191–204.

Bergeron, A., Matsuo, S. and Guitton, D. (2003) Superior colliculus encodes distance to target, not saccade amplitude, in multi-step gaze shifts. Nat. Neurosci., 6: 404–413.

Bridgeman, B. and Palca, J. (1980) The role of microsaccades in high acuity observational tasks. Vision Res., 20: 813–817.

Brockmann, D. and Geisel, T. (2000) The ecology of gaze shifts. Neurocomputing, 32–33: 643–650.

Ciuffreda, K.J. and Tannen, B. (1995) Eye Movement Basics for the Clinician. St. Louis, Mosby.

Collins, J.J. and De Luca, C.J. (1993) Open-loop and closed-loop control of posture: a random-walk analysis of center-of-pressure trajectories. Exp. Brain Res., 95: 308–318.

Collins, J.J., Imhoff, T.T. and Grigg, P. (1996) Noise-enhanced tactile sensation. Nature, 383: 770.

Cornsweet, T.N. (1956) Determination of the stimuli for involuntary drifts and saccadic eye movements. J. Opt. Soc. Am., 46: 987–993.

Cox, D.R. and Miller, H.D. (1965) The Theory of Stochastic Processes. London, Methuen.

Dayan, P. and Abbott, L.F. (2001) Theoretical Neuroscience. Computational and Mathematical Modeling of Neural Systems. Cambridge, MA, The MIT Press.

Deaner, R.O. and Platt, M.L. (2003) Reflexive social attention in monkeys and humans. Curr. Biol., 13: 1609–1613.

Ditchburn, R.W. (1980) The function of small saccades. Vision Res., 20: 271–272.

Ditchburn, R.W. and Ginsborg, B.L. (1952) Vision with a stabilized retinal image. Nature, 170: 36–37.

Ditchburn, R.W. and Ginsborg, B.L. (1953) Involuntary eye movements during fixation. J. Physiol. (Lond.), 119: 1–17.

Douglass, J.K., Wilkens, L., Pantazelou, E. and Moss, F. (1993) Noise enhancement of information-transfer in crayfish mechanoreceptors by stochastic resonance. Nature, 365: 337–340.

Einstein, A. (1905) Über die von der molekularkinetischen Theorie der Wärme geforderte Bewegung von in ruhenden Flüssigkeiten suspendierten Teilchen. Ann. d. Physik, 322: 549–560.

Engbert, R. and Kliegl, R. (2003a) Microsaccades uncover the orientation of covert attention. Vision Res., 43: 1035–1045.

Engbert, R. and Kliegl, R. (2003b) Binocular coordination in microsaccades. In: Hyönä, J., Radach, R. and Deubel, H. (Eds.), The Mind's Eye: Cognitive and Applied Aspects of Eye Movements. Oxford, Elsevier, pp. 103–117.

Engbert, R. and Kliegl, R. (2004) Microsaccades keep the eyes' balance during fixation. Psychol. Sci., 15: 431–435.

Engbert, R. and Mergenthaler, K. (2006) Microsaccades are triggered by low retinal image slip. Proc. Natl. Acad. Sci. USA, 103: 7192–7197.

Findlay, J.M. and Walker, R. (1999) A model of saccade generation based on parallel processing and competitive inhibition. Behav. Brain Sci., 22: 661–721.

Galfano, G., Betta, E. and Turatto, M. (2004) Inhibition of return in microsaccades. Exp. Brain Res., 159: 400–404.

Hafed, Z.M. and Clark, J.J. (2002) Microsaccades as an overt measure of covert attention shifts. Vision Res., 42: 2533–2545.

Helmholtz, H. v. (1866) Handbuch der Physiologischen Optik. Hamburg/Leipzig, Voss.

Klein, R.M. (2000) Inhibition of return. Trends Cogn. Sci., 4: 138–147.

Kowler, E. and Steinman, R.M. (1980) Small saccades serve no useful purpose: reply to a letter by R. W. Ditchburn. Vision Res., 20: 273–276.

Krauskopf, J., Cornsweet, T.N. and Riggs, L.A. (1960) Analysis of eye movements during monocular and binocular fixation. J. Opt. Soc. Am., 50: 572–578.

Krauzlis, R., Basso, M. and Wurtz, R. (2000) Discharge properties of neurons in the rostral superior colliculus of the monkey during smooth-pursuit eye movements. J. Neurophysiol., 84: 876–891.

Laubrock, J., Engbert, R. and Kliegl, R. (2005) Microsaccade dynamics during covert attention. Vision Res., 45: 721–730.

Lettvin, J.Y., Maturana, H.R., McCulloch, W.S. and Pitts, W.H. (1959) What the frog's eye tells the frog's brain. Proc. Inst. Radio Eng., 47: 1940–1951.

Lord, M.P. (1951) Measurement of binocular eye movements of subjects in the sitting position. Br. J. Ophthalmol., 35: 21–30.

Mandelbrot, B.B. and Van Ness, J.W. (1968) Fractional Brownian motions, fractional noises and applications. SIAM Rev., 10: 422–436.

Martinez-Conde, S., Macknik, S.L. and Hubel, D.H. (2000) Microsaccadic eye movements and firing of single cells in the striate cortex of macaque monkeys. Nat. Neurosci., 3: 251–258.

Martinez-Conde, S., Macknik, S.L. and Hubel, D.H. (2002) The function of bursts of spikes during visual fixation in the awake primate lateral geniculate nucleus and primary visual cortex. Proc. Natl. Acad. Sci. USA, 99: 13920–13925.

Martinez-Conde, S., Macknik, S.L. and Hubel, D.H. (2004) The role of fixational eye movements in visual perception. Nat. Rev. Neurosci., 5: 229–240.

Metzler, R. and Klafter, J. (2000) The random walk's guide to anomalous diffusion: a fractional dynamics approach. Phys. Rep., 339: 1–77.

Munoz, D.P. and Everling, S. (2004) Look away: the antisaccade task and the voluntary control of eye movements. Nat. Rev. Neurosci., 5: 218–228.

Posner, M.I. (1980) Orientation of attention The VIIth Sir Frederic Bartlett lecture. Quart. J. Exp. Psychol., 32A: 3–25.

Priplata, A.A., Niemi, J.B., Harry, J.D., Lipsitz, L.A. and Collins, J.J. (2003) Vibrating insoles and balance control in elderly people. Lancet, 362: 1123–1124.

Priplata, A., Niemi, J., Salen, M., Harry, J., Lipsitz, L.A. and Collins, J.J. (2002) Noise-enhanced human balance control. Phys. Rev. Lett., 89: 238101.

Ratliff, F. and Riggs, L.A. (1950) Involuntary motions of the eye during monocular fixation. J. Exp. Psychol., 40: 687–701.

Riggs, L.A., Ratliff, F., Cornsweet, J.C. and Cornsweet, T.N. (1953) The disappearance of steadily fixated test objects. J. Opt. Soc. Am., 43: 495–501.

Robinson, D.A. (1972) Eye movements evoked by collicular stimulation in the alert monkey. Vision Res., 12: 1795–1808.

Rolfs, M., Engbert, R. and Kliegl, R. (2004) Microsaccade orientation supports attentional enhancement opposite a peripheral cue. Psychol. Sci., 15: 705–707.

Rolfs, M., Engbert, R. and Kliegl, R. (2005) Crossmodal coupling of oculomotor control and spatial attention in vision and audition. Exp. Brain Res., 166: 427–439.

Rolfs, M., Laubrock, J. and Kliegl, R. (2006) Shortening and prolongation of saccade latencies following microsaccades. Exp. Br. Res., 169: 369–376.

Schiller, P.H. and Stryker, M. (1972) Single-unit recording and stimulation in superior colliculus of the alert rhesus monkey. J. Neurophysiol., 35: 915–924.

Sparks, D.L. (2002) The brainstem control of saccadic eye movements. Nat. Rev. Neurosci., 3: 952–964.

Steinman, R.M., Cunitz, R.J., Timberlake, G.T. and Herman, M. (1967) Voluntary control of microsaccades during maintained monocular fixation. Science, 155: 1577–1579.

Trappenberg, T.P., Dorris, M.C., Munoz, D.P. and Klein, R.M. (2001) A model of saccade initiation based on the competitive integration of exogenous and endogenous signals in the superior colliculus. J. Cogn. Neurosci., 13: 256–271.

Tse, P.U., Sheinberg, D.L. and Logothetis, N.K. (2003) Attentional enhancement opposite a peripheral flash revealed using change blindness. Psychol. Sci., 14: 91–99.

Viswanathan, G.M., Afanasyev, V., Buldyrev, S.V., Murphy, E.J., Prince, P.A. and Stanley, H.E. (1996) Lévy flight search patterns of wandering albatrosses. Nature, 381: 413–415.

Viswanathan, G.M., Buldyrev, S.V., Havlin, S., Da Luz, M.G.E., Raposo, E.P. and Stanley, H.E. (1999) Optimizing the success of random searches. Nature, 401: 911–914.

Wiesenfeld, K. and Moss, F. (1995) Stochastic resonance and the benefits of noise — from ice ages to crayfish and squids. Nature, 373: 33–36.

Winterson, B.J. and Collewijn, H. (1976) Microsaccades during finely guided visuomotor tasks. Vision Res., 16: 1387–1390.

Zhou, W. and King, W.M. (1998) Premotor commands encode monocular eye movements. Nature, 393: 692–695.

Zuber, B.L., Stark, L. and Cook, G. (1965) Microsaccades and the velocity — amplitude relationship for saccadic eye movements. Science, 150: 1459–1460.

Martinez-Conde, Macknik, Martinez, Alonso & Tse (Eds.)
Progress in Brain Research, Vol. 154
ISSN 0079-6123

CHAPTER 10

Fixational eye movements and motion perception

Ikuya Murakami*

Department of Life Sciences, University of Tokyo, 3-8-1 Komaba, Meguro-ku, Tokyo 153-8902, Japan

Abstract: Small eye movements are necessary for maintained visibility of the static scene, but at the same time they randomly oscillate the retinal image, so the visual system must compensate for such motions to yield the stable visual world. According to the theory of visual stabilization based on retinal motion signals, objects are perceived to move only if their retinal images make spatially differential motions with respect to some baseline movement probably due to eye movements. Motion illusions favoring this theory are demonstrated, and psychophysical as well as brain-imaging studies on the illusions are reviewed. It is argued that perceptual stability is established through interactions between motion-energy detection at an early stage and spatial differentiation of motion at a later stage. As such, image oscillations originating in fixational eye movements go unnoticed perceptually, and it is also shown that image oscillations are, though unnoticed, working as a limiting factor of motion detection. Finally, the functional importance of non-differential, global motion signals are discussed in relation to visual stability during large-scale eye movements as well as heading estimation.

Keywords: small eye movement; motion perception; illusion; perceptual stability; aftereffect; motion detection; visual jitter

Introduction

While we maintain fixation, our eyes are incessantly moving in tiny oscillations (Steinman et al., 1973). This type of eye movements are called "fixational eye movements" or "small eye movements" (for extensive review, see Martinez-Conde et al., 2004). Accordingly, the images of objects projected onto the retina are always moving randomly, even though the objects themselves are static.

There is ample evidence for the functional significance of such retinal image motions. One of the most striking demonstrations comes from stabilized retinal images. If the image of a static object is artificially stabilized on the retina, the visual image tends to fade away in tens of seconds (Ditchburn and Ginsborg, 1952; Riggs et al., 1953; Yarbus, 1967). Even if stimulus duration is less than 1 s, image stabilization is reported to decrease sensitivity for shape discrimination in low-contrast and noisy environment (Rucci and Desbordes, 2003). In a neurophysiological study also, retinal ganglion cells in the turtle retina show greatly increased activities when the stimulus is wobbled by simulated eye movements of fixation (Greschner et al., 2002). These findings tell us that the visual system requires retinally moving points and edges, or temporal changes of light intensity, as valid visual inputs. Those that do not fulfill this requirement in normal conditions include the blind spot and the cast shadow of retinal blood vessels, which should be invisible from an ecological viewpoint and are actually unnoticed in one's life (Ramachandran, 1992; Komatsu and Murakami, 1994; Coppola and Purves, 1996; Murakami et al., 1997).

*Corresponding author. Tel.: +81-3-5454-4437;
Fax: +81-3-5454-6979; E-mail: ikuya@fechner.c.u-tokyo.ac.jp

DOI: 10.1016/S0079-6123(06)54010-5

Thus, small eye movements are believed to counteract neural adaptation and noise and to maintain visibility of objects by oscillating their retinal images all the time (Greschner et al., 2002). However, small eye movements produce non-negligible velocity noise in the retinal image. How large is the noise? In the three major classes of small eye movements, "tremors," "microsaccades," and "drifts," tremors are extremely small in amplitude (<1 arcmin) and rapid (>30 Hz), and thus seem to interact minimally with visual processing. Microsaccades are the largest of them (>10 arcmin), but also the rarest (roughly one saccade per second or less), and their frequency is reducible by instruction (Steinman et al., 1967). Drifts, in contrast, occur constantly between every couple of consecutive microsaccades and have relatively large positional random walk. Their position time series can be approximated by the "$1/f$" amplitude spectra in frequency domain (Eizenman et al., 1985), and result in a velocity distribution obeying the zero-centered Gaussian (Murakami, 2003). The statistics of a typical untrained observer's eye drifts are illustrated in Fig. 1. As small eye movements are known to get smaller over fixation training (Di Russo et al., 2003), Fig. 1 is not meant to show the lower-bound performance of human observers (e.g., my own eye movements are less than 1/3 as small as this example). This particular observer might get better after fixation training in laboratory. Importantly, however, her visual world does not appear to oscillate, even without fixation training, although her retinal images always contain this large velocity noise due to fixational eye drifts. Also, I often encounter subjects with random fixational noise twice as large as Fig. 1, but with normal vision otherwise — with normal reading and driving skills as well. Thus, we must understand how retinal image instability is transformed to perceptual stability of visual scenes.

The easiest scheme for visual stability would be to reduce motion sensitivity to tiny oscillation. For tremors this would be the case, because their amplitudes are as small as visual acuity limit and their frequency range is more or less comparable to the critical fusion frequency (Gerrits and Vendrik, 1970; Eizenman et al., 1985; Spauschus et al.,

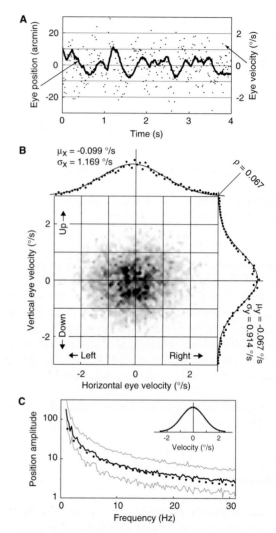

Fig. 1. Small eye movements; data from an untrained naive subject. While the observer was passively viewing a stationary random-dot pattern with a fixation spot, horizontal gaze position was recorded by an infrared limbus eye tracker (Iota Orbit 8) with the sampling resolution of 1 kHz and was bandpass-filtered (1–31 Hz). (A) Horizontal eye position during fixation (thick curve) and instantaneous velocity (dots). (B) The two-dimensional (horizontal × vertical) histogram of eye-drift velocity. Darker pixels correspond to more frequent occurrence. (C) Position amplitude spectral density calculated by Fourier transform of eye-position data during 4 s of fixation (thick and thin curves, mean ± 1 SD). The function $y = a f^b$ yielded the best-fit result when $b = -1.067$ ($R^2 = 0.98$), indicating that the amplitude is well approximated by $y = a/f$, consistently with previous estimation (Eizenman et al., 1985). The amplitude spectra of an example of this theoretical function are overlaid (dots). Inset: The velocity histogram of the eye-position time series synthesized from this theoretical function by using inverse Fourier transform. The histogram obeys Gaussian with σ linearly related to a. Reprinted from Murakami (2004), with permission from Elsevier.

1999). Microsaccades and drifts can, however, move the retinal image much faster than the threshold speed at which one would detect external object motion (Nakayama and Tyler, 1981; McKee et al., 1990), and in fact, visual neurons in area V1 and higher do respond to image oscillations originating in such eye movements (Bair and O'Keefe, 1998; Leopold and Logothetis, 1998; Martinez-Conde et al., 2000, 2002; Snodderly et al., 2001). It may be the case that image slip by a microsaccade goes unnoticed because of "saccadic suppression," i.e., transient decrease of displacement detectability, as seen in large-scale saccades (for review, see Ross et al., 2001). But eye drifts cannot be cancelled the same way, because they occur constantly, unlike saccades and microsaccades.

Since the retinal image motions, due to small eye movements, are indeed registered in early cortical representation, some computation must be needed to cancel them. For large-scale eye movements such as smooth pursuit, the visual system may use extraretinal estimation as to how the eye is currently moving: efference copy of oculomotor commands (Helmholtz, 1866) and proprioceptive signals from extraocular muscles (Sherrington, 1918). By subtracting the estimated eye-movement vectors from retinal image flow, image motion components due to eye movements would ideally be cancelled out. For fixational eye movements, however, it is doubtful that these extraretinal signals work effectively. First, cancellation of random image oscillation would require temporally precise and accurate synchronization between visual and extraretinal processes, but actually this may be difficult because each of them, or at least visual processing, may be running on its own clock with poor temporal resolution and wide latency fluctuation (Murakami, 2001a,b; Kreegipuu and Allik, 2003). Second, eye drifts are partly derived from random neuronal activities in the periphery of the oculomotor system (Cornsweet, 1956; Eizenman et al., 1985), to which the efference-copy system may be blind. Third, eyes can be moved by skull oscillation such as in chewing behavior and by head and body movements, to which the oculomotor proprioception system may be blind (Skavenski et al., 1979). Fourth, images can also be moved by imperfect rigidity between artificial optical devices (e.g., spectacles and head-mounted display goggles) and the observer's head, to which the neural system may be blind.

The remaining source of visual stabilization, then, would be nothing but the visual inputs themselves — sometimes called reafference signals — which exhibit systematic image changes each time the eye moves (for review, see Wertheim, 1994).

Visual stability based on visual motion

For the processing of lightness and color constancy, the visual system does not measure the light source by an extraretinal colorimeter, but uses input images themselves to estimate light-source intensity and albedo variations together. Similarly, the visual system may use retinal image motions themselves to estimate eye movements and external motions.

When the outer world is stationary and the eye rotates, the frontal visual field contains approximately pure translation, or a field of spatially common velocities (Harris, 1994). When there is a moving object on a stationary background and the eye rotates, the retinal image now contains spatially common motions and spatially differential motions as well. Therefore, the visually based theory of visual stability (Murakami and Cavanagh, 1998, 2001) posits that the visual system uses this relationship as the natural constraint (Fig. 2); the visual system constantly dismisses spatially common image motions, as they are most probably derived from eye movements, and interprets spatially differential motions as coming from external object motion.

We have recently found a dramatic illusion (Murakami and Cavanagh, 1998, 2001) that reveals such a visual-motion-based mechanism of noise reduction in the visual system. A possible neural counterpart of such a mechanism has also been identified by using the brain-imaging technique (Sasaki et al., 2002). Next, as converging evidence of the mechanism, another compelling illusion has been developed and demonstrated (Murakami, 2003). I have further examined the effects of small eye movements on visual motion sensitivity in the perithreshold velocity range, and

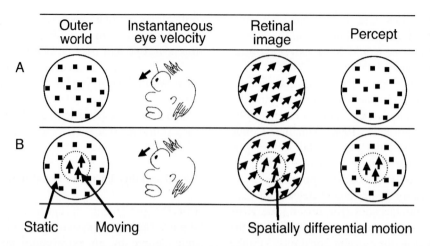

	Outer world	Instantaneous eye velocity	Retinal image	Percept

A

B

Static   Moving   Spatially differential motion

Fig. 2. Illustration of the visually based theory of perceptual stability. (A) The case in which the observer is viewing a stationary pattern. When an instantaneous eye movement occurs to the left and down, the retinal image of the stationary pattern moves to the right and up altogether. As the visual input contains no differential motion, the visual system interprets this velocity field as spurious motion due to eye movements, so that the resulting percept is stationary. (B) The case in which the observer is viewing a moving disk in a stationary background. When the same instantaneous eye movement occurs, the retinal image moves to the right and up but simultaneously the central disk moves differently from its surround. As the visual system uses this spatially differential motion as a cue of external object motion, a moving object in a stationary background is perceived.

detection threshold has been found to correlate positively with fixation instability if differential motion signals between target and background are not available (Murakami, 2004). Finally, the mechanism to discount small eye movements is discussed in relation to other visual functions mediated by the motion processing system.

The visual jitter aftereffect

While I was coding computer graphics programs for preliminary observation of various illusions related to noise adaptation (e.g., Ramachandran and Gregory, 1991), a strange thing was observed: prolonged viewing of dynamic random noise transiently disrupted perceptual stability. That is, after adaptation, some part of a physically stationary stimulus appeared to "jitter" in random directions. We named this new illusion the "visual jitter" and started systematic psychophysical experiments on it (Murakami and Cavanagh, 1998, 2001). A typical experimental setup consists of a concentric disk and annulus, both of which are filled with dense (50% of the dots black, 50% white) random-dot texture. In the example shown in Fig. 3C, the noise

in the disk is static, whereas the noise in the annulus is re-generated every computer frame (75 frames/s) so that it appears like snowstorm. The observer passively views this adapting stimulus with steady fixation. After adaptation, static noise patterns are presented in both regions. During observation of this test stimulus with steady fixation, oscillatory random movements are perceived within the physically stationary *disk* region — i.e., in the previously *unadapted* region — whereas the previously adapted annulus region seems stationary (Murakami and Cavanagh, 1998). The illusion lasts for only 2–15 s, but is quite vigorous and well repeatable for hundreds of casual observers. Also, changing stimulus configurations as shown in Fig. 3 do not affect this relationship: the previously unadapted region appears to move, whereas the previously adapted region appears static.

A number of observations indicate that the jitter perception is related to eye movements during fixation (Figs. 3, 4). (1) When two or more unadapted regions are configured at spatially remote locations, the jitter aftereffect that occurs in these regions are synchronized in direction and speed, suggesting that the source of the illusion is global through the visual field. (2) When the illusion is observed with smooth pursuit of some tracking target, the jitter is not random but biased toward the direction opposite to the pursuit direction. (3) When the illusion is observed while the eye is

artificially moved (either by post-rotational nystagmus or by mechanical vibration), the jitter perception is biased toward the directions as predicted by these artificial movements of the eye. (4) When the test stimulus is retinally stabilized, the illusion never occurs.

This illusion therefore suggests that retinal image motions due to small eye movements are not negligibly small but substantial enough to be seen as a compelling illusion, when noise adaptation somehow confuses our biological mechanism of noise reduction.

To reiterate, it is proposed that in a normal environment, the brain assumes common image motions over the visual field to be originating in eye movements, so that we are normally unaware of image jitter of the whole visual field. The brain only interprets differential motions with respect to a surrounding reference frame as originating in

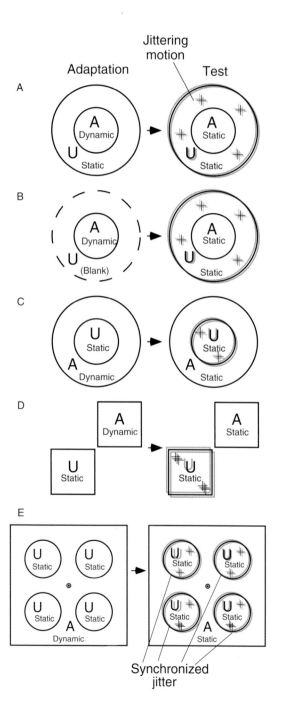

Fig. 3. Schematics of the visual jitter aftereffect. The left-hand column indicates the adapting stimulus, whereas the right-hand column indicates the test stimulus and the illusory jitter perception in a specific part of the stimulus. A fixation point is typically provided at the center of the stimulus, but the illusion occurs for peripheral viewing as well. A and U stand for adapted and unadapted (static or blank) regions, respectively. The blur of circles and crosses in the test stimuli depicts the visual jitter schematically. (A) The typical experimental setup is as follows. In the inner disk subtending $6.67°$ in diameter, black and white dynamic random noise was presented; each dot (8 arcmin × 8 arcmin) was randomly assigned black ($0.18\,cd/m^2$) or white ($52.4\,cd/m^2$) every frame (75 Hz). In the outer annulus with an outer diameter of $13.33°$, static random noise was presented. There was a uniform gray surround ($23.5\,cd/m^2$) outside the stimuli. After 30 s of adaptation, the two regions were changed to a new pattern of static random noise. The illusory jitter was perceived within the previously unadapted annulus region. (B) During adaptation, the annulus region was left blank. The same illusion occurred. (C) The center–surround relationship was reversed in the adapting stimulus. The illusory jitter was perceived in the previously unadapted disk region. (D) Even though the adapted and unadapted regions were separated squares, the previously unadapted square region appeared to jitter. (E) When the adapting stimulus consisted of four separate patches of static noise embedded within a larger region of dynamic noise, the jitter aftereffect was synchronized across these remote locations, namely they appeared to move in the same direction at the same speed at the same time. Reprinted from Murakami and Cavanagh (1998), with permission from Nature Publishing Group.

198

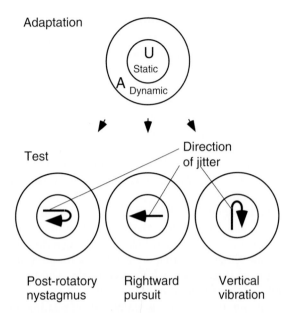

Fig. 4. The visual jitter aftereffect that is contingent on various kinds of eye movement during the test period. The black arrows in the central unadapted regions schematically illustrate the direction of illusory motion, which is consistent with the direction of retinal image slip due to eye movements indicated below each figure. In the condition of post-rotatory nystagmus, the subject's body was rotated about vertical with eyes closed for several seconds in the interval between a standard adaptation and test. When the body was suddenly stopped, back-and-forth horizontal oscillations of gaze was induced by the vestibulo-ocular reflex mechanism. The direction of the visual jitter seen in the unadapted area during the test was almost always horizontal with quick and slow phases, in accordance with the nature of the post-rotatory nystagmus. In the condition of rightward pursuit, a smooth pursuit eye movement was made to track a moving spot during test. The unadapted region perceptually slid in the opposite direction. This is the direction of the retinal image slip produced by smooth pursuit. In the condition of vertical vibration, up-and-down eye displacements were induced externally by a rapid alternation of extension/relaxation in the nearby skin by a mechanical vibrator. At amplitudes of vibration that did not produce noticeable motions of the world, vertical jitter nevertheless occurred in the unadapted area of the static test with a temporal profile synchronized with this vibration. Reprinted from Murakami and Cavanagh (1998), with permission from Nature Publishing Group.

object motions in the outer world. When noise adaptation lowers motion sensitivity of the adapted region (annulus in Fig. 3C), the representation of the same image motion due to eye movement becomes greater in the unadapted region (disk) and lesser in the adapted region, artificially

creating relative motion between disk and annulus in the topographic neural representation. This results in the perception of illusory jitter in the unadapted region only. As the source of this artificial relative motion is one's own fixational eye movements, the visual jitter illusion is one of the rare cases in which one becomes aware of one's own eye movements during fixation.

Assuming that relative motion between disk and annulus is represented in the brain, does not answer the question as to why the unadapted region always appears to move while the adapted region looks stationary and not the other way around. We suggest that a baseline velocity is estimated from the region of the retina that has the slowest instantaneous velocity. (After adaptation to dynamic noise, the adapted region in the cortical topography would have the lowest sensitivity to motion and thus would register the slowest velocity.) If this baseline velocity is subtracted from the velocities of all points on the retina, the adapted region (having an artificially slowed velocity representation) will be zero velocity, whereas all other previously unadapted regions will have residual, under-compensated velocities, when every part of the retina is in fact oscillated equally by small eye movements. Many observers experience that after adaptation, synchronized jitter is perceived in a variety of stationary parts of the visual field, including the static noise pattern in the unadapted region per se, the fixation spot, and the frame of the display monitor. Accordingly, velocity subtraction using the artificially slowed baseline velocity seems to take place in a relatively large spatial scale.

The mechanisms of the jitter aftereffect

According to the above explanation, at least two distinct mechanisms are involved in the jitter aftereffect. One is the processing stage that is adaptable by prolonged viewing of dynamic noise, and the other is the processing stage that detects differential motions in the cortical representation of visual motion. Let us call these first and second hypothetical stages the "adaptable stage" and "compensation stage" respectively.

Psychophysical findings as listed below indicate that the first adaptable stage should be as early as V1 (Murakami and Cavanagh, 1998, 2001). (1) There is no interocular transfer of the effect of adaptation, so the adapted neural site must be monocular. (2) After adaptation to directionally limited noise (e.g., only vertical motion energy exists), the jitter aftereffect is also directionally biased toward the adapted axis (e.g., vertical jitter is seen stronger). Hence the adapted site must be directionally selective. (3) When the adapting and test stimuli are band-limited in spatial frequency, the jitter aftereffect is greatest when the passbands of the adapting and test stimuli match. Hence the adapted site must have spatial-frequency tuning. It seems that, in the brain, area V1 (layer 4B, in particular) is the most represent-ative locus where a majority of cells possess all these characteristics: ocular dominance, direc-tional selectivity, and spatial-frequency tuning (Hubel and Wiesel, 1968). In addition, subcortical streams may also be involved in neural adaptation, since dynamic random noise contains broadband spatial and temporal frequencies suitable for driv-ing cells in the retinal ganglion cells and LGN cells, especially the magnocellular subsystem (Livingstone and Hubel, 1987).

Where is the second stage, namely the compen-sation stage, in which differential motions are to be segregated from common motions? To elucidate the spatial operation range of such a neural mech-anism, the stimulus size as well as the separation between the adapting and test stimuli were ma-nipulated. The results as listed below altogether suggest the involvement of high-level motion processing stages later than V1. (1) The perceived jitter strength decreases with increasing spatial gap between the adapting and test stimuli, but the effect persists even if the gap is as wide as 5–7°. (2) The effect transfers between hemifields, such that the adaptation in the left hemifield yields the jitter perception in the test stimulus in the right hemifield. (3) As the size of the concentric stimulus field (Fig. 3C) is scaled larger, the jitter strength increases up to a particular stimulus size and then levels off, so there seems to be a certain stimulus size specificity. (4) The inflection point (optimal stimulus size) of this stimulus-size function

systematically shifts toward larger size with in-creasing eccentricity, showing a linear relationship between optimal stimulus size and eccentricity. These characteristics suggest that the underlying mechanism has a spatially localized receptive field, but it is much larger than typical receptive field sizes of V1 neurons, and that the receptive field can span across hemifields and becomes larger with eccentricity. We consider that area MT is the most likely candidate for the neural counterpart of the compensation stage. According to studies done on monkeys, the receptive-field size specificity in this cortical area, especially its dependence on eccen-tricity, looks quantitatively similar to the stimulus-size specificity of the visual jitter that we obtained psychophysically (Gattass and Gross, 1981; Tanaka et al., 1986; Tanaka et al., 1993). A large proportion of neurons in this area also exhibit surround suppression, such that they are relatively insensitive to common motions presented both in-side and outside the classical receptive field, but are strongly driven by a moving stimulus confined within the classical receptive field; the spatial range of the suppressive surround also account for our findings of the gap-size effect and the inter-hemi-field transfer (Allman et al., 1985a,b, 1990; Tanaka et al., 1986; Lagae et al., 1989; Born and Tootell, 1992; Xiao et al., 1995; Hupé et al., 1998). Thus, the single neuron's behavior looks similar to the model description of the compensation stage, in which common image motions are discounted and only differential motions are interpreted as exter-nal object motions (Born et al., 2000).

Brain activity during the jitter aftereffect

The previous psychophysical investigations on the neural mechanisms of the visual jitter suggest that at least two distinct neural sites, namely the adapt-able stage and the compensation stage, are actually involved. To get more direct evidence for the neu-ral substrates, we further proceed to record human brain activity by using high-field functional mag-netic resonance imaging (fMRI). As a result, we obtained different patterns of brain activity de-pending on adaptation conditions (Sasaki et al., 2002), as predicted by the proposed framework of adaptable and compensation stages.

As before, the stimulus was presented within the concentric regions, the disk and annulus (Fig. 5). In the jitter-disk condition, dynamic noise was presented within the annulus, so that after adaptation, the disk appeared to jitter during the test period. The relationship was reversed in the jitter-annulus condition, so the annulus appeared to jitter. In the control-static condition, the noise pattern was static all the time. In the control-dynamic, the disk and annulus were both dynamic during adaptation, and there was no jitter perception afterward. We focused on the cortical activity during the test period, monitoring the blood-oxygenation-level-dependent (BOLD) signals from several visual areas that had been identified by the standard mapping protocol (Sereno et al., 1995). Two distinct cortical activity patterns emerged at different processing stages: activity decrease after adaptation to dynamic noise was observed in early visual areas such as V1, whereas activity increase was observed in higher areas such as MT when the visual jitter was actually perceived.

Figure 6 shows the time course of brain activity with different visual areas plotted separately. As these areas exhibit retinotopic organization, BOLD signal change can be recorded separately for the two regions of interest (ROI): the disk and annulus representations. Clearly, the cortical representations stimulated by dynamic noise showed positive activity during adaptation, and negative activity after adaptation, relative to the baseline (i.e., the signal level in the control-static condition). (A control study confirmed that this negativity was distinguishable from the effect of BOLD undershoot.) Interestingly, this was always the case whether the observer perceived jitter or not; the transient decrease of activity therefore seems to reflect neuronal desensitization due to prolonged exposure to dynamic noise, rather than the neural correlate of the illusory motion. The activity change was also greater for lower visual areas, with area V1 showing the most dramatic response peaks and troughs.

Figure 7 shows the signal changes in the same time course, but here we set the ROI to cover both the disk and annulus regions, because we want to include area MT+, to which the retinotopic analysis is not easily applicable (Tootell et al.,

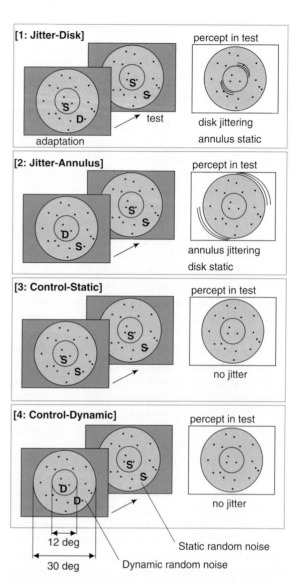

Fig. 5. The four stimulus conditions in the fMRI experiment. In each condition, a trial consisted of an adaptation period (32 s) and a subsequent test period (32 s). The adapting stimulus varied across conditions. "D" indicates dynamic random noise and "S" indicates static random noise. In the test period, the visual stimulus was identical throughout the four conditions (i.e., static random noise occupied both regions). However, perception in the test period differed across conditions: in Jitter-Disk, illusory jitter was perceived in the disk region; in the Jitter-Annulus, jitter was perceived in the annulus region; no jitter was perceived in the other two conditions. For illustrative purpose, noise is shown as if sparse, but actually it was 50% density. Reprinted from Sasaki et al. (2002), with permission from Elsevier.

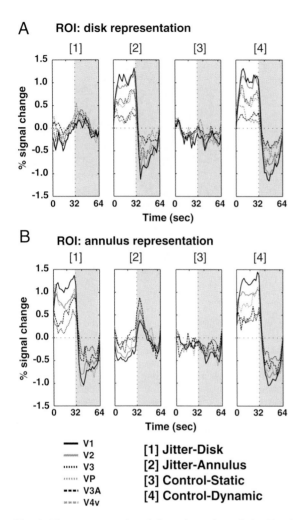

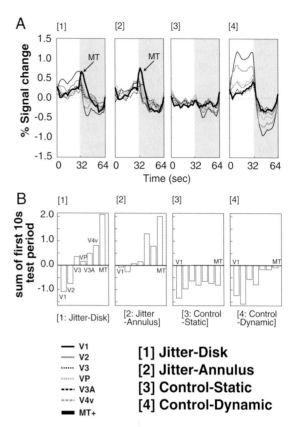

Fig. 6. Time course results of the retinotopic analysis. Signal changes are plotted in separate columns for the four conditions. The abscissa indicates time (adaptation for 0–32 s and test for 32–64 s). The ordinate indicates the signal change relative to the average activity level across all the conditions; thus its zero level had no functional meaning. Instead, the virtually flat profiles in condition Control-Static were considered to reflect the baseline activity, relative to which signal changes in other conditions were assessed. (A) The latency-corrected time course of the signal change in each visual area in the disk representation. (B) The same analysis for the annulus representation. No subject showed activation in the annulus representation in V4v. Reprinted from Sasaki et al. (2002), with permission from Elsevier.

Fig. 7. Time course results of the non-retinotopic analysis. (A) The latency-corrected time course of signal change in each visual area. The conventions are identical to those in Fig. 6. (B) Signal integration over the first 10 s of the test period (dark gray strip in A), during which the subject perceived illusory jitter in conditions 1 and 2. Reprinted from Sasaki et al. (2002), with permission from Elsevier.

1998). Hence the pattern of signal change is roughly the average of disk and annulus data that have appeared in Fig. 6. In the control-dynamic condition, the effect of adaptation is still observed.

In the jitter-disk and jitter-annulus conditions, the effect looks much less clear because of summation of opposite response tendencies, but the same pattern as before can still be seen. For MT+, however, a qualitatively different pattern is evident: just after the adapting stimulus was changed to the test stimulus, there was a transient increase of BOLD signals — when the observer perceived jitter. This activity change is distinct from the above mentioned desensitization effect, because there was no such response peak in the control-dynamic condition.

From these results, we argue that there are indeed two functionally distinct mechanisms in the human brain, namely the adaptable stage and the compensation stage, and these mechanisms are implemented in two anatomically distinct loci, namely early retinotopic areas (e.g., V1) and higher-tier motion processing areas (e.g., MT +).

The on-line jitter illusion

Adaptation, or desensitization after prolonged exposure to intense stimuli, alters neural sensitivity in multiple processing stages and, in a sense, functionally damages some of brain mechanisms for a few seconds — a considerably long term in neuronal time scale. Thus, although the adaptation paradigm is a useful psychophysical tool to tap specific brain mechanisms, the phenomenology of aftereffects can only provide indirect evidence of a mechanism in a normally functioning system. It is important to show converging evidence of the mechanism of interest using a different, adaptation-free paradigm. Is there a variant of the visual jitter illusion that does not require noise adaptation?

In developing the new illusion, my design rule was quite simple: make the stimulus so that in the observer's brain, some artificial differential motion is immediately created between disk and annulus, without adaptation. I tried to "shut up" motion detectors in the annulus, keeping the detectors intact in the disk region, so that the same retinal image motion due to small eye movements is detected in the disk region but not in the annulus region. Finally I came up with the flicker stimulation as the most effective stimulus to neutralize the motion detectors in the annulus region (Murakami, 2003).

In this new version of the jitter illusion (Fig. 8), the motion sensitivity in the annulus region is affected by such synchronous flicker: the overall random-dot pattern is periodically turned on (80 ms) and off (27 ms) in a typical experimental setup. The central disk region is filled with a stationary random-dot pattern that is constantly visible. When the observer looks at this kind of concentric stimulus with steady fixation, the disk region appears to "jitter" in random directions. Unlike the jitter aftereffect, this illusion occurs immediately and lasts as long as the annulus

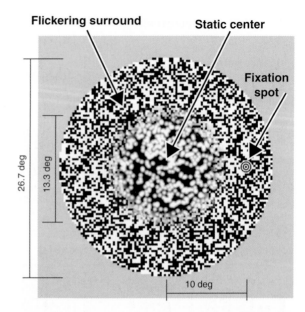

Fig. 8. Typical stimulus configuration for the on-line jitter illusion. The central pattern was static, whereas the surrounding annulus region was synchronously flickering, i.e., periodically turned on (80 ms) and off (27 ms). (While it was off, the annulus region was filled with the same uniform gray as the background.) Perceptually the central pattern appeared to move in random directions. The illusion was more salient with peripheral viewing of the stimulus, presumably because the stationary center–surround border as a frame of reference would become perceptually more obscure. However, the illusion persisted if viewed centrally or if the stimulus was enlarged to cover tens of degrees. Reprinted from Murakami (2003), with permission from Elsevier.

region continues to flicker. Thus I tentatively called this illusion the "on-line jitter." In this case also, psychophysical tests indicated that the source of the illusion is indeed retinal image motions originating in fixational eye movements, and again, the most likely interpretation is that motions due to eye movements are visible because relative motions between disk and annulus are artificially created in the brain.

Psychophysics as well as computer simulation have revealed that the temporal parameters mentioned above are optimal for confusing elementary motion-sensing neural units. The space–time plot of the simulated synchronous flicker on the moving retina is shown in Fig. 9A. This stimulus pattern was submitted to the standard motion-energy computation algorithm. The simulation output, i.e., bipolar motion-energy contrast (Georgeson and Scott-Samuel, 1999), is shown in Fig. 9B as a color plot. The simulation revealed that, although more than half the time they correctly reported "right," the motion-energy

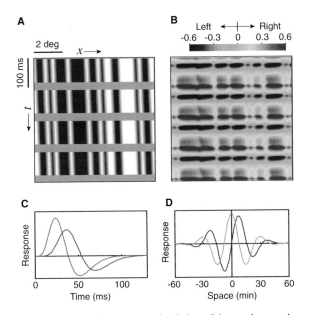

Fig. 9. Results of the computer simulation of the synchronously flickering pattern. Under the assumption that the eye was constantly moving to the left at 0.625°/s, the retinal image of a flickering random-dot pattern was rendered on a space–time surface (A). The retinal image was filtered by biphasic temporal impulse response functions (C), and was also filtered by Gabor-shaped spatial impulse response functions (D), using biologically plausible parameters (Watson, 1982; McKee and Taylor, 1984; Watson and Ahumada, 1985; Pantle and Turano, 1992; Takeuchi and De Valois, 1997). Appropriate spatiotemporal combinations of the filtered outputs were squared and summed to yield motion-energy responses (Adelson and Bergen, 1985). The final bipolar motion responses (plotted by color scale in B) indicate the motion-energy contrast, i.e., the difference of leftward and rightward responses divided by their sum (Georgeson and Scott-Samuel, 1999). (A) Spatiotemporal (horizontal × time) plot of the retinal image motion of the flickering surround. Patterns are made oblique by eye velocity and are periodically interrupted by the mean-luminance gray due to synchronous flicker. The input image actually had the size of 512 × 512 simulation-pixels (one pixel subtended 2 arcmin × 1.667 ms), but only its central region of 256 × 256 simulation-pixels is shown. (B) Spatiotemporal plot of the outputs of motion-energy units. (C) Temporal impulse response functions used as a part of motion-energy computation. (D) Spatial impulse response functions used as a part of motion-energy computation. Reprinted from Murakami (2003), with permission from Elsevier.

processing units occasionally reported "left" also, slightly (≈ 50 ms) after the onset of each gray interval. This result is consistent with previous studies on the behavior of motion-energy processing units during the blank interstimulus interval (Shioiri and Cavanagh, 1990; Pantle and Turano, 1992; Takeuchi and De Valois, 1997). These are critical occasions when center and surround are reported oppositely (because the non-flickering center moving at the same velocity is constantly reported "*right*"). Therefore, the simulation demonstrates that biological motion-processing units artificially create spatially differential motion between the disk and annulus regions when eye movements actually produce common image motions in these regions on the retina. Furthermore, it has also been confirmed that the temporal parameters (duty cycle and frequency) of the synchronous flicker that yield the maximally negative motion-energy output in the simulation quantitatively match the parameters that yield the strongest illusion in a psychophysical experiment (Murakami, 2003).

This phenomenon has a conceptual similarity to visual effects of stimulus blinking on perceptual space constancy during large, small, and artificial eye movements (MacKay, 1958; Deubel et al., 1996, 1998; Spillmann et al., 1997; Peli and García-Pérez, 2003), in which the observer experiences perceptual dissociation between continuously illuminated parts and flickering parts of the visual field. As the motion processing system in our brain has evolved under continuous illumination, it is conceivable that artificial flicker stimulation at a certain frequency will confuse motion detectors, leading to computational errors of perceptual stability.

Motion detection thresholds

It is generally considered that a finding obtained in highly suprathreshold realm (e.g., illusion) cannot be immediately generalized to perithreshold behavior (e.g., detection performance). That is, different mechanisms with different principles might be operating at different points on a response function of stimulus intensity. As empirical testing was necessary in order to elucidate the

204

operating range of the proposed scheme, I moved my research interest from jitter illusions to absolute threshold measurement for motion detection.

According to these illusions and the theory of visual stability described so far, external object motions are segregated from their background motions because there are differential motions between object and background; other retinal image motions lacking differential information tend to be dismissed. The theory therefore raises testable predictions about motion detection performance as follows: (1) Motion without a surrounding reference frame should be hard to detect. (2) The detection of motion with a surrounding reference frame should be easy if the surround is static, but should be hard if the surrounding reference provides only unreliable information as the synchronous flicker does. (3) An external motion without a surrounding frame would be indistinguishable from eye-movement noise and thus should have a positive relation to fixation instability. (4) An external motion with a static reference frame should be easiest to detect and independent of fixation instability.

All these predictions were found to be actually the case (Murakami, 2004). In concentric regions, the central disk contained a slowly translating random-dot pattern, accompanied by another pattern in the annulus region (the stimulus configuration was identical to Fig. 8). The annulus could contain a static random-dot pattern (with-surround condition), a synchronously flickering random-dot pattern (flicker-surround condition), or a mean-luminance, uniform gray field (no-surround condition). The subject's task was to identify the movement direction of the central pattern while maintaining fixation. Lower threshold for motion was determined for each condition (Fig. 10A). In no-surround, the threshold was significantly higher than with surround. In flicker-surround, the threshold was even higher. Thus, predictions 1 and 2 were fulfilled. Also, the pattern of results is consistent with earlier investigations on the detection threshold for comparable situations: referenced motion is easier to detect than unreferenced motion (Levi et al., 1984; Tulunay-Keesey and VerHoeve, 1987; Whitaker and MacVeigh, 1990; Shioiri et al., 2002).

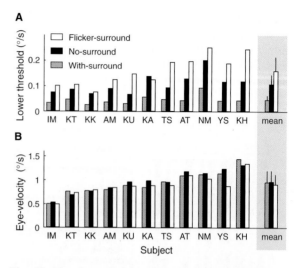

Fig. 10. The data of the motion-detection experiment for the 11 observers and their mean (error bar, 1 SD). Results in the three conditions (with-surround, no-surround, and flicker-surround) are plotted in separate bars. The lower threshold for motion was determined by the method of constant stimuli. In each trial, the surround (if visible) first appeared and remained for 2 s, within which the center appeared at a randomized timing and remained for 0.85 s. The subject had to indicate the direction of the translation by choosing one out of eight possible directions differing by 45°. The correct rate was plotted against translation speed and was fit with a sigmoidal psychometric function. The motion detection threshold was determined as the speed that yielded 53.3% correct identification. (A) Lower threshold for motion. (B) Fixation instability estimated from eye-movement records. See the legend of Fig. 1 for experimental details of eye-movement recording. Instantaneous horizontal eye velocity with the temporal resolution of 13 ms was recorded while the observer was steadily fixating at a noise pattern with a fixation spot, and was plotted in a velocity histogram with 0.1°/s bin (see Fig. 1B). The histogram was fit with a Gaussian, and the best-fit σ was taken as the index of fixation instability for each observer. Reprinted from Murakami (2004), with permission from Elsevier.

Fixation instability (standard deviation of instantaneous eye velocity during fixation) of each observer was also determined by monitoring small eye movements with a high-speed eye tracker (Fig. 10B), and inter-observer correlograms between motion detection threshold and fixation instability were plotted for 11 observers (Fig. 11). In no-surround, threshold positively correlated with fixation instability. A similar positive correlation was found in flicker-surround. With a static surround, however, threshold did not systematically vary with

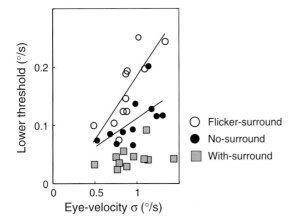

Fig. 11. Correlations between psychophysical data and eye-movement data. Each point represents each observer. The solid lines indicate linear regressions for the correlations that reached statistical significance. Results in the three conditions (with-surround, no-surround, and flicker-surround) are plotted in separate symbols. Reprinted from Murakami (2004), with permission from Elsevier.

fixation instability. Thus, predictions 3 and 4 were fulfilled.

When the observer's task is to detect differential motion between disk and annulus, the disk contains motion signal (to be detected) and eye-jitter noise, whereas the annulus contains only eye-jitter noise. As the noises in both regions are perfectly correlated, the brain ignores such spatially common eye-movement noise and accomplishes fine detectability of target motion irrespective of fixation instability by detecting a spatially differential motion. This explanation accounts for the lowest detection threshold in the with-surround condition.

In the detection of uniform motion without a reference frame, however, the brain is not very certain as to whether the origin of incoming image motion is the movement of the eye or an object. The visual system's task, then, is to estimate whether the current samples of motion signals in the disk region come from the distribution of "motion signal plus eye-originating noise" $(S + N)$ or the distribution of "eye-originating noise only" (N). The signal detection theory dictates that the performance in such a situation is limited by noise variance — fixation instability. The detection threshold in no-surround indeed showed a linear relationship with the size of eye-movement noise.

The detection of motion with a flickering reference frame is even harder, because the flicker not only makes the reference frame unreliable but also induces perceptual jitter in the central disk (Murakami, 2003), within which the detection target was presented. The results indeed indicate higher thresholds that also show dependence on fixation instability. Previously it was reported that motion detection in the central patch was harder with a retinally stabilized pattern in the surround than without a structured surround (Tulunay-Keesey and VerHoeve, 1987). In this case also, the presence of the stabilized surround was detrimental presumably because it enhanced the spatially differential motions between eye-originating jitter in the center and the jitter-free surround.

Roles of global motions

The psychophysical findings described above have convincingly shown that our motion perception is in fact largely mediated by local differential-motion sensors. In primate studies, there is neuro-physiological evidence of segregated processing between differential and common motions in the dorsal visual pathway (Tanaka et al., 1986; Tanaka and Saito, 1989; Born and Tootell, 1992). The retinal ganglion cells of the rabbit and the salamander also exhibit greater excitation to spatially differential motions than common ones (Ölveczky et al., 2003). There is also a line of psychophysical evidence for differential motion detectors in the human (Murakami and Shimojo, 1993, 1995, 1996; Watson and Eckert, 1994; Sachtler and Zaidi, 1995; Tadin et al., 2003). These kinds of neuronal architecture are probably suitable for implementing the rule that spatially common image motions should be omitted from object motion perception. However, is this rule applicable to general movements of the eye other than fixational eye movements? Do spatially non-local and non-differential, globally coherent image motions have anything to do with visual information processing?

I would argue that non-differential, global motions are usually removed from one's consciousness, but at the same time, they are playing major roles in other important tasks related to visually based orienting. In natural environment, the spatially global velocity field on the retina conveys ecologically meaningful information as to the physical relationship between the eye and world that made the particular retinal velocity field happen at the particular instant (e.g., Gibson, 1966). Also, there is evidence for neuronal processes specialized to such global motions as uniform translation, expansion/contraction, and rotation, presented in a spatially large velocity field spanning over tens of degrees (e.g., Sakata et al., 1986; Tanaka and Saito, 1989; Orban et al., 1992).

It will be useful to classify global velocity fields into two major types of motions that are generated by distinct sources: one is the retinal image slip caused by eye rotation with respect to the orbit, and another type is the optic flow caused by the observer's body movement with respect to outer environment (Harris, 1994). When a stationary observer looks at a stationary scene, orbit-relative eye movements cause approximately pure translation in the retinal image. Conversely, when the visual system simultaneously receives pure image translation and extraretinal information (e.g., efference copy) reporting eye velocity, it is possible to recover the stationary visual field. Even if it is not used to counteract fixational eye movements, this extraretinal scheme must be used to compensate for pure translation caused by large-scale eye movements, such as smooth pursuit (e.g., Macknik et al., 1991; Freeman and Banks, 1998) and saccades (Ross et al., 1997). Were pure translation simply ignored, we would not be able to understand why the whole visual field has been displaced after each gaze shift. Only by combining retinal and extraretinal signals at each occurrence of saccadic eye movements, it would be possible for the visual system to transform the retina-zcentered spatial coordinate system to the orbit-centered one. Also, if head movements and body movements are similarly monitored by proprioception and used in conjunction with the orbit-centered representation, it would be possible to transform visual events in reference to the observer-centered spatial coordinate system that no longer depends on eye/head/ body movements (Wade and Swanston, 1996). Once the observer-centered representation is established, the next important task is to describe objects and the observer with respect to the environment-centered frame of reference. The optic flow can be used as a strong cue of the observer's heading direction. The observer's pure translation and pure rotation with respect to the environment yield radial and solenoidal, respectively, optic flow (Warren, 1998). Conversely, it seems that the neural system heavily depends on the optic flow information in computing body orienting and heading with respect to the environment. For example, driving in fog leads to serious speed underestimation of car speed (Snowden et al., 1998), presumably because the optic flow, on which the driver heavily relies in estimating driving speed, becomes low-contrast and thus it is registered as slower than actual flow (Thompson, 1982).

In light of the above empirical evidence and theoretical consideration, it can be argued that there are more than one neural processes to handle perceptual stability of the world and oneself. Precise coding of object velocity would be based on differential-motion detectors, whereby common image motions are filtered out. The process that is responsible for coordinate transformation from retinal to egocentric would rather rely on common image motions, assuming that they may originate in eye movements, and compares them to extraretinal information of eye velocity. The process that is responsible for self-movement in the environmental reference frame would be monitoring patterns of optic flow as a direct cue of orienting and heading. These processes may be located independently and connected serially, either in the above-mentioned order or differently, but it may also be possible that they operate in parallel and share a common mechanism. Therefore, my current investigations to answer future questions include: (1) Is there a strict principle as to when common motions are perceptually dismissed and when they are not? (2) Can the illusory jitter perception lead to an error of orienting or heading of oneself with respect to the environment? (3) Do patients with motion-related deficits in vision perceive the visual jitter? (4) Is the neural correlate of perceptual stability observable in single neurons? We are obtaining a few empirical data in relation to these questions (Kitazaki et al., 2004; Murakami et al., 2004), but they are still open to extensive investigations.

Acknowledgments

I would like to acknowledge financial support from the Center for Evolutionary Cognitive Sciences at the University of Tokyo during the writing of this manuscript. I thank Drs. Susana Martinez-Conde and Stephen L. Macknik for their expertise and stimulating discussions on the topics reviewed here. Part of the experimental research was done at Vision Sciences Laboratory, Harvard University, and at Massachusetts General Hospital, while I was a post-doctoral fellow under the supervision of Dr. Patrick Cavanagh, Harvard University, and part of the research was conducted at Human and Information Science Laboratory, NTT Communication Science Laboratories, while I was working for NTT Corporation.

References

Adelson, E.H. and Bergen, J.R. (1985) Spatiotemporal energy models for the perception of motion. J. Opt. Soc. Am. A, 2: 284–299.

Allman, J., Miezin, F. and McGuinness, E. (1985a) Direction- and velocity-specific responses from beyond the classical receptive field in the middle temporal visual area (MT). Perception, 14: 105–126.

Allman, J., Miezin, F. and McGuinness, E. (1985b) Stimulus specific responses from beyond the classical receptive field: neurophysiological mechanisms for local–global comparisons in visual neurons. Annu. Rev. Neurosci., 8: 407–430.

Allman, J., Miezin, F. and McGuinness, E. (1990) Effects of background motion on the responses of neurons in the first and second cortical visual areas. In: Edelman, G.M., Gall, W.E. and Cowan, W.M. (Eds.), Signal and Sense: Local and Global Order in Perceptual Maps. Wiley, New York, pp. 131–141.

Bair, W. and O'Keefe, L.P. (1998) The influence of fixational eye movements on the response of neurons in area MT of the macaque. Visual Neurosci., 15: 779–786.

Born, R., Groh, J.M., Zhao, R. and Lukasewycz, S.J. (2000) Segregation of object and background motion in visual area MT: effects of microstimulation on eye movements. Neuron, 26: 725–734.

Born, R.T. and Tootell, R.B.H. (1992) Segregation of global and local motion processing in primate middle temporal visual area. Nature, 357: 497–499.

Coppola, D. and Purves, D. (1996) The extraordinarily rapid disappearance of entoptic images. Proc. Natl. Acad. Sci. USA, 93: 8001–8004.

Cornsweet, T.N. (1956) Determination of the stimuli for involuntary drifts and saccadic eye movements. J. Opt. Soc. Am., 46: 987–993.

Deubel, H., Bridgeman, B. and Schneider, W.X. (1998) Immediate post-saccadic information mediates space constancy. Vision Res., 38: 3147–3159.

Deubel, H., Schneider, W.X. and Bridgeman, B. (1996) Post-saccadic target blanking prevents saccadic suppression of image displacement. Vision Res., 36: 985–996.

Di Russo, F., Pitzalis, S. and Spinelli, D. (2003) Fixation stability and saccadic latency in élite shooters. Vision Res., 43: 1837–1845.

Ditchburn, R.W. and Ginsborg, B.L. (1952) Vision with a stabilized retinal image. Nature, 170: 36–37.

Eizenman, M., Hallett, P.E. and Frecker, R.C. (1985) Power spectra for ocular drift and tremor. Vision Res., 25: 1635–1640.

Freeman, T.C.A. and Banks, M.S. (1998) Perceived head-centric speed is affected by both extra-retinal and retinal errors. Vision Res., 38: 941–945.

Gattass, R. and Gross, C.G. (1981) Visual topography of striate projection zone (MT) in posterior superior temporal sulcus of the macaque. J. Neurophysiol., 46: 621–638.

Georgeson, M.A. and Scott-Samuel, N.E. (1999) Motion contrast: a new metric for direction discrimination. Vision Res., 39: 4393–4402.

Gerrits, H.J.M. and Vendrik, A.J.H. (1970) Artificial movements of a stabilized image. Vision Res., 10: 1443–1456.

Gibson, J.J. (1966) The Senses Considered as Perceptual Systems. Houghton Mifflin, Boston.

Greschner, M., Bongard, M., Rujan, P. and Ammermüller, J. (2002) Retinal ganglion cell synchronization by fixational eye movements improves feature estimation. Nat. Neurosci., 5: 341–347.

Harris, L.R. (1994) Visual motion caused by movements of the eye, head and body. In: Smith, A.T. and Snowden, R.J. (Eds.), Visual Detection of Motion. Academic Press, London, pp. 397–435.

Helmholtz, H.v. (1866) Handbuch der physiologischen Optik. Voss, Leipzig.

Hubel, D.H. and Wiesel, T.N. (1968) Receptive fields and functional architecture of monkey striate cortex. J. Physiol., 195: 215–243.

Hupé, J.M., James, A.C., Payne, B.R., Lomber, S.G., Girard, P. and Bullier, J. (1998) Cortical feedback improves discrimination between figure and background in V1, V2 and V3 neurons. Nature, 394: 784–787.

Kitazaki, M., Kubota, M. and Murakami, I. (2004) Effects of the visual jitter aftereffect on the control of posture. Eur. Conf. Visual Perception, 33(Suppl.): 104.

Komatsu, H. and Murakami, I. (1994) Behavioral evidence of filling-in at the blind spot of the monkey. Visual Neurosci., 11: 1103–1113.

Kreegipuu, K. and Allik, J. (2003) Perceived onset time and position of a moving stimulus. Vision Res., 43: 1625–1635.

Lagae, L., Gulyás, B., Raiguel, S. and Orban, G.A. (1989) Laminar analysis of motion information processing in macaque V5. Brain Res., 496: 361–367.

Leopold, D.A. and Logothetis, N.K. (1998) Microsaccades differentially modulate neural activity in the striate and extrastriate visual cortex. Exp. Brain Res., 123: 341–345.

Levi, D.M., Klein, S.A. and Aitsebaomo, P. (1984) Detection and discrimination of the direction of motion in central and peripheral vision of normal and amblyopic observers. Vision Res., 24: 789–800.

Livingstone, M.S. and Hubel, D.H. (1987) Psychophysical evidence for separate channels for the perception of form, color, movement, and depth. J. Neurosci., 7: 3416–3468.

MacKay, D.M. (1958) Perceptual stability of a stroboscopically lit visual field containing self-luminous objects. Nature, 181: 507–508.

Macknik, S.L., Fisher, B.D. and Bridgeman, B. (1991) Flicker distorts visual space constancy. Vision Res., 31: 2057–2064.

Martinez-Conde, S., Macknik, S.L. and Hubel, D.H. (2000) Microsaccadic eye movements and firing of single cells in the striate cortex of macaque monkeys. Nat. Neurosci., 3: 251–258.

Martinez-Conde, S., Macknik, S.L. and Hubel, D.H. (2002) The function of bursts of spikes during visual fixation in the awake primate lateral geniculate nucleus and primary visual cortex. Proc. Natl. Acad. Sci. USA, 99: 13920–13925.

Martinez-Conde, S., Macknik, S.L. and Hubel, D.H. (2004) The role of fixational eye movements in visual perception. Nat. Rev. Neurosci., 5: 229–240.

McKee, S.P. and Taylor, D.G. (1984) Discrimination of time: comparison of foveal and peripheral sensitivity. J. Opt. Soc. Am. A, 1: 620–627.

McKee, S.P., Welch, L., Taylor, D.G. and Bowne, S.F. (1990) Finding the common bond: stereoacuity and the other hyperacuities. Vision Res., 30: 879–891.

Murakami, I. (2001a) The flash-lag effect as a spatiotemporal correlation structure. J. Vision, 1: 126–136.

Murakami, I. (2001b) A flash-lag effect in random motion. Vision Res., 41: 3101–3119.

Murakami, I. (2003) Illusory jitter in a static stimulus surrounded by a synchronously flickering pattern. Vision Res., 43: 957–969.

Murakami, I. (2004) Correlations between fixation stability and visual motion sensitivity. Vision Res., 44: 751–761.

Murakami, I. and Cavanagh, P. (1998) A jitter after-effect reveals motion-based stabilization of vision. Nature, 395: 798–801.

Murakami, I. and Cavanagh, P. (2001) Visual jitter: evidence for visual-motion-based compensation of retinal slip due to small eye movements. Vision Res., 41: 173–186.

Murakami, I., Kitaoka, A. and Ashida, H. (2004) The amplitude of small eye movements correlates with the saliency of the peripheral drift illusion. Society for Neuroscience Annual Meeting Abstract, 30: 302.

Murakami, I., Komatsu, H. and Kinoshita, M. (1997) Perceptual filling-in at the scotoma following a monocular retinal lesion in the monkey. Visual Neurosci., 14: 89–101.

Murakami, I. and Shimojo, S. (1993) Motion capture changes to induced motion at higher luminance contrasts, smaller eccentricities, and larger inducer sizes. Vision Res., 33: 2091–2107.

Murakami, I. and Shimojo, S. (1995) Modulation of motion aftereffect by surround motion and its dependence on stimulus size and eccentricity. Vision Res., 35: 1835–1844.

Murakami, I. and Shimojo, S. (1996) Assimilation-type and contrast-type bias of motion induced by the surround in a random-dot display: evidence for center–surround antagonism. Vision Res., 36: 3629–3639.

Nakayama, K. and Tyler, C.W. (1981) Psychophysical isolation of movement sensitivity by removal of familiar position cues. Vision Res., 21: 427–433.

Ölveczky, B.P., Baccus, S.A. and Meister, M. (2003) Segregation of object and background motion in the retina. Nature, 423: 401–408.

Orban, G.A., Lagae, L., Verri, A., Raiguel, S., Xiao, D., Maes, H. and Torre, V. (1992) First-order analysis of optical flow in monkey brain. Proc. Natl. Acad. Sci. USA, 89: 2595–2599.

Pantle, A. and Turano, K. (1992) Visual resolution of motion ambiguity with periodic luminance- and contrast-domain stimuli. Vision Res., 32: 2093–2106.

Peli, E. and García-Pérez, M.A. (2003) Motion perception during involuntary eye vibration. Exp. Brain Res., 149: 431–438.

Ramachandran, V.S. (1992) The blind spot. Sci. Am., 266: 86–91.

Ramachandran, V.S. and Gregory, R.L. (1991) Perceptual filling in of artificially induced scotomas in human vision. Nature, 350: 699–702.

Riggs, L.A., Ratliff, F., Cornsweet, J.C. and Cornsweet, T.N. (1953) The disappearance of steadily fixated visual test objects. J. Opt. Soc. Am., 43: 495–501.

Ross, J., Morrone, M.C. and Burr, D.C. (1997) Compression of visual space before saccades. Nature, 386: 598–601.

Ross, J., Morrone, M.C., Goldberg, M.E. and Burr, D.C. (2001) Changes in visual perception at the time of saccades. Trends Neurosci., 24: 113–121.

Rucci, M. and Desbordes, G. (2003) Contributions of fixational eye movements to the discrimination of briefly presented stimuli. J. Vision, 3: 852–864.

Sachtler, W.L. and Zaidi, Q. (1995) Visual processing of motion boundaries. Vision Res., 35: 807–826.

Sakata, H., Shibutani, H., Ito, Y. and Tsurugai, K. (1986) Parietal cortical neurons responding to rotary movement of visual stimulus in space. Exp. Brain Res., 61: 658–663.

Sasaki, Y., Murakami, I., Cavanagh, P. and Tootell, R.B.H. (2002) Human brain activity during illusory visual jitter as revealed by functional magnetic resonance imaging. Neuron, 35: 1147–1156.

Sereno, M.I., Dale, A.M., Reppas, J.B., Kwong, K.K., Belliveau, J.W., Brady, T.J., Rosen, B.R. and Tootell, R.B.H. (1995) Borders of multiple visual areas in humans revealed by functional magnetic resonance imaging. Science, 268: 889–893.

Sherrington, C.S. (1918) Observations on the sensual role of the proprioceptive nerve supply of the extrinsic ocular muscles. Brain, 41: 332–343.

Shioiri, S. and Cavanagh, P. (1990) ISI produces reverse apparent motion. Vision Res., 30: 757–768.

Shioiri, S., Ito, S., Sakurai, K. and Yaguchi, H. (2002) Detection of relative and uniform motion. J. Opt. Soc. Am. A, 19: 2169–2179.

Skavenski, A.A., Hansen, R.M., Steinman, R.M. and Winterson, B.J. (1979) Quality of retinal image stabilization during small natural and artificial body rotations in man. Vision Res., 19: 675–683.

Snodderly, D.M., Kagan, I. and Gur, M. (2001) Selective activation of visual cortex neurons by fixational eye movements: implications for neural coding. Visual Neurosci., 18: 259–277.

Snowden, R.J., Stimpson, N. and Ruddle, R.A. (1998) Speed perception fogs up as visibility drops. Nature, 392: 450.

Spauschus, A., Marsden, J., Halliday, D.M., Rosenberg, J.R. and Brown, P. (1999) The origin of ocular microtremor in man. Exp. Brain Res., 126: 556–562.

Spillmann, L., Anstis, S., Kurtenbach, A. and Howard, I. (1997) Reversed visual motion and self-sustaining eye oscillations. Perception, 26: 823–830.

Steinman, R.M., Cunitz, R.J., Timberlake, G.T. and Herman, M. (1967) Voluntary control of microsaccades during maintained monocular fixation. Science, 155: 1577–1579.

Steinman, R.M., Haddad, G.M., Skavenski, A.A. and Wyman, D. (1973) Miniature eye movement. Science, 181: 810–819.

Tadin, D., Lappin, J.S., Gilroy, L.A. and Blake, R. (2003) Perceptual consequences of centre-surround antagonism in visual motion processing. Nature, 424: 312–315.

Takeuchi, T. and De Valois, K.K. (1997) Motion-reversal reveals two motion mechanisms functioning in scotopic vision. Vision Res., 37: 745–755.

Tanaka, K., Hikosaka, K., Saito, H., Yukie, M., Fukada, Y. and Iwai, E. (1986) Analysis of local and wide-field movements in the superior temporal visual areas of the macaque monkey. J. Neurosci., 6: 134–144.

Tanaka, K. and Saito, H. (1989) Analysis of motion of the visual field by direction, expansion/contraction, and rotation cells clustered in the dorsal part of the medial superior temporal area of the macaque monkey. J. Neurophysiol., 62: 626–641.

Tanaka, K., Sugita, Y., Moriya, M. and Saito, H. (1993) Analysis of object motion in the ventral part of the medial superior temporal area of the macaque visual cortex. J. Neurophysiol., 69: 128–142.

Thompson, P. (1982) Perceived rate of movement depends on contrast. Vision Res., 22: 377–380.

Tootell, R.B.H., Mendola, J.D., Hadjikhani, N.K., Liu, A.K. and Dale, A.M. (1998) The representation of the ipsilateral visual field in human cerebral cortex. Proc. Natl. Acad. Sci. USA, 95: 818–824.

Tulunay-Keesey, U. and VerHoeve, J.N. (1987) The role of eye movements in motion detection. Vision Res., 27: 747–754.

Wade, N.J. and Swanston, M.T. (1996) A general model for the perception of space and motion. Perception, 25: 187–194.

Warren, W.H.J. (1998) The state of flow. In: Watanabe, T. (Ed.), High-Level Motion Processing. MIT Press, Cambridge, MA, pp. 315–358.

Watson, A.B. (1982) Derivation of the impulse response: comments on the method of Roufs and Blommaert. Vision Res., 22: 1335–1337.

Watson, A.B. and Ahumada, A.J.J. (1985) Model of human visual-motion sensing. J. Opt. Soc. Am. A, 2: 322–342.

Watson, A.B. and Eckert, M.P. (1994) Motion-contrast sensitivity: visibility of motion gradient of various spatial frequencies. J. Opt. Soc. Am. A, 11: 496–505.

Wertheim, A.H. (1994) Motion perception during self-motion: the direct versus inferential controversy revisited. Behav. Brain Sci., 17: 293–355.

Whitaker, D. and MacVeigh, D. (1990) Displacement thresholds for various types of movement: effect of spatial and temporal reference proximity. Vision Res., 30: 1499–1506.

Xiao, D.-K., Raiguel, S., Marcar, V., Koenderink, J. and Orban, G.A. (1995) Spatial heterogeneity of inhibitory surrounds in the middle temporal visual area. Proc. Natl. Acad. Sci. USA, 92: 11303–11306.

Yarbus, A.L. (1967) Eye Movements and Vision. Plenum Press, New York.

Martinez-Conde, Macknik, Martinez, Alonso & Tse (Eds.)
Progress in Brain Research, Vol. 154
ISSN 0079-6123

CHAPTER 11

A cholinergic mechanism underlies persistent neural activity necessary for eye fixation

José M. Delgado-García[1,*], Javier Yajeya[2] and Juan de Dios Navarro-López[3]

[1]*División de Neurociencias, Universidad Pablo de Olavide, 41013-Seville, Spain*
[2]*Departamento de Fisiología y Farmacología, Facultad de Medicina, Instituto de Neurociencias de Castilla y León, Universidad de Salamanca, Salamanca, Spain*
[3]*Department of Physiology, University College of London, London WC1E 6BT, UK*

Abstract: It is generally accepted that the prepositus hypoglossi (PH) nucleus is the site where horizontal eye-velocity signals are integrated into eye-position ones. However, how does this neural structure produce the sustained activity necessary for eye fixation? The generation of the neural activity responsible for eye-position signals has been studied here using both in vivo and in vitro preparations. Rat sagittal brainstem slices including the PH nucleus and the paramedian pontine reticular formation (PPRF) rostral to the abducens nucleus were used for recording intracellularly the synaptic activation of PH neurons from the PPRF. Single electrical pulses applied to the PPRF showed a monosynaptic projection on PH neurons. This synapse was found to be glutamatergic in nature, acting on alpha-amino-3-hydroxy-5-methylisoxazole propionate (AMPA)/kainate receptors. Train stimulation (100 ms, 50–200 Hz) of the PPRF evoked a depolarization of PH neurons, exceeding (by hundreds of ms) the duration of the stimulus. Both duration and amplitude of this long-lasting depolarization were linearly related to train frequency. The train-evoked sustained depolarization was demonstrated to be the result of the additional activation of cholinergic fibers projecting onto PH neurons, because it was prevented by slice superfusion with atropine sulfate and pirenzepine (two cholinergic antagonists), and mimicked by carbachol and McN-A-343 (two cholinergic agonists). These results were confirmed in alert behaving cats. Microinjections of atropine and pirenzepine evoked an ipsilateral gaze-holding deficit consisting of an exponential-like, centripetal eye movement following saccades directed toward the injected site. These findings suggest that the sustained activity present in PH neurons carrying eye-position signals is the result of the combined action of PPRF neurons and the facilitative role of cholinergic terminals, both impinging on PH neurons. The present results are discussed in relation to other proposals regarding integrative properties of PH neurons and/or related neural circuits.

Keywords: prepositus hypoglossi; eye movements; eye fixation; persistent activity; cholinergic neurons; muscarinic receptors; glutamate receptors; short-term potentiation

Introduction

The eye moves in the horizontal plane under the action of two antagonist extraocular muscles: the lateral and medial recti. The lateral rectus muscle is innervated by motoneurons located in the pontine abducens nucleus, while the medial rectus muscle is innervated by motoneurons located in the mesencephalic oculomotor complex (Büttner-Ennever and Horn, 1997). As illustrated in Figs. 1 and 2, these extraocular motoneurons are capable of

*Corresponding author. Tel.: +34-954-349374;
Fax: +34-954-349375; E-mail: jmdelgar@upo.es

DOI: 10.1016/S0079-6123(06)54011-7

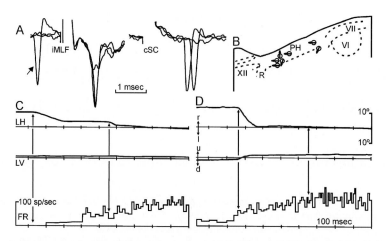

Fig. 1. Comparative analysis of eye-position signals present in abducens (ABD) motoneurons and in PH (PH) neurons. (A) Examples of the firing rate (FR, in spikes/s) of representatives (1) ABD motoneurons and (2) *position*, (3) *position-velocity*, and (4) *velocity-position* PH neurons, during eye fixations following on-directed saccades. Mean firing rate followed by its standard deviation (SD) is shown for each neuron. Note that the coefficient of variation of the mean (indicated between brackets) increased five-fold from (1) to (4). (B) At the top is shown a plot of the SD of the interspike intervals (ISIs, in ms) against mean ISI for a PH neuron (open circles, continuous line) and an ABD motoneuron (solid circles and discontinuous line). The ISIs were measured during eye fixations of 500–1000 ms through the whole oculomotor range. Coefficients of correlation (*r*) are indicated. The steeper slope for the PH neuron indicates a higher variability in the firing rate of this neuron for a given mean ISI. At the bottom are shown examples corresponding to three ABD motoneurons (discontinuous lines) and eight PH neurons (continuous lines). Coefficients of correlation for these regression lines were ⩾0.7. (Reproduced with permission from Delgado-García et al., 1989.)

evoking phasic firing (i.e., high-frequency bursts of action potentials lasting ≈100 ms) that will produce a strong muscular contraction which is able to generate a fast eye displacement — that is, a saccade or a fast phase of the vestibulo-ocular or opto-kinetic reflexes (Robinson, 1981; Moschovakis et al., 1996; Delgado-García, 2000). This fast muscular activation is necessary to overcome the viscous drag of the orbit. In order to maintain a stable position of the eye in the orbit, extraocular motoneurons are also capable of a sustained tonic firing, necessary to counteract the restoring elastic components of orbital tissues (Robinson, 1981; Escudero et al., 1992; Fukushima et al., 1992; Moschovakis, 1997; Delgado-García, 2000; Major and Tank, 2004). Thus, horizontal motoneurons encode the necessary velocity and position signals to rotate the eye toward the appropriate visual target and to hold the eye stable in the orbit. In fact, the firing properties of ocular motoneurons can be precisely represented by a first-order linear model (Robinson, 1981). In cats, horizontal motoneurons increase their mean firing rate by ≈7 spikes/s per degree of eye position, and by 1 spike/s per degree/s of eye velocity in

the pulling direction of the involved muscle (see references in Delgado-García, 2000).

In the next few pages we will concentrate on experiments carried out by our group regarding the firing activities of prepositus hypoglossi (PH) neurons during eye movements, and on recent in vitro studies on the functional properties of reticular afferents to these neurons. More-detailed and comparative reviews regarding the integrative properties of PH neurons for the generation of eye-position signals can be found elsewhere (Robinson, 1981; Cannon and Robinson, 1987; Fukushima et al., 1992; Moschovakis et al., 1996; Moschovakis, 1997; Delgado-García, 2000; Major and Tank, 2004).

The final common pathway for horizontal eye movements

Abducens and medial rectus motoneurons represent the final common neural pathway interposed between eye-movement-related brainstem centers and extraocular muscles in the horizontal plane. Thus, abducens and medial rectus motoneurons

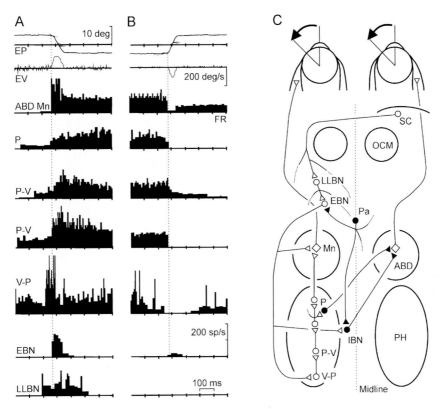

Fig. 2. A plausible hypothesis regarding the generation of eye-position signals in the PH nucleus. (A, B) Firing rate (FR, in spikes/s) of seven different types of neuron during eye fixations before and after on- and off-directed saccades. From bottom to top are illustrated the FRs of a long-lead burst neuron (LLBR), an excitatory burst neuron (EBN), four PH neurons showing *velocity-position* (V-P), *position-velocity* (P-V), or pure *position* (P) signals, and an abducens motoneuron (ABD Mn). The horizontal eye position (EP) and velocity (EV) corresponding to these neural activities are illustrated at the top. (C) Diagram showing the possible pathways generating eye-position signals following a saccade triggered from the superior colliculus (SC). Other abbreviations: OCN, oculomotor nucleus; IBN, inhibitory burst neuron; Pa, omnipause cells. (Modified from Escudero et al. (1992), and reproduced with permission of the Physiological Society.)

must be able to translate to the lateral and medial recti muscles the precise neural motor commands corresponding to each type of eye movement (Robinson, 1981; Escudero and Delgado-García, 1988; Fukushima et al., 1992; Büttner-Ennever and Horn, 1997; Moschovakis, 1997). As indicated above, abducens and medial rectus motoneurons occupy widely separated sites in the brainstem, but they have similar firing profiles and gain for eye-position and eye-velocity motor commands (de la Cruz et al., 1989). In this regard, it is well known that horizontal conjugate eye movements are generated in the abducens nucleus (Delgado-García et al., 1986). Abducens internuclear neurons convey a signal similar to that present in abducens motoneurons to

the medial rectus motoneurons located in the contralateral oculomotor nucleus. That is, besides eye accommodation and vergence signals produced in midbrain regions close to the oculomotor nucleus, neural signals present in abducens motoneurons represent horizontal oculomotor commands to the corresponding extraocular muscles.

Abducens motoneurons fire a burst of action potentials slightly preceding and during eye movements in the on-direction, and decrease, or even stop, their firing during off-directed saccades. Moreover, abducens motoneurons present a tonic firing proportional to eye positions in the orbit. This tonic firing increases or decreases linearly with eye positions further in the on- or off-direction, respectively. Thus, abducens moto-neurons are multi-stable, in the sense that they are capable of producing a stable firing for every position of the eye in the orbit (Major and Tank, 2004). Recently, we have shown that the firing variability of abducens motoneurons in alert cats during ocular positions of fixation depends on the balance between inhibitory and excitatory synaptic innervation, but also on intrinsic mechanisms capable of stabilizing motoneuron firing (see Fig. 1; Delgado-García et al., 1989; González-Forero et al., 2002).Whereas, the neural origin and functional properties of oculomotor subsystems generating eye saccades and the vestibulo-ocular and opto-kinetic reflexes are relatively well known, the site and (mainly) the mechanism by which eye position is generated has always been somewhat of a mystery (Robinson, 1981; Escudero and Delgado-García, 1988; Fukushima et al., 1992; Moschovakis, 1997; Aksay et al., 2001; Major and Tank, 2004). In an initial seminal study, Robinson (1981) proposed the presence of a common eye-position-signal neural integrator subserving all of the eye-movement subsystems. However, available experimental information suggests that there are several integrators, depending on the origin of extraocular motor commands and on the plane of movement (horizontal or vertical). For example, horizontal and vertical eye-position signals are integrated separately in the PH nucleus and in the interstitial nucleus of Cajal (Fukushima et al., 1992; Moschovakis, 1997; Delgado-García, 2000). Moreover, other brainstem (medial vestibular nucleus) and cerebellar (fastigial nucleus) structures seem to carry eye-position signals, at least in the cat (Escudero et al., 1992; Gruart and Delgado-García, 1994). Integration is assumed to take place within those nuclei and/or as a result of the functional interactions they establish by their reciprocal connections and connections with other brainstem (vestibular nuclei) and cerebellar (flocculus) structures. That is, neural integration subserving eye positions of fixation could be the result of a network property.

Firing properties of prepositus hypoglossi neurons

Neurons located in the paramedian pontine reticular formation (PPRF), in particular those called excitatory burst neurons (EBN; Fig. 2), are able to generate bursts of action potentials that encode the amplitude, peak velocity, and duration of eye saccades and fast phases of the vestibulo-ocular and opto-kinetic reflexes (Igusa et al., 1980; Escudero and Delgado-García, 1988; Fukushima et al., 1992; Moschovakis et al., 1996). These neurons project monosynaptically onto abducens motoneurons and PH neurons. The question is: how do PH neurons transform the eye-velocity signals provided by EBN into eye-position ones?

Permanent or transient blockage of the functional processes taking place in the PH nucleus in both cats and monkeys supports the assumption that this structure is the site where eye-velocity signals are integrated into eye-position motor commands. It has also been suggested that eye-position signals are the result of functional interactions established by the reciprocal connections of PH nucleus with the vestibular nuclei, the contralateral PH, and the cerebellum (Cannon and Robinson, 1987; Cheron and Godaux, 1987; Moreno-López et al., 1996, 1998, 2001; Kaneko, 1997). Here, we will firstly consider the firing activities related to eye movements that can be recorded in the PH nucleus of alert cats, and later we will address the issue of how PH neurons generate eye-position signals.

Depending on their linear relationships with eye position and/or velocity, PH neurons can be classified as *position*, *position-velocity* and *velocity-position* cells (Figs. 1 and 2; López-Barneo et al., 1982; Delgado-García et al., 1989; Escudero et al., 1992). Pure *position neurons* are activated during ipsilateral eye fixations. Their mean position gain is ≈ 7 spikes/s per degree (i.e., similar to position-signal values present in abducens motoneurons), and they present no noticeable eye-velocity signals. It has been shown in alert cats that position neurons project monosynaptically onto abducens motoneurons (Escudero and Delgado-García, 1988; Escudero et al., 1992). *Position-velocity neurons* located in the PH nucleus seem to encode both eye-position and eye-velocity signals in the horizontal plane, and are activated by eye movements in the ipsilateral direction. The mean position and velocity sensitivity of these neurons are ≈ 5 spikes/s per degree and 0.6 spikes/s per degree/s respectively, with correlation coefficients, $r \geq 0.6$ for both parameters. Position-velocity neurons project onto abducens motoneurons and to the vicinity of the oculomotor complex. Finally, *velocity-position* neurons present a rather irregular tonic firing, not related significantly with the position of the eye in the orbit, but they present a significant relationship with ipsi- or contra-lateral eye movements (0.75 spikes/s per degree/s, $r \geq 0.6$; Figs. 1 and 2). Velocity-position neurons related to ipsilateral eye movements seem to project to the peri-oculomotor area, while those related to contralateral saccades project mainly to the cerebellum (Delgado-García et al., 1989).

The variability of interspike intervals for a similar eye position in the orbit decreases from velocity-position to position-velocity and to pure position neurons in relation to the increase in eye-position signals (Figs. 1 and 2). However, it can be noticed in these two figures that the firing rate of abducens motoneurons appears to be more stable than that presented by pure position neurons. In accord to this, extraocular motoneurons have already been reported as having some intrinsic mechanisms that might contribute to a persisting and stable firing rate (Delgado-García et al., 1989; González-Forero et al., 2002). Thus, the stabilizing role of the intrinsic membrane properties of extraocular motoneurons in the generation of eye-position motor commands should also be taken into account. The algebraic addition (i.e., another form of integration) of many different sources of eye-position signals upon the distal dendrites of ocular motoneurons (Baker and Spencer, 1981) could be further elaborated by the intrinsic active properties of the motoneuron membrane to produce the stable firing rate they display during eye positions of fixation (Delgado-García, 2000).

The cascade model for the generation of eye-position signals

Using available data collected from extracellular recordings of firing activities of PH neurons during eye movements in alert cats, Delgado-García et al. (1989) have proposed a neural circuit *in cascade* to explain the generation of eye-position signals. In this circuit, the three neuronal types described above (velocity-position, position-velocity, and position neurons) were assumed to receive similar inputs from vestibular and reticular origins. This early proposal was modified following data obtained later (Fig. 2; Escudero et al., 1992) showing that only position neurons seem to project monosynaptically onto abducens motoneurons. Nevertheless, it has also been shown that position-velocity neurons project onto extraocular motoneurons, at least in the vertical motor system (Moschovakis et al., 1996; Moschovakis, 1997). In short, the cascade model proposes that eye-velocity signals coming from EBN (Fig. 2) terminate preferentially on velocity-position neurons. Each component of the rest of the chain (position-velocity and position neurons) projects on both the following and the preceding one, forming an integrated loop, and these chains of neurons would be superimposed upon the shorter, direct pathways carrying eye-velocity motor commands (Escudero et al., 1992). Evidence supporting this proposal is the following: (i) The larger latency of activation of abducens motoneurons presented by position-velocity versus position neurons (Escudero et al., 1992). (ii) The decreasing onset time with respect to the triggering saccade presented by velocity-position, position-velocity, and position neurons. Indeed, the latency to the beginning of the saccade was higher for those neurons exhibiting larger velocity signals. Peri-event time histograms of the spike activity of identified abducens motoneurons showed activation latencies in the monosynaptic range when triggered by premotor PH position neurons. In contrast, the activation of abducens motoneuron discharge by position-velocity and (mainly) velocity-position PH neurons was in the di- or polysynaptic range (Escudero et al., 1992). (iii) The progressive disappearance of velocity signals, as opposed to the increasing position signal, from velocity-position to position neurons (Delgado-García et al., 1989; Escudero et al., 1992). (iv) The presence of cascade-like, polysynaptic connections could explain the high susceptibility of eye-position neuronal systems to drugs, anesthetics, conscious state, and attention level (Delgado-García, 2000).

Experiments carried out in goldfish (Pastor et al., 1994) and cats (Moschovakis et al., 1996) suggest the presence of separate integrator mechanisms to store eye-velocity signals and to generate pure eye-position ones. It is possible, with the help of selected pharmacological tools, to dissociate the two mechanisms related to neural integration in the oculomotor system (Moreno-López et al., 1996, 1998). For example, the injection of nitric oxide synthase inhibitors into the PH nucleus in alert cats produces significant changes in eye velocity, but no noticeable effect on eye positions of fixation. Thus, the unilateral injection of the nitric oxide synthase inhibitor L-nitro-arginine methyl ester (L-NAME; Fig. 3) produces ramp-like eye movements in the contralateral direction, interrupted by fast phases directed ipsilaterally

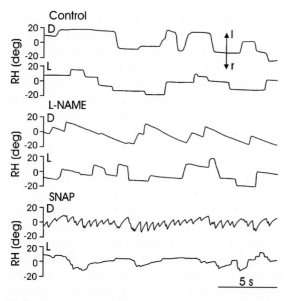

Fig. 3. Records of the right eye position in the horizontal plane (RH, in degrees) obtained from control alert cats and following the injection of the indicated drugs in the left PH nucleus. Recordings were carried out in conditions of darkness (D) and light (L). Control recordings are illustrated at the top. Doses and time after the injection of the illustrated recordings were as follows: L-nitro-arginine methyl ester (L-NAME; an inhibitor of the nitric oxide synthase), 28 nmol, 4 (D) and 5 (L) min (middle set of records); S-nitroso-N-acetylpenicillamine (SNAP; a nitric oxide donor), 20 nmol, 3 (D) and 5 (L) min (bottom set of records). Vertical arrows indicate the direction of eye movements: l, left; r, right. (Reproduced from Moreno-López et al. (1998), with permission of Cell Press.)

(Moreno-López et al., 1996). In contrast, a group of PH neurons, located laterally in the so-called marginal zone close to the medial vestibular nucleus, seems to be involved in the generation of eye-position signals (Kaneko, 1997; Moreno-López et al., 1998, 2001). Neurons located in the marginal zone are characterized by having a nitric-oxide-sensitive guanylyl cyclase, and the injection of nitric oxide donors in this area significantly affects the neural integrator for eye position. As indicated, the injection of the nitric oxide donor S-nitroso-N-acetylpenicillamine (SNAP; Fig. 3) in the marginal zone produces an exponential ipsilaterally directed drift of eye position.

The sustained firing reported here as a characteristic of PH position neurons seems to be necessary for the generation of a stable eye position in the orbit (Robinson, 1981; Moschovakis, 1997; Delgado-García, 2000). In vivo studies carried out in goldfish also reported a persistent firing of selective neuronal brainstem populations related to the generation of eye-position signals (Aksay et al., 2001, 2003). It should be pointed out that the long-lasting neuronal responses surpassing the duration of the triggering stimulus described here have evident relationships with the persistent firing underlying working memory and other cognitive processes (Fuster, 1997; Goldman-Rakic, 1995; Major and Tank, 2004). Two recent papers have made it possible to bridge the gap between these two apparently different neural mechanisms. Thus, Egorov et al. (2002) have reported that a muscarinic cholinergic mechanism underlies persisting firing observed in slices of rat entorhinal cortex, while Navarro-López et al. (2004) have suggested that a similar muscarinic cholinergic action could be responsible (at least in part) for the neuronal processes that is able to generate eye-position signals. The rest of the present report will be devoted to analyzing the relationships between cholinergic mechanisms and neural integration in the PH nucleus.

In search of a synaptic mechanism for eye fixation

As shown by Aksay et al. (2001), the sustained firing rate observed in the neural integrator subserving eye position does not depend on neuronal intrinsic properties, but has to be ascribed to the amplitude and rate of the synaptic inputs arriving at the integrator (brainstem area I, where position-related neurons are located in goldfish). It has also been proposed that synaptic feedback among neurons located in the brainstem area I is still necessary for temporal integration (Aksay et al., 2003). Recently, we have reported a slightly different mechanism possibly underlying the generation of eye-position signals (Navarro-López et al., 2004). In this case, the interaction between those PPRF neurons (mainly EBN, Igusa et al., 1980) projecting to the PH nucleus and meso-pontine reticular cholinergic neurons (Semba et al., 1990) is an additional mechanism, occurring at the synaptic level, necessary for the generation of persistent firing in PH neurons.

Experiments reported by Navarro-López et al. (2004) were carried out in brainstem slices collected from newborn rats. As illustrated in Fig. 4, single pulses applied to the PPRF evoked graded responses in identified PH neurons. If the stimulus reached threshold, prepositus neurons fired a synaptically triggered action potential presenting a characteristic biphasic appearance of the after-hyperpolarization (Fig. 4A, B). The short latency ($\approx 2.5\,ms$) of the evoked excitatory postsynaptic potential and its negligible variability when evoked at a high rate (up to 200 Hz) suggested its monosynaptic nature, confirming that EBN (Fig. 2) located at the rostral PPRF project directly onto prepositus neurons.

Train stimulation (100 ms) of the same reticular formation area evoked sustained depolarizations in the recorded PH neurons, exceeding the end of the stimulus by hundreds of milliseconds (Fig. 4C, D). This is indeed an example of persisting activity, evoked in this case on neurons supposedly endowed with the capacity of neural integration. Both the amplitude and the duration of the evoked sustained depolarization were linearly related with

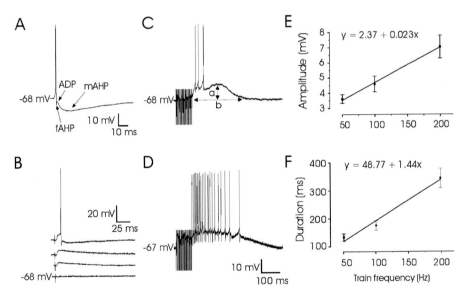

Fig. 4. Differential effects of single and train stimulation of the PPRF (PPRF) on PH (PH) neurons. Recordings were carried out in sagittal brainstem slices from 1-month-old rats. (A) Example of typical action potentials recorded in PH neurons. Note the biphasic appearance of the afterhyperpolarization, presenting a fast (fAHP) and a medium (mAHP) component, separated by an afterde-polarization (ADP). (B) Graded nature of the excitatory postsynaptic potentials (EPSPs) evoked in the same PH neuron by single-pulse (100 µs) stimulation of the PPRF at increasing intensities (200–400 µA) until reaching the threshold intensity necessary to evoke an action potential. (C, D) Effects of PPRF train (200 Hz, 250 µA) on two PH cells. Note the sustained depolarization after the train of stimuli. Calibration in (D) is also for (C). (E, F) Plots of train frequency during PPRF stimulation (abscissas, in Hz) against the amplitude (E, in mV) and duration (F, in ms) of the EPSPs evoked by the train (ordinates). Values corresponding to EPSP amplitude (a) and duration (b) were measured as indicated in (C). (Taken from Navarro-López et al. (2004), with permission of the Society for Neuroscience.)

train frequency (Fig. 4E, F). This latter result is very important, since firing frequency in EBN encodes saccade velocity, and the neural signal elaborated at the integrator needs to be proportional to it (Robinson, 1981; Moschovakis, 1997). On many occasions, the sustained depolarization was able to reach threshold and evoke a train of action potentials (Fig. 4C, D). Interestingly, the evoked burst of action potentials decay with a constant time, similar to that measured for orbital mechanics and to that present in abducens and medial rectus motoneurons (i.e., 120–150 ms; Delgado-García et al., 1986; de la Cruz et al., 1989).

We have also shown that the synapse of EBN on PH neurons is glutamatergic in nature, acting probably on alpha-amino-3-hydroxy-5-methylisoxazole propionate (AMPA)/kainate receptors (Fig. 5A; Navarro-López et al., 2004). However, it is important to indicate here that the sustained depolarization evoked by train stimulation of the PPRF was produced by a different synaptic mechanism, involving cholinergic receptors (Fig. 5B). As shown in Fig. 6, these cholinergic receptors are located postsynaptically, i.e., on PH neurons, because the application of carbachol (a non-specific cholinergic agonist) was able to depolarize recorded neurons even in the presence of TTX (to remove all possible action potentials spontaneously present in the preparation).

Moreover, pirenzepine (a selective blocker of muscarinic M1 receptors) was able to block both the sustained depolarization evoked by train stimulation of the PPRF and the long-lasting depolarization produced by the addition of carbachol to the bathing solution. In accordance with these results, we have proposed that the interaction between AMPA/kainate and muscarinic M1 receptors underlies a plausible mechanism able to generate persistent activity in PH neurons.

It is known that extraocular muscles present a greater strength for the same length when relaxing than when contracting. For this reason, abducens motoneurons and PH neurons present a higher firing rate for a similar eye position when the position follows on-directed saccades than when it follows off-directed saccades (Delgado-García et al. 1986,

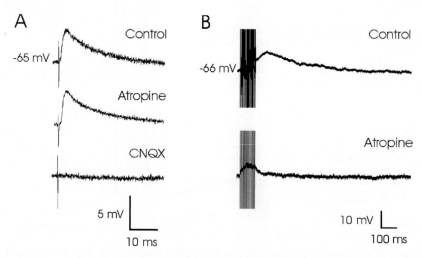

Fig. 5. Glutamatergic nature of excitatory burst neuron (EBN) synapses on PH neurons, and the depolarizing effect of cholinergic inputs on the same postsynaptic PH neuron. (A) At the top (control) is illustrated the excitatory postsynaptic potential (EPSP) evoked in a PH neuron by a single subthreshold stimulus applied to the PPRF where EBN are located. Note (middle record) that this EPSP was not affected by atropine sulfate (a non-specific antagonist of cholinergic receptors, 1.5 μM), but that the superfusion of the recording slice with CNQX (a specific blocker of AMPA/kainate receptors, 10 μM) completely removed the evoked EPSP (bottom record). (B) A train stimulation of the same PPRF site evoked a sustained post-train depolarization of the same PH neuron (top record). This sustained depolarization was impossible to evoke in the presence of atropine (1.5 μM; bottom record), suggesting the involvement of cholinergic terminals in its generation. (Taken from Navarro-López et al. (2004), with permission of the Society for Neuroscience.)

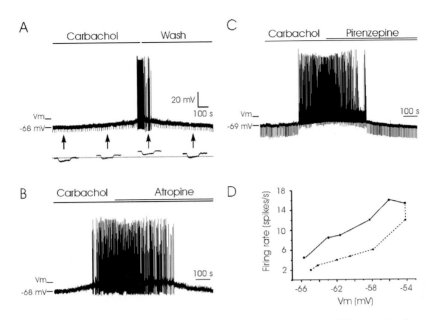

Fig. 6. Depolarizing effects of carbachol on PH neurons. (A) Intracellular recording of a PH neuron in the presence of carbachol (25 μM). Carbachol evokes a slow depolarization of the cell, with a non-significant decrease in membrane input resistance. At threshold, the neuron started to fire. Carbachol effects disappeared with washing. Arrows indicate points at which records have been expanded in time to show membrane potential during the presentation of hyperpolarizing pulses (0.3 nA, 300 ms). The dotted line indicates membrane resting potential. (B, C) The depolarization evoked by carbachol was blocked by atropine (a non-specific antagonist of cholinergic receptors, 1.5 μM) and by pirenzepine (a specific agonist of M1 receptors, 0.5 μM). In (C), the recorded neuron was hyperpolarized with current pulses (0.3 mA, 300 ms) at a frequency of 0.2 Hz. (D) A plot of the membrane potential (VM, in mV) against neuron firing rate (spikes/s) for data shown in (C). The continuous line indicates that firing frequencies reached when the membrane potential was changing in the depolarizing direction. The dotted line indicates neuron firing when the membrane potential was going in the hyperpolarizing direction. Vm, maximum depolarizing level evoked by carbachol. (Taken from Navarro-López et al. (2004), with permission of the Society for Neuroscience.)

1989). It is important to note (Fig. 6D) that when PH neurons are depolarized and repolarized in vitro, they still present the same hysteresis phenomenon (i.e., a different firing rate for the same resting potential, depending on the changing direction of membrane potential values). As shown in vivo, PH neurons presented a higher firing rate when the membrane potential was going in the depolarizing direction than when being repolarized.

From the results reported here, PH neurons recorded in an in vitro preparation presented similar functional properties (sustained firing proportional to train stimulation of the PPRF area, hysteresis, etc.) to those recorded in alert mammals (Delgado-García et al., 1989). In order to confirm the cholinergic nature of the synaptic mechanisms involved in the generation of the persistent activity observed in PH neurons, we decided

to carry out a pharmacological study in alert cats. For this aim, we localized the PH nucleus using electrophysiological techniques (Escudero and Delgado-García, 1988), and then carried out microinjections of selected drugs during the performance of spontaneous eye movements (Navarro-López et al., 2004).

In the light of this view, the administration of non-specific (carbachol) and specific (McN-A-343) agonists of the muscarinic M1 receptors did not produce any noticeable effect on eye movements (Fig. 7B, D). In contrast, the pharmacological blockage of muscarinic receptors by atropine sulfate and, especially, by pirenzepine (a specific antagonist of the M1 receptor) produced postsaccadic, centripetal drifts of the eye in the horizontal plane (Fig. 7A, C). In the dark, agonists induced a nystagmus with slow phases directed to the

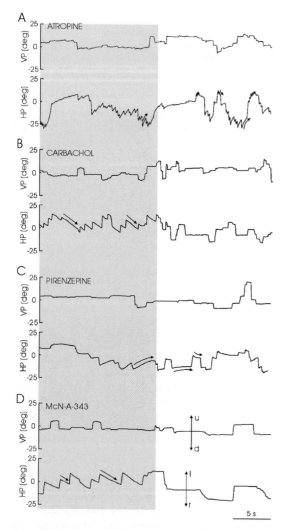

Fig. 7. Records of left eye position in the vertical (VP) and horizontal (HP) planes, from alert cats, following the injection of the indicated drugs in the ipsilateral PH nucleus. Recordings were carried out in conditions of darkness (gray lane) and light. Doses and time of recording after injection were as follows: atropine sulfate (A) (a non-specific antagonist of cholinergic receptors), 8 mM and 22.3 min; carbachol (B) (a cholinergic agonist), 1.5 pM and 12.8 min; pirenzepine (C) (a specific antagonist of M1 receptors), 0.12 pM and 4.9 min; and McN-A-343 (D) (a specific agonist of muscarinic M1 receptors), 20 pM and 4.5 min. Eye position is plotted as degrees of rotation in the horizontal plane, positive to the left (l) and up (u), and negative to the right (r) and down (d). Zero (0) indicates the central, resting position of the eye in the orbit. Curved and straight arrows indicate important peculiarities of eye movements evoked by the drugs. (Taken from Navarro-López et al. (2004), with permission of the Society for Neuroscience.)

contralateral side, while cholinergic antagonists produced a nystagmus with ipsilaterally curved slow phases. Results obtained with cholinergic agonists (ramp-like displacement of the eyes in the horizontal plane to the contralateral side in darkness) are similar to the effects evoked by blockage of GABA and glycine receptors (Moreno-López et al., 2002). The absence of effects of cholinergic agonists in the light can be explained by the fact that in this situation, and because of the activation of the visual system, there is a massive availability of acetylcholine in the PH nucleus (Chan and Galiana, 2005). In contrast, results obtained with cholinergic antagonists (curved slow phases in darkness and postsaccadic drifts in the light) suggest a loss of eye-position signals in the horizontal plane. These results are in agreement with those reported by Moreno-López et al. (2002), following the administration of antagonists of glutamate receptors. Thus, these experiments confirm results obtained in vitro suggesting that the activation of muscarinic M1 receptors is necessary for the generation of eye-position signals (Navarro-López et al., 2004).

The cholinergic connection

The diagrams illustrated in Figs. 8–10 attempt to summarize the results obtained by our group in a recent series of in vitro and in vivo experiments (Navarro-López et al., 2004, 2005).

The electrical stimulation of the PPRF (i.e., of EBN; Fig. 8) by single pulses evokes a monosynaptic depolarization of PH neurons (Igusa et al., 1980). The PPRF synapse is glutamatergic in nature, acting on AMPA/kainate receptors. It has been shown (Navarro-López et al., 2004) that the presentation of a single pulse to the PPRF area is unable to open the N-methyl-D-aspartate (NMDA) receptors also present on the membrane of PH neurons. However, in this situation, cholinergic (ACh; Fig. 8) terminals are able to modulate glutamate release by the presynaptic activation of muscarinic receptors. This presynaptic mechanism could act as a high-pass filter (Lisman, 1997), canceling out the disturbing effects of a low-rate firing of EBN.

Train stimulation of the PPRF evokes a sustained depolarization of PH neurons surpassing

SINGLE PULSES

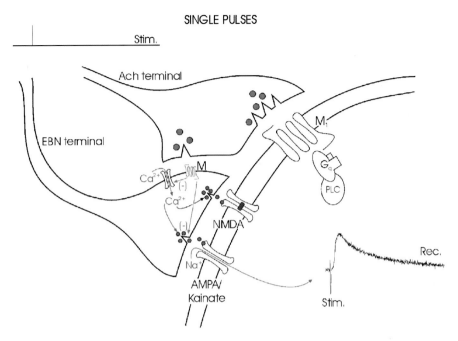

Fig. 8. Projections from the PPRF onto PH neurons are represented mainly by excitatory burst neurons (EBN). EBN projections are glutamatergic in nature (green circles represent glutamate-containing vesicles, acting on AMPA/kainate receptors. When single pulses are applied to the PPRF, the evoked excitatory postsynaptic potential (EPSP) can be modulated at the presynaptic level by the activation of muscarinic receptors (M), decreasing neurotransmitter release. However, this type of stimulation is not able to generate an NMDA current or to activate cholinergic (Ach) terminals acting on postsynaptic M1 receptors (red circles represent Ach-containing vesicles). Other abbreviations: G_q, G-protein associated to the M1 receptor; PLC, phospholipase C.

the duration of the train (Fig. 9). This depolarization is able to reach threshold and evoke a persistent firing of the PH neuron. The depolarization is apparently the result of the joint activation of EBN and of meso-pontine cholinergic axons (Igusa et al., 1980; Navarro-López et al., 2004). Train stimuli (>30 Hz) seem to be necessary in order to activate cholinergic axons (Moises et al., 1995). It has been shown that cholinergic terminals act on muscarinic M1 receptors located in the membrane of PH neurons (Navarro-López et al., 2004), and it can be suggested that cholinergic effects on PH neurons are produced by the generation of a Ca^{2+}-dependent unspecific cationic current (Klink and Alonso, 1997a, b; Haj-Dahmane and Andrade, 1999). Thus, the joint activation of glutamatergic and cholinergic receptors located in the membrane of PH neurons seems to be one of the mechanisms involved in the generation of the persistent activity necessary for eye-position signals (Navarro-López et al., 2004).

According to unpublished data from our laboratory (Navarro-López et al., 2005), train stimulation of the PPRF is also able to open postsynaptic NMDA receptors located in PH neurons. This effect is probably mediated by the simultaneous activation of muscarinic M1 receptors by the train. The activation of M1 receptors increases the production of inositoltriphosphate and diacylglycerol by a molecular mechanism involving a G_p-protein, coupled to the M1 receptor, which activates the phospholipase C. The increase in available diacylglycerol seems to activate endogenous protein kinase C (Fig. 10). Furthermore, it has been proposed that the activation of G-proteins coupled to certain receptors, such as muscarinic M1, is able to potentiate NMDA-responses, at least in hippocampal neurons (Marino et al., 1998; Sur et al., 2003). This potentiation can be blocked by drugs inhibiting protein kinase C (Lu et al., 1999). In turn, protein kinase C seems to enhance the activity of NMDA

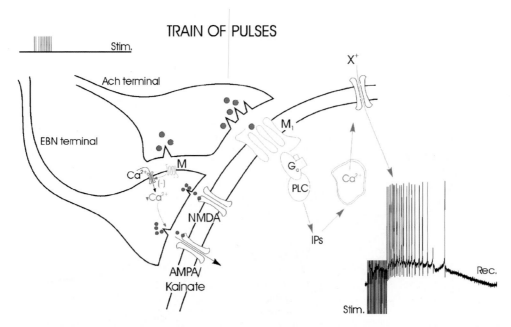

Fig. 9. A workable hypothesis is able to explain the generation of the sustained depolarization observed in PH neurons following train stimulation of the PPRF The stimulation (Stim.) of excitatory burst neurons (EBN), located in the PPRF, with a train of pulses produces a sustained depolarization of PH neurons. The duration and amplitude of the evoked post-train depolarization is proportional to train frequency. At the same time, this sustained depolarization is cholinergic in nature, acting on M1 muscarinic receptors. Other abbreviations: G_q, G-protein associated to the M1 receptor; IPs, inositolphosphates; PLC, phospholipase C; X^+, Ca^{2+}-dependent unspecific cationic conductance.

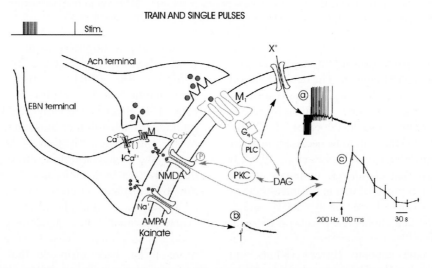

Fig. 10. A further hypothesis proposing a mechanism for stabilization of the oculomotor integrator following on-directed saccades. Electrical stimulation of the PPRF with a train of pulses activates both glutamatergic and cholinergic axons terminating on PH neurons. Specifically, (a) cholinergic axons seem to be responsible for evoking the sustained depolarization necessary for the generation of the persistent neural activity underlying eye fixation. Moreover, (b) train stimulation of the PPRF area also seems to evoke a short-term potentiation of PH neurons, probably by the joint activation of AMPA/kainate and muscarinic M1 receptors, with the participation of NMDA receptors. In this situation, the amplitude of excitatory postsynaptic potentials evoked by single pulses applied to the PPRF (c) increases after train stimulation of the same area. Green circles, vesicles with glutamate; red circles, vesicles with acetylcholine. G_q, G-protein associated to the M1 receptor; DAG, diacylglycerol; IPs, inositolphosphates; PKC, protein-kinase C; PLC, phospholipase C; X^+, Ca^{2+}-dependent unspecific cationic conductance.

receptors through different intermediate steps, including the activation of tyrosine kinases such as Src (Salter and Kalia, 2004). Thus, the final result is an activation of the NMDA receptors by the joint activation of AMPA/kainate and cholinergic receptors located on PH neurons. In fact, some preliminary results from our laboratory (Navarro-López et al., 2005) indicate that a short-term potentiation of the PPRF synapse on PH neurons is produced after train stimulation of the same area (c, in Fig. 10). It is difficult to envisage at the moment the role that a short-term potentiation could have on oculomotor performance, but it could be related with the facilitation of the synchronous activation of PH neurons, increasing the signal-to-noise ratio (Lisman, 1997) during alertness.

Abbreviations

AMPA	alpha-amino-3-hydroxy-5-methylisoxazole propionate
NMDA	N-methyl-D-aspartate
PH	prepositus hypoglossi
PPRF	paramedian pontine reticular formation

Acknowledgments

We acknowledge the editorial help of Mr. R. Churchill. The authors thank the help of Dr. Agnès Gruart in the edition of the figures. This work was supported by grant BFI2000-00939 from the Spanish Ministry of Science.

References

Aksay, E., Baker, R., Seung, H.S. and Tank, D.W. (2003) Correlated discharge among cell pairs within the oculomotor horizontal velocity-to-position integrator. J. Neurosci., 23: 10852–10858.

Aksay, E., Gamkrelidze, G., Seung, H.S., Baker, R. and Tank, D.W. (2001) In vivo intracellular recording and perturbation of persistent activity in a neural integrator. Nat. Neurosci., 4: 184–193.

Baker, R. and Spencer, R.F. (1981) Synthesis of horizontal conjugate eye movement signals in the abducens nucleus. Jap. J. EEG EMG Suppl., 31: 49–59.

Büttner-Ennever, J.A. and Horn, A.K.E. (1997) Anatomical substrates of oculomotor control. Curr. Opin. Neurobiol., 7: 872–879.

Cannon, S.C. and Robinson, D.A. (1987) Loss of the neural integrator of the oculomotor system from brain stem lesions in monkey. J. Neurophysiol., 57: 1383–1409.

Chan, W.W. and Galiana, H.L. (2005) Integrator function in the oculomotor system is dependent on sensory context. J. Neurophysiol., 93: 3709–3717.

Cheron, G. and Godaux, E. (1987) Disabling of the oculomotor neuronal integrator by kainic acid injections in the prepositus-vestibular complex of the cat. J. Physiol. (Lond.), 394: 267–290.

De la Cruz, R.R., Escudero, M. and Delgado-García, J.M. (1989) Behaviour of medial rectus motoneurons in the alert cat. Eur. J. Neurosci., 1: 288–295.

Delgado-García, J.M. (2000) Why move the eyes if we can move the head? Brain Res. Bull., 52: 475–482.

Delgado-García, J.M., del Pozo, F. and Baker, R. (1986) Behavior of neurons in the abducens nucleus of the alert cat. II. Internuclear neurons. Neuroscience, 17: 953–973.

Delgado-García, J.M., Vidal, P.P., Gómez, C. and Berthoz, A. (1989) A neurophysiological study of PH neurons projecting to oculomotor and preoculomotor nuclei in the alert cat. Neuroscience, 29: 291–307.

Egorov, A.V., Hamam, B.N., Fransén, E., Hasselmo, M.E. and Alonso, A.A. (2002) Graded persistent activity in entorhinal cortex neurons. Nature, 420: 173–178.

Escudero, M. and Delgado-García, J.M. (1988) Behavior of reticular, vestibular and prepositus neurons terminating in the abducens nucleus of the alert cat. Exp. Brain Res., 71: 218–222.

Escudero, M., de la Cruz, R.R. and Delgado-García, J.M. (1992) A physiological study of vestibular and PH neurons projecting to the abducens nucleus in the alert cat. J. Physiol. (Lond.), 458: 539–560.

Fukushima, K., Kaneko, C.R.S. and Fuchs, A.F. (1992) The neuronal substrate of integration in the oculomotor system. Prog. Neurobiol., 39: 609–639.

Fuster, J.M. (1997) Network memory. Trends Neurosci., 20: 451–459.

Goldman-Rakic, P.S. (1995) Cellular basis of working memory. Neuron, 14: 477–485.

González-Forero, D., Álvarez, F.J., de la Cruz, R.R., Delgado-García, J.M. and Pastor, A.M. (2002) Influence of afferent synaptic innervation on the discharge variability of cat abducens motoneurones. J. Physiol. (Lond.), 541: 283–299.

Gruart, A. and Delgado-García, J.M. (1994) Signalling properties of identified deep cerebellar nuclear neurons related to eye and head movements in the alert cat. J. Physiol. (Lond.), 478: 37–54.

Haj-Dahmane, S. and Andrade, R. (1999) Muscarinic receptors regulate two different calcium-dependent non-selective cation current in rat prefrontal cortex. Eur. J. Neurosci., 11: 1973–1980.

Igusa, A., Sasaki, S. and Shimazu, H. (1980) Excitatory premotor burst neurons in the cat pontine reticular formation related to the quick phase of vestibular nystagmus. Brain Res., 182: 451–456.

Kaneko, C.R.S. (1997) Eye movement deficits after ibotenic acid lesions of the nucleus PH in monkeys. I. Saccades and fixation. J. Neurophysiol., 78: 1753–1768.

Klink, R. and Alonso, A. (1997a) Muscarinic modulation of the oscillatory and repetitive firing properties of entorhinal cortex layer II neurons. J. Neurophysiol., 77: 1813–1828.

Klink, R. and Alonso, A. (1997b) Ionic mechanism of muscarinic depolarization in entorhinal cortex layer II neurons. J. Neurophysiol., 77: 1829–1843.

Lisman, J.E. (1997) Burst as a unit of neuronal information: making unreliable synapses reliable. Trends Neurosci., 20: 38–43.

López-Barneo, J., Darlot, C., Berthoz, A. and Baker, R. (1982) Neuronal activity in prepositus nucleus correlated with eye movement in alert cat. J. Neurophysiol., 47: 329–352.

Lu, W.Y., Xiong, Z.G., Lei, S., Orser, B.A., Dudek, E., Browning, M.D. and MacDonald, J.F. (1999) G-protein-coupled receptors act via protein kinase C and Src to regulate NMDA receptors. Nat. Neurosci., 2: 231–338.

Major, G. and Tank, D. (2004) Persistent neural activity: prevalence and mechanisms. Curr. Opin. Neurobiol., 14: 675–684.

Marino, M.J., Rouse, S.T., Levey, A.I., Potter, L.T. and Conn, P.J. (1998) Activation of the genetically defined m1 muscarinic receptor potentiates N-methyl-D-aspartate (NMDA) receptor currents in hippocampal pyramidal cells. Proc. Natl. Acad. Sci. USA, 95: 11465–11470.

Moises, H.C., Womble, M.D., Washburn, M.S. and Willians, L.R. (1995) Nerve growth facilitates cholinergic neurotransmission between nucleus basalis and the amygdale in the rat: an electrophysiological analysis. J. Neurosci., 15: 8131–8142.

Moreno-López, B., Escudero, M., Delgado-García, J.M. and Estrada, M. (1996) Nitric oxide production by brainstem neurons is required for normal performance of eye movements in alert animals. Neuron, 17: 739–745.

Moreno-López, B., Escudero, M., de Ventge, J. and Estrada, C. (2001) Morphological identification of nitric oxide sources and targets in the cat oculomotor system. J. Comp. Neurol., 435: 311–324.

Moreno-López, B., Escudero, M. and Estrada, M. (2002) Nitric oxide facilitates GABAergic neurotransmission in the cat oculomotor system: a physiological mechanism in eye movement control. J. Physiol. (Lond.), 540: 295–306.

Moreno-López, B., Estrada, C. and Escudero, M. (1998) Mechanisms of action and targets of nitric oxide in the oculomotor system. J. Neurosci., 18: 10672–10679.

Moschovakis, A.K. (1997) The neural integrators of the mammalian saccadic system. Front. Biosci., 2: D552–D577.

Moschovakis, A.K., Scudder, C.A. and Highstein, S.M. (1996) The microscopic anatomy and physiology of the mammalian saccadic system. Prog. Neurobiol., 50: 133–254.

Navarro-López, J.D., Alvarado, J.C., Márquez-Ruíz, J., Escudero, M., Delgado-García, J.M. and Yajeya, J. (2004) A cholinergic synaptically triggered event participates in the generation of persistent activity necessary for eye fixation. J. Neurosci., 24: 5109–5118.

Navarro-Lopez, J.deD., Delgado-Garcia, J.M. and Yajeya, J. (2005) Cooperative glutamatergic and cholinergic mechanisms generate short-term modifications of synaptic effectiveness in prepositus hypoglossi neurons. J. Neurosci., 25: 9902–9906.

Pastor, A.M., de la Cruz, R.R. and Baker, R. (1994) Eye position and eye velocity integrators reside in separate brainstem nuclei. Proc. Natl. Acad. Sci. USA, 91: 807–811.

Robinson, D.A. (1981) The use of control systems analysis in the neurophysiology of eye movements. Ann. Rev. Neurosci., 4: 463–503.

Salter, M.W. and Kalia, L.V. (2004) SRC kinases: a hub for NMDA receptor regulation. Nat. Rev. Neurosci., 5: 317–328.

Semba, K., Reiner, P.B. and Fibigere, H.C. (1990) Single cholinergic mesopontine tegmental neurons project to both the pontine reticular formation and the thalamus in the rat. Neuroscience, 38: 643–654.

Sur, C., Mallorga, P.J., Wittmann, M., Jacobson, M.A., Pascarella, D., Williams, J.B., Brandish, P.E., Pttibone, D.J., Scolnick, E.M. and Conn, P.J. (2003) N-desmethylclozapine, an allosteric agonist at muscarinic 1 receptor, potentiates N-methyl-D-aspartate receptor activity. Proc. Natl. Acad. Sci. USA, 100: 13674–13679.

Perceptual Completion

Introduction

Some diseases of the visual system result in a paradoxical situation: patients may present multiple blind regions or scotomas within their visual field, and yet be unaware of their existence, owing to the brain process of "perceptual completion", or "filling-in". Filling-in also takes place near the center of vision in normal healthy retinas, in the blind spot. In 1804, Troxler discovered that during strict visual fixation of a target, a small perimetric stimulus would quickly fade from perception and fill in with the visual texture surrounding it, so that it becomes indistinguishable from the background. This fading stimulus can be thought of as an "artificial scotoma". These facts suggest that filling-in is a brain process of great importance to surface perception in normal vision as well as in clinical patients. Interestingly, filling-in is usually stronger when the artificial scotoma is presented on a dynamic noise background.

Peter De Weerd's chapter reviews psychophysical and electrophysiological studies on filling-in of dynamic textures, and discusses the importance of both low-level factors and high-order processes such as the role of attention. Akiyoshi Kitaoka and colleagues describe how several visual illusions can be explained in terms of surface completion.

Susana Martinez-Conde

Martinez-Conde, Macknik, Martinez, Alonso & Tse (Eds.)
Progress in Brain Research, Vol. 154
ISSN 0079-6123

CHAPTER 12

Perceptual filling-in: more than the eye can see

Peter De Weerd*

Neurocognition Group, Psychology Department, University of Maastricht, 6200 MD Maastricht, The Netherlands

Abstract: When a gray figure is surrounded by a background of dynamic texture, fixating away from the figure for several seconds will result in an illusory replacement of the figure by its background. This visual illusion is referred to as perceptual filling-in. The study of filling-in is important, because the underlying neural processes compensate for imperfections in our visual system (e.g., the blind spot) and contribute to normal surface perception. A long-standing question has been whether perceptual filling-in results from symbolic tagging of surface regions in higher order cortex (ignoring the absence of information), or from active neural interpolation in lower order visual areas (active filling-in of information). The present chapter reviews a number of psychophysical studies in human subjects and physiological experiments in monkeys to evaluate the above two hypotheses. The data combined show that there is strong evidence for neural interpolation processes in retinotopically organized, lower order areas, but that there is also a role for higher order perceptual and cognitive factors such as attention.

Keywords: vision; filling-in; completion; Troxler effect; illusions; attention

Introduction

Visual perceptual filling-in refers to the interpolation of information across a region in the visual environment in the absence of any physical evidence for that information in that region. Filling-in can be triggered under a number of different conditions, and some types of perceptual filling-in occur very fast. An example of fast, quasi-instantaneous filling-in is the perceptual filling-in of the blind spot with the information surrounding it. Similar types of perceptual filling-in have been reported for pathological scotomas (Bender and Teuber, 1946; Sergent, 1988). Another form of fast filling-in (within 80 ms) has been observed across entopic images of vasculature (Coppola and Purves, 1996). Slower filling-in of retinal images (within a few seconds) has been reported under conditions of artificial retinal stabilization (Ratliff,

1958; Gerrits et al., 1966; Yarbus, 1967), and during the stabilization of peripheral images through maintained fixation (Troxler, 1804; Riggs et al., 1953). Perceptual filling-in during maintained fixation away from a figure is often referred to as the 'Troxler effect', and it can also be observed during fixation in the middle of a disk with boundaries sufficiently far from the fixation point. Depending on the exact stimulus conditions, the time required to achieve Troxler fading can range from a few to many seconds. It has been demonstrated for color, brightness, and (dynamic) texture (Ramachandran and Gregory, 1991; Spillmann and Kurtenbach, 1992; Ramachandran et al., 1993; Fujita, 1993; Friedman et al., 1999; Hamburger et al., 2006).

A distinction can be made between surface filling-in and contour completion. Many studies have been devoted to the completion of contours, and several have investigated the neural correlates underlying contour completion (von der Heydt et al., 1984; Peterhans and von der Heydt, 1989; Grosof et al., 1993). This chapter, however, will focus on

*Corresponding author. Tel.: +31-43-388-4513;
Fax: +31-43-388-4125; E-mail: P.deweerd@Psychology.unimaas.nl

DOI: 10.1016/S0079-6123(06)54012-9

surface filling-in. Note that in this chapter, the terms 'completion' and 'filling-in' will be used synonymously, although some authors exclusively use the term completion in association with contours, and filling-in in association with surfaces (e.g., Grossberg, 2003a). Furthermore, there is a distinction made in the literature between modal and amodal completion. Amodal completion refers to the perception of continuity of a surface and its contours behind an object positioned in front. Modal completion refers to the perceptible (but illusory) effects of contour and surface completion when local cues suggest that one surface is in front of other surfaces or objects in the background (as in the Kanizsa square; Kanizsa, 1955). Despite differences in perceptual effects between modal and amodal completion, both types of completion also share common mechanisms (for review, see Davis and Driver, 2003). The type of filling-in studied in this chapter can be best considered as a modal type of filling-in. Finally, it is important to draw a conceptual distinction between perceptual filling-in (the illusory perception of a feature in a region where it is physically absent), and neural filling-in (the neuronal mechanism that produces the perceptual illusion).

In the last few decades, the illusion of the perceptual filling-in of surfaces has been studied extensively for several reasons. First, the perception of continuous surfaces (and contours) despite design features of the visual system that might interfere with these percepts, strongly suggests the existence of filling-in mechanisms. In this chapter, it will be argued that the perceptual filling-in processes that generate the illusory completion of backgrounds across regions physically occupied by a figure are relevant for normal surface perception. Second, given the likely existence of filling-in mechanisms, it is unavoidable to ask what might be the specific neural correlate underlying perceptual filling-in, and where in the visual system it might be found. Third, the perceptual filling-in of a figure by its background implies that the same physical stimulus can be perceived in two different ways (similar to bi-stable stimuli). This is suggestive of neural mechanisms that permit fast perceptual reorganization and that might be related to plastic mechanisms underlying other types of visual reorganization and

learning. These will be the three main topics in this chapter.

The paradigmatic example of filling-in upon which this chapter will focus is the slowest type of perceptual filling-in, in which figures presented away from the center of gaze become filled-in perceptually only after many seconds of maintained fixation. The processes underlying this type of perceptual filling-in may therefore, at first sight, be unlikely to contribute to normal surface perception, but a closer investigation suggests that the opposite may be true. Because of its focus on one type of perceptual filling-in, the chapter does not offer a complete review of the literature related to this phenomenon. Instead, the chapter provides the logic behind the work from our group on perceptual filling-in, with reference only to directly related work from other investigators.

Perceptual filling-in and the design of the visual apparatus

There are several design features of the visual system that provide reasons for suspecting the existence of a filling-in process. One such feature is the anatomical design of the retina, which interferes with the continuity of the retinal image. The retina is constructed such that light finding its way to the photoreceptors has to pass first through a mesh of ganglion cells, amacrine cells, bipolar cells, and blood vessels (Sterling, 1998). The fact that we do perceive the world in an uninterrupted continuous fashion points to the existence of interpolation mechanisms that complete the image. An even stronger example of the existence of such interpolation mechanisms is given by the perceptual filling-in of the blind spot. The blind spot is a scotoma corresponding to the place in the retina where the optic nerve leaves the eye, and where there are no photoreceptors. When one eye is closed, the blind spot might be expected to reveal itself as a 'black hole' in the visual image of the world. Instead, the 'black hole' is perceptually filled in by the surrounding background, and therefore goes unnoticed. Neural mechanisms of perceptual filling-in across the blind spot (Fiorani et al., 1992; Komatsu and Murakami, 1994; Komatsu et al., 2000), and across scotomas induced by a small retinal lesion (Murakami et al., 1997) have been described in V1 of the monkey. The perceptual filling-in of a scotoma is experienced as instantaneous, while perceptual filling-in of figures on a background during maintained fixation of a point away from the figure takes time. Nevertheless, both types of filling-in may be related, as will be discussed in the next section. A second plausible reason to suspect the existence of filling-in processes is the fact that processing in the early visual system (retina, lateral geniculate nucleus (LGN)), and especially cortical processing in early visual areas is heavily biased toward the detection of local contrasts and boundaries (Kuffler, 1953; Hubel and Wiesel, 1959, 1962). Most neurons in the early cortical stages of the visual system are highly responsive to local contrasts in luminance or other features within their receptive field (RF), and fairly unresponsive to homogeneous stimulation in the RF. Ever since Hubel and Wiesel described the responses of neurons in the early visual areas to oriented lines, the contribution of visual cortex to the detection of boundaries and contours in the visual image has been heavily emphasized. Nevertheless, the world is visible only owing to our vivid perception of surfaces (Grossberg 1987a, b, 2003a). How could surface perception be accomplished given the scarcity of neurons that directly and efficiently code surface properties?

The locus of perceptual filling-in

Isomorphic representation of surfaces: some background

The debate on the nature of a neural correlate of perceptual filling-in, and on the locus of such a correlate in the visual system has been waged not only in the neuroscientific community, but also in philosophical circles (for review, see Pessoa et al., 1998; Pessoa and Neumann, 1998). Dennett (1991) in particular has taken the strong stance that perceptual filling-in is *not* associated with isomorphic neural filling-in processes in the brain. An isomorphic cortical representation of a perceptually filled-in visual surface would imply a point-to-point correspondence between the distribution of neural activity in the cortex triggered by the surface and

the spatial distribution of the surface in the visual field (Todorovic, 1987). Dennett (1991) has argued that this is unlikely, and that surfaces would be coded in the brain in analogy with the way in which digitized images are coded in graphical environments. A surface with a specific color in a digitized image will be represented by its boundaries and a single value that represents the color (symbolic coding). This would avoid the waste associated with a 'pixel-by-pixel' code in the brain. According to Dennett (1991), a cortical pixel-by-pixel code to represent surface features would be tantamount to creating a 'Cartesian theatre' in the brain that would have to be viewed by a separate conscious entity (a 'homunculus') in order to produce surface perception. A potential neurophysiological translation of the symbolic coding hypothesis of Dennett (1991) would be that surface features are coded in higher-order visual areas, which are also involved in high-level representations of objects, and which are not retinotopically organized (Desimone and Ungerleider, 1989; Wang et al., 1996; Biederman, 2000; Haxby et al., 2001; Kayaert et al., 2003). According to Dennett (1991), perceptual filling-in corresponds to ignoring the absence of information rather than the active interpolation of existing information across regions in the visual field where that information is absent.

Although physiological investigations of perceptual filling-in of surfaces had not been conducted at the time Dennett published his book, his resistance to the idea of an isomorphic neural interpolation process associated with perceptual filling-in was remarkable in view of the known organization of the visual system, and in view of existing studies demonstrating isomorphic neural interpolation processes during contour perception. The visual system consists of a number of hierarchical levels, wherein light that enters the retina reaches the cortex through the intermediary of a thalamic relay station (the LGN). Primary visual cortex and other early visual areas each consist of a sheet of neurons with small RFs that are organized in a retinotopic fashion. A classical RF is the region in visual space where a single, small bar stimulus can influence the firing rate of a neuron. A cortical area is retinotopically organized if neighboring points in the retinal image are represented by physically neighboring neurons in the cortical sheet with neighboring RFs. Placing stimuli inside a neuron's RF often reveals multiple tuning properties, including orientation, disparity, velocity, direction, or color tuning, in various combinations (for review, see Desimone and Ungerleider, 1989). Placing stimuli outside the RF (in the RF surround) usually does not produce a change in activation, but can modulate activity to stimuli presented inside the classical RF. For example, an enhancement of neural responses has been reported in V1 neurons when a central line in the classical RF was flanked by a collinear line segment in the RF surround (Kapadia et al., 1995; Polat et al., 1998). Influences from the surround may be related in part to lateral, long-range connections, which show a bias to link neurons with similar orientation tuning properties (Gilbert and Wiesel, 1989; McGuire et al., 1991; Lund et al., 1993). In addition to a contribution of horizontal connections, feedback influences contribute to the extent and properties of the surround (Lamme, 1995; Zipser et al., 1996; Levitt and Lund, 2002; Bair et al., 2003).

The retinotopic organization of early visual cortex as well as the availability of circuitry that could propagate activity laterally along the cortical sheet suggests that isomorphic filling-in processes might indeed exist. In the domain of contour perception, it has been claimed that the above-described circuitry forms the basis for a tendency of the visual system to complete fragmentary information. For example, a line segment will be detected more readily if it is flanked by one or two collinear line segments (Dresp, 1993; Field et al., 1993; Polat and Sagi, 1994; Wehrhahn and Dresp, 1998). It is as if the visual system 'expects' that two aligned fragments are likely to come from a single contour. This expectation reveals itself in an enhanced capability to detect a third aligned line segment placed in the gap in-between the flanking line segments, and an enhanced physiological response to the middle line segment from neurons with RFs covering the gap. Neural activity in response to illusory contours is another striking example of the bias of the visual system to interpret fragments as belonging to a contour (von der Heydt et al., 1984). These types of 'hypotheses' are built into the architecture of visual circuitry, which is shaped through the statistical regularities of visual stimulation during early development (Hirsch and Spinelli, 1971; Pettigrew and Freeman, 1973; Blakemore, 1976; Blasdel et al., 1977). Given the evidence for isomorphic neural mechanisms that interpolate contour information in retinotopically organized cortex (von der Heydt et al., 1984; Peterhans and von der Heydt, 1989), the question can be asked why such neural interpolation processes would not exist for surface properties as well.

Ramachandran (2003) has linked perceptual filling-in phenomena to the philosophical idea of sensory 'qualia'. The idea of qualia in visual perception can be related to the undeniable and vivid experience of a particular percept, and the vividness of that percept can be linked to the probability that it corresponds to physical reality. This hypothesis is compatible with the idea that the architecture of the visual system is shaped in early development to reflect statistical regularities in the visual environment (e.g., Blasdel et al., 1977). If particular configurations of local elements in a 2D display are statistically likely to be linked to real contours or (occluding) surfaces in 3D environment, the adult visual system will generate the (illusory) perception of contours and surfaces even in the absence of physical evidence for it in the display (as in the Kanizsa square; Kanizsa, 1955). The strong bias or 'certainty' of the visual system to interpret appropriately spaced and aligned local cues as belonging to real contours and surfaces produces illusory percepts of contours and surfaces, which could be considered as qualia. The convincing percepts (qualia) of contours or surfaces based on fragmentary information may reflect the action of isomorphic interpolation processes in retinotopically organized visual cortex. Because local elements that are not properly aligned or too widely spaced are less likely to be associated with a single contour or surface, they will not produce illusory contours or surfaces. Neural interpolation process therefore should have a limited spatial range.

Psychophysical evidence

Independent of physiological measurements, psychophysical experiments have suggested that neural processes in retinotopic areas do contribute to some types of surface filling-in. In one such experiment (De Weerd et al., 1998), observers were instructed to fixate a red fixation spot, while a gray figure was presented away from fixation on a dynamic texture background (Fig. 1). The texture consisted of a black background densely filled with jittering white

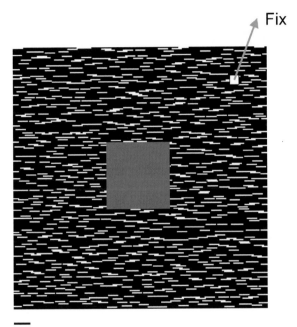

Fix

1°

Fig. 1. Typical stimulus used in perceptual filling-in experiments. The picture shows a single frame of a dynamic texture stimulus used by De Weerd et al. (1999) (see Fig. 2), and similar to the stimuli used in other experiments discussed in this chapter. The homogeneous region in the center was approximately equiluminous with the average luminance of the surrounding dynamic texture, and with the gray background upon which the texture was presented (23 cd/m²). The dynamic texture was a 'movie' made of five frames, each of which consisted of horizontal, white line segments ($0.7° \times 0.1°$) on a dark background, spaced $0.4°$ apart on average. Since the position of the line elements was randomized in each frame of the movie, playing the movie (at 20 Hz) created a stimulus with continuously jittering line elements on the dark texture background. In physiological recordings, the orientation of the bars and square hole typically were chosen to match the preferred orientation of the cell. The small white square indicates the position of the fixation point (Fix). The illustration is approximately to scale, except for the fixation spot which is exaggerated in size for clarity (and which in reality was red). In this example, the texture was $16° \times 16°$ in size, square size was $4°$, and the eccentricity of the square's center relative to fixation was $8°$.

line segments, and the gray square (of which the size and eccentricity was varied) was equiluminant with the background. Subjects were instructed to report with a button press when the gray figure 'disappeared'. The time required for filling-in varied from about 4 to 10 s, depending on the exact stimulus conditions (Fig. 2). At an eccentricity of $4°$, increases in the size of the square produced very steep increases in the time required for subjects to fill in, whereas at an eccentricity of $8°$, the same increases in square size produced much smaller increases in the time required for filling-in. This pattern of results suggested that cortical magnification of central vision was a critical factor determining the time required for perceptual filling-in. Because the cortical magnification of the projection of a stimulus in retinotopically organized cortex increases with decreasing eccentricity, the same increase in square size will produce much larger increases in cortical projection at the fovea than in the periphery. This parallels the psychophysical effects of eccentricity and figure size on the time required to perceive filling-in. Indeed, re-plotting the time required for perceptual filling-in as a function of cortical projection size (after converting all figure sizes at all eccentricities into their cortical projection size), revealed a linear relationship between the two variables. Figure 2 shows filling-in times plotted as a function of cortical projection size in human V3, but equivalent results were obtained in V1 and V2. The reason for this equivalence is that the reduction of cortical magnification with eccentricity follows a similar function in different areas, except for a scaling factor to take account of the differences in size among those cortical areas (Sereno et al., 1995). Although the data in Fig. 2, therefore, could not pinpoint to what extent different early visual area(s) contribute differentially to perceptual filling-in, the data did establish a role for retinotopically organized cortex in the perception of filling-in. Other experiments in which a larger set of eccentricities and square sizes was used, confirmed the observations shown in Fig. 2 (De Weerd et al., 1998). Furthermore, the observation that two identical squares at identical eccentricities often filled in at different moments (De Weerd et al., 1998), fits with the idea that perceptual filling-in reflects localized interpolation mechanisms in retinotopic areas.

232

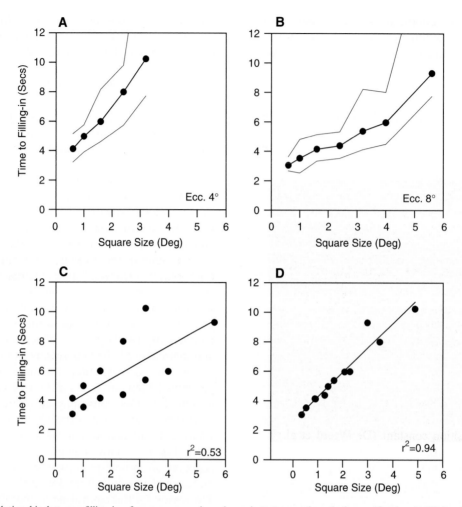

Fig. 2. The relationship between filling-in of a gray square in a dynamic texture and cortical magnification in V3 (modified from De Weerd et al., 1998). (A) Median time from stimulus onset to filling-in as a function of square size at an eccentricity of 4°. (B) Median time to filling-in as a function of square size at an eccentricity of 8°. Thin lines below and above each data symbol in (A) and (B) indicate the 25th and 75th percentile, respectively (pooled data from four subjects). (C) Linear regression through the filling-in times plotted as a function of square size, ignoring the fact that the data were collected at different eccentricities. (D) Linear regression through filling-in times plotted as a function of the square root of the square's cortical projection area in human V3, computed for all square sizes and both eccentricities. In this experiment, there were four participants who performed about 20 trials each in each condition.

At first sight, the finding that time to filling-in is predicted by cortical projection size might suggest that perceptual filling-in reflects a slow neural filling-in process that takes many seconds to interpolate the cortical projection area. Such a slow neural interpolation process would be unlikely to contribute to normal surface perception, and an experiment was designed to reject this hypothesis. Different stimuli of which the projection areas in retinotopical cortex were rectangular with different widths and lengths

were devised (De Weerd et al., 1998). If filling-in proceeded slowly across the rectangular projection area, one would predict that the time from stimulus onset to filling-in would correlate with the shortest distance across. However, the data showed that the shortest distance across the rectangular projection area was *not* a good predictor of the time required for filling-in. Among the factors that were correlated with the time required for filling-in was the total length of the cortical projection of the figure's boundary. Computational work suggests that the representation of a longer boundary is more robust than that of a shorter boundary, because of the larger recurrent network that would be involved in encoding a longer boundary (Francis et al., 1994; Francis and Grossberg, 1996). The stabilization of a boundary inside the RF of boundary detectors might lead, therefore, to slower adaptation the longer the boundary is. Assuming that perceptual filling-in occurs because of a weakening of boundary representations owing to a stabilization of boundaries on the retina and adaptation of boundary detectors, the robustness of the boundaries of figures with longer cortical boundary projections would explain why such figures take longer to fill in. This proposal goes back to the original ideas by Gerrits and Vendrik (1970), who suggested that boundary detectors inhibit the spread of activity related to the surfaces, and models from Grossberg (1987a, b; 2003a, b) in which interactions between a boundary system and a surface feature system form the basis of perception. The idea that the adaptation of an inhibitory signal related to boundary representations is required for filling-in was supported by a psychophysical experiment in which the gray figure was moved by a few tenths of a degree at a rate of 1 Hz, keeping the average position constant (De Weerd et al., 1998). This subtle manipulation was sufficient to significantly increase the time required to observe perceptual filling-in, suggesting that neurons with small RFs may contribute to the boundary representation, and that the jittering of the boundaries delayed the adaptation of neurons encoding the boundaries. Thus, any condition in which the inhibitory signal from the boundary would remain intact longer would delay perceptual filling-in, while any condition that would promote adaptation and weakening of the inhibitory signal (e.g., very accurate fixation) would make filling-in happen sooner. This implies that the time that elapses before filling-in is related to the adaptation of boundary detectors, while the time required for the filling-in process itself might be much briefer. This hypothesis is compatible with the idea that the neural filling-in processes producing the filling-in illusion might also be relevant for normal surface perception.

Evidence for the existence of fast perceptual filling-in processes came from several elegant experiments by Paradiso and colleagues. In one such experiment (Paradiso and Nakayama, 1991), the presentation of a homogeneous bright disk on a dark background was followed by a dark 'masking' display with a bright circle of a diameter smaller than that of the bright disk. When the masking display with the circle was presented at an appropriate, very brief time interval after presentation of the bright disk, observers temporarily perceived the inner part of the circle as dark, as if the circle had blocked the spread of brightness from the inner side of the disk's contour toward the middle. By manipulating the size of the circle and the time interval between the disk and masking stimulus, Paradiso and colleagues arrived at a speed of brightness spread on the order of $100°/s$. A number of other experiments confirmed that brightness spread, though fast, is in fact a time-consuming event that takes place on a time scale of milliseconds for objects that span few visual degrees on the retina (Todorovic, 1987; Paradiso and Hahn, 1996; Rossi and Paradiso, 1996). Biologically plausible, computational models have provided theoretical support for fast brightness interpolation (Grossberg and Todorovic, 1988; Arrington, 1994; Neumann et al., 2001).

Although the previous experiments suggest that brightness perception is related to an interpolation process that proceeds from the inner side of a surface's edges toward its middle, the perceived brightness is related to luminance ratios between neighboring surfaces (Rossi and Paradiso, 1996). Thus, in a display consisting of a gray patch flanked by two dark patches, the gray patch will be perceived as brighter compared to the same gray patch flanked by two bright patches. Accordingly, ramping up the luminance values of the flanking patches from dark to bright will decrease the perceived brightness of the gray patch, and the opposite will occur when the luminance values of the flanking patches are ramped down, in the absence of any physical changes inside the gray patch. This shows that perceptual filling-in represents more than just an interpolation of physical features in the image, and that what is interpolated depends on global computations of brightness ratios (contrasts) in the image. These computations come down to a discounting of the illuminant, and have been mathematically described by Grossberg (1997, 2003a, b). Grossberg's models include an on-center off-surround input network, which normalizes incoming luminance signals and discounts the illuminant. Boundary signals are constructed by a boundary-processing system based on the discounted signals, which are also used as a seed for a diffusion-like interpolation process in a surface-processing system.

Neurophysiological evidence

If one could find neurons with RFs inside the gray patch whose firing rate would be correlated with the perceived brightness of the patch while the luminance values of the flanking patches in the RF surround would be ramped up and down, then one would have evidence for neural brightness interpolation being directly related to brightness perception. Exactly that type of responses was found in a small proportion of V1 neurons recorded in cat visual cortex (Rossi and Paradiso, 1999). A strict correspondence between variations in brightness perception and variations in neuronal responses across a large set of experimental conditions was found in 10% of the recorded neurons. These data indicate that V1 may be involved in the interpolation of brightness, but do not exclude that other visual areas getting input from V1 may participate as well. In addition, the brightness-related responses observed in V1 could at least in part reflect feedback from extrastriate cortex.

The experiments from Paradiso and colleagues show that global computations during which the illuminant is discounted determine the brightness values that are interpolated between surface contours. Likewise, the adaptation process preceding filling-in of a figure by its background during maintained fixation depends in part on global computations determining what region in the image is a figure, and what region is the background. In an experiment in which the amount of texture surrounding a gray region was manipulated (De Weerd et al., 1998), two observations were made. First, to produce effective filling-in of the gray region, a minimal amount of surrounding texture

was required, and up to a certain limit, adding texture increased the percentage of trials with filling-in and reduced the time before filling-in took place. Second, when the amount of texture was reduced, such that the stimulus started to look like a narrow frame made of texture on a homogeneously gray background (a narrow, textured frame with the gray background showing on the inside and on the outside), it was the frame that became filled-in by the gray background, rather than the opposite. For square sizes up to 4°, the ratio of the width of the surrounding frame to the size of the square had to exceed about 0.25 before the gray square would be filled in by the texture on a majority of trials. This suggests that what becomes filled-in is always what is perceived as the figure. These observations show that the direction of filling-in is fundamentally influenced by global computations in the image that determine what is perceived as a foreground and what as a background.

The above studies suggest that perceptual filling-in may involve a slow adaptation of figure-ground segregation mechanisms (in which both the boundaries and the size of the figure play a role) followed by a fast neural interpolation of the figure by its background. To directly test the presence of active, neural interpolation in retinotopic visual cortical areas, neurophysiological recordings were carried out in areas V1, V2, and V3 of the Rhesus monkey. Two monkeys were trained to fixate a point away from a gray square surrounded by dynamic texture, which in humans would produce perceptual filling-in, while recording from neurons with RFs that overlapped with the gray square. The monkeys were trained to maintain fixation for up to 14 s. Average population histograms of neurons in V2 and V3 (showing activity averaged over trials and neurons from the two monkeys) showed an initial onset transient followed by reduced activity. After a number of seconds, however, neuronal activity increased and reached a level that matched activity measured when the RF was physically filled in with texture (Fig. 3a). The activity increases occurred in the absence of any physical changes to the stimuli. Increasing square size delayed the time point at which the increasing activity reached a level that was indistinguishable from the activity produced by physical filling-in of the RF. From that time point on, one might expect that the neurons would not be able to distinguish the stimulus with the square from the stimulus with the square physically filled in with texture. Accordingly, across a range of square sizes, that time point accurately predicted when human subjects exposed to the same stimuli viewed by the monkey, on average, perceived perceptual filling-in (Fig. 3b). In addition, stimulus conditions that prevented perceptual filling-in in humans did not produce activity increases in V2 and V3 neurons, or produced activity increases that did not reach the activity level obtained when the square was physically filled in (Fig. 3a, 12.8° panel). These observations strongly suggested that V2 and V3 contribute to neural interpolation processes associated with perceptual filling-in.

While about half of the recorded neurons in V2 and V3 showed activity that correlated with filling-in, no such neurons were found in V1. A conclusion that V1 neurons did not contribute to perceptual filling-in would be flawed if the V1 recordings had been done at the same eccentricities used for the recordings in V2 and V3. At matched eccentricities, RFs in V1 are much smaller than those in V2 or V3, such that small fixation errors would affect visual processing in V1 much more than in V2 or V3. However, the V1 recordings were carried out at an average eccentricity of 24.5°, where human subjects experienced strong filling-in, and where the average RF size (2.5°) was comparable to that of cells in V2 (2°) and V3 (3.7°), at the smaller eccentricities studied in these areas (9.1° and 7.6°, respectively). This suggests that the absence of activity increases related to filling-in in V1 was not due to smaller RF size, which could have interfered with the adaptation of boundary detectors in V1. Thus, because the relationships between the size of fixation errors and RF size were similar in V1, V2, and V3, it is reasonable to conclude that V1 plays little role in the filling-in of dynamic texture (see De Weerd et al., 1995).

The human psychophysical data from De Weerd et al. (1998) suggest the importance of accurate fixation for filling-in to occur, and so do the data collected in the monkeys. The average standard deviation of the fixation errors was around 0.4° in one monkey and 0.8° in the other (De Weerd et al., 1995). The activity increases related to filling-in were less prevalent in the second monkey. Further, an analysis of eye movements in a subset of trials indicated that small deviations away from fixation within the electronic fixation window abolished the activity increases associated with filling-in.

It would be interesting to repeat the above experiments after training monkeys to signal the moment of filling-in. The correlation between perceptual filling-in responses in humans and neuronal responses in monkeys measured under identical stimulus conditions, however appealing, remains an indirect way of linking a neural response to a percept. It has been shown that monkeys can be trained to report the filling-in of the blind spot (Komatsu and Murakami, 1994), so there is no reason in principle why monkeys could not also be trained to report filling-in produced by prolonged stabilization of the image (through fixation). Measuring the monkey's ability to fill in would require that the monkey be trained first to respond to the physical filling-in of a gray square embedded in texture, choosing stimulus parameters that would make perceptual filling-in impossible (e.g., choosing a square that is too large). Interspersed among a majority of these trials, a minority of test trials would present stimuli that do produce filling-in, to which the monkey would then be expected to respond as if the perceived filling-in were physical. Unfortunately, a test trial approach is difficult to combine with recording experiments such as those performed by De Weerd et al. (1995), which yielded no more than a single successfully completed trial per minute (due to fixation errors that prematurely ended the trial).

The slowly increasing activity level visible in the average histograms in neurons with their RFs centered on the gray square suggests that the neural interpolation process is slow, in contradiction to what was argued earlier. This raises doubts again as to whether such a slow process could be relevant for normal surface perception. However, when considering neurons with strong activity increases (and strong responses to physical filling-in), individual trials suggest that the activity increases in fact occur at a discrete moment in time. In a typical trial in an example of such neurons

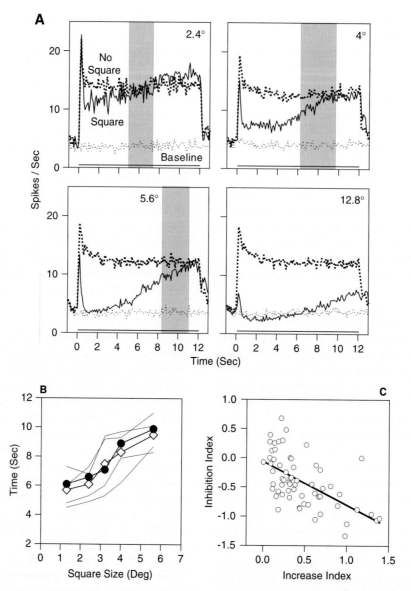

Fig. 3. Correlation between increased neural activity in V2/V3 and perceptual filling-in, and underlying disinhibition (modified from De Weerd et al., 1995). (A) Average activity in stimulus conditions with a gray square over or inside the RF (solid line) and in conditions in which the square, and thus the RF were physically filled in with texture (heavy dotted line) in subsets of V2 and V3 neurons with significant activity increases. Fine dots show baseline activity recorded without any stimulus. Square size is given at the top of the panels. The horizontal line on top of each abscissa indicates the 12 s stimulus presentation time, preceded and followed by a 1 s period in which activity was recorded in the absence of a stimulus. Human observers reported filling-in for the same stimuli as used during physiological recordings in the monkey in a time range indicated by the shaded zones. Activity increases were evaluated in 93 V2/V3 neurons that were recorded in all square size condition with paired t-tests (2-tailed, $P < 0.05$), in which firing rates in a 2.5 s interval starting 1.5 s after stimulus onset were compared with the rates in the last 2.5 s of stimulus presentation. A few cells that showed a response increase in the no-square condition were excluded from the count of cells with increased activity. For cells that initially responded similarly in the square and no-square conditions, there was less opportunity to show an activity increase. This was particularly true at square sizes of 3.2° and smaller (which were also not as conspicuous to the subjects as the larger square), for which we found significant activity increases in only 12% of V2/V3 cells. For larger square sizes, significant activity increases were found for about a third of V2 neurons, and about half of V3 neurons. Thus, the number of neurons and trials (up to 20 trials per condition per

(Fig. 4), a number of seconds with limited activity was followed by one or two discrete periods of increased activity, very much similar to filling-in reports. Thus, the gradual activity increases in average histograms correspond to an increase in the probability of perceptual filling-in rather than to a slow neural filling-in process. The activity in individual trials is in agreement with the idea that a long process of adaptation (in which boundary representations and figure-ground segregation weaken) is followed by a sudden filling-in event. The sudden onset of the increased activity suggests the presence of a fast filling-in process, which might be similar to the processes that can be inferred from the psychophysical experiments on brightness filling-in. This confirms that the neural interpolation process associated with dynamic texture filling-in revealed in V2 and V3 might also play a role in normal surface perception.

Is there a correlate of the adaptation process in the data? An interesting hint is given by the neuronal recordings obtained with square sizes that were too large to produce perceptual filling-in (12.8° panel in Fig. 3a), but which still produced moderate activity increases toward the end of stimulus presentation. In that stimulus condition, stimulus onset was followed by several seconds of activity below the baseline, indicating that the low activity level after stimulus onset reflected the presence of an active inhibitory process. It is possible that the increase of activity observed toward the end of the trial with the 12.8° square was made possible only

because of a decline in inhibition. For smaller square sizes (e.g., 4°), the decrease in activity immediately following the stimulus onset transient did not decrease below the baseline. Nevertheless, it is possible that this activity decrease reflected inhibitory inputs partially counteracting strong excitatory inputs from the surrounding texture, rather than just a passive decline in activity. Thus, activity increases after a number of seconds of fixation might be enabled by an adaptation of this inhibitory signal (related to mechanisms of figure-ground segregation). To test this hypothesis, we quantified the amount of inhibition revealed in the large-square condition (12.8°; Fig. 3b), and correlated it with the magnitude of the activity increases during prolonged fixation for a smaller square size (4°; Fig. 3b), both stimulus conditions having been recorded in each neuron. We found that the more inhibition there was in the beginning of the trial in the 12.8° square condition, the stronger the activity increases were later in the trial in the 4° condition. This indicates that a reduction of inhibition that permits existing excitatory inputs from the texture to become effective in driving neurons with RFs in the gray square is a plausible model for the adaptation process that precedes filling-in. Based on the psychophysical data outlined above, one could speculate that this inhibitory signal is related to several mechanisms that maintain figure-ground segregation, including signals related to boundary representations. One interesting question related to this interpretation is whether neurons were encountered that showed an

neuron) included in the different histograms was different at different square sizes. (B) Comparison of the average time required by humans to report filling-in (open symbols) with the average time it took neurons with significant activity increases to minimize the response difference in the square and no-square conditions (solid symbols). Data from individual observers are shown with thin, solid lines. The slopes of linear regression lines fit though the observers' data ranged from 0.8 to 1.1 ($r^2 > 0.90$ for three subjects and $r^2 = 0.73$ for the fourth) compared to 0.9 for the cells ($r^2 = 0.94$). For the neuronal data, the smallest difference between the firing rates in the square and no-square conditions was determined for each 1 s epoch beginning 1 s after stimulus onset, for each cell and square size. The time at which the difference was the smallest was considered as the 'filling-in time' for that cell at that square size. Data at the smallest square size were pooled from square sizes of 1.0 and 1.6°. (C) Correlation between inhibition and activity increases. The activity increase index was calculated from data in the 4° square condition, by subtracting the activity in the 4° square condition in a 1.5–4 s time window from activity in a 9.5–12 s time window, and dividing the result by the sum of those activities. The inhibition index was calculated from data in the 12.8° square condition, by subtracting the activity in the square condition in a 1.5–4 s time window from activity in the baseline condition, and dividing the result by the sum of those activities. Inhibition and increase indices were calculated individually in 62 V2/V3 neurons with positive increase indices for the 4° square, also tested with the 12.8° square, and a negative correlation between both indices was obtained (the more negative the inhibition index, the stronger the inhibition). The data indicate that larger activity increases with the 4° square were associated with stronger inhibition with the 12.8° square. Hence, activity increases associated with perceptual completion result from the adaptation of an inhibitory mechanism.

238

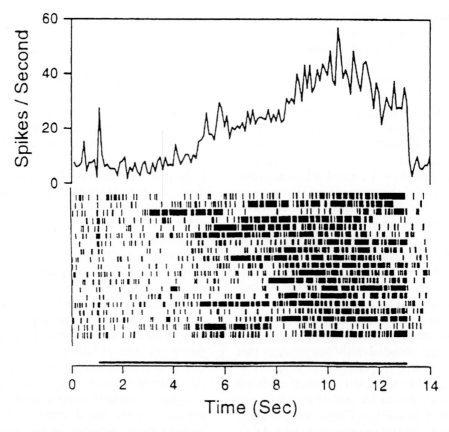

Fig. 4. Response of a single V3 cell (Fi0177-2) to the presentation of a gray square surrounded by a dynamic texture as a function of time. The square was 4°, and was shown at an eccentricity of 7°. A trial started with the recording of 1 s of baseline activity in the absence of a stimulus. Stimulus presentation is indicated by the solid bar at the bottom of the graph. The cell's RF was centered over the square, and straddled the square's edges, causing a brief onset response. The top panel shows the cumulative histogram demonstrating increased activity toward the end of stimulus presentation. Activity is expressed as spikes (action potentials) per second, calculated in bins of 100 ms. The bottom panel shows individual response traces for the individual trials upon which the cumulative histogram is based. Each small vertical mark indicates an action potential (and at high firing rates the vertical marks tend to blend). Individual trials show discrete episodes of increased activity, which were more likely to occur toward the end of each trial. The histogram only includes trials in which the monkey maintained fixation till the end of the trial.

adaptation of their signal around the time observers perceive perceptual filling-in. We have encountered only a single neuron in which activity was high starting at stimulus onset, and then showed a strong decline at the time where other neurons would show activity increases. This neuron could have been an inhibitory neuron, whose signal could have been directly related to declines in the strength of boundary representations.

To interpret the activity increases of neurons with RFs over the gray square as a neural interpolation process associated with perceptual filling-in does not answer the question as to what aspect of the dynamic texture surface is actually carried by that neural signal. Is it the brightness, the general statistics of the line texture, the motion signal, or all aspects of the stimulus together? This is a relevant question, because of the multiple features present in the texture surface used, and because behavioral data from Ramachandran and Gregory (1991) suggest that different features of a background might spread across a figure at slightly different times during peripheral fixation of the figure. It has been suggested (Gattass et al., 1998) that interactions between boundaries and surface features might take place in different areas depending on the features that define the boundaries between surface areas (e.g., luminance, color, motion, or texture). However, although a differential contribution of different areas to different types of interpolation processes is likely, activity increases in single neurons correlated with perceptual filling-in are probably related to the interpolation of several features at once. Experiments in which neural responses are measured in different cortical areas during perceptual filling-in of figures defined in different feature domains will be required to address this issue.

During the adaptation process preceding perceptual filling-in and neural interpolation, a competition may occur between the texture background and the figure's boundaries. Specifically, the dynamic texture could contribute to the weakening of boundary information, which is predominantly given by the static black/gray luminance boundary. In computational models from Grossberg and colleagues (e.g., see Grossberg, 2003a), cooperative groupings of iso-oriented and collinear line segments compete with orthogonally oriented surface contours through inhibitory mechanisms, which are compatible with anatomical and physiological data (McGuire et al., 1991; Kisvarday and Eysel, 1993). Thus, in the stimulus used to test perceptual filling-in, the grouping of line segments in the background may weaken the representation of square boundaries that are orthogonal to the background elements. Furthermore, the competition between groupings in the background and the square's contours may be biased in favor of the elements in the background, because they are constantly refreshed and the square's boundary is not.

In addition, surface features might compete or interact with each other during filling-in. In the domain of color filling-in, the perceptual result of filling-in across a figure has been reported to be determined by both the figure and the background color. Fujita (1993) has demonstrated that prolonged fixation of the middle of a colored disk on a background of another color does not lead to a replacement of the figure's color with that of the background, but rather with the appearance of a color mix throughout the entire stimulus. This suggests that filling-in is a bi-directional process,

which might be revealed only when the two features that fill the figure and background region are about equally strong.

Finding a neural interpolation response associated with perceptual filling-in for one feature does not imply the existence of a similar interpolation process for another feature. Von der Heydt et al. (2003) recorded in V1 and V2 of Rhesus monkeys that were trained to respond to color filling-in (see also Friedman et al., 2003). The monkeys were trained to respond to physical color filling-in, and appeared to respond in a similar fashion to conditions in which the stimulus remained physically identical, but which led to perceptual filling-in in human observers. This experiment was possible because stimulus conditions were chosen to lead to sufficient perceptual filling-in in trials lasting only 6 s. Recordings were performed from two types of neurons: recordings were made from neurons that were sensitive to large color stimuli filling the RF (surface-sensitive neurons), which existed in limited numbers in V1 and V2. For those cells, the color figure to be filled in was positioned inside the RF. Recordings were also made from neurons with optimal responses to oriented bars or edges (border-sensitive neurons), which are often sensitive to (color) contrast polarity, and which can also signal border ownership (Zhou et al., 2000). For those cells, the RF was placed along the border of the figure. In 92% of trials, the trials presented physical filling-in of the disk by the surrounding annulus. In 8% of the trials (test trials), monkeys responded, while there was no physical change suggesting that the responses were triggered by perceptual filling-in. It was found that during test trials, there were no changes in activity in the surface-response neurons, but there were declines in activity in the border-sensitive neurons which, with some assumptions, could be related to the filling-in responses of the monkeys. By assuming that border detectors sensitive to polarity and border ownership in fact have the information to trigger a symbolic 'filling-in' of spaces between borders (potentially with help from higher-order visual cortex), von der Heydt et al. (2003) interpreted the link between activity changes in border detectors and the filling-in reports as support for a symbolic filling-in theory. Thus, even if the surface-sensitive cells contributed to normal perception of surface color, they would not contribute to perceptual filling-in of color induced by maintained fixation away from a figure. The experiments from von der Heydt et al. (2003), however, may not be entirely conclusive for several reasons. The correlation between filling-in responses and neural data was not tested under a variety of conditions that would lead to large differences in the timing of filling-in responses. In trials without physical filling-in, monkeys might have timed their responses to occur just before the end of the trial (after 6 s). It is, therefore, not entirely established to what extent the monkeys' responses correlated with a percept of color filling-in. Furthermore, the activity of border-sensitive neurons was only recorded with their RF on the edge of the figure, and it is possible that activity increases during filling-in would have been present if the neurons' RFs had been placed in the middle of the figure. In addition, the physical filling-in that stood as a model for the percept of perceptual filling-in was a very gradual transition from the figure color to the annulus color in the figure region. One could question whether this is an appropriate model for perceptual filling-in. If the perceptual change in the figure during maintained fixation was indeed so slow, one might wonder whether the conditions were optimal to induce a strong illusion. Despite these considerations, the data of von der Heydt et al. (2003) suggest the interesting possibility that color filling-in is a result of a high-level symbolic operation, rather than the result of an isomorphic interpolation process. Color perception (and not brightness perception) is strongly affected by lesions in infero-temporal cortex, which is also strongly involved in object recognition (Heywood et al., 1995; Huxlin et al., 2000; Cowey et al., 2001). Further, color (and not brightness) is an important surface feature that permits the evaluation of important qualities of an object (e.g., whether a fruit is ripe and eatable). Color may be an integral part of object meaning, more than brightness or other surface features, and may thus be coded at high levels in the visual system.

For surface features that do show a neural correlate of perceptual filling-in in retinotopic areas, such a finding does not exclude a contribution of

higher-level areas. When neurons in retinotopically organized visual areas signal that a region of visual space that was occupied by a given surface now belongs to another surface, this new 'interpretation' of the visual scene might be enhanced by feedback, similar to the way in which feedback mechanisms can enhance the interpolation of fragmentary contour information. Furthermore, even when there is evidence for a neural interpolation process in retinotopical areas associated with filling-in, it is possible that the activity is not linked with perception *per se*, but rather, is used as a signal by higher-order areas to 're-interpret' the arrangement of (perceived) surfaces. Thus, the experience of perceptual filling-in might rely on cooperative activity in a network of visual areas at several levels in the visual system. This point is reinforced by other findings reviewed above, showing that global computations across the image strongly contribute to the outcome of the filling-in process.

It is an interesting finding that, for the stimulus used in our experiments, a correlate of perceptual filling-in was present in extrastriate cortex, but not in V1. Although V1 lesions at best leave the possibility for unconscious processing of visual information ('blind sight'; Weiskrantz, 2004), and although V1 therefore could be considered necessary for conscious perception, the presently discussed data suggest that V1 activity is closer to 'physical' reality than to perception. Furthermore, while V1 may be a required component to produce conscious perception, we may not be 'aware' of the presence or absence of V1 activity (Crick and Koch, 1995). This illustrates further that perception is a reconstructive process that hinges on interactions between multiple areas and processing streams. It is not surprising then (in retrospect) that an attempt to replicate the activity increases associated with perceptual filling-in in anesthetized monkeys failed. Recordings in a number of neurons in V2 and V3 of the anesthetized Cebus monkey, using stimulus conditions comparable to those used in Fig. 3, showed initial onset transients associated with stimulus presentation, followed by a lower, constant level of activity (De Weerd, Gattass, Desimone and Ungerleider, unpublished data). These data suggest that lateral connectivity within areas V2 and V3 is insufficient to generate the neuronal processes associated with perceptual filling-in, and that interaction with higher levels of the system is required. In other words, the neural interpolation of surfaces might be tightly linked with conscious perception. Other studies in V1 have shown that anesthesia abolishes neural responses related to signaling figure ground segregation (Lamme et al., 1998), and to signaling the direction of plaid motion (Guo et al., 2004); responses that likely involve feedback from the extrastriate cortex.

Relationship of perceptual filling-in to attention, learning, and cortical plasticity

Given the role of feedback speculated about in the previous section, it is a natural question to ask what the role of attention might be during perceptual filling-in. In a previous study of filling-in during maintained fixation, Lou (1999) had reported that instructing subjects to pay attention to a set of peripherally presented figures made it more likely to perceive filling-in of the attended figures, compared to other unattended figures. This was interpreted as an inhibitory effect of attention on the perception of peripheral figures during maintained fixation. In one of our experiments (De Weerd et al., 2006), subjects fixated a red dot on a large dynamic texture background, while to the left and to the right of fixation ($7°$ eccentricity), a gray square was presented. Fixation was measured, and trials aborted when fixation errors occurred. The luminance of the two squares as a whole or of a small cue in the middle of the squares was modulated slightly in an independent fashion on the left and on the right side, at a rate of $1\,Hz$ on average. Subjects were instructed to report all of these small, regularly occurring luminance changes with button presses, and they were cued to do so on the left or on the right in blocks of about 20 trials. Furthermore, subjects were told that the disappearance of one (or sometimes both) of the squares ($1.5° \times 1.5°$) would signal the end of the trial. Subjects were instructed to press the space bar when this occurred, and then to indicate with the left and right arrow keys of the keyboard

which square(s) had disappeared. Subjects ($N = 5$) were not aware that the filling-in of the squares was in fact the measurement of interest in this experiment. We found that filling-in was reported more frequently at the attended side (71%) than at the unattended side (23%). However, when filling-in happened to be perceived at the unattended side, it occurred after a time period that was indistinguishable from the time period required to perceive filling-in at the attended side (Fig. 5). This indicates that attention changed the probability of detecting filling-in, without modifying the disinhibitory processes that lead to filling-in. Hence, although conscious perception seems to be required to trigger the disinhibitory processes preceding filling-in, selective attention does not modulate the time course of disinhibition. Instead, attention might simply boost the response to inputs provided by the background texture to neurons with RFs over the gray square, similar to the boosting of responses to real stimuli placed inside the RF. Once neural interpolation has started, attention might have the effect of increasing the magnitude of the neural responses related to interpolation, in turn increasing the probability of detecting perceptual filling-in of the attended square.

There is ample evidence for the effects of selective attention on stimuli presented within the classical RF (e.g., Moran and Desimone, 1985; McAdams and Maunsell, 1999; Reynolds et al., 1999), and there is also evidence for attentional modulation of the effects of surround stimuli on stimuli in the RF (e.g., Ito et al., 1998; Ito and Gilbert, 1999). The former effects are related to mechanisms of selective attention, while the latter effects to attentional modulation of mechanisms of boundary perception and grouping. The findings reported in Fig. 5 are related to yet another effect of attention: the modulation of the strength of interpolation mechanisms contributing to surface perception. Computational models from Grossberg have proposed how the same cortical circuitry might explain all these attentional phenomena in an integrative manner. In an original study by Grossberg (1973), shunting inhibition was used to implement multiplicative dynamics in neuronal connectivity. This idea was the basis for adaptive resonance models in which

A

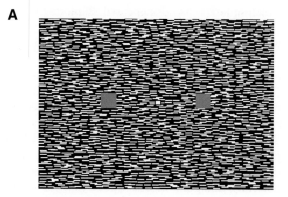

B

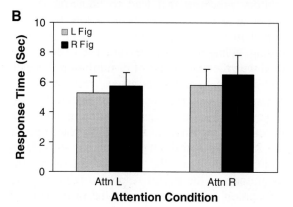

C

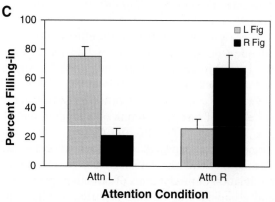

Fig. 5. Effects of attention on perceptual filling-in. (A) Stimulus used in Experiment 3, which contained two shapes each at 7° from a central fixation spot. (B) Response times of trials with perceptual filling-in in trials in which the left (gray bars) or right (black bars) figure filled in first, under two attention conditions: attention to the left (Attn L), and attention to the right (Attn R). Error bars are standard errors. Five subjects each carried out an average of 56 trials in each condition. (C) Percentage of trials with perceptual filling-in (conventions as in (B)).

multiplicative modulation of neuronal connectivity was used to explain attention phenomena (Grossberg, 1980), and led recently to a neural model of a biased competition model of selective attention (Reynolds et al., 1999). The same basic ideas of neural connectivity had also been the basis of a theory for boundary completion and surface filling-in (Grossberg, 1987a, b). Both lines of work have been combined in a model in which specific laminar circuits have been proposed to explain grouping, boundary completion, surface filling-in, and effects of attention on these processes (Grossberg, 1999, 2003a, b; Grossberg and Raizada, 2000; Raizada and Grossberg, 2001). In this model, feedback from layer VI to IV is used to modulate the on-center off-surround mechanisms that produce boundaries, and that produce input to surface filling-in mechanisms. This feedback (which will modulate the interaction between RF center and surround) provides access for various higher order variables (including attention) to influence perceptual processing. Within this theoretical framework, effects of attention on perceptual filling-in are entirely expected.

Perceptual filling-in and its associated neural interpolation response is a phenomenon that complicates the concept of a RF. A RF is typically subdivided into the 'classical' RF, and its surround. As defined in the Introduction, the classical RF is the region in visual space within which a small oriented bar must be placed to drive a recorded neuron. The same small bar placed in the region immediately surrounding the classical RF will not drive the neuron, but can modulate responses to stimuli placed inside the classical RF. However, the onset responses of textures surrounding the RF, and far outside the classical RF, show that when the surround stimulus is large enough, it can in fact be sufficient to drive both striate and extrastriate neurons. In addition, in extrastriate cortex, the stabilization of a stimulus on the retina is sufficient to change the interactions between classical RF and surround within seconds. The finding that within a few seconds, dynamic texture placed outside the classical RF of a recorded neuron starts driving that neuron (De Weerd et al., 1995) can be interpreted as a fast, temporary remapping of the visual field onto retinotopic cortex. The finding that attention can boost that response, and thus could also boost correlated activity between neurons with RFs in the figure and neurons with RFs in the surrounding background, suggests that perceptual filling-in could be enhanced by learning (in part, perhaps, through Hebbian mechanisms enhancing lateral connectivity). One might therefore speculate that repeated exposure to perceptual filling-in during a task that requires directing attention to the locus of filling-in might reduce the time required for perceptual filling-in.

There is a similarity between fast, temporary re-mapping during perceptual filling-in, and slower re-mapping phenomena after infliction of localized damage to the retina (see Kaas et al., 2003), which suggests that both might be linked. The neural responses in extrastriate cortex associated with perceptual filling-in might be similar to the responses immediately following a retinal lesion. Because of the permanent nature of a retinal scotoma, filling-in responses might become permanent, ultimately leading to neural plasticity and more permanent re-mapping. Disinhibition has been proposed to be a permissive mechanism for neural interpolation responses (De Weerd et al., 1995), and likewise, disinhibition owing to removal of afferent input after retinal injury has been proposed to be a mechanism facilitating cortical plasticity underlying re-mapping (Kaas et al., 2003). It is possible that under particular conditions the responses we have associated with perceptual filling-in represent a beginning of plastic changes in cortex (Tremere et al., 2003).

References

Arrington, K.F. (1994) The temporal dynamics of brightness filling-in. Vis. Res., 34: 3371–3387.

Bair, W., Cavanaugh, J.R. and Movshon, J.A. (2003) Time course and time-distance relationships for surround suppression in macaque V1 neurons. J. Neurosci., 23: 7690–7701.

Bender, M.B. and Teuber, L.H. (1946) Phenomena of fluctuation, extinction and completion in visual perception. Arch. Neurol. Psychiat. (Chicago), 55: 627–658.

Biederman, I. (2000) Recognizing depth-rotated objects: a review of recent research and theory. Spatial Vis., 13: 241–253.

Blakemore, C. (1976) The conditions required for the maintenance of binocularity in the kitten's visual cortex. J. Physiol., 261(2): 423–444.

Blasdel, G.G., Mitchell, D.E., Muir, D.W. and Pettigrew, J.D. (1977) A physiological and behavioural study in cats of the effect of early visual experience with contours of a single orientation. J. Physiol. (Lond.), 265: 615–636.

Coppola, D. and Purves, D. (1996) The extraordinarily rapid disappearance of entopic images. Proc. Natl. Acad. Sci. USA, 93: 8001–8004.

Cowey, A., Heywood, C.A. and Irving-Bell, L. (2001) The regional cortical basis of achromatopsia: a study on macaque monkeys and an achromatopsic patient. Eur. J. Neurosci., 14: 1555–1566.

Crick, F. and Koch, C. (1995) Are we aware of neural activity in primary visual cortex? Nature, 375: 121–123.

Davis, G. and Driver, J. (2003) Effects of modal and amodal completion upon visual attention: a function for filling-in. In: Pessoa, L. and De Weerd, P. (Eds.), Filling-In: From Perceptual Completion to Cortical Reorganization. Oxford University Press, Oxford, pp. 128–150.

Dennett, D. (1991) Consciousness Explained. Little Brown, Boston.

Desimone, R. and Ungerleider, L.G. (1989) Neural mechanisms of visual processing in monkeys. In: Boller, F. and Grafman, J. (Eds.), Handbook of Neuropsychology, Vol. 2. Elsevier, Amsterdam, pp. 267–299.

De Weerd, P., Desimone, R. and Ungerleider, L.G. (1998) Perceptual filling-in: a parametric study. Vis. Res., 38: 2721–2734.

De Weerd, P., Gattass, R., Desimone, R. and Ungerleider, L.G. (1995) Responses of cells in monkey visual cortex during perceptual filling-in of an artificial scotoma. Nature, 377: 731–734.

De Weerd, P., Smith, E. and Greenberg, P. (2006) Effects of selective attention on perceptual filling-in. J. Cogn. Neurosci., 18(3): 335–347.

Dresp, B. (1993) Bright lines and edges facilitate the detection of small light targets. Spatial Vis., 7: 213–225.

Field, D.J., Hayes, A. and Hess, R.F. (1993) Contour integration by the human visual system: evidence for local association field. Vis. Res., 37: 173–193.

Fiorani, M., Rosa, M.G.P., Gattass, R. and Rocha-Miranda, C.E. (1992) Dynamic surrounds of receptive fields in primate striate cortex: a physiological basis for perceptual completion. Proc. Natl. Acad. Sci. USA, 89: 8547–8551.

Francis, G. and Grossberg, S. (1996) Cortical dynamics of boundary segmentation and reset: persistence, afterimages, and residual traces. Perception, 35: 543–567.

Francis, G., Grossberg, S. and Mingolla, E. (1994) Cortical dynamics of feature binding and reset: control of visual persistence. Vis. Res., 34: 1089–1104.

Friedman, H.S., Zhou, H. and von der Heydt, R. (1999) Color filling-in under steady fixation: behavioral demonstration in monkeys and humans. Perception, 28(11): 1383–1395.

Friedman, H.S., Zhou, H. and von der Heydt, R. (2003) The coding of uniform colour figures in monkey visual cortex. J. Physiol., 548(2): 593–613.

Fujita, M. (1993) Color filling-in in foveal vision. Soc. Neurosci. Abstract, 19(3): 1802.

Gattass, R., Pessoa, L.A., De Weerd, P. and Fiorani, M. (1998) Filling-in in topographically organized distributed networks. Proc. Brazil. Acad. Sci., 70: 1–11.

Gerrits, H.J.M., De Haan, B. and Vendrik, A.J.H. (1966) Experiments with retinal stabilized images. Relations between the observations and neural data. Vis. Res., 6: 427–440.

Gerrits, H.J.M. and Vendrik, A.J.H. (1970) Simultaneous contrast, filling-in process and information processing in man's visual system. Exp. Brain Res., 11: 411–440.

Gilbert, C.D. and Wiesel, T.N. (1989) Columnar specificity of intrinsic cortico-cortical connections in cat visual cortex. J. Neurosci., 9: 2432–2442.

Grosof, D.H., Shapley, R.M. and Hawken, M.J. (1993) Macaque V1 neurons can signal 'illusory' contours. Nature, 365: 550–552.

Grossberg, S. (1973) Contour enhancement, short-term memory, and constancies in reverberating neural networks. Stud. App. Math., 52: 217–257.

Grossberg, S. (1980) How does a brain build a cognitive code? Psychol. Rev., 87: 1–51.

Grossberg, S. (1987a) Cortical dynamics of three-dimensional form, color, and brightness perception: I. Monocular theory. Percept. Psychophys., 41: 87–116.

Grossberg, S. (1987b) Cortical dynamics of three-dimensional form, color, and brightness perception: II Binocular theory. Percept. Psychophys., 41: 17–158.

Grossberg, S. (1997) Cortical dynamics of 3D figure ground perception of 2D pictures. Psychol. Rev., 104: 618–658.

Grossberg, S. (1999) How does the cerebral cortex work? Learning, attention and grouping by the laminar circuits of visual cortex. Spatial Vis., 12: 163–186.

Grossberg, S. (2003a) Filling-in the forms: surface and boundary interaction in visual cortex. In: Pessoa, L. and De Weerd, P. (Eds.), Filling-In: From Perceptual Completion to Cortical Reorganization. Oxford University Press, Oxford, pp. 13–37.

Grossberg, S. (2003b) How does the cerebral cortex work? Development, learning, attention, and 3D vision by laminar circuits of visual cortex. Behav. Cogn. Neurosci. Rev., 2: 47–76.

Grossberg, S. and Raizada, R. (2000) Contrast-sensitive perceptual grouping and object-based attention in the laminar circuits of primary visual cortex. Vis. Res., 40: 1413–1432.

Grossberg, S. and Todorovic, D. (1988) Neural dynamics of 1-D and 2-D brightness perception: a unified model of classical and recent phenomena. Percept. Psychophys., 43: 241–277.

Guo, K., Benson, P.J. and Blakemore, C. (2004) Pattern motion is present in V1 of awake but not anaesthetized monkeys. Eur. J. Neurosci., 19: 1055–1066.

Hamburger, K., Prior, H., Sarris, V. and Spillmann, L. (2006) Filling-in with colour: different modes of surface completion. Vision Res., 46(6–7): 1129–1138.

Haxby, J.V., Gobbini, M.I., Furey, M.L., Ishai, A., Schouten, J.L. and Pietrini, P. (2001) Distributed and overlapping representations of faces and objects in ventral temporal cortex. Science, 293: 2425–2430.

Heywood, C.A., Gaffan, D. and Cowey, A. (1995) Cerebral achromatopsia in monkeys. Eur. J. Neurosci., 7: 1064–1073.

Hirsch, H.V. and Spinelli, D.N. (1971) Modification of the distribution of receptive field orientation in cats by selective visual exposure during development. Exp. Brain Res., 12(5): 509–527.

Hubel, D.H. and Wiesel, T.N. (1959) Receptive fields of single neurons in the cat's striate cortex. J. Physiol. (Lond.), 148: 574–591.

Hubel, D.H. and Wiesel, T.N. (1962) Receptive fields, binocular interaction and functional architecture in the cats's visual cortex. J. Physiol. (Lond.), 160: 106–154.

Huxlin, K.R., Saunders, R.C., Marchionini, D., Pham, H.A. and Merigan, W.H. (2000) Perceptual deficits after lesions of inferotemporal cortex in macaques. Cereb. Cortex, 10: 671–683.

Ito, M. and Gilbert, C.D. (1999) Attention modulates contextual influences in the primary visual cortex of alert monkeys. Neuron, 22: 593–604.

Ito, M., Westheimer, G. and Gilbert, C.D. (1998) Attention and perceptual learning modulate contextual influences on visual perception. Neuron, 20: 1191–1197.

Kaas, J.H., Collins, C.E. and Chino, Y.M. (2003) The reactivation and reorganization of retinotopic maps in visual cortex of adult mammals after retinal and cortical lesions. In: Pessoa, L. and De Weerd, P. (Eds.), Filling-In: From Perceptual Completion to Cortical Reorganization. Oxford University Press, Oxford, pp. 187–206.

Kanizsa, G. (1955) Margini quasi-percettivi in campi con stimulazioni omagenea (Conditions and effects of apparent transparency). Rivista di Psicologia, 49: 7–30.

Kapadia, M.K., Ito, M., Gilbert, C.D. and Westheimer, G. (1995) Improvement in visual sensitivity by changes in local context: parallel studies in human observers and in V1 of alert monkeys. Neuron, 15(4): 843–856.

Kayaert, G., Biederman, I. and Vogels, R. (2003) Shape tuning in macaque inferior temporal cortex. J. Neurosci., 23: 3016–3027.

Kisvarday, Z.F. and Eysel, U.T. (1993) Functional and structural topography of horizontal inhibitory connections in cat visual cortex. Eur. J. Neurosci., 5: 1558–1572.

Komatsu, H., Kinoshita, M. and Murakami, I. (2000) Neural responses in the retinotopic representation of the blind spot in the macaque V1 to stimuli for perceptual filling-in. J. Neurosci., 20: 9310–9319.

Komatsu, H. and Murakami, I. (1994) Behavioral evidence of filling-in at the blind spot of the monkey. Vis. Neurosci., 11: 1103–1113.

Kuffler, S.W. (1953) Discharge patterns and functional organization of mammalian retina. J. Neurophysiol., 16: 37–68.

Lamme, V.A., Zipser, K. and Spekreijse, H. (1998) Figure-ground activity in primary visual cortex is suppressed by anesthesia. Proc. Natl. Acad. Sci. USA, 95: 3263–3268.

Lamme, V.A.F. (1995) The neurophysiology of figure-ground segregation in primary visual cortex. J. Neurosci., 10: 649–669.

Levitt, J.B. and Lund, J.S. (2002) The spatial extent over which neurons in macaque striate cortex pool visual signals. Vis. Neurosci., 19: 439–452.

Lou, L. (1999) Selective peripheral fading: evidence for inhibitory sensory effect of attention. Perception, 28: 519–526.

Lund, J.S., Yoshioka, T. and Levitt, J.B. (1993) Comparison of intrinsic connectivity in different areas of macaque monkey cerebral cortex. Cerebr. Cortex, 3: 148–162.

McAdams, C.J. and Maunsell, J.H. (1999) Effects of attention on orientation-tuning functions of single neurons in macaque cortical area V4. J. Neurosci., 19: 431–441.

McGuire, B.A., Gilbert, C.D., Rivlin, P.K. and Wiesel, T.N. (1991) Targets of horizontal connections in macaque primary visual cortex. J. Comp. Neurol., 305: 370–392.

Moran, J. and Desimone, R. (1985) Selective attention gates visual processing in the extrastriate cortex. Science, 229: 782–784.

Murakami, I., Komatsu, H. and Kinoshita, M. (1997) Perceptual filling-in at the scotoma following a monocular retinal lesion in the monkey. Vis. Neurosci., 14: 89–101.

Neumann, H., Pessoa, L. and Hansen, T. (2001) Visual filling-in for computing perceptual surface properties. Biol. Cybern., 85: 355–369.

Paradiso, M.A. and Hahn, S. (1996) Filling-in percepts produced by luminance modulation. Vis. Res., 36: 2657–2663.

Paradiso, M.A. and Nakayama, K. (1991) Brightness perception and filling-in. Vis. Res., 31: 1221–1236.

Pessoa, L. and Neumann, H. (1998) Why does the brain fill-in? TICS, 2: 422–424.

Pessoa, L., Thompson, E. and Noë, A. (1998) Finding out about filling-in: a guide to perceptual completion for visual science and the philosophy of perception. Behav. Brain Sci., 21: 723–748.

Peterhans, E. and von der Heydt, R. (1989) Mechanisms of contour perception in monkey visual cortex. II. Contours bridging gaps. J. Neurosci., 9: 1749–1763.

Pettigrew, J.D. and Freeman, R.D. (1973) Visual experience without lines: effect on developing cortical neurons. Science, 182(112): 599–601.

Polat, U. and Sagi, D. (1994) The architecture of perceptual spatial interactions. Vis. Res., 34: 73–78.

Polat, U., Mizobe, K., Pettet, M.W., Kasamatsu, T. and Norcia, A.M. (1998) Collinear stimuli regulate visual responses depending on cell's contrast threshold. Nature, 5:391(6667): 580–584.

Raizada, R. and Grossberg, S. (2001) Context-sensitive bindings by the laminar circuits of V1 and V2: a unified model of perceptual grouping, attention, and orientation contrast. Vis. Cogn., 8: 431–466.

Ramachandran, V.S. (2003) Foreword. In: Pessoa, L. and De Weerd, P. (Eds.), Filling-In: From Perceptual Completion to Skill Learning. Oxford University Press, Oxford, pp. xi–xxii.

Ramachandran, V.S. and Gregory, R.L. (1991) Perceptual filling-in of artificially induced scotomas in human vision. Nature, 350: 699–702.

Ramachandran, V.S., Gregory, R.L. and Aiken, W. (1993) Perceptual fading of visual texture borders. Vis. Res., 33: 717–721.

Ratliff, F. (1958) Stationary retinal images requiring no attachments to the eye. J. Opt. Soc. Am., 48: 274–275.

Reynolds, J.H., Chelazzi, L. and Desimone, R. (1999) Competitive mechanisms subserve attention in macaque areas V2 and V4. J. Neurosci., 19: 1736–1753.

Riggs, L.A., Ratliff, F., Cornsweet, J.C. and Cornsweet, T.N. (1953) The disappearance of steadily fixated visual test objects. J. Opt. Soc. Am., 43(6): 495–501.

Rossi, A.F. and Paradiso, M.A. (1996) Temporal limits of brightness induction and mechanisms of brightness perception. Vis. Res., 36: 1391–1398.

Rossi, A.F. and Paradiso, M.A. (1999) Neural correlates of perceived brightness in the retina, lateral geniculate nucleus, and striate cortex. J. Neurosci., 19: 6145–6156.

Sereno, M.I., Dale, A.M., Reppas, J.B., Kwong, K.K., Belliveau, J.W., Brady, T.J., Rosen, B.R. and Tootell, R.B.H. (1995) Borders of multiple visual areas in humans revealed by functional magnetic resonance imaging. Science, 268: 889–893.

Sergent, J. (1988) An investigation into perceptual completion in blind areas of the visual field. Brain, 111: 347–373.

Spillmann, L. and Kurtenbach, A. (1992) Dynamic noise backgrounds facilitate target fading. Vis. Res., 32: 1941–1946.

Sterling, P. (1998) Retina. In: Sheperd, G.M. (Ed.), The Synaptic Organization of the Brain. Oxford University Press, Oxford, pp. 205–252.

Todorovic, D. (1987) The Craik-O'Brien-Cornsweet effect: new varieties and their theoretical implications. Percept. Psychophys., 42: 545–560.

Tremere, L., Pinaud, R. and De Weerd, P. (2003) Contributions of inhibitory mechanisms to perceptual completion and cortical reorganization. In: Pessoa, L. and De Weerd, P. (Eds.), Filling-In: From Perceptual Completion to Cortical Reorganization. Oxford University Press, Oxford, pp. 295–322.

Troxler, D. (1804) Ueber das Verschwinden gegebener Gegenstände innerhalb unseres Gesichtskreises. In: Himly, K. and Schmidt, J.A. (Eds.) Ophthalmolgische Bibliothek II, Vol. 2. Jena, Frommann, F., pp. 1–119.

von der Heydt, R., Friedman, H.S. and Zhou, H. (2003) Searching for the mechanism of color filling-in. In: Pessoa, L. and De Weerd, P. (Eds.), Filling-in: From Perceptual Completion to Cortical Reorganization. Oxford University Press, pp. 106–127.

von der Heydt, R., Peterhans, E. and Baumgartner, G. (1984) Illusory contours and cortical neuron responses. Science, 224: 1260–1262.

Wang, G., Tanaka, K. and Tanifuji, M. (1996) Optical imaging of functional organization in the monkey inferotemporal cortex. Science, 272: 1665–1668.

Wehrhahn, C. and Dresp, B. (1998) Detection facilitation by collinear stimuli in humans: dependence on strength and sign of contrast. Vis. Res., 38: 423–428.

Weiskrantz, L. (2004) Roots of blind sight. Prog. Brain Res., 144: 229–241.

Yarbus, A.L. (1967) Eye Movements and Vision. Plenum Press, New York.

Zhou, H., Friedman, H.S. and von der Heydt, R. (2000) Coding of border ownership in monkey visual cortex. J. Neurosci., 20: 6594–6611.

Zipser, K., Lamme, V.A.F. and Schiller, P.H. (1996) Contextual modulation in the primary visual cortex. J. Neurosci., 16: 7376–7389.

Martinez-Conde, Macknik, Martinez, Alonso & Tse (Eds.)
Progress in Brain Research, Vol. 154
ISSN 0079-6123

CHAPTER 13

The visual phantom illusion: a perceptual product of surface completion depending on brightness and contrast

Akiyoshi Kitaoka[1,*], Jiro Gyoba[2] and Kenzo Sakurai[3]

[1]*Department of Psychology, Ritsumeikan University, 56-1 Toji-in Kitamachi, Kita-ku, Kyoto 603-8577, Japan*
[2]*Department of Psychology, Graduate School of Arts & Letters, Tohoku University, Kawauchi 27-1, Aoba-ku, Sendai 980-8576, Japan*
[3]*Department of Psychology, Tohoku Gakuin University, 2-1-1 Tenjinzawa, Izumi-ku, Sendai 981-3193, Japan*

Abstract: The visual phantom illusion was first discovered by Rosenbach in 1902 and named 'moving phantoms' by Tynan and Sekuler in 1975 because of its strong dependence on motion. It was later revealed that phantoms can be generated by flickering the grating (flickering phantoms) or by low-luminance stationary gratings under dark adaptation (stationary phantoms). Although phantoms are much more visible at scotopic or mesopic adaptation levels (scotopic phantoms) than at photopic levels, we proposed a new phantom illusion which is fully visible in photopic vision (photopic phantoms). In 2001, we revealed that the visual phantom illusion is a higher-order perceptual construct or a Gestalt, which depends on the mechanism of perceptual transparency. Perceptual transparency is known as a perceptual product based upon brightness and contrast. We furthermore manifested the shared mechanisms between visual phantoms and neon color spreading or between visual phantoms and the Petter effect. In our recent study, the visual phantom illusion can also be seen with a stimulus of contrast-modulated gratings. We assume that this effect also depends on perceptual transparency induced by contrast modulation. Moreover, we found that the Craik–O'Brien–Cornsweet effect and other brightness illusions can generate the visual phantom illusion. In any case, we explain the visual phantom illusion in terms of surface completion, which is given by perceptual transparency.

Keywords: visual phantoms; perceptual transparency; surface completion; illusion; neon color spreading; grating induction

What is the visual phantom illusion?

The visual phantom illusion refers to the completion phenomenon in which something like mist appears to lie over a physically homogenous surface (Tynan and Sekuler, 1975). Specifically, when part of a sinusoidally modulated luminance grating is transversely occluded by another surface, the two separate gratings appear to be continual *in front of* the occluder (Weisstein et al., 1982; Brown and Weisstein, 1988, 1991). If the luminance of the occluder is the same as the darkest one of the grating, the darker parts of the gratings appear to be continual (dark phantoms: Fig. 1a), while if the luminance of the occluder is the same as the brightest one of the grating, the brighter parts of the gratings appear to be continual (Sakurai and Gyoba, 1985) (light phantoms: Fig. 1b).

*Corresponding author. Tel.: +81-75-466-3402;
Fax: +81-75-465-8188; E-mail: akitaoka@lt.ritsumei.ac.jp

DOI: 10.1016/S0079-6123(06)54013-0

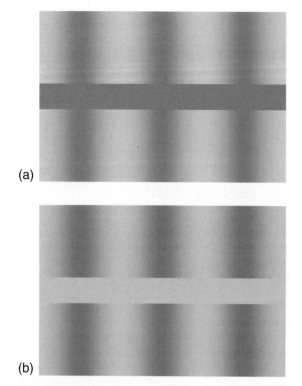

(a)

(b)

Fig. 1. Visual phantoms, which appear to be continual in front of the physically homogeneous occluder. (a) The dark parts of the upper grating appear to be continual with those of the lower grating in front of the occluder, when the occluder is dark. We hereafter call them 'dark phantoms'. (b) The light parts of the upper grating appear to be continual with those of the lower grating in front of the occluder, when the occluder is bright. We hereafter call them 'light phantoms'. Note that the inducing gratings are identical while only the occluder luminance is different between the two images.

(a)

(b)

Fig. 2. Misty appearance of visual phantoms. (a) It appears as if there were two spirals of white mist floating in front of a black-and-white spiral stripes, though the white stripes are physically homogenous. (b) There appears a sea of clouds with tops of mountains protruding. This 'sea' is visual phantoms.

This illusion was first discovered by Rosenbach (1902) and developed by Tynan and Sekuler (1975) as 'moving phantoms' because of its strong dependence on motion. It was later revealed that phantoms can be generated by flickering the grating (flickering phantoms) (Genter and Weisstein, 1981) as well as by low-luminance stationary gratings under dark adaptation (stationary phantoms) (Gyoba, 1983). Moreover, we found that dark adaptation is not necessary for stationary phantoms, because many people see stationary phantoms in photopic vision when inducing gratings are of low spatial frequency and of low contrast (Fig. 1). Actually, stationary phantoms are ubiquitous, e.g. the perception of mist (Fig. 2).

The visual phantom illusion is a phenomenon of brightness induction

It had been believed that the brightness induction in visual phantoms is *in phase* with the inducing grating (Mulvanny et al., 1982; Weisstein et al., 1982), as exaggerated in Fig. 3a, because the induced phantom grating appears to be continual with the inducing grating. This belief was questioned by McCourt (1994) who pointed out that the induced brightness in the occluder is *counterphase* with the inducing grating, as exaggerated in Fig. 3b. This finding was confirmed in subsequent research (May et al., 1999).

According to McCourt (1994), this counterphase induction is a kind of grating induction

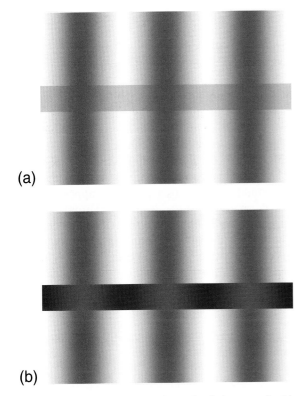

(a)

(b)

Fig. 3. (a) Exaggeration of 'in-phase' visual phantoms. In this figure, luminance modulation is given to the occluder in phase with the flanking gratings. (b) Exaggeration of 'counterphase' visual phantoms. In this figure, luminance modulation is given to the occluder out of phase with the flanking gratings. In both cases, three dark columns appear to be continual in front of the occluder.

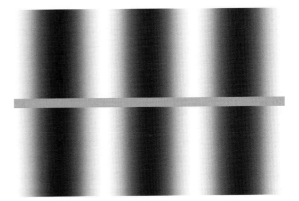

Fig. 4. Grating induction. The homogenously gray occluder appears to be modulated in brightness, counterphase with the inducing gratings. In grating induction, inducing gratings do not appear to be continual in front of the occluder.

The visual phantom illusion is a phenomenon of perceptual transparency

Grating induction, however, is quite different from visual phantoms in its appearance since the standard grating induction pattern never gives perceptual continuation of the gratings in front of the occluder. What is responsible for this difference?

We directed our attention to a phenomenon in perceptual transparency (Kitaoka et al., 2001a), in which adjacent areas that are similar in luminance tend to appear in front (Fuchs, 1923; Oyama and Nakahara, 1960; Metelli, 1974). We considered the mechanism of perceptual continuation of visual phantoms in terms of different types of transparency classified by Anderson (1997). Our idea is that visual phantoms reflect 'unique' transparency whereas McCourt's grating induction is characterized by 'no' transparency.

Anderson (1997) phenomenologically classified perceptual transparency into two types: unique and bistable transparency. In the former, a transparent surface is always perceived in front of the other surface, while in the latter, the perceived depth of two surfaces alternates and the surface in front appears to be transparent. These two types of perceptual transparency depend on the type of X-junctions. Unique transparency appears when contrast polarity along one edge is reversed over the X-junction, while contrast polarity along the other

(McCourt, 1982), which is one of the strongest brightness illusions (Fig. 4). He proposed an explanation of dark phantoms in which the dark phase of the inducing grating is combined with the light phase of induced brightness on the black occluder because of brightness similarity between them. Then, he thought that counterphase brightness induction sponsors this perceptual continuation.[1]

---

[1]We have recently found that in brief presentation (e.g. 50 ms) visual phantoms are in phase with the inducing gratings. This leaves the possibility that moving or flickering phantoms are in-phase brightness induction (Mulvanny et al., 1982); this point has not been argued by McCourt (1994).

edge is preserved over the X-junction (Fig. 5a). On the other hand, bistable transparency appears when contrast polarity along both edges is preserved over the X-junction (Fig. 5b). If contrast polarity along both edges is reversed over the X-junction, no or invalid transparency appears (Fig. 5c). According to Anderson (1997), this idea was first proposed by Adelson and Anandan (1990). For further research on this issue, see Kitaoka (2005).

It is evident that the configuration of grating induction (Fig. 4) corresponds to that of no transparency. This is the main reason why the grating induction never shows perceptual continuation of the grating, since those edges are not grouped by transparency.

On the other hand, the configuration of visual phantoms (Fig. 1) renders a series of configurations of unique transparency, as demonstrated in Fig. 6. In unique transparency, adjacent regions of the lower contrast always appear to be continual in front of the background when they are aligned with edges of different contrast polarities. This depth order is irreversible. Thus we (Kitaoka et al., 2001a) conjectured that this characteristic should give the two pieces of appearance of visual phantoms. One is that phantoms always appear to be continual in front of the occluder. The other is that the phantom visibility is reduced by giving crossed (= near) disparity to the occluder (Weisstein et al., 1982; Brown and Weisstein, 1991) because this binocular cue comes into conflict with the monocular cue (= unique transparency).

Furthermore, we have recently devised a new type of visual phantoms called 'mixed phantoms' in which dark gratings and light gratings cooperatively generate visual phantoms (Fig. 7a). The phases of these inducing gratings are opposite.

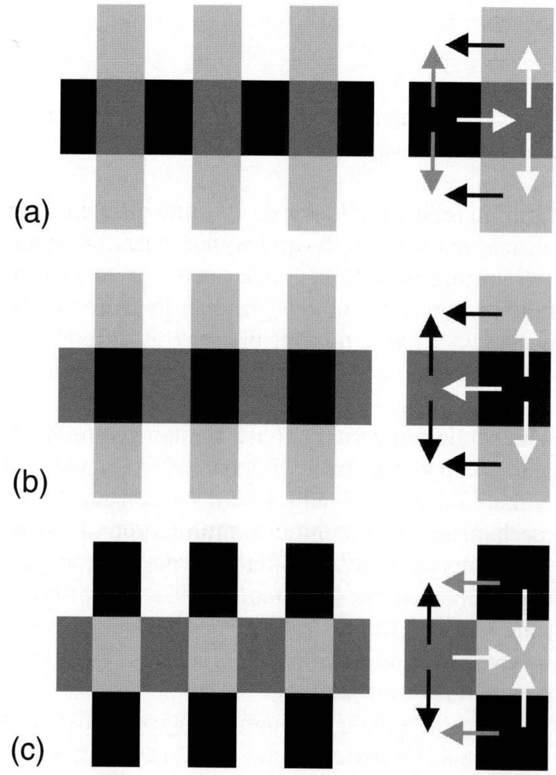

(a)

(b)

(c)

Fig. 5. The phenomenological classification of perceptual transparency depending on contrast polarity along edges over X-junctions. (a) Unique transparency, in which vertical gray rectangles appear to be transparent or translucent in front of the horizontal black rectangle. (b) Bistable transparency, in which vertical gray rectangles appear to be transparent in front of the horizontal gray rectangle, or the latter appears to be transparent in front of the former. (c) No transparency, in which transparency is not perceived. Arrows indicate which region is brighter between two adjacent regions.

Valid X-junctions giving unique transparency

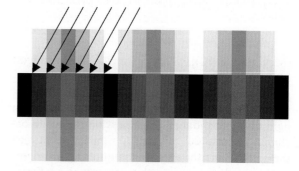

Fig. 6. A schematic explanation (Kitaoka et al., 2001a) of perceptual continuation of visual phantoms in terms of unique transparency. In this figure, every series of vertical edges changes contrast polarity when crossing horizontal edges while every series of the horizontal edges keeps contrast polarity. Thus the appearance is stratiform, i.e. the dark-gray, narrowest rectangle is in front of the middle-gray, middle-size rectangle, which in turn is in front of the light-gray, widest rectangle. These layers appear to be transparent and the order in apparent depth is fixed. This figure corresponds to Fig. 3b.

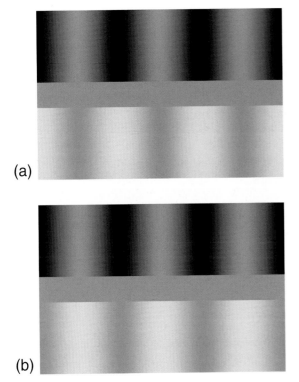

(a)

(b)

Fig. 7. 'Mixed' phantoms. The upper half gives the visual phantoms with the 'bright' occluder, i.e. light phantoms (Fig. 1b), while the lower half renders those with 'dark' occluder, i.e. dark phantoms (Fig. 1a). In this image, light phantoms and dark ones share the same gray occluder. (a) Three gray vertical columns or phantoms appear to be continual in front of the occluder when the luminance phases between the upper and lower gratings are shifted for half a cycle. (b) Phantoms disappear or bridge obliquely when the luminance phases are aligned.

This variant cannot be explained simply with brightness illusion but is quite consistent with the idea that the visual phantom illusion is characterized by perceptual transparency. If the phases of the inducers are aligned, grating induction becomes obvious but phantoms disappear or bridge obliquely (Fig. 7b).

Finally, we have recently devised an image that simultaneously shows visual phantoms and grating induction, as shown in Fig. 8. These findings, combined, indicate that the visual phantom illusion is different from grating induction though they share the same characteristics in brightness induction in the early stage.

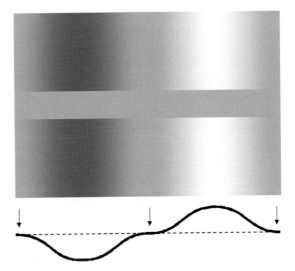

Fig. 8. An image where visual phantoms and grating induction are simultaneously observed. The luminance modulation of the inducing gratings is 'double-sinusoidal' as shown in the lower, where the broken line shows the luminance of the occluder. Visual phantoms are seen around the intersection between the luminance of the inducing gratings and that of the occluder, as pointed by arrows. For grating induction, the left half of the occluder appears to be brighter than the right half.

The visual phantom illusion is a phenomenon of figure–ground segregation

When the luminance of the occluder is low, the dark gratings appear to form phantoms or the *figure* (dark phantoms: Fig. 1a). This indicates that the lighter parts of the gratings become the *ground*. Inversely, when the luminance of the occluder is high, the light gratings appear to be phantoms or the figure (light phantoms: Fig. 1b) and the darker parts of the gratings become the ground. This appearance is enhanced by adding binocular disparity between the inducing gratings and the occluder (Brown and Weisstein, 1991) (Fig. 9).

Although visual phantoms are characterized by visual interpolation between two gratings, visual extrapolation from one grating shares the properties with visual phantoms (Gyoba, 1996). Figure 10 shows this stereogram. When the occluder or the surround is dark, dark gratings appear to be the figure in front while the rest bright parts become the background (Fig. 10a). Inversely, when

252

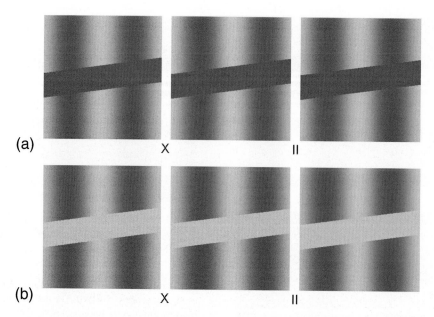

Fig. 9. A stereogram of visual phantoms. When observers cross-fuse the left and middle panels or uncross-fuse the middle and right panels, dark phantoms (a) or light phantoms (b) appear to float in front of the occluder, where the background appears to be bright (a) or dark (b). Note that inducing gratings are identical between the two stereograms and that only the occluder luminance is different. Oblique occluders are depicted to avoid their 'stereo capture' to the depth of phantoms.

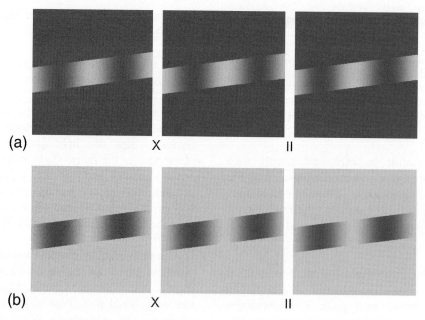

Fig. 10. A stereogram of visual extrapolation. When observers cross-fuse the left and middle panels or uncross-fuse the middle and right panels, (a) dark columns appear to float in front of the bright background or (b) light columns appear to float in front of the dark background. Note that inducing gratings are identical between the two stereograms and that only the occluder (= surround) luminance is different.

the occluder is bright, light gratings appear to be the figure in front, while the rest dark parts become the background (Fig. 10b). This reversibility was pointed out by Anderson (1999, 2003), who did not mention visual phantoms, however.

The visual phantom illusion includes the square-wave version

Although the inducing gratings are usually sinusoidal-wave luminance gratings, it is known that square-wave luminance gratings also give visual phantoms (Gyoba, 1983), as shown in Fig. 11. Different from the visual phantoms induced by sinusoidal-wave gratings, the square-wave phantoms do not give misty or hazy appearance but look like 'opaque' columns. In this regard, this version is not characterized by perceptual transparency.

The square-wave phantom illusion shares many properties with the Kanizsa square (Kanizsa, 1976) (Fig. 12). The illusory square corresponds to visual phantoms, and the four 'pac-men' and their wedges correspond to inducing gratings while the rest or the surround corresponds to the occluder.

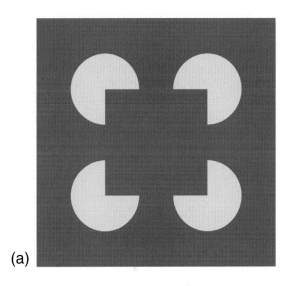

(a)

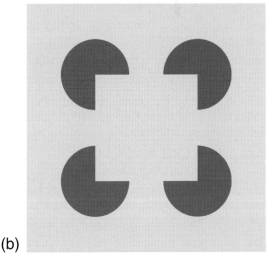

(b)

Fig. 12. The Kanizsa square. (a) A dark square appears to be in front of four light circles and the dark background. (b) A light square appears to be in front of four dark circles and the bright background.

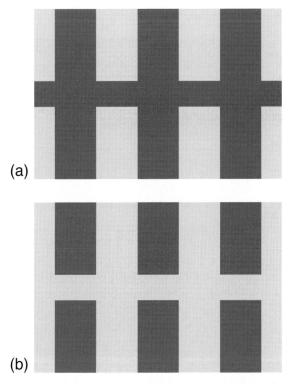

(a)

(b)

Fig. 11. The square-wave version of visual phantoms. (a) Dark columns appear to be continual in front of the occluder, where the background appears to be bright. (b) Light columns appear to be continual in front of the occluder, where the background appears to be dark. Note that inducing gratings are identical between the two images and that only the occluder luminance is different.

Inversely, the Kanizsa square can be changed to the sinusoidal-wave or misty version, as shown in Fig. 13. The inducers are circles with sinusoidally modulated luminance radials. Dark 'mist' appears when the surround (i.e. corresponding to the occluder) is dark whereas light 'mist' appears when the surround is bright. This 'mist' corresponds to visual phantoms.

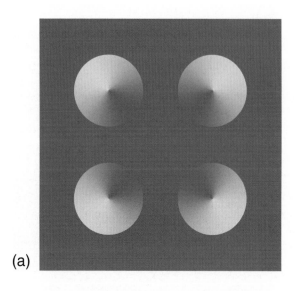

(a)

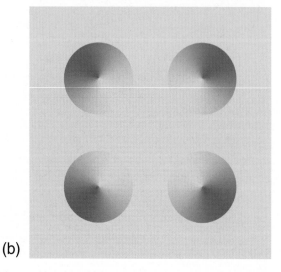

(b)

Fig. 13. The 'sinusoidal-wave' or misty version of the Kanizsa square. (a) Dark mist appears to be in front of four light circles and the dark background. (b) Light mist appears to be in front of four dark circles and the light background.

The visual phantom illusion depends on spatial frequency

Visual phantoms can be seen clearly when the inducing gratings are of low spatial frequency (Tynan and Sekuler, 1975; Genter and Weisstein, 1981; Gyoba, 1983). The higher the spatial frequency, the lower the visibility of phantoms. Moreover, the taller the occluder height, the lower the visibility of phantoms (Fig. 14).

This characteristic quite resembles Petter's effect (Kitaoka et al., 2001c). When two objects of the same brightness and of different sizes overlap, the perceived period of the thicker region that appears in front of the thinner region is longer than that of the thinner region that appears in front of the thicker region (Petter, 1956; Kanizsa, 1979; Shipley and Kellman, 1992; Masin, 1999; Singh

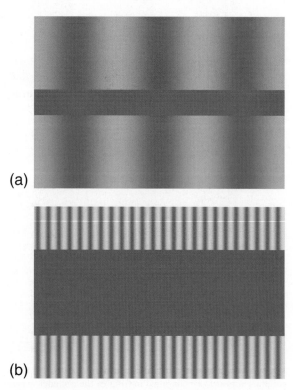

(a)

(b)

Fig. 14. Dependence of the visual phantom illusion on spatial frequency. (a) Phantoms are visible when the spatial frequency of the inducing gratings is low or the occluder height is short. (b) Phantoms are invisible when the spatial frequency of the inducing gratings is high or the occluder height is tall.

et al., 1999). The preferred explanation of Petter's effect is that an object completed by a shorter interpolating contour tends to be perceived in front of an object completed by a longer one, since modal completion requires more 'energy' than amodal completion and the larger object in front is usually accompanied by shorter interpolating contours than does the smaller one (Petter, 1956; Tommasi et al., 1995; Takeichi et al., 1995; Forkman and Vallortigara, 1999).

This explanation agrees with the dependence on spatial frequency in visual phantoms.

Aligned phantoms versus misaligned phantoms

As one of visual completion phenomena, the visual phantom illusion displays two rules. One is that completion prefers linearity, while the other is that completion prefers proximity. Figure 15 shows that oblique gratings may be linearly completed (aligned phantoms) while instead they may appear to be connected vertically between the nearest neighbors (misaligned phantoms) (Gyoba, 1994a; Brown et al., 2001). This bistable characteristic of visual phantoms is not observed in grating induction, in which the proximity rule is dominant.

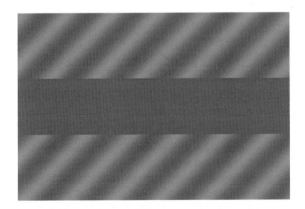

Fig. 15. Two pieces of appearance of visual phantoms. One is that phantoms are obliquely aligned with the inducing gratings (aligned phantoms). The other is that phantoms vertically connect the nearest neighbors of the gratings (misaligned phantoms).

New types of the visual phantom illusion

Envelope phantoms

Regions of low contrast can generate visual phantoms (Sakurai et al., 2000), as shown in Fig. 16. This phenomenon is explained as follows. First, regions of low contrast appear to include transparent layers or surfaces in front of carriers (Langley et al., 1998). Then, the perceived surfaces are completed over the occluder, thus displaying visual phantoms. This appearance is enhanced by adding crossed disparity to the envelope, as shown in Fig. 17.

In psychophysics, contrast modulation is often applied to examine the second-order processing. Here is an open question: Does this envelope phantom illusion disappear when the contrast-modulated

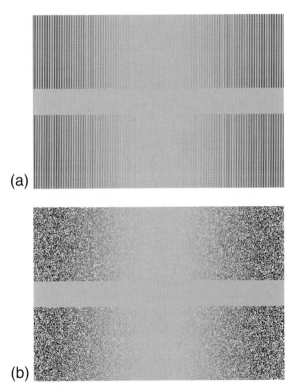

Fig. 16. Contrast-modulation-dependent phantoms or envelope phantoms. The envelope grating renders the phantom grating. In particular, regions of the lower contrast form phantoms. (a) The carriers are luminance gratings of high spatial frequency. (b) These are random dots.

256

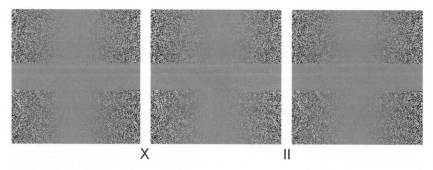

X ||

Fig. 17. A stereogram of envelope phantoms. When observers cross-fuse the left and middle panels or uncross-fuse the middle and right panels, regions of the lower contrast appear to be continual in front of the occluder.

Fig. 18. 'Chimera' phantoms. The upper half is a luminance grating while the lower half is an envelope grating. These different gratings cooperatively render visual phantoms.

stimulus is perfectly isoluminant? This question is important because visual phantoms have been thought to be the first-order processing. We speculate that envelope phantoms would remain in the isoluminant conditions because surface can be obtained from regions of low contrast even if the contrast-modulated stimulus is isoluminant (Langley et al., 1999), though visual phantoms cannot be produced by the stimuli of isoluminant colors (Gyoba, 1994b)[2].

Anyway, the envelope phantom illusion is not a special one. Indeed, it can render phantoms in co-operation with the standard phantoms (Fig. 18). Moreover, the envelope phantom illusion can yield the contrast-modulation version of Varin's figure (Fig. 19). This relationship is parallel to the close

relationship between standard phantoms and the Kanizsa Square.

Photopic phantoms

The standard configuration of visual phantoms is that the luminance of the occluder is the same as the darkest or brightest luminance of the inducing gratings (Fig. 1). When the occluder luminance is placed between them, grating induction appears (Fig. 4). What happens if the occluder luminance is higher or lower than the range of the gratings? Then, another type of visual phantoms appears (Kitaoka et al., 1999).

When the grating of high luminance is occluded by a dark occluder, the occluder appears to be transparent through which an illusory grating in phase with the surrounding gratings is observed, accompanied by clear illusory contours (Fig. 20a). On the other hand, when the grating of dark luminance is occluded by a bright occluder, the same illusion occurs (Fig. 20b). These illusions are called 'photopic' phantoms because we had thought that the standard visual phantoms are visible only in scotopic vision (Kitaoka et al., 1999), which was wrong because standard phantoms are fully visible in photopic vision, too (Fig. 1).

Photopic phantoms are characterized by the in-phase appearance in phantoms and the transparent appearance in the occluder. It is true that the photopic phantom illusion shares the mechanism of perceptual transparency with the standard phantom illusion, but it depends on bistable transparency (Anderson, 1997) (Fig. 5b). This type of transparency allows that any of the intersected

---

[2]According to Brown (2000), isoluminant color can generate visual phantoms in the moving condition.

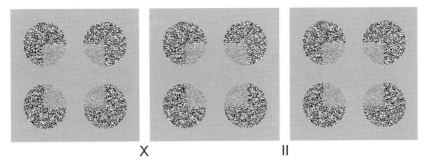

Fig. 19. A stereogram of the contrast-modulation version of Varin's figure. When observers cross-fuse the left and middle panels or uncross-fuse the middle and right panels, they can see a translucent square in front of the background. For Varin's figure, see Fig. 22.

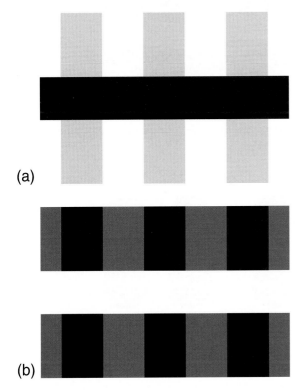

(a)

(b)

Fig. 20. Photopic phantoms. In-phase phantoms accompanied by clear illusory contours are observed behind the occluder. The occluder thus appears to be transparent. (a) The inducing gratings are of high luminances while the occluder is dark. (b) The inducing gratings are of low luminances while the occluder is bright. Some observers have difficulty seeing photopic phantoms.

surfaces can be perceived in front of the other. In the monocular condition, phantoms are faintly seen behind the occluder that appears to be transparent. However, phantoms appear to be much more vivid when crossed disparity is added to the inducing

gratings and phantoms are perceived in front of the occluder (Kitaoka et al., 1999) (Fig. 21).

This enhancement is called 'stereoscopic enhancement' (Harris and Gregory, 1973; Gregory and Harris, 1974; Lawson et al., 1974; Whitmore et al., 1976; Fujita, 1993). Stereoscopic enhancement is much more frequently mentioned in Varin's figure (Varin, 1971; Nakayama et al., 1990) (Fig. 22), which we think can be a variant of the photopic phantom illusion.

In relation to stereoscopic enhancement, there is an accepted idea on contour completion that illusory contours are modal (= visible) when they appear in front of the background while they are amodal (= invisible) when they appear behind the occluder (Michotte et al., 1991; Kanizsa, 1974, 1976, 1979; Shimojo and Nakayama, 1990; Kellman and Shipley, 1991; Tommasi et al., 1995; Ringach and Shapley, 1996; Kellman et al., 1998; Liu et al., 1999; Singh et al., 1999). This idea does not agree with the characteristics of photopic phantoms because they are visible behind the occluder. It is therefore suggested that the visibility of illusory contours is not characterized by a dichotomy (i.e. visible or invisible) depending on the depth order but is continuous in magnitude.

Actually, the first visual phantoms that Rosenbach (1902) demonstrated were photopic phantoms. More precisely, they were moving photopic phantoms.

Neon phantoms

Neon color spreading is a striking visual illusion which produces apparent completion of color or

258

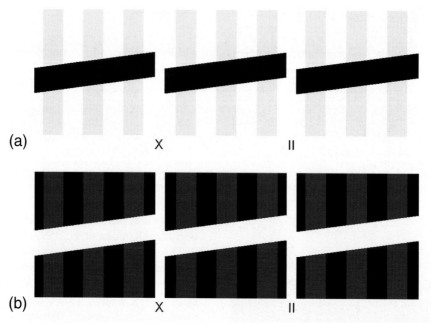

(a)　　　X　　　II

(b)　　　X　　　II

Fig. 21. A stereogram of photopic phantoms. When observers cross-fuse the left and middle panels or uncross-fuse the middle and right panels, three light-gray columns (a) or three dark-gray columns (b) appear to bridge in front of the occluder, in which the illusory contours appear to be clearer than observed monocularly.

X　　　　　II

Fig. 22. A stereogram of Varin's figure. When observers cross-fuse the left and middle panels or uncross-fuse the middle and right panels, they can see a translucent square in front of the background.

lightness (Varin, 1971; Van Tuijl, 1975; Van Tuijl and de Weert, 1979; Redies and Spillmann, 1981; Redies et al., 1984; Bressan et al., 1997) (Fig. 23). The brightness induction is in phase with the inducer. This character is the same as that of photopic phantoms. The only difference in stimulus configurations is the difference in the height of the inducing grating. That is short in neon color spreading while that is tall in photopic phantoms. What happens if the latter is shortened? A new in-phase phantom illusion appears (Fig. 24). This

new type or the neon phantom illusion shares the same characteristic in spatial frequency as those of the other phantoms (Kitaoka et al., 2001b).

Counterphase photopic phantoms

Photopic and neon phantoms are characterized by in-phase brightness induction, but there is a counterphase version, as shown in Fig. 25a. This version looks like standard stationary phantom illusion, giving a misty appearance (Fig. 1b). The

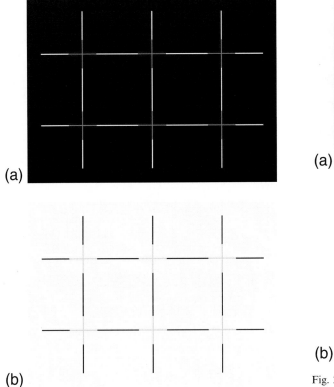

(a)

(b)

(a)

(b)

Fig. 23. Neon color spreading. (a) Illusory, dark 'diamonds' are seen in front of gray crosses though the background is homogeneously dark. (b) Illusory, light 'diamonds' are seen in front of gray crosses though the background is homogeneously bright.

Fig. 24. Neon phantoms. In-phase phantoms accompanied by clear illusory contours are observed in front of, behind, or flush with the gap. (a) The inducing gratings are of high luminances while the gap or background is dark. (b) The inducing gratings are of low luminances while the gap or background is bright.

difference between them is that for the former the brightest parts in the inducing gratings are a little bit darker than the bright occluder, whereas for the latter they are the same in luminance. This counterphase photopic phantom illusion is different from the 'in-phase' photopic phantom illusion (Fig. 20) at the point that counterphase phantoms are always seen in front of the occluder, which is based upon unique transparency (Figs. 5b and 6) like the standard stationary phantoms.

This counterphase photopic phantom illusion is closely related to the glare effect as proposed by Zavagno (1999) or Zavagno and Caputo (2001). When the gratings are separated at the brightest parts, the glare effect is manifest (Fig. 25b). It should be noted that the glare is brighter at a glance than the nonglare white parts, but the glare actually appears to be darker when analytically

observed. This inconsistency reminds us of the interaction between high-order mechanisms and low-order ones in brightness perception and might bring some fruitful hints to its study in the future.

Craik–O'Brien–Cornsweet phantoms

Our latest finding is that the stimulus of the Craik–O'Brien–Cornsweet effect (O'Brien, 1958; Craik, 1966; Cornsweet, 1970) can yield in-phase phantoms like photopic phantoms or neon phantoms, where apparent lightness gratings induce phantoms (Fig. 26a). In the Craik–O'Brien–Cornsweet image, there are no luminance gratings physically. This means that the Craik–O'Brien–Cornsweet phantoms do not depend on luminance but *lightness*. Figure 26b shows a variant, in which the apparent lightness induces in-phase phantoms in the occluder though the whole luminance profiles are

260

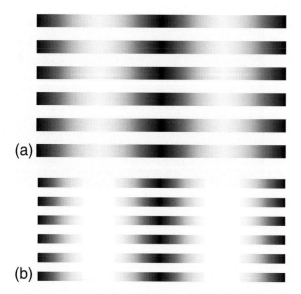

Fig. 25. Counterphase photopic phantoms. (a) Illusory mist like light phantoms (Fig. 1b) can be seen. In the bright gaps, counterphase brightness induction is observed. (b) When the brightest parts of inducing gratings are separated, Zavagno's glare effect appears.

counterphase sinusoidal waves. These new types also support our idea that the visual phantom illusion is based upon some high-order visual processing.

Possible neural mechanisms underlying the visual phantom illusion

It had been discussed that visual phantoms might depend on transient channels (Kulikowski and Tolhurst, 1973; Tolhurst, 1973) because phantoms had been visible in motion or flicker before Gyoba (1983) who first demonstrated stationary phantoms. Gyoba (1994b) suggested the involvement of the magnocellular pathway (Livingstone and Hubel, 1987) because phantoms are vivid in low luminance and disappear with isoluminant colors. Brown (1993, 2000) reported the involvement of fundus pigmentation, in which lightly pigmented observers saw phantoms more clearly than did darkly pigmented ones (e.g. African-Americans). We speculate that the pigmentation might affect the perception of brightness or contrast, which plays a critical role in the visual phantom illusion.

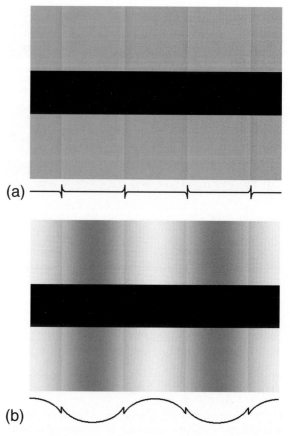

Fig. 26. (a) Craik–O'Brien–Cornsweet phantoms. (b) Sine-to-square phantoms. For both panels, phantoms are in phase with apparent lightness gratings, not luminance gratings. That is, the occluder appears to be dark, light, dark, light and dark from the left to the right. The luminance profile of the inducers is shown below in each panel, where the luminance modulation given at the edges is similar to each other.

Recent evidence supports the idea that the visual phantom illusion originates from high-order visual areas because it involves the figure-ground processing (Weisstein et al., 1982; Brown and Weisstein, 1991), illusory contours (Kitaoka et al., 1999, 2001b), or perceptual transparency (Kitaoka et al., 1999, 2001a,b). Physiology has started to support this idea, in which visual phantoms activated the extrastriate cortex as well as V1 (Sasaki and Watanabe, 2004; Meng and Tong, 2004; Meng et al., 2005). However, there remains an open question why moving or flickering phantoms are much more vivid than stationary phantoms.

Finally, it should be stressed again that the visual phantom illusion is ubiquitous, not a special percept. For example, Kawabata et al. (1999) have suggested that infants younger than one month of age can see visual phantoms.

References

Adelson, E. H. and Anandan, P. (1990) Ordinal characteristics of transparency. Paper presented at the AAAI-90 Workshop on Qualitative Vision, July 29, 1990, Boston, MA.

Anderson, B.L. (1997) A theory of illusory lightness and transparency in monocular and binocular images: the role of contour junctions. Perception, 26: 419–453.

Anderson, B.L. (1999) Stereoscopic surface perception. Neuron, 24: 919–928.

Anderson, B.L. (2003) The role of occlusion in the perception of depth, lightness, and opacity. Psychol. Rev., 110: 785–801.

Bressan, P., Mingolla, E., Spillmann, L. and Watanabe, T. (1997) Neon color spreading: a review. Perception, 26: 1353–1366.

Brown, J.M. (1993) Moving phantom visibility as a function of fundus pigmentation. Percept. Psychophys., 53: 367–371.

Brown, J.M. (2000) Fundus pigmentation and equiluminant moving phantoms. Percept. Mot. Skills, 90: 963–973.

Brown, J.M. and Weisstein, N. (1988) A phantom context effect: visual phantoms enhance target visibility. Percept. Psychophys., 43: 53–56.

Brown, J.M. and Weisstein, N. (1991) Conflicting figure-ground and depth information reduces moving phantom visibility. Perception, 20: 155–165.

Brown, J.M., Gyoba, J. and May, J.M. (2001) Stationary phantoms and grating induction with oblique inducing gratings: implications for different mechanisms underlying the two phenomena. Psychon. Bull. Rev., 8: 278–283.

Cornsweet, T. (1970) Visual Perception. Academic Press, New York.

Craik, K.J.W. (1966) The Nature of Psychology. Cambridge University Press, Cambridge.

Forkman, B. and Vallortigara, G. (1999) Minimization of modal contours: an essential cross-species strategy in disambiguating relative depth. Animal Cogn., 2: 181–185.

Fuchs, W. (1923) Untersuchungen über das simultane Hintereinandersehen auf derselben Sehrichtung. Z. Psychol., 91: 145–235.

Fujita, N. (1993) An occluded contour becomes visible with reversal of disparity. Perceptual Mot. Skills, 77: 271–274.

Genter II, C.R. and Weisstein, N. (1981) Flickering phantoms: a motion illusion without motion. Vision Res., 21: 963–966.

Gregory, R.L. and Harris, J.P. (1974) Illusory contours and stereo depth. Percept. Psychophys., 15: 411–416.

Gyoba, J. (1983) Stationary phantoms: a completion effect without motion and flicker. Vision Res., 23: 205–211.

Gyoba, J. (1994a) The visual phantom illusion under oblique inducing gratings. Jpn. Psychol. Res., 36: 182–187.

Gyoba, J. (1994b) Disappearance of stationary visual phantoms under high luminant or equiluminant inducing gratings. Vision Res., 34: 1001–1005.

Gyoba, J. (1996) Perceptual extrapolation of a stereoscopically raised surface depends upon its spatial frequency content. Invest. Ophthalmol. Vis. Sci., 37(3): 284.

Harris, J.P. and Gregory, R.L. (1973) Fusion and rivalry of illusory contours. Perception, 2: 235–247.

Kanizsa, G. (1974) Contours without gradients or cognitive contours? Ital. J. Psychol., 1: 93–113.

Kanizsa, G. (1976) Subjective contours. Sci. Am., 234: 48–52.

Kanizsa, G. (1979) Organization in Vision: Essay on Gestalt Perception. Praeger, New York.

Kawabata, H., Gyoba, J., Inoue, H. and Ohtsubo, H. (1999) Visual completion of partly occluded grating in infants under 1 month of age. Vision Res., 39: 3586–3591.

Kellman, P.J. and Shipley, T.F. (1991) A theory of visual interpolation in object perception. Cognit. Psychol., 23: 141–221.

Kellman, P.J., Yin, C. and Shipley, T.E. (1998) A common mechanism for illusory and occluded object completion. J Exp. Psychol.: Hum. Percept. Perform., 24: 859–869.

Kitaoka, A. (2005) A new explanation of perceptual transparency connecting the X-junction contrast-polarity model with the luminance-based arithmetic model. Jpn. Psychol. Res., 47: 175–187.

Kitaoka, A., Gyoba, J. and Kawabata, H. (1999) Photopic visual phantom illusion: Its common and unique characteristics as a completion effect. Perception, 28: 825–834.

Kitaoka, A., Gyoba, J., Kawabata, H. and Sakurai, K. (2001a) Perceptual continuation and depth in visual phantoms can be explained by perceptual transparency. Perception, 30: 959–968.

Kitaoka, A., Gyoba, J., Kawabata, H. and Sakurai, K. (2001b) Two competing mechanisms underlying neon color spreading, visual phantoms and grating induction. Vision Res., 41: 2347–2354.

Kitaoka, A., Gyoba, J., Sakurai, K. and Kawabata, H. (2001c) Similarity between Petter's effect and visual phantoms. Perception, 30: 519–522.

Kulikowski, J.J. and Tolhurst, D.J. (1973) Psychophysical evidence for sustained and transient detectors in human vision. J. Physiol., 232: 149–162.

Langley, K., Fleet, D.J. and Hibbard, P.B. (1998) Linear and nonlinear transparencies in binocular vision. Proc. R. Soc. Lond. B, 265: 1837–1845.

Langley, K., Fleet, D.J. and Hibbard, P.B. (1999) Stereopsis from contrast envelopes. Vision Res., 39: 2313–2324.

Lawson, R.B., Cowan, E., Gibbs, T.D. and Whitmore, C.G. (1974) Stereoscopic enhancement and erasure of subjective contours. J. Exp. Psychol., 103: 1142–1146.

Liu, Z., Jacobs, D.W. and Basri, R. (1999) The role of convexity in perceptual completion: beyond good continuation. Vision Res., 39: 4244–4257.

Livingstone, M.S. and Hubel, D.H. (1987) Psychophysical evidence for separate channels for the perception of form, color, movement and depth. J. Neurosci., 7: 3416–3468.

Masin, S.C. (1999) Test of Petter's rule for perceived surface stratification. Perception, 28: 1147–1154.

May, J.G., Brown, J.M. and Roberts, S. (1999) Afterimages, grating induction and illusory phantoms. Vision Res., 39: 3025–3031.

McCourt, M. (1994) Grating induction: a new explanation for stationary phantom gratings. Vision Res., 34: 1609–1617.

McCourt, M.E. (1982) A spatial frequency dependent grating-induction effect. Vision Res., 22: 119–134.

Meng, M., Remus, D.A. and Tong, F. (2005) Filling-in of visual phantoms in human brain. Nat. Neurosci., 8: 1248–1254.

Meng, M. and Tong, F. (2004) Binocular rivalry and perceptual filling-in of visual phantoms in human visual cortex. J. Vision, 4(8): 63.

Metelli, F. (1974) The perception of transparency. Sci. Am., 230: 90–98.

Michotte, A., Thinès, G. and Crabbé, G. (1991) Amodal completion of perceptual structures. In: Thinès, G., Costall, A. and Butterworth, G. (Eds.), Michotte's Experimental Phenomenology of Perception. Lawrence Erlbaum Associates, Hillsdale, NJ, pp. 140–167 (translated from the original paper published in 1964).

Mulvanny, P., Macarthur, R. and Sekuler, R. (1982) Thresholds for seeing visual phantoms and moving gratings. Perception, 11: 35–46.

Nakayama, K., Shimojo, S. and Ramachandran, V.S. (1990) Transparency: relation to depth, subjective contours, luminance, and neon color spreading. Perception, 19: 497–513.

O'Brien, V. (1958) Contour perception, illusion and reality. J. Opt. Soc. Am., 48: 112–119.

Oyama, T. and Nakahara, J. (1960) The effect of lightness, hue and area upon the apparent transparency. Jpn. J. Psychol., 31: 35–48.

Petter, G. (1956) Nuove ricerche sperimentali sulla totalizzazione percettiva. Rivista di Psicologia, 50: 213–227.

Redies, C. and Spillmann, L. (1981) The neon color effect in the Ehrenstein illusion. Perception, 10: 667–681.

Redies, C., Spillmann, L. and Kunz, K. (1984) Colored neon flanks and line gap enhancement. Vision Res., 24: 1301–1309.

Ringach, D.L. and Shapley, R. (1996) Spatial and temporal properties of illusory contours and amodal boundary completion. Vision Res., 36: 3037–3050.

Rosenbach, O. (1902) Zur Lehre von den Urtheilstäuschungen. Z. Psychologie, 29: 434–448.

Sakurai, K. and Gyoba, J. (1985) Optimal occluder luminance for seeing stationary visual phantoms. Vision Res., 25: 1735–1740.

Sakurai, K., Kawabata, H., Sasaki, H. and Kitaoka, A. (2000) Effects of occluder luminance on appearance of moving visual phantoms induced by second-order components. Invest. Ophthalmol. Vis. Sci., 41(4): S228 Abstract 1199.

Sasaki, Y. and Watanabe, T. (2004) The primary visual cortex fills in color. Proc. Natl. Acad. Sci., USA, 101: 18251–18256.

Shimojo, S. and Nakayama, K. (1990) Amodal representation of occluded surfaces: role of invisible stimuli in apparent motion correspondence. Perception, 19: 285–299.

Shipley, T.F. and Kellman, P.J. (1992) Perception of partly occluded objects and illusory figures: evidence for an identity hypothesis. J. Exp. Psychol.: Hum. Percept. Perform., 18: 106–120.

Singh, M., Hoffman, D.D. and Albert, M.K. (1999) Contour completion and relative depth: Petter's rule and support ratio. Psychol. Sci., 10: 423–428.

Takeichi, H., Nakazawa, H., Murakami, I. and Shimojo, S. (1995) The theory of the curvature-constraint line for amodal completion. Perception, 24: 373–389.

Tolhurst, D.J. (1973) Separate channels for the analysis of the shape and the movement of a moving visual stimulus. J. Physiol., 231: 385–402.

Tommasi, L., Bressan, P. and Vallortigara, G. (1995) Solving occlusion indeterminacy in chromatically homogeneous patterns. Perception, 24: 391–403.

Tynan, P. and Sekuler, R. (1975) Moving visual phantoms: a new contour completion effect. Science, 188: 951–952.

Van Tuijl, H.F.J.M. (1975) A new visual illusion: neonlike color spreading and complementary color induction between subjective contours. Acta Psychol., 39: 441–445.

Van Tuijl, H.F.J.M. and de Weert, C.M.M. (1979) Sensory conditions for the occurrence of the neon spreading illusion. Perception, 8: 211–215.

Varin, D. (1971) Fenomeni di contrasto e diffusione cromatica nell'organizzazione spaziale del campo percettivo. Riv. Psicol., 65: 101–128.

Weisstein, N., Maguire, W. and Williams, M.C. (1982) The effect of perceived depth on phantoms and the phantom motion aftereffect. In: Beck, J. (Ed.), Organization and Representation in Perception. Lawrence Erlbaum, Hillsdale, NJ, pp. 235–249.

Whitmore, C.L.G., Lawson, R.B. and Kozora, C.E. (1976) Subjective contours in stereoscopic space. Percept. Psychophys., 19: 211–213.

Zavagno, D. (1999) Some new luminance-gradient effects. Perception, 28: 835–838.

Zavagno, D. and Caputo, G. (2001) The glare effect and the perception of luminosity. Perception, 30: 209–222.

Form, Object and Shape Perception

Introduction

Although seeing 3D shapes seems effortless, a tremendous amount of computation underlies form perception. The following chapters explore various aspects of these computations at various levels of analysis. A fundamental unsolved problem in visual neuroscience is how the human visual system recovers the 3D form of objects and scenes from inherently ambiguous 2D retinal images. The core computational problem is one of correctly and rapidly interpreting inherently ambiguous patterns of retinal activation. The following chapter by Mamassian most explicitly explores the nature of the inferences underlying perception, but inference is a recurring theme in all the chapters.

The processing of form information begins in the retina with center–surround and color-opponent receptive fields that confer upon ganglion cells a sensitivity to edges. The information the retina transmits to the brain for further processing is a compressed version of the image that emphasizes border information at multiple spatial scales. Ganglion cells with similar tuning characteristics are distributed throughout the retina, and can be thought of as bandpass channels that accomplish a decomposition of the image using something analogous to Fourier analysis or in more recent conceptualizations, a wavelet decomposition. The fact that retinal processing emphasizes edge information suggests that contours play a crucial role in generating later representations of 3D form.

A keystone in thinking about the neural mechanisms of visual perception is the concept of hierarchical processing of the details of the visual image. A widely held view is that this processing occurs in a number of stages, the first of which performs an analysis or filtering of the retinal image by extracting different, elementary features (primitives) or classes of image "energy". Early cortical processing appears to consist of a neural description of various image primitives and their locations within the scene. This description is a simplified version of the original retinal image, but is still far from explicit identification of the 3D structure of the visible world. One basic form of feature energy is motion energy. How form information constrains the interpretation of this energy so that correct 3D trajectories can be computed is considered in the following chapter by Tse and Caplovitz.

It is commonly held that "higher" stages of visual processing combine aggregates of primitive features into progressively more complex representations. Two generally dichotomous characterizations of these elements should be mentioned. One is that contours are primitives, which are used to define surfaces and object boundaries. The other is that receptive fields operate as localized, spatial frequency filters. It is not known which of these two general formulations (edge detection or spatial frequency analysis) is more accurate. Both types of detectors appear to exist and may constitute the extremes of a spectrum of processing types.

It is widely believed that early visual cortical areas are involved in grouping local information across the image into aggregate wholes. Because grouping involves decisions about what belongs with what in an image, it is tantamount to an inference about the state of the world. As such, grouping procedures do not merely extract information from the image; they create or construct new information. Thus, detected types of orientation, motion and other types of energy are used to construct an experience of a 3D world.

Implicit in either the "primitive feature extraction" or "spatial frequency decomposition" views is the idea that individual cells can "code for" or are "tuned to" the presence of a specific feature or spatial frequency component. A literal interpretation of this idea is that activity in individual visual neurons signifies the presence of the feature to which that neuron is tuned. In order for neural activity to code for the presence of particular features, the cells would have to respond to that particular feature (such as orientation) independently of other properties (such as contrast). However, it has been recognized that at least at the earliest stages of neural processing, individual neurons cannot uniquely encode specific features because their patterns of discharge vary across several dimensions simultaneously (for example, contrast, orientation, direction of motion, stereo disparity and/or spatial frequency). This has led to a greater appreciation of the fact that the neural representation of images and visible objects must somehow involve the profile of activity across populations of neurons. Computational studies have illustrated that this kind of "population coding" can generate very precise representations — much more precise than the tuning characteristics of any of the individual neurons contributing to the profile. The possibility that shape is coded in terms of contour curvature cues at the population level in V4 is explored in the following chapter by Pasupathy.

The problems inherent in determining an object's 3D form are inseparable from the problems associated with recognizing that object. The perception of form does not necessarily involve making matches to representations of things seen in the past, but form recognition does. This is a theme considered in the following chapter by Bülthoff and Newell, who explore how familiarity with objects influences our ability to recognize them and how they may be encoded. Once even a part of an unfamiliar object has been recognized, the object's representation in memory can be used to constrain possible shape solutions for its other, still unrecognized parts. Multiple representations of an object appear to be formed, each specialized for the purpose and coordinate transformation required for some perceptual or behavioral task.

Solving for shape and matching to stored representations using multiple strategies and cues has numerous advantages. Multiple circuits can compute shape solutions in parallel, reducing the overall computational time. If one subsystem should reach a solution before the others, that shape solution can constrain and be constrained by the computations being carried out by the other subsystems. Parallel and concurrent processing also reduces the likelihood of attaining non-veridical solutions because all shape subsystems must come to mutually consistent solutions, permitting a form of error checking and redundancy.

While we know that 3D form and motion are computed extremely rapidly, we do not know how they are coded or decoded from image cues. While some things can be agreed upon such as the necessity of parallel, population level coding or the importance of contour cues such as curvature and junctions in recovering 3D form, we are very far from a complete understanding of how form is constructed by the brain. The following chapters offer a useful cross section of current work in the field of human visual form processing.

Peter U. Tse

Martinez-Conde, Macknik, Martinez, Alonso & Tse (Eds.)
Progress in Brain Research, Vol. 154
ISSN 0079-6123

CHAPTER 14

Bayesian inference of form and shape

Pascal Mamassian*

CNRS UMR 8581, LPE Université Paris 5, 71 ave. Edouard Valliant, 92100 Boulogne-Billancourt, France

Abstract: The ability to visually perceive two-dimensional (2D) form and three-dimensional (3D) shape is one of our most fundamental faculties. This ability relies on considerable prior knowledge about the way edge elements in an image are likely to be connected together into a contour as well as the way these 2D contours relate to 3D shapes. The interaction of prior knowledge with image information is well modeled within a Bayesian framework. We review here the experimental evidence of shape perception seen as a Bayesian inference problem.

Keywords: visual perception; Bayesian inference; form; shape; contour; three-dimensional perception

Introduction

What are shapes good for? Imagine you are interested in identifying a tree in your local park. Neither its size nor its color is a reliable cue for its name because the tree's size is dependent on its age and its color on the current season. In contrast, a very robust cue is the shape of the tree, be it pyramidal, columnar, V-shaped, round, or oval. Thus for example, a pin oak will be easily distinguished from a white oak because the former is pyramidal and the latter is round in shape. Not only is the global shape important, but the local shapes also carry critical information. To pursue our example, the pin oak has leaves that contain lobes with bristle-tipped teeth whereas the leaves of the white oak have rounded lobes.

In addition to the difference in scale, the whole tree and the individual leaves also differ in another fundamental way. The tree is described as a 3D shape whereas the leaves are 2D. While the distinction between 2D and 3D shapes is clear in the world, it is often much more subtle for the visual system. Figure 1 illustrates how a simple line

drawing can appear 3D. Interestingly, once the 3D percept arises, it is difficult to simultaneously hold a planar interpretation.

In this chapter, we discuss how 2D and 3D object shape can be inferred from an image. It is well known that a single image is consistent with an infinite number of scenes and yet our perception does not reflect these ambiguities. Therefore, it seems that additional information is used to disambiguate the image. This extra information has

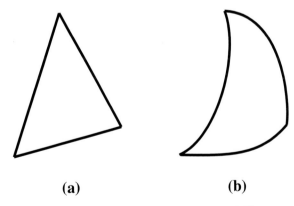

(a)   **(b)**

Fig. 1. Distinction between 2D and 3D shapes. (a) Three vertices connected by straight lines produce the impression of a flat triangle. (b) In contrast, when those three vertices are connected by curved lines, a curved 3D surface can be perceived.

*Corresponding author. Tel.: +33-155-205-930,
Fax: +33-155-205-854; E-mail: pascal.mamassian@univ-paris5.fr

DOI: 10.1016/S0079-6123(06)54014-2

been variously called a Gestalt principle, a rule, a law, or a property. In contemporary Bayesian models, this extra information is called prior knowledge. We start this chapter with a brief reminder of the way the Bayesian framework can be used in visual perception, and then describe how this framework has been used to understand contour groupings. We then review some studies on 2D and 3D shape from contours that can be cast within this Bayesian framework.

Basic Bayes

The Bayesian framework has the merit to make a clear distinction between the information available at the level of the receptors and the information specific to the organism. The former is called the likelihood and the latter the prior knowledge. Priors are often seen as supplementary information brought in to disambiguate sensory information.

Imagine that you are interested in inferring the shape of a closed contour from an image such as the outline of a sculpture from Constantin Brancusi (Fig. 2). The sculpture is egg-shaped and sliced by a plane on its left side thus forming an elliptical section. The section is discontinuous in the image because of the particular illumination conditions when the image was taken but the elliptical contour does exist on the sculpture. How can we infer the closed contour in the world given

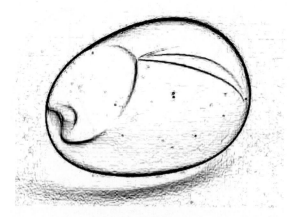

Fig. 2. Contour image of the sculpture "The Newborn" by Constantin Brancusi (1915). The original sculpture can be seen at the Philadelphia Museum of Art.

the broken contour in the image? We come back to this question in the next section; suffice to say at this stage that the image impinging on our retinas should be complemented by additional information if we want to conclude that the object does contain an elliptical cut.

The Bayesian framework offers the optimal way to combine multiple sources of information. The likelihood is combined with the prior knowledge thanks to Bayes' theorem, thus producing a posterior distribution. The posterior represents the relative probability of all possible interpretations of the stimulus, taking into account all the information available. When one is interested in the choice of the organism for one particular interpretation, we also need to take into account a decision rule that is applied on the posterior distribution (Fig. 3). More details about the Bayesian approach to visual perception can be found in Mamassian et al. (2002) and Kersten et al. (2004).

Bayesian models of contour grouping

Let us come back to our example of the broken elliptical contour in the picture of Brancusi's sculpture. If we want to infer the closed contour, we need some prior knowledge about contours. This extra knowledge is likely to be similar to what Gestalt psychologists referred to as the principle of good continuation: two nearby line segments in an image can be thought to belong to the same contour in the world if there is a smooth curve of minimal curvature that joins them.

Gestalt principles have often been criticized for being qualitative descriptions of phenomena rather than proper explanations. Recent work in experimental psychology has however allowed us to quantify these principles. For instance, Kubovy and Wagemans (1995) have quantified the principle of proximity with the elegant use of dot lattice stimuli. This kind of experimental approach will allow us to measure the relative strength of Gestalt principles and thus to make quantitative predictions on how these principles can act as prior knowledge in the interpretation of contour shape.

If human observers use a prior assumption that contours are smooth, then we should be able to

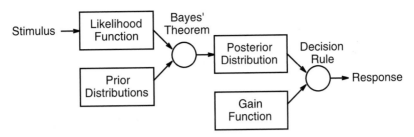

Fig. 3. Flowchart of the Bayesian framework. Reprinted from Mamassian et al. (2002), with permission from MIT Press.

predict when a series of dots are perceived as a single contour or as two contours connected at a corner. Feldman (2001) measured the probability to perceive one smooth contour versus a corner in a variety of conditions where one dot was more or less aligned with the dots on either side (see also Warren et al., 2002). He could account for the relative probability to perceive a corner thanks to prior probabilities on the angle between successive dots along a smooth contour.

The prior in Feldman's study had a general shape (it favored straight contours) and had a degree of freedom adjusted to the data (the belief that most contours are straight). Other studies have attempted to extract the prior probabilities from the statistics of natural scenes. Two groups of researchers have independently found that scenes in our environment do indeed contain a predominance of straight contours. Geisler et al. (2001) extracted the edges of 20 natural scenes and measured how nearby edges varied in orientation. They found a prevalence not only of parallel contours, but also of co-circular contours (edges that would fall on a circle). This latter property is a physical property that could be at the origin of the Gestalt principle of good continuation and the existence of association fields in the human visual cortex (Field et al., 1993). From their statistics, Geisler et al. (2001) extracted a likelihood ratio that was a unit-free estimate that two edges belonged to the same contour and found that this measure was a good descriptor of humans' ability to detect a contour in noise. In a similar vein, Elder and Goldberg (2002) extracted the edges of nine natural scenes and measured statistics on proximity, good continuation and luminance similarity. They found that proximity was by far the most reliable cue to group

elements along a contour and that human observers were in good agreement with a model based on the proximity cue. Finally, one should also note the effort of Howe and Purves (2005) to relate the statistics of oriented edges to contours in the 3D world instead of just the 2D retinal image. They also found a predominance of aligned contours in their analysis.

From contours to 2D shape

We have seen in the previous section how a fragmented contour could be virtually reconnected into a continuous contour. Once the contour has been revealed, its shape can be inferred. Here again, at the level of the 2D shape of the contour, the Bayesian framework has proved useful. One elegant demonstration of the role of prior knowledge on shape perception is the work of Liu et al. (1999) on amodal completion. When the contour of an object is occluded by another object, the missing part of the contour can be interpolated via a process called amodal completion (Kellman and Shipley, 1991). For instance in Fig. 4, the upper parts of two objects are perceptually completed behind the occluder and linked to their respective lower parts. Liu et al. (1999) placed the occluder either in front or behind the object parts thanks to some binocular disparities and found that convex objects were completed more easily than concave ones. They reached this conclusion by placing the upper and lower parts at different depths and found higher depth discrimination thresholds when the object underwent a convex completion.

As we discussed in the introduction, there are a few objects in the world, like leaves, that are

268

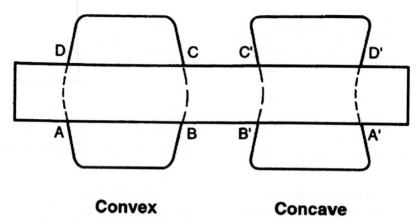

Convex   **Concave**

Fig. 4. Convex and concave amodal completion. Reprinted from Liu et al. (1999), with permission from Elsevier.

basically 2D. There are also quasi-objects that have a shape and that are 2D. The most common example is probably the shadow cast by an object on a remote surface. Interestingly, we are very tolerant of the relationship between the shape of the casting object and the shape of the cast shadow, so much so that we seem insensitive to impossible shadows (Mamassian, 2004). Other 2D quasi-objects are holes in a surface (Casati and Varzi, 1994; Bertamini & Croucher, 2003). The extent to which the shapes of cast shadows and holes are processed the same way as other physical objects is still an empirical question.

From contours to 3D shape

Contours in an image are not only informative about the outline of an object, but also about the 3D shape of the object. Perhaps, the most celebrated demonstration of the 3D shape information available from contours is Attneave's cat shown in Fig. 5 (Attneave, 1954). Attneave argued that most of the information along a contour was concentrated in parts of high curvature. This argument has since received considerable support from both computational and psychophysical studies (Norman et al., 2001; Feldman and Singh, 2005). However, proving that information is concentrated at points of high curvature does not prove that the rest of the contour is uninformative. Koenderink (1984) showed that for a smooth surface, there is a one-to-one relationship between the

Fig. 5. Attneave's cat. Reproduced from Attneave (1954), with permission.

sign of curvature of the occluding contour and the sign of curvature of the surface. More specifically, a convex contour is the signature of a convex surface patch, and a concave contour indicates a hyperbolic (saddle-shaped) surface patch. Moreover, the curvature magnitude of the object can be inferred from the occluding contour when the observer is active (Mamassian and Bülthoff, 1996). Once the shape of the occluding contour is inferred, the 3D shape can be extrapolated thanks to a surface filling-in mechanism (Tse, 2002).

There are also other sources of information about 3D shape than those present on the occluding contour. For instance, parallel structure in a solid (like in marble) will create stripes on the surface of the object and the deformation of the resulting texture will be informative about the shape of the surface (Li and Zaidi, 2004). Lines that appear painted on the surface on an object tend to be perceived as geodesics (contours of minimum curvature) and are therefore also informative about the 3D shape of the object (Knill, 1992). In addition, these contours tend to be interpreted in such a way that the surface that supports them is seen from above (rather than from below), thereby constraining further the 3D shape of the object (Mamassian and Landy, 1998).

Other contours are also informative about the 3D shape of the surface on which these contours appear to be painted. For instance, alternating dark and bright parallel contours tend to be perceived as shaded bevelled patterns on the surface (Mamassian and Goutcher, 2001). These contours are informative about the 3D shape of the surface only if you know the illumination position, namely above the observer. In summary, contours are informative about 3D shape provided a number of assumptions about the illumination position, the viewer position, or the alignment of the surface contours with geometric properties of the object. Without these assumptions, the contours are ambiguous about the 3D shape of the object. The fact that human observers are able to consistently perceive the same 3D shape from these images is consistent with a mechanism that integrates image information with prior knowledge. Bayesian models are precisely models of this type.

Discussion

The state-of-the-art we have presented in this chapter on shape perception can be schematized by the diagram in Fig. 6. Once basic features such as oriented edges are extracted from an image, they can be grouped together to form continuous contours. These contours can then be used to infer the shape of an object in various ways. First, if the object is 2D, the shape of the contour will simply be related to the shape of the object. Second, if the contour is identified as the contour of the object against the background, there are mechanisms that will allow us to fill-in the surface from this occluding contour. Finally, if a contour is near other similar contours, these can be grouped together into a texture pattern, and a 3D shape can then be inferred from this texture. Even though we have drawn uni-directional arrows in this figure, it is clear that feedback mechanisms do influence early representations (Murray et al., 2004).

One critical aspect of this way to understand shape perception is that prior knowledge is brought in at all stages of visual processing. For instance, features are grouped into contours following preferences that contours are straight. Contours are grouped into texture patterns following some principles of similarity. 3D shape is inferred from texture following homogeneity assumptions on the texture. Each stage can be seen as a Bayesian inference process where information

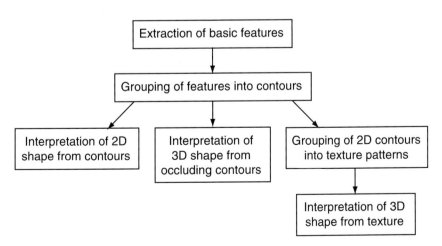

Fig. 6. A schematic diagram of the perception of 2D and 3D shape from an image.

from the earlier stage is combined with the relevant prior knowledge for the current stage. While there are a growing number of studies on the individual stages, little work has been done to integrate these stages into a hierarchical Bayesian model of shape perception.

We conclude this chapter with some outstanding questions. First, what are the Bayesian aspects that are equivalent to generic principles such as "simplicity" or "smoothness"? We have argued that such principles would intuitively translate as priors in a Bayesian model, but the precise nature of these priors and their origin remains a serious empirical question. Theoretical works showing the equivalence between simplicity and some statistical quantity are very important to explore this question (Chater, 1996).

A second issue is whether priors necessarily reflect regularities of the natural world. Some works within the Bayesian framework do indeed take this assumption as a premise, but some believe that this is not a requirement. For instance, we found that human observers behaved as if their assumption on the light source position was biased to the left (Mamassian and Goutcher, 2001), an assumption that is difficult to relate to statistics of the natural world. Future work should investigate the origin of such priors.

Finally, we may wonder what are the neural bases of Bayesian processing? The empirical determination of prior assumptions used by human observers when they perceive shape opens the door to fascinating investigations on the way these priors are implemented.

References

Attneave, F. (1954) Some informational aspects of visual perception. Psychol. Rev., 61: 184–193.

Bertamini, M. and Croucher, C.J. (2003) The shape of holes. Cognition, 87: 33–54.

Casati, R. and Varzi, A.C. (1994) Holes and Other Superficialities. MIT Press, Cambridge, MA.

Chater, N. (1996) Reconciling simplicity and likelihood principles in perceptual organization. Psychol. Rev., 103: 566–581.

Elder, J.H. and Goldberg, R.M. (2002) Ecological statistics of Gestalt laws for the perceptual organization of contours. J. Vision, 2: 324–353.

Feldman, J. (2001) Bayesian contour integration. Perception Psychophys., 63: 1171–1182.

Feldman, J. and Singh, M. (2005) Information along contours and object boundaries. Psychol. Rev., 112: 243–252.

Field, D.J., Hayes, A. and Hess, R.F. (1993) Contour integration by the human visual system: evidence for a local 'association field'. Vision Res., 33: 173–193.

Geisler, W.S., Perry, J.S., Super, B.J. and Gallogly, D.P. (2001) Edge co-occurrence in natural images predicts contour grouping performance. Vision Res., 41: 711–724.

Howe, C.Q. and Purves, D. (2005) Natural-scene geometry predicts the perception of angles and line orientation. Proc. Nat. Acad. Sci., USA, 102: 1228–1233.

Kellman, P.J. and Shipley, T.F. (1991) A theory of visual interpolation in object perception. Cogn Psychol., 23: 141–221.

Kersten, D., Mamassian, P. and Yuille, A. (2004) Object perception as Bayesian inference. Annu. Rev. Psychol., 55: 271–304.

Knill, D.C. (1992) Perception of surface contours and surface shape: from computation to psychophysics. J. Opt. Soc. Am. A, 9: 1449–1464.

Koenderink, J.J. (1984) What does the occluding contour tell us about solid shape? Perception, 13: 321–330.

Kubovy, M. and Wagemans, J. (1995) Grouping by proximity and multistability in dot lattices: a quantitative Gestalt theory. Psychol. Sci., 6: 225–234.

Li, A. and Zaidi, Q. (2004) Three-dimensional shape from non-homogeneous textures: carved and stretched surfaces. J. Vis., 4: 860–878.

Liu, Z., Jacobs, D. and Basri, R. (1999) The role of convexity in perceptual completion: beyond good continuation. Vision Res., 39: 4244–4257.

Mamassian, P. (2004) Impossible shadows and the shadow correspondence problem. Perception, 33: 1279–1290.

Mamassian, P. and Bülthoff, H. H. (1996) Active kinetic depth effect. Technical Report No. 27. Max-Planck-Institut für biologische Kybernetik, Tübingen, Germany.

Mamassian, P. and Goutcher, R. (2001) Prior knowledge on the illumination position. Cognition, 81: B1–B9.

Mamassian, P. and Landy, M.S. (1998) Observer biases in the 3D interpretation of line drawings. Vision Res., 38: 2817–2832.

Mamassian, P., Landy, M.S. and Maloney, L.T. (2002) Bayesian modelling of visual perception. In: Rao, R., Olshausen, B. and Lewicki, M. (Eds.), Probabilistic Models of the Brain: Perception and Neural Function. MIT Press, Cambridge, MA, pp. 13–36.

Murray, S.O., Schrater, P. and Kersten, D. (2004) Perceptual grouping and the interactions between visual cortical areas. Neural Netw., 17: 695–705.

Norman, J.F., Phillips, F. and Ross, H.E. (2001) Information concentration along the boundary contours of naturally shaped solid objects. Perception, 30: 1285–1294.

Tse, P.U. (2002) A contour propagation approach to surface filling-in and volume formation. Psychol. Rev., 109: 91–115.

Warren, P.E., Maloney, L.T. and Landy, M.S. (2002) Interpolating sampled contours in 3D: analyses of variability and bias. Vision Res., 42: 2431–2446.

Martinez-Conde, Macknik, Martinez, Alonso & Tse (Eds.)
Progress in Brain Research, Vol. 154
ISSN 0079-6123

CHAPTER 15

Contour discontinuities subserve two types of form analysis that underlie motion processing

Peter Ulric Tse* and Gideon P. Caplovitz

H.B. 6207, Moore Hall, Department of Psychological and Brain Sciences, Dartmouth College, Hanover, NH 03755, USA

Abstract: Form analysis subserves motion processing in at least two ways: first, in terms of figural segmentation dedicated to solving the problem of figure-to-figure matching over time, and second, in terms of defining trackable features whose unambiguous motion signals can be generalized to ambiguously moving portions of an object. The former is a primarily ventral process involving the lateral occipital complex and also retinotopic areas such as V2 and V4, and the latter is a dorsal process involving V3A. Contour discontinuities, such as corners, deep concavities, maxima of positive curvature, junctions, and terminators, play a central role in both types of form analysis. Transformational apparent motion will be discussed in the context of figural segmentation and matching, and rotational motion in the context of trackable features. In both cases the analysis of form must proceed in parallel with the analysis of motion, in order to constrain the ongoing analysis of motion.

Keywords: Motion; Contour discontinuities; Form; Vision.

The importance of contour discontinuities

Contour discontinuities are more informative about object shape and motion than other portions of contour. They function in part as invariants in the image that permit shape and motion to be extracted and constructed from information available in the visual array. One of the first to realize the importance of contour discontinuities was Attneave (1954). He realized that the visual system is more sensitive to the presence of local maxima of positive or negative contour curvature (CC) than it is to inflection points. Attneave argued on the basis of information theory that contour discontinuities such as highly curved portions of a contour provide more information than straight sections of contour because the location of neighboring points is more unpredictable

around contour discontinuities than straight portions of contour. This insight was extended by Biederman (1987), who showed that objects are unrecognizable if contour discontinuities are removed, but are easily recognizable if an equal amount of contour is removed that spares contour discontinuities. Neuropsychological work has also demonstrated the importance of contour discontinuities. Humphreys et al. (1994) found that Balint's patients, who can only perceive one object at a time, tended to perceive the version of a square or diamond defined by corner segments of contour over those defined by the straight sections of contour between corners.

Because contour discontinuities are so informative, it is not surprising that classes of detectors have evolved to detect them. Fast search rates in the visual search paradigm have generally been regarded as evidence that the visual system is predisposed to rapidly detect a given image feature, event, or configuration. It is well known that

*Corresponding author. Tel.: +1-603-646-4014;
E-mail: Peter.U.Tse@dartmouth.edu

DOI: 10.1016/S0079-6123(06)54015-4
271

tangent and curvature "pop out" (i.e., are rapidly detected, as determined by flat search slopes as a function of distractor set size) when the distractors are straight lines (Treisman and Gormican, 1988; Wolfe et al., 1992; see also Zucker et al., 1989). Kristjansson and Tse (2001) showed that CC discontinuities pop out among continuous curvature distractors. In contrast, straight lines do not pop out among curves, and continuous curves do not pop out among discontinuous ones. The search asymmetry for curved targets among straight-line distractors attests to the visual system's bias to preferentially detect information about curvature rapidly. Similar results using other contour discontinuities have led to the conclusion that contour discontinuities in general, including tangent discontinuities, junctions, corners, deep concavities, terminators, curvature, and even CC discontinuities, are basic features to which the visual system is highly sensitive.

Contour discontinuities and figural matching

Here we will focus on transformational apparent motion (TAM), because it is a phenomenon that makes apparent the importance of form processing in determining perceived motion. When an object discretely and instantaneously changes its shape, observers typically do not perceive the abrupt transition between shapes that in fact occurs. Rather, a continuous shape change is perceived. Although TAM is a faulty construction of the visual system, it is not arbitrary. From the many possible shape changes that could have been inferred, usually just one is perceived because only one is consistent with the shape-based rules that the visual system uses to (1) segment figures from one another within a scene, and (2) match figures to themselves across successive scenes. TAM requires an interaction between neuronal circuits that process form relationships with circuits that compute motion trajectories. In particular, this form–motion interaction must happen before TAM is perceived, because the direction of perceived motion is dictated by form relationships among figures in successive images. Overall, the occurrence of TAM is consistent with a view of motion processing

comprised of two subsystems: one low level, driven by the processing of motion-energy extracted from changes in various features, and the other high level, driven by global contour-based form analysis and the attentional tracking of figures.

Historical background to TAM

In recent decades, motion perception has predominantly been studied from the perspective of neurophysiology, occasioned by the dramatic discovery of motion-sensitive neurons (e.g., Barlow and Levick, 1965). Psychophysical and physiological observations support the notion that motion perception arises from the responses of populations of neurons tuned to various spatiotemporal offsets. Recent formal models of motion perception emphasize "motion-energy," which could be captured by the receptive fields of such neurons (e.g., Hassenstein and Reichardt, 1956; Reichardt, 1957, 1961; Barlow and Levick, 1965; Adelson and Bergen, 1985; van Santen and Sperling, 1985). According to such motion-energy detection models, a given motion-sensitive neural "comparator" signals motion when it registers some feature, most commonly relative luminance, at one location, and then later registers that feature at a second location. A simplified low-level motion unit has two receptive fields with a particular spatial offset, and their activity is compared upon a given temporal offset (e.g., a "Reichardt detector"; Hassenstein and Reichardt, 1956). If the two receptive regions become successively active in the preferred order, the unit responds, signaling that something has moved in a direction and speed consistent with the spatiotemporal offset that the unit is wired up to detect. Such "low-level" motion processing does not segment the image, but rather decomposes the image pattern into components of motion-energy, allowing separate responses of many different comparators to any one location at the same time. This property of registering the motion-energy of multiple, superimposed components at a location, rather than treating the pattern at that location as a single entity, is the critical difference between low-level and high-level motion-processing systems.

High-level motion processing, in contrast, has been described as a process of identifying forms and then matching those forms across time intervals (Anstis, 1980; Braddick, 1980). Grouping and segmentation procedures would treat a pattern at a single location as a single entity that is not decomposable into Fourier components. The high-level motion-processing system would not detect spatiotemporal displacements of luminance or any other image feature per se. Rather, the high-level system would track a single thing as it moved, even when the features defining that thing changed, and, indeed, would do so even if motion-energy implied motion in a direction opposite that of the tracked figure.

Motivated by early speculation about the possibility of a form-sensitive motion-processing stream, researchers went in search of form–motion interactions but were unable to demonstrate strong effects of form on motion. What studies using translational apparent motion have generally shown is that few of the elements of scene or form analysis have much impact on matching in apparent motion (see e.g., Kolers and Pomerantz, 1971; Kolers and von Grünau, 1976; Navon, 1976; Burt and Sperling, 1981; Baro and Levinson, 1988; Cavanagh et al., 1989; Victor and Conte, 1990; Dawson, 1991). Indeed, low spatial frequencies appear to count for more than shape (Ramachandran et al., 1983), and the most potent factor is found to be spatiotemporal proximity. There are some examples of object properties — shape, color, depth — playing a minor role in matches (e.g., Green, 1986a, b, 1989). But the effect of these factors is only revealed when the much stronger factor of proximity has been carefully controlled for. This tendency to match elements on the basis of proximity in successive displays has been called "the nearest neighbor principle" (Ullman, 1979). Relative to proximity factors, color and form factors seem insignificant in determining matches between scenes using translational apparent motion as a probe. The conclusion that form played little role in matching elements in apparent motion displays seemed to undermine the defining characteristic of high-level motion processing, namely that form extraction precedes motion perception. Results were not that

different from what would have been predicted on the basis of low-level mechanisms.

If versions of the standard low-level motion units were available with very large receptive fields, they could respond to isolated items of two frames even over large spatial separations. However, they could not resolve the details of each item's shape. They would signal motion of the nearest neighbor pairs and these simple receptive fields would ignore the form and color details of the items. There are examples of apparent motion that deviate from the expected properties of low-level detectors, such as matching opposite-contrast items or matching across eyes, but these examples do not rule out contributions from low-level units in most apparent motion displays. They only suggest that low-level mechanisms cannot be solely responsible for all motion phenomena.

The seeming indifference of motion processing to form information in apparent motion meshed well with neurophysiological findings that the low-spatial, high-temporal resolution magnocellular system, specialized for the processing of motion and spatial relationships, responds poorly to contours and boundaries defined only by color contrast (e.g., Ramachandran and Gregory, 1978; Livingstone and Hubel, 1987), whereas the high-spatial, low-temporal resolution parvocellular/interblob system, specialized for form analysis responds poorly to motion (e.g., Livingstone and Hubel, 1987; Schiller, 1991).

Because both psychophysical and neurophysiological evidence seemed to indicate that motion analysis was not dependent on form-based matching, many in the visual neuroscience community reached the conclusion by the early 1990s that the presumed stage of form extraction was either not present or not significant in high-level motion processing. Many researchers in the recent past, at least by implication, therefore attempted to reduce the high-level motion-processing stream to a variant of the low-level motion-processing stream. For example, instances of motion phenomena that were not easily reducible to luminance-defined motion-energy processing were deemed to be processed by detectors that had additional properties, such as rectification prior to the stage where motion-energy would be detected by more or less traditional motion-energy detectors. A common claim was that most types of image motion could be detected and analyzed using elaborations of motion-energy detectors that were not explicitly tuned to form or contour relationships. (See, e.g., Braddick, 1974; Marr and Ullman, 1981; Adelson and Movshon, 1982; Watson and Ahumada, 1983; van Santen and Sperling, 1984; Adelson and Bergen, 1985; Burr et al., 1986; Chubb and Sperling, 1988; Cavanagh and Mather, 1989; Lu and Sperling, 1995a, 2001.) The emergence of TAM in the past 10 years has, however, changed this debate by offering a reassessment of the role that form analysis plays in high-level motion processing, and by calling into question recent attempts to reduce all motion processing to the analysis of various orders of motion-energy processing.

Transformational apparent motion

Over the past several years authors have investigated a new type of motion phenomenon that occurs when two spatially overlapping shapes presented discretely in time appear to transform smoothly and illusorily from the first shape into the second as if the sequence were animated using a succession of intermediate shapes, as shown in Fig. 1. The illusion of apparently smooth and continuous shape change was termed TAM (Tse et al., 1998) in order to contrast it with the translational apparent motion first discovered by Gestalt psychologists (e.g., Wertheimer, 1912; Kenkel, 1913). TAM occurs when an object is abruptly flashed on next to an abutting static object, causing the new object to appear to smoothly extend from the static object. Upon offset, it appears to smoothly retract into the static object. Abutment or near abutment appears to be a necessary condition for the occurrence of TAM. A precedent to TAM was first described by Kanizsa (1951, 1971) and termed "polarized gamma motion." This phenomenon was rediscovered in a more compelling form by Hikosaka et al. (1991, 1993a, b). They showed that when a horizontal bar is presented shortly after a cue, the bar appears to shoot away from the cue. Calling this phenomenon "illusory line motion," they hypothesized that the

effect was due to the formation of an attentional gradient around the cue. Because an attentional gradient would fall off with distance from the cue, and because it has been shown that attention increases the speed of stimulus detection (Titchener, 1908; Sternberg and Knoll, 1973; Stelmach and Herdman, 1991, 1994), Hikosaka et al. hypothesized that illusory line motion occurs because of the asynchronous arrival of visual input to a motion detector such as human area hMT+. Shortly after, other authors (e.g., von Grünau and Faubert, 1994; Faubert and von Grünau, 1995) discovered low-level contributions to this phenomenon that could not easily be attributed to a gradient of attention. For example, a red line flashed on all at once between two existing spots, one red and one green, would always appear to move away from the same color spot, regardless of how attention was allocated. Soon after this, other authors (Downing and Treisman, 1995, 1997; Tse and Cavanagh, 1995; Tse et al., 1996, 1998) rejected the attentional gradient account and suggested that this phenomenon was actually a type of apparent motion where the initial shape appears to smoothly transform into the second shape. These authors have shown that TAM arises even when attention is paid to the opposite end of the cued location, implying that there must be other contributors to the motion percept than a gradient of attention. The attentional gradient account has now been disproved in direct psychophysical tests (Hsieh et al., 2005), meaning that an alternative explanation of TAM is required.

Tse and colleagues (Tse et al., 1998; Tse and Logothetis, 2002) have shown that figural parsing plays an essential role in determining the perceived direction of TAM. Figural parsing involves a comparison of contour and surface relationships among successive scenes. In particular, the visual system appears to infer which figures at time 2 are derived from which figures at time 1 on the basis of contour and surface relationships. If a given figure has a different shape at times 1 and 2, a continuous deformation between those shapes is constructed and perceived, presumably because the new figure is inferred to be a change in the shape of an already existing figure. The key point here is that form processing must temporally precede or at the

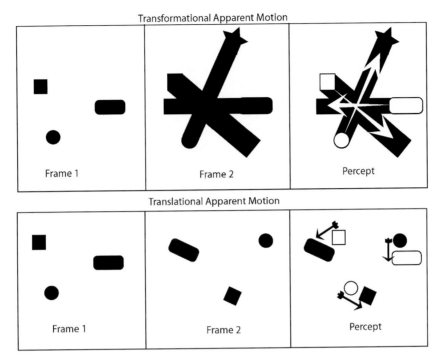

Fig. 1. Transformational apparent motion (TAM) (top row) is contrasted with translational apparent motion (bottom row). When the figures in frame 1 (leftmost column) are replaced instantaneously with the figures in frame 2 (middle column), then a percept (rightmost column) of TAM (i.e., smooth deformations) results when figures spatiotemporally abut, but of rigid translations results when they do not abut. Note that even though frame 2 may consist of a single contiguous shape, as in the top middle frame, multiple motions can be perceived to occur simultaneously, as depicted by the arrows on the top right.

very least accompany the motion processing that subserves the percept of TAM. This is because the motion that is seen depends on how figures at time 1 have been matched to figures at time 2. Determining that something has moved requires that the "something" be identified in the first instant and then paired off with what is presumed to be the same thing in the next instant. The first component of this processing is to identify candidates at both instants, and the second is to match them. We can call the first component a parsing step and the second component a matching step. By "parsing" is meant the spatial isolation and specification of individual objects, including any necessary segmentation away from overlapping objects or background elements of the image, as well as the discounting of noise.

A representative example of TAM is shown in Fig. 1. TAM can occur behind partially occluding objects and over illusory surfaces that have not been cued to attract attention to one location over

another, and TAM can appear to emerge out of the picture plane, implying that it can take place over 3D representations after a stage of modal and amodal completion (Tse and Logothetis, 2002). Whereas translational apparent motion is generally insensitive to shape and color constraints so long as the two stimuli presented remain within the optimal range of spatiotemporal offsets (see, e.g., Cavanagh et al., 1976), TAM is sensitive to such shape and color constraints, because these can be used by the parser to disambiguate figures in scenes that can only be ambiguously parsed otherwise. Moreover, TAM can occur over figures that have been defined by changes in successive subregions of a field of random dots (Hsieh and Tse, 2006).

TAM reveals that parsed figures are matched to parsed figures from scene to scene in high-level motion processing, in some cases violating nearest-neighbor principles. In TAM, new image data appearing closer to one figure than another, can still get matched as comprising a shape change in the more distant figure. Tse and colleagues (Tse et al., 1998; Tse and Logothetis, 2002) showed that a set of parsing and matching principles based on analyzing contour relationships among abutting and successive figures aids in determining figural identity within and between scenes holds for TAM. These principles approximately reduce to the nearest-neighbor principle for cases of translational apparent motion. Their behavioral research to date demonstrates that a stage of figural parsing and matching precedes the perception of motion in TAM displays. This stage of segmentation takes place on the basis of Gestalt-like grouping principles that process the spatiotemporal relationships among figures. Figure formation and matching appear to primarily take place on the basis of good contour continuity (indicating figural identity between scenes) and contour discontinuity (indicating possible figural difference). In particular, the high-level motion-processing system receives segmented and completed contours, and perhaps surfaces as input. The percept of motion is therefore in part the perception of how figures have translated and transformed their shape.

The importance of figural parsing could not have been revealed by research into translational apparent motion, because in translational apparent motion displays, the parsing of each successive scene is generally given unambiguously in the sense that figures are spatially distinct. In translational apparent motion displays, a figure seems to disappear at one location and reappear at a different, nonoverlapping location some time later. The problem in translational apparent motion experiments has generally been the match between figures, not the parsing of figures. However, in TAM displays, there is usually ambiguity in determining which figure in one scene has become which figure in the following scene, because of the spatiotemporal overlap of succeeding figures. That is, in the case of the apparent shape transformations of figures, new image data generally appear without the disappearance of the figure (s) that existed in the previous scene (although brief "figureless" intervals are tolerated, as in translational apparent motion). In TAM, the parsing problem has to be solved before the problem of matching or

maintaining figural identity across successive scenes can be attempted. Since the image itself is not parsed, the visual system faces a problem of ambiguity in its efforts to correctly parse the image so as to coincide with the actual segmentation of the world into independent, but abutting or overlapping, objects. Since many possible parsings are consistent with a single image, the visual system has evolved default processes for solving the parsing problem, realized, in part by contour-based completion and segmentation of figures. This contour-based completion and segmentation is driven in large part by an analysis of contour continuity and contour discontinuity relationships among successive, abutting, or overlapping figures.

The neuronal correlates of TAM

Recent functional magnetic resonance imaging (fMRI) work examined the neuronal correlates of TAM (Tse, 2006). The blood oxygen level dependent (BOLD) signal in several cortical areas was found to increase during the perception of TAM relative to the perception of control stimuli in which TAM was not observed. In particular, a region of interest analysis revealed that the BOLD signal in areas V1, V2, V3, V4, V3A/B, hMT+ (the human homologue of macaque MT) and LOC (lateral occipital complex) increased during the perception of TAM relative to the control. An additional, whole brain analysis revealed an additional area in the posterior fusiform gyrus that was also more active during the percept of TAM than during control. We can therefore conclude that the neural basis of TAM resides in these and perhaps other areas. Because TAM is thought to invoke high-level motion-processing mechanisms, it is likely that high-level motion processing occurs in at least these areas.

The LOC has been implicated in the processing of form (Malach et al., 1995; Kanwisher et al., 1996; Grill-Spector et al., 2001; Haxby et al., 2001). Evidence is emerging that LOC processes global 3D object shape, rather than local 2D shape features (Grill-Spector et al., 1998, 1999; Malach et al., 1998; Mendola et al., 1999; Gilaie-Dotan et al., 2001; Moore and Engel, 2001; Avidan et al.,

2002; Kourtzi and Kanwisher, 2000b, 2001; Kourtzi et al., 2003), and may even mediate aspects of object recognition (Grill-Spector et al., 2000). Indeed TAM itself occurs over 3D representations of form (Tse and Logothetis, 2002). The fMRI data concerning suggest that the LOC plays a greater role in TAM than the control. While both the TAM condition and the control condition involve global forms, the TAM condition may place a greater workload upon the LOC because the global form or global figural relationships that are believed to be computed there are presumably fed into motion-processing areas in the TAM case, whereas this output is not required for the control case. While fMRI data cannot specify the temporal dynamics of interactions among these areas, a reasonable model would place contour-based form analysis in the LOC and in retinotopic areas. The results of parsing would then be sent to hMT+ where motion trajectories would be computed in light of both figural matching and motion-energy cues.

These findings contrast with recent fMRI findings that used translational apparent motion as a probe. Liu et al. (2004) found no difference between apparent motion and flicker conditions in any retinotopic area, although they did see greater activation for apparent motion than flicker in hMT+. They conclude from this that there is no evidence for the filling-in of features along the path of perceived motion in early retinotopic areas. Other fMRI studies have found greater activation for apparent motion than flicker in hMT+, but not in V1 (Goebel et al., 1998; Muckli et al., 2002). While both hMT+ and V1 contain motion sensitive cells, the types of motion to which these cells respond need not be the same, perhaps accounting for this difference. For example, Mikami et al. (1986) found that neurons in MT but not in V1 responded to long-range apparent motion. Since the present data reveal differences between TAM and flicker in early retinotopic areas, it is possible that TAM, unlike translational apparent motion, does involve filling-in of features in early retinotopic areas.

These findings suggest that hMT+ should be thought of as part of a form/motion-processing stream of analysis, rather than as an area just dedicated to the processing of motion, and in particular, just motion-energy. Similarly, these findings suggest that the LOC should be thought of as part of a form/motion-processing stream of analysis, rather than as an area just dedicated to form processing. Indeed, a number of recent papers have come to the conclusion that there is potential anatomical and functional overlap between hMT+ and the LOC (Ferber et al., 2003; Kourtzi et al., 2003; Murray et al., 2003; Zhuo et al., 2003; Liu et al., 2004) supported by behavioral data as well (Stone, 1999; Liu and Cooper, 2003). Because the form processing that underlies TAM must be extremely rapid, it would appear that LOC, hMT+, and other areas involved in form/motion processing operate in conjunction to solve the problem of what went where in TAM before TAM is perceived.

The unification of hMT+ and LOC into interacting motion–from–form analyzers blurs the distinction between the "what" vs. "where" pathways. Traditionally it has been maintained that form processing is a ventral or "what" stream process (Ungerleider and Mishkin, 1982; Goodale and Milner, 1992), whereas motion processing is a dorsal or "where/whence" stream process. While the ventral/dorsal distinction is a useful heuristic for understanding gross-scale information-processing architecture, the current data imply that ventral and dorsal processing are not independent, and interact in a bottom-up manner in the few hundred milliseconds between sensory input and the conscious perception of TAM.

Implications for models of TAM

Baloch and Grossberg (1997) modeled TAM to involve three interacting subprocesses: (1) a boundary completion process (V1 → interstripe V2 → V4); (2) a surface filling-in process (blob V1 → thin stripe V2 → V4); and (3) a long-range apparent motion process (V1 → MT → MST), adding a link between V2 and MT in their model to allow the motion-processing stream to track emerging boundaries and filled-in surface colors. The model is foremost a bottom-up account of boundary completion, where existing boundaries enhance growth of neighboring collinear boundaries and inhibit growth of neighboring dissimilarly oriented boundaries. Color is then filled in from existing color regions to new color boundaries. The data from our recent fMRI experiments cannot distinguish BOLD responses in V2 due to thin stripe vs. thick stripe activity, and can therefore not precisely test the model. However, the present fMRI data implicate all the areas forecast by this model. In addition to areas of activation predicted by this theory, V3 V, V3A/B, and LOC were found to be more responsive to TAM than the control, suggesting that the model is incorrect, or at best incomplete. Another difficulty for the model emerges from the local nature of the boundary interactions they describe. TAM is influenced by global configural relationships among stimuli. For example, in the stimuli in our fMRI experiments (Fig. 2), the direction of perceived motion in the central bar is influenced by the location of the square many degrees away. This means that the mechanism underlying TAM must be sensitive to global configural relationships within and between the two images. Purely local lateral excitation and inhibition are not sufficient to account for TAM (this applies also to Zanker, 1997).

TAM highlights a central problem in visual neuroscience: how does local feature information become integrated into a global representation of spatiotemporal figural relationships, which in turn appears to influence how local features are interpreted? TAM should prove a useful probe in future studies that attempt to answer this difficult question first raised by the Gestalt psychologists nearly a century ago (e.g., Wertheimer, 1912; Kenkel, 1913). TAM requires contour integration over large scales and a global analysis of contour relationships in order to determine the correct correspondence of figures over time, such that TAM can take place over corresponding figures (see Fig. 1). fMRI studies that have examined how contours are integrated into global shapes reveal that contour integration activates V1 and V2 in humans and monkeys but produces strongest activation in lateral occipital areas (Altmann et al., 2003; Kourtzi et al., 2003). That striate cortex shows BOLD activation for a task that requires global contour integration is consistent with recent work in neurophysiology. However, this is perhaps surprising, given that to date

V2 is the first area in the visual hierarchy where illusory contours have been shown to have a direct effect using single-unit recording (von der Heydt et al., 1984). While processing in V1 has traditionally (Hubel and Wiesel, 1968) been thought to be limited to the processing of local features, recent evidence has implicated early visual areas such as V1 and V2 in the processing of global shape (Allman et al., 1985; Gilbert, 1992, 1998; see also Fitzpatrick, 2000; Lamme et al., 1998, for reviews). In light of these findings, it is possible that the global form processing that underlies TAM begins as early as V1. The recent fMRI data examining the neural basis of TAM (Tse, 2006) indeed does find greater V1 activation to TAM than a no-TAM control. However, this cannot be used to establish, for example, that global processing takes place in V1. These data are also not sufficient to determine whether the activity seen in V1 arises because of bottom-up or top-down activation. The BOLD activity seen is consistent with both possibilities and indeed may arise from both mechanisms.

It is widely recognized that Fourier decomposition of motion-energy and pooling of motion signals at a single location are not capable of solving all motion-processing problems (e.g., Born and Bradley, 2005). More than one moving object can appear in a single region in space, and, on the retina, motions from multiple depth planes can project to the same retinal location, as occurs in cases of transparency. Pooling motions across depths or objects would create spurious motion signals, since the true motions can be independent. Thus, parsing mechanisms must exist that segregate motion signals from distinct objects or depth layers from each other for separate motion trajectory computations (Hildreth et al., 1995; Nowlan and Sejnowski, 1995). One possibility is that the parsing takes place within MT (hMT +) itself. There is evidence that motion-processing cells may contribute to such parsing procedures in that MT cells are highly tuned to depth and the response suppression that usually occurs with nonpreferred motion is attenuated when nonpreferred motion lies on a different depth plane from the preferred motion (Bradley et al., 1995). Moreover, whether overlapping sine waves are manipulated to look like a plaid or two overlapping transparent layers

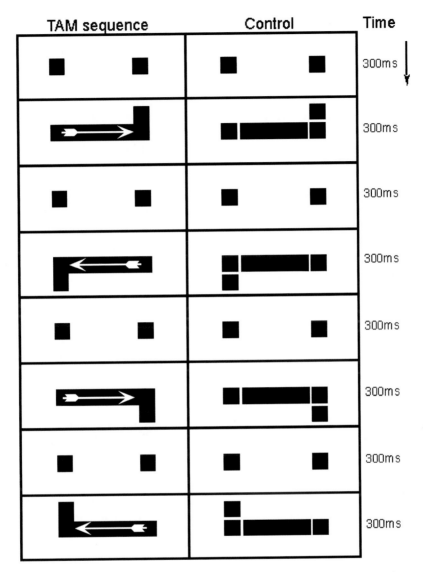

Fig. 2. Two conditions were tested, one where TAM was seen, and the other, where no TAM or other motion was seen. In both cases, the stimuli consisted of identically sized white squares (1.68° × 1.68°) and bars (5.40° × 1.68°) on a black background. In the TAM case, the bars and squares abutted. In the control case they were separated by a small gap (0.32°). The direction of perceived motion is indicated with black arrows (no arrows were actually shown to observers). A small fixation point (0.2° × 0.2°, not shown) was located just below (0.30°) the lower contour of the bar.

by changing the luminance of x-junctions effects whether individual MT cells respond with pattern- or component-like directional tuning (Stoner and Albright, 1992, 1996). The typical motion opponency observed in MT cells may play a role in segregating figures moving relative to a moving background. However, the present fMRI data imply that it is more likely that at least some aspects

of parsing, particularly those involved in segmenting figures on the basis of global contour and other form relationships, do not take place in hMT +. It is more likely that areas specialized in such form processing, such as the LOC, carry out this type of global figural segmentation and then constrain motion processing in hMT + to be consistent with figural identity matches over time.

Finally, the steps of parsing figures in space and matching figures in time are not separable. While it is conceivable that form analysis would operate on each image independently, followed by an independent stage of matching based on form correspondence among elements in successive images, form analysis and correspondence matching are not independent or serial processes. Because the form interpretation reached for a given figure takes into account the form of figures in the previous scene, form analysis and matching are part and parcel of the same spatiotemporal process. Matching occurs because figures are defined spatiotemporally, rather than just spatially in each successive scene. In other words, matching is subsumed by a figure formation process that operates over a certain range of spatiotemporal extents, and what gets matched across scenes is the corresponding 3D figures. Thus, form as processed by the visual system, is spatiotemporal rather than just spatial. Therefore, the existing dichotomy between (spatial) form or figure analysis and the (temporal) matching of figures across images is a misleading one. The first proponent of an entirely spatiotemporal analysis of form and motion may have been Gibson (1979). Recent work emphasizing the importance of spatiotemporal form and motion processing can be found in Gepshtein and Kubovy (2000) and Wallis and Bülthoff (2001).

Contour discontinuities as trackable features

So far we have been emphasizing that contour cues, in particular cues associated with contour continuity and discontinuity, are essential to figural segmentation, matching, and tracking, and that this stage of form analysis is central to high-level motion processing. In this section we will discuss how contour cues such as regions of high curvature and curvature discontinuities can serve as trackable features (TFs), disambiguating the aperture problem as it arises in the processing of rotational motion.

Motion perception is beset with the problem that motion signals are inherently ambiguous. There are many possible motions in the world that could have given rise to any particular motion that could be measured at the level of the retina or later. A key

cause of this ambiguity is the so-called "aperture problem," which arises because receptive fields in early stages of the visual processing hierarchy are small. Because they are small, they can only recover the component of motion perpendicular to the orientation of a contour. At the heart of the problem is the fact that an infinite number of 3D velocity fields can generate the same 2D retinal sequence. The local motion information at any point along a contour is consistent with an infinite number of possible motions that all lie on a "constraint line" in velocity space (Adelson and Movshon, 1982) for the 2D case. The problem of interpreting this many-to-one mapping is commonly termed the "aperture problem" (Fennema and Thompson, 1979; Adelson and Movshon, 1982; Marr, 1982; Nakayama and Silverman, 1988).

Explaining how the aperture problem is solved is perhaps the most basic challenge that must be met by any model of motion perception. While there are several theoretical solutions to the aperture problem that account for many aspects of motion perception, no single general theory has yet emerged that can explain how the visual system actually processes motion in every instance. Many authors have argued that the aperture problem can be solved by integrating component motion signals along the contour (Bonnet, 1981; Burt and Sperling, 1981; Adelson and Movshon, 1982; Watson and Ahumada, 1985). These models are based on the assertion that ambiguous motion signals can, via integration, be disambiguated. Several models that provide reasonable solutions to the aperture problem for the case of translational motion, such as "intersection of constraints" and "vector summation" models, fail to provide unique solutions in the case of rotational motion.

An account that does provide a solution to the aperture problem in the case of both translational and rotational motion is that based upon the tracking of TFs (Ullman, 1979). TFs disambiguate ambiguous component motion signals that arise along portions of contour distant from TFs (Ullman, 1979) because these locations along a contour such as corners, terminators, and junctions do not move ambiguously when they are intrinsically part of the moving object (Shimojo et al., 1989; i.e., terminators that arise from endpoints belonging to the moving line vs. those arising from points where the moving line is occluded). Recent work has shown how such terminator motions influence processes such as amodal completion and global integration of local motion signals (Lorenceau and Shiffrar, 1992; Shiffrar et al., 1995).

Relying solely on such form features, however, creates new problems that are in some ways as problematic as the aperture problem they are meant to solve. For one, corners, terminators, and junctions can arise in the image for reasons of occlusion that have nothing to do with the motion of the stimulus. These "extrinsic" form features give rise to spurious motion signals that can lead to incorrect conclusions about what motion is actually taking place in the world (Shimojo et al., 1989). The human brain appears to get around the ambiguities of motion stimuli by having multiple motion systems that process different characteristics of the moving stimulus, namely,

one based upon motion-energy processing and another that processes form cues.

Contour curvature is a vital cue for the analysis of both form and motion. We recently conducted a series of fMRI experiments (Caplovitz and Tse, 2006) designed to isolate the neural circuitry underlying the processing of CC as a TF for the perception of motion in general and rotational motion in particular. Using stimuli, called bumps (Tse and Albert, 1998; Kristjansson and Tse, 2001), in each of the experiments the degree of CC was parametrically modulated across stimulus group, as shown in Fig. 3. Each of the three experiments controlled for a different aspect of the visual stimulus, and together isolated CC as the sole form cue, which varied across stimulus condition. A region of interest analysis was applied to the data from each experiment identify brain areas in which the BOLD signal was systematically modulated as a function of CC. We found that the

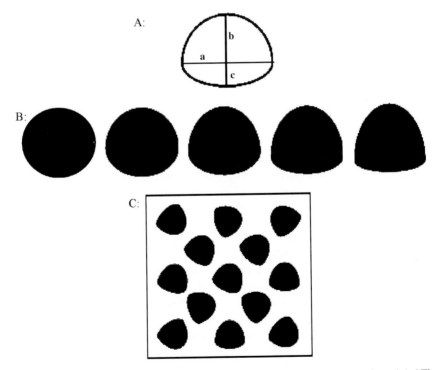

Fig. 3. (A) The bump stimulus is constructed by combining two half-ellipses along their common major axis 'a.' The degree of contour curvature and severity of contour curvature discontinuities can be controlled for by carefully selecting the relative sizes of the two minor axes 'b' and 'c' while still preserving the overall area of the stimulus. (B) In each of the fMRI experiments contour curvature was parametrically varied across each of the five stimulus groups shown here. (C) Stimuli as presented in the scanner: in each stimulus block, 13 rotating bumps were presented continually for 20 s.

BOLD signal in retinotopic area V3A varied parametrically with the degree of CC. The convergent results of the three fMRI experiments ruled out the possibility that these modulations resulted from changes in the area of the stimuli, the velocity with which contour elements were translating, and perceived angular velocity. We concluded that neurons within V3A process continuously moving CC as a TF, and this form-based feature is used to determine the speed and direction of motion of the objects as a whole. In this context, it is suggested that V3A contains neural populations that process form, not to solve the "ventral problem" of determining object shape, but in order to solve the "dorsal problem" of what is going where.

V3A is an area that several studies have found to be motion selective (Tootell et al., 1997; Vanduffel et al., 2002). Schira et al. (2004) demonstrated that %BOLD signal in V3A is also correlated with contour and figural processing, even in the absence of conscious perception. Figural processing is central to the TF argument, as the motion signal derived from the TF must be generalized to the rest of the contour.

V3A has also been identified as a neural correlate of form–motion interactions. Several groups (Braddick et al., 2000, 2001; Vaina et al., 2003; Moutoussis et al., 2005) have shown that %BOLD signal change in V3A was greater for coherent than for random motion. Koyama et al. (2005) showed V3A is more responsive to radial than for translational motion. These findings suggest a role for V3A in the generation of global motion percepts. It is notable that the fMRI investigation of TAM described earlier in this chapter found, in addition to other areas, V3A to be more active during the TAM percept relative to control.

The recent finding that BOLD signal in area V3A is modulated by CC in the context of rotational motion expanded upon this previous work by suggesting a specific mechanism concerning how form and motion may interact to construct global motion percepts. Namely, we hypothesize that neural activity within V3A serves to extract reliable motion information from regions of high CC. Such TF motion information may then be propagated to the entire moving object, resulting in the global motion percept.

How might TFs be processed? Recent neurophysiological data have shown that neurons in MT in the macaque respond more to terminator motion in a barber pole stimulus than to the ambiguous signals generated by contours. Furthermore, they respond more to intrinsically owned terminators than to extrinsic terminators (Pack et al., 2004). It has also been shown that neurons in MT in the macaque will initially respond to the direction of motion that is perpendicular (component direction) to a moving line independent of the actual direction of motion (Pack and Born 2001). These same neurons will, over a period of approximately 60 ms, shift their response properties so that they respond to the true motion of the line independent of its orientation, suggesting that the unambiguously moving endpoints of the line are quickly but not instantaneously exploited to generate a veridical motion solution. The response properties of these neurons match behavioral data that show initial pursuit eye movements will be in the direction perpendicular to the moving line, and then rapidly adapt to follow the direction of veridical motion as defined by the terminators of the lines (Pack and Born, 2001). There is also neurophysiological evidence of end-stopped neurons in V1 that respond to the motion of line terminators independently of the line's orientation (Pack et al., 2003), suggesting that form-based TFs such as line terminators can be directly extracted from the image as early as V1. Such cells are largely immune to the aperture problem.

In line with this view, features to which such end-stopped cells would respond have been shown psychophysically to be processed both rapidly and in parallel across the visual scene. Visual search studies have found several form-based features, including certain types of contour junctions (Enns and Rensink, 1991), contour concavities (Hulleman et al., 2000), corners (Humphreys et al., 1994), CCs (Wolfe et al., 1992) and curvature discontinuities (Kristjansson and Tse, 2001), which will pop out among a set of distractors. It is commonly believed that features that exhibit pop out during visual search are processed rapidly and in parallel across the visual field (Treisman and Gelade, 1980), suggesting the existence of hardwired contour discontinuity detectors in V1 or later. Indeed, contour discontinuity information may be extracted even before V1, since circular center-surround receptive fields will respond more to corners than to edges, and more to bar terminators than corners (Troncoso et al., 2005).

However, simply because local features pop out in visual search experiments does not mean that they pop out because there is a dedicated local detector for this feature that is uninfluenced by global context. For example, relative maxima of positive curvature are inherently relational in character. They do not count as maxima because of their local curvature but because of their curvature relative to other CCs in a global figure. Furthermore, such features may arise due to occlusion, and the determination of whether a feature is intrinsic to the moving object or not requires nonlocal processes. Only a global analysis of form can adequately specify which contour discontinuities are intrinsically owned by a moving object and which are extrinsic, and therefore potential sources of spurious motion signal (Lee et al., 1998). Thus, even if end-stopped hypercomplex cells in V1 play an essential role in detecting CC, as first suggested by Hubel and Wiesel (1965; Dobbins et al., 1987), they cannot directly or locally isolate these contour elements as TFs of interest in moving figures. Similarly, the form features that underlie figural segmentation are inherently global and relational in character. It therefore appears that a more global analysis of form than that available in V1 or hMT+ must interact with the motion-processing system to account for TAM and to account for the extraction and analysis of intrinsically owned TFs.

Implications for the architecture of motion processing

The data from the experiments described shed light on a debate concerning the basic architecture of the motion-processing system that has concerned the field for more than 30 years. Early studies implied that there are two motion-processing subsystems, one called the "short-range" system and the other called the "long-range" system (Braddick, 1974). The short-range process was initially touted to operate over short distances and durations and

emerge from the responses of low-level motion-energy detectors that respond automatically, passively, and in parallel across the whole visual field, while the long-range process was described as operating over long distances and durations and was thought to emerge from the responses of a cognitive mechanism that identifies forms and then tracks them over time (Anstis, 1980; Braddick, 1980). Subsequent research showed that the presumed distinctions between these processing systems did not hold. Contrary to the initial delineation of the two subsystems, short-range motion could occur over large distances and color stimuli, and long-range motion over short distances (Cavanagh et al., 1985). Contrary to predictions, it was found that long-range motion can produce motion aftereffects (von Grünau, 1986).

The reason the short- vs. long-range motion-processing distinction foundered was that it confounded differences in processing type with differences in stimulus type. Cavanagh and Mather (1989) tried to remedy this confusion by arguing that while there was only one type of motion processing, namely, the detection of motion-energy, there were two types of motion stimuli, first order (luminance and color defined) and second order (texture, relative motion, and disparity defined). There were first-order motion-energy detectors that detected first-order stimuli and second-order motion-energy detectors that detected second-order stimuli.

This argument belonged to a tradition that is still alive today that attempts to reduce all of motion perception, ultimately, to the detection of motion-energy (Lelkens and Koenderink, 1984; Chubb and Sperling, 1988; Cavanagh and Mather, 1989; Johnston et al., 1992; Wilson et al., 1992). When certain stimuli cannot be detected by luminance-defined (first-order) motion-energy detectors, a "front end" of neuronal circuitry is posited that converts the signal into a change that a more or less straightforward motion-energy detector sitting on top of the front-end circuitry can detect, whether using a front end that carries out some form of rectification (as, supposedly, occurs in second-order motion processing; Chubb and Sperling, 1988; Solomon and Sperling, 1994), or salience mapping (as is posited to occur in third-order motion processing; Lu and Sperling, 1995, 2001). Note that Sperling and colleagues have taken Cavanagh and Mather's (1989) initial distinction between first-order and second-order stimulus characteristics, and turned them into first-, second-, and third-order motion-processing systems, complicating the debate because terms like "second-order motion" are now used to mean different things by different authors.

Several authors (Mather and West, 1993; Ledgeway and Smith, 1994, 1995; Seiffert and Cavanagh, 1998) have provided evidence that first- and second-order motion are not processed by a common motion detector type. Rather, they are processed by separate low-level detectors, each insensitive to motion of the other class. Edwards and Badcock (1995), however, provided evidence that unlike first-order motion detectors, second-order motion detectors were sensitive to motion of the other class. This would be expected if second-order filters detect or are influenced by contour, texture, or form information, because form can be defined using first-order cues. Wilson et al. (1992; see also Derrington et al., 1993; Zanker and Huepgens, 1994) suggested that first-order motion is processed foremost in V1, whereas second-order motion also requires processing in V2, with both streams converging in MT. This was supported by the finding that 87% of sampled cells in MT in the alert macaque are tuned to second-order motion as well as first-order motion (Albright, 1992, see also Olavarria et al., 1992; but compare O'Keefe and Movshon, 1998 who found <25% such cells in MT in the anaesthetized macaque; for second-order motion tuning in cat cortex see Mareschal and Baker, 1998, 1999; Zhou and Baker, 1993, 1994, 1996). Albright (1992) suggested that these cells may underlie form–cue invariant motion processing (see also, Buracas and Albright, 1996). Form–cue invariant TAM, and TAM more generally, may be processed in MT (cf. Kawamoto et al., 1997) by such cells, and may therefore have much in common with second-order motion. Some authors have reported areas of the brain that respond more to second-order motion than first-order motion. Smith et al. (1998) reported that V3 was more activated by second-order motion, and Dumoulin et al. (2003) found a region

posterior to hMT+ that was preferentially activated by second-order motion. Other authors (Seiffert et al., 2003) have found no difference in retinotopic areas, hMT+, or LOC in the processing of first- vs. second-order motion.

Several authors have argued that second-order motion processing requires a stage of texture extraction (Derrington and Henning, 1993; Stoner and Albright, 1993; Werkhoven et al., 1993). This stage of texture extraction (presumably carried out in V2, V3, and/or V4; compare Smith et al., 1998) may play a role in the stage of form analysis that must precede the assignment of motion paths in TAM. Of course, there need not be a single form processor underlying TAM just as there need not be a single stage of form extraction underlying second-order motion. There may exist several shape-from-x systems, just as there may be multiple types of second-order motion detectors (Petersik, 1995). But for form–cue invariant TAM to be possible (e.g., between a luminance-defined square and color-defined rectangle; Tse and Logothetis, 2002), the various shape-from-x systems and/or second-order motion detectors must converge on a common representation of moving shape. Texture features, once extracted, could play a role in the formation of the figural representations that are transformed in TAM.

Reducing all types of motion processing, ultimately, to the processing of motion-energy fails to account for the importance of spatiotemporal segmentation and tracking of figures in motion processing. TAM reveals a role of figural tracking in motion perception that cannot be readily reduced to motion-energy accounts. There are cases where TAM can be perceived in stimulus configurations for which there is no known motion-energy detector. For example, TAM is seen when replacing a square section of a random dot field with new random dots, and then replacing this square with a rectangle of new random dots 200 ms later (Hsieh and Tse, 2006). On each frame there is nothing but random dots, so no figure seems to exist when each frame is considered in isolation. The square and rectangular figures emerge in the relationship among successive random dot fields, and must involve a parser that can extract these figures on the basis of inferred contour or surface information. This and other examples of TAM imply that there is a type of motion processing that cannot be reduced to low-level processing of local motion-energy.

TAM requires that a stage of figural segmentation and matching on the basis of global form cues exists that can determine the perceived direction of motion, even when motion-energy would predict the opposite direction of motion. For example, in Fig. 4, all motion-energy accounts would predict that motion should be perceived to the left, whereas TAM is perceived to proceed to the right, as indicated by the white arrow. Although there are many different motion-energy models, all have in common that the centroid of a luminance blob remains the centroid regardless of spatial frequency.

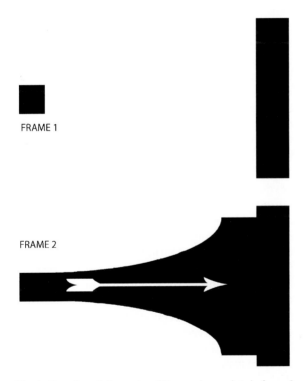

FRAME 1

FRAME 2

Fig. 4. Even though the center of the new image data in frame 2 is closer to the right-hand cue, motion proceeds away from the left-hand cue, implying that more is involved than the matching of luminance blobs, as is commonly assumed in motion-energy models. In particular, perceived TAM proceeds from left to right in this example because the new image data are matched as comprising a shape change in the left-hand cue, based on the location of continuous contours and deep concavities, which permit figural parsing and matching.

Moreover, all models of motion-energy agree that motion-energy is stronger from one centroid in the first frame to the nearest centroid in the subsequent frame (The "nearest neighbor principle"; Ullman, 1979). Centroid-to-centroid matching would imply motion to the left, whereas inferred motion that matches on the basis of figural identity across images would imply motion to the right. Because this type of high-level motion processing can violate a basic premise of all motion-energy models, it can be concluded that there are at least two mutually irreducible types of motion processing, one built upon the detection of local motion-energy (low-level, passive, parallel, monocular, automatic) and the other built upon the tracking (parsing and matching) of global figures based upon form relationships within and between successive images. A second type of form analysis that subserves motion processing is the analysis of TFs.

Reducing all types of motion processing, ultimately, to the processing of motion-energy also fails to take into account the importance of TFs in disambiguating low-level motion signals. It is likely that the paramount TFs used by the visual system to overcome the aperture problem are contour discontinuities, including terminators, junctions, corners, and curvature. These are manifestly not a motion-energy cue and cannot be reduced to motion-energy. Rather, like the stage of contour-based segmentation that underlies TAM, which apparently involves retinotopic areas and the LOC, the stage of TF analysis that we argue takes place in V3A must proceed in parallel with a motion-energy analysis. The neural computations that underlie the conscious experience of motion, which presumably take place in hMT +, combine inputs from the low-level, motion-energy-based system, and the two form-based processes described here. It is as if there is a ventral process of global contour-based form segmentation that culminates in the LOC, and a dorsal process that tracks key contour discontinuities that culminates in V3A. Both these form analyses apparently feed into hMT + where the final motion trajectory is thought to be computed in light of the constraints that they provide.

Although Cavanagh (Cavanagh and Mather, 1989) initially argued that all motion processing could be reduced to motion-energy processing, Cavanagh (1992) subsequently reported the existence of a new attention-based or high-level motion process that could not be reduced to an account in terms of motion-energy. Cavanagh (1995) has called the low-level and high-level systems the "passive" (low-level, motion-energy based for color, texture, and luminance) and "active" (based upon attentional tracking) motion-processing systems. Since TAM cannot be accounted for by the responses of the passive system, this raises the question of whether TAM requires not just form-based parsing, but also attention in order to be seen. It has been shown that TAM does not arise because of attentional gradients (Hsieh et al., 2005). However, other evidence is emerging that TAM requires the specification and tracking of a figure over time. It could be that TAM does indeed require attention in order to be perceived, not in the sense of attentional gradients, but in the sense that attention is required for the tracking of figural changes over successive scenes. In essence, attention converts what would otherwise be a mere succession of disjoint shapes (Horowitz and Wolfe, 1998) into a bound object representation that can undergo changes in location, shape, and other features, and that can be tracked over time (Kahneman et al., 1992).

To test whether attention is needed for the perception of TAM in the sense of figural tracking, a simple psychophysical experiment was carried out on TAM in the context of a standard change blindness paradigm. An array of squares and rectangles was continuously flashed on and off. On alternating flashes one of the squares would be replaced by a rectangle, which in turn would be replaced by the square on the next flash, and so on. While maintaining fixation, observers were asked to localize the figure undergoing shape changes. This is slow and requires a serial search of the whole array, as has been shown previously (Rensink et al., 1997; Rensink, 2000; Tse et al., 2003). However, if the location is cued, the change is perceived immediately even when the figure is located as far as fifteen degrees from fixation, in line with past findings (Rensink et al., 1997) and the general notion of attentional "pop-out." What is new is that as soon as a change is seen, the figure

is seen to undergo TAM from a square into a rectangle or vice versa. This implies that TAM indeed does require attention because attention is required to bind successive shape representations into the representation of a single spatiotemporal figure that can undergo a shape change.

What does this tell us about motion processing? It tells us that the high-level motion-processing stream is one that tracks figural changes on the basis of changes in position or shape of various features, and it tells us that such figural tracking requires attention to bind successive disjoint (Horowitz and Wolfe, 1998) shape representations into a representation of a single object (Kahneman et al., 1992) undergoing a shape change. Thus the high-level motion-processing stream is inherently driven by form analysis and attentional tracking, and the low-level motion-processing stream is driven by the analysis of various classes of motion-energy. This is not to say that low-level motion processing is stimulus driven whereas high-level processing is not. Once attention has marked an object as a figure to be tracked, the fact that TAM can be processed so rapidly implies that form-based processes parse, match, and track figures over time as they change shape and position in a largely automatic and stimulus-driven manner.

How does the low-level/high-level dichotomy advocated here mesh with past views? TAM vindicates the initial speculations of Braddick (1980) and Anstis (1980) that the essence of the high-level motion-processing system involves matching on the basis of form. During the 1980s researchers found little evidence to support this view of high-level motion processing, and many came to the conclusion that all motion processing could be reduced to the processing of motion-energy detectors of various types, none of which were sensitive to form relationships per se. The present view is consistent with Cavanagh's post-1992 view (Cavanagh, 1992, 1995) of passive and active motion-processing systems. The high-level motion-processing system described here is the same as his active vision system, as long as it is understood that the motion signal is not necessarily or primarily driven by the top-down movements of an attentional window, but rather by the outputs of a stimulus-driven analysis of form. Once a figure is defined by attention (i.e., by its binding into an object representation; Kahneman et al., 1992), its tracking becomes largely stimulus driven.

The present view of motion processing is not entirely consistent with the hierarchy of motion-energy types posited by Sperling and colleagues (e.g., Lu and Sperling, 1995) because there is no place in their models for matching based upon the explicit processing of global form relationships, or for TFs. The saliency mapping posited to underlie third-order motion processing resembles attention, but nonetheless feeds into a more or less standard motion-energy detector. Even though the motion-energy detector detects shifts in the peaks of salience, such a detector would still map peaks in salience to their nearest neighbor. Such a detector would presumably predict leftward motion for the stimulus in Fig. 4. But because rightward motion is what observers actually perceive, matching must take place on the basis of something other than peak to peak or centroid-to-centroid matching, whether of luminance or of salience.

Lu and Sperling (2001) justify their claim that third-order motion involves motion-energy detection over shifts in salience on the basis of an evolutionary argument:

> An important reason for assuming that third-order motion is computed by an algorithm similar to that for first- and second-order motion is that the genetic code needed to instantiate a computation in the brain is quite complex. The likelihood that a new gene for a motion computation would evolve separately vs. the original motion computation gene being spliced from one location to another is negligibly small. (p. 2335)

In contrast, a central claim of the present paper is that the figural tracking underlying high-order motion processing is not based upon motion-energy detection at all, but is based upon a very different algorithm premised upon spatiotemporal figural parsing, matching, and binding on the basis of global form cues.

In addition, it may not be true that only third-order motion is salience- or attention based, whereas second-order motion is not. While Lu and Sperling (1995) argue that neither first- nor second-order motion depends on tracking, there is evidence that at low contrasts, second-order motion is dependent on figural tracking (Seiffert and Cavanagh, 1998; Derrington and Ukkonen, 1999). In addition, it is likely that some kinds of stimuli classified as second-order in the past have unwittingly relied on the type of form analysis that underlies TAM. Thus the trichotomy into first-, second-, and third-order motion-processing systems may complicate matters by viewing all three systems as involving motion-energy detection, by segregating attentional mechanisms from second-order mechanisms, and by ignoring the contribution of figural analysis based on global form relationships to motion processing. Any model of motion processing that fails to explicitly incorporate matching on the basis of global form-relationships within and between successive scenes will be incomplete. A simpler architecture may be adequate to explain all motion data, where a dichotomy exists between a low-level system which processes motion in terms of motion-energy derived from various features such as luminance, color, and texture, and a high-level system that processes motion in terms of changes over figures that are parsed and matched on the basis of their global form relationships rather than salience.

Conclusion

Contour discontinuities, such as corners, deep concavities, maxima of positive curvature, junctions, and terminators, are the most informative portions of an object's contour used in determining perceived object shape (e.g., Tse, 2001) and motion (e.g., Tse, 2006, Caplovitz and Tse, 2006; Caplovitz, Hsieh and Tse, 2006). Although contour discontinuities are a form cue, they play at least two important roles in determining the direction of perceived motion. First, they play a central role in a stage of figural segmentation and matching that must precede the perception of motion because it can determine the direction of perceived motion. TAM was discussed in this context. This is primarily a ventral process involving the LOC and probably also V2 and V4, and perhaps even V1. Second, contour discontinuities play an important role as TFs that can solve the aperture problem. Their role in the perception of rotating figures was considered. This TF analysis is a dorsal process involving at least V3A (Caplovitz and Tse, submitted). In this context, it is suggested that area V3A contains neural populations that process form, not to solve the "ventral problem" of determining object shape, but in order to solve the "dorsal problem" of what is going where. To put it more succinctly, form is involved in motion processing at least two ways, in terms of figural segmentation dedicated to solving the problem of figure to figure matching over time, and in terms of defining TFs whose unambiguous motion signals can be generalized to ambiguously moving portions of an object. In both cases, the analysis of form must proceed in parallel with the analysis of motion, in order to constrain the ongoing analysis of motion. The neural computations that underlie the conscious experience of motion, which presumably take place in hMT+, combine inputs from the low-level, motion-energy based system, and the two form-based processes described here. There is a ventral process of global contour-based form segmentation that culminates in the LOC, and a dorsal process that tracks key contour discontinuities that culminates in V3A. Both these form analyses apparently feed into hMT+ where the final motion trajectory is presumably computed in light of the constraints that they provide.

References

Adelson, E.H. and Bergen, J.R. (1985) Spatiotemporal energy models for the perception of motion. J. Opt. Soc. Am. A, 2(2): 284–299.

Adelson, E.H. and Movshon, J.A. (1982) Phenomenal coherence of moving visual patterns. Nature, 30: 523–525.

Albright, T.D. (1992) Form-cue invariant motion processing in primate visual cortex. Science, 255: 1141–1143.

Allman, J.M., Miezin, F. and McCuiness, E. (1985) Stimulus specific responses from beyond the classical receptive field: neurophysiological mechanisms for local-global comparisons in visual neurons. Ann. Rev. Neurosci., 8: 407–430.

Altmann, C.F., Bülthoff, H.H. and Kourtzi, Z. (2003) Perceptual organization of local elements into global shapes in the human visual cortex. Curr. Biol., 13(4): 342–349.

Anstis, S.M. (1980) The perception of apparent motion. Philos. Trans. R. Soc. Lond. B, 290: 153–168.

Attneave, F. (1954) Some informational aspects of visual perception. Psychol. Rev., 61: 183–193.

Avidan, G., Harel, M., Hendler, T., Ben-Bashat, D., Zohary, E. and Malach, R. (2002) Contrast sensitivity in human visual areas and its relationship to object recognition. J. Neurophysiol., 87: 3102–3116.

Baloch, A.A. and Grossberg, S. (1997) A neural model of high-level motion processing: line motion and form-motion dynamics. Vision Res., 37(21): 3037–3059.

Barlow, H.B. and Levick, W.R. (1965) The mechanism of directionally selective units in rabbit's retina. J. Physiol., 178(3): 477–504.

Baro, J.A. and Levinson, E. (1988) Apparent motion can be perceived between patterns with dissimilar spatial frequencies. Vision Res., 28: 1311–1313.

Biederman, I. (1987) Recognition-by-components: a theory of human image understanding. Psychol. Rev., 94(2): 115–147.

Bonnet, C. (1981) Processing configurations of visual motion. In: Long, J. and Bradley, A. (Eds.), Attention and Performance IX. Erlbaum, Hillsdale, NJ.

Born, R.T. and Bradley, D.C. (2005) Structure and function of visual area MT. Ann. Rev. Neurosci., 28: 157–189.

Braddick, O. (1974) A short-range process in apparent motion. Vision Res., 14(7): 519–527.

Braddick, O. (1980) Low-level and high-level processes in apparent motion. Philos. Trans. R. Soc. Lond. B, Biol. Sci., 290(1038): 137–151.

Braddick, O.J., O'Brien, J.M., Wattam-Bell, J., Atkinson, J. and Turner, R. (2000) Form and motion coherence activate independent but not dorsal/ventral segregated, networks in the human brain. Curr. Biol., 10: 731–734.

Braddick, O.J., O'Brien, J.M., Wattam-Bell, J., Atkinson, J., Hartley, T. and Turner, R. (2001) Brain areas sensitive to coherent visual motion. Perception, 30: 61–72.

Bradley, D.C., Qian, N. and Andersen, R.A. (1995) Integration of motion and stereopsis in middle temporal cortical area of macaques. Nature, 373(6515): 609–611.

Buracas, G.T. and Albright, T.D. (1996) Contribution of area MT to perception of three-dimensional shape: a computational study. Vision Res., 36(6): 869–887.

Burr, D.C., Ross, J. and Morrone, M.C. (1986) Seeing objects in motion. Proc. R. Soc. Lond., B227: 171–195.

Burt, P. and Sperling, G. (1981) Time, distance, and feature trade-offs in visual apparent motion. Psychol. Rev., 88: 171–195.

Caplovitz, G.P. and Tse, P.U. (2006) V3A processes contour curvature as a trackable feature for the perception of rotational motion. Cereb. Cortex, in press.

Caplovitz, G.P., Hsieh, P.-J. and Tse, P.U. (2006) Mechanisms underlying the perceived angular velocity of a rigidly rotating object. Vision Res., Apr. 27 [Epub ahead of print].

Cavanagh, P. (1992) Attention-based motion perception. Science, 257: 1563–1565.

Cavanagh, P. (1995). Is there low-level motion processing for non-luminance-based stimuli? In: Thomas, V. Papathomas, Charles Chubb, Andrei Gorea and Eileen Kowler (Eds.),

Early Vision and Beyond. MIT Press, Cambridge, MA, pp. 113–120.

Cavanagh, P., Arguin, M. and von Grünau, M. (1989) Inter-attribute apparent motion. Vision Res., 29(9): 1197–1204.

Cavanagh, P. and Mather, G. (1989) Motion: the long and short of it. Spatial Vis., 4: 103–129.

Cavanagh, P., Boeglin, J. and Favreau, O.E. (1985) Perception of motion in equiluminous kinematograms. Perception, 14(2): 151–162.

Chubb, C. and Sperling, G. (1988) Drift-balanced random stimuli: a general basis for studying non-Fourier motion perception. J. Opt. Soc. Am. A, 5(11): 1986–2007.

Dawson, M.R.W. (1991) The how and why of what went where in apparent motion: modeling solutions to the motion correspondence problem. Psychol. Rev., 33(4): 569–603.

Derrington, A.M., Badcock, D.R. and Henning, G.B. (1993) Discriminating the direction of second-order motion at short stimulus durations. Vision Res., 33(13): 1785–1794.

Derrington, A.M. and Henning, G.B. (1993) Detecting and discriminating the direction of motion of luminance and colour gratings. Vision Res., 33(5-6): 799–811.

Derrington, A.M. and Ukkonen, O.I. (1999) Second-order motion discrimination by feature-tracking. Vision Res., 39(8): 1465–1475.

Dobbins, A., Zucker, S.W. and Cynader, M.S. (1987) End-stopped neurons in the visual cortex as a substrate for calculating curvature. Nature, 329(6138): 438–441.

Dumoulin, S., Baker, C.L., Hess, R.F. and Evans, A.C. (2003) Cortical specialization for processing first- and second-order motion. Cereb. Cortex, 13(12): 1375–1385.

Downing, P. and Treisman, A. (1995) The shooting line illusion: attention or apparent motion? Invest. Ophthalmol. Vis. Sci., 36: S856.

Downing, P. and Treisman, A. (1997) The line motion illusion: attention or impletion? J. Exp. Psychol. Hum. Percept. Perform., 23(3): 768–779.

Edwards, M. and Badcock, D.R. (1995) Global motion perception: no interaction between the first- and second-order motion pathways. Vision Res., 35(18): 2589–2602.

Enns, J.T. and Rensink, R.A. (1991) Preattentive recovery of three-dimensional orientation from line drawings. Psychol Rev., 98(3): 335–351.

Faubert, J. and von Grünau, M. (1995) The influence of two spatially distinct primers and attribute priming on motion induction. Vision Res., 35(22): 3119–3130.

Fennema, C. and Thompson, W. (1979) Velocity determination in scenes containing several moving objects. Comput. Graph. Image Process., 9: 301–305.

Ferber, S., Humphrey, G.K. and Vilis, T. (2003) The lateral occipital complex subserves the perceptual persistence of motion-defined groupings. Cereb. Cortex., 13: 716–721.

Fitzpatrick, D. (2000) Seeing beyond the receptive field in primary visual cortex. Curr. Opin. Neurobiol., 10: 438–443.

Gepshtein, S. and Kubovy, M. (2000) The emergence of visual objects in spacetime. Proc. Natl. Acad. Sci. U S A, 97(14): 8186–8191.

Gibson, J.J. (1979) The Ecological Approach to Visual Perception. Houghton Mifflin, Boston.

Gilaie-Dotan, S., Ullman, S., Kushnir, T. and Malach, R. (2001) Shape-selective stereo processing in human object-related visual areas. Hum. Brain Map., 15: 67–79.

Gilbert, C.D. (1992) Horizontal integration and cortical dynamics. Neuron, 9: 1–13.

Gilbert, C.D. (1998) Adult cortical dynamics. Physiol. Rev., 78: 467–485.

Goebel, R., Khorram-Sefat, D., Muckli, L., Hacker, H. and Singer, W. (1998) The constructive nature of vision: direct evidence from functional magnetic resonance imaging studies of apparent motion and motion imagery. Eur. J. Neurosci., 10: 1563–1573.

Goodale, M. and Milner, A. (1992) Separate visual pathways for perception and action. Trends Neurosci., 15: 20–25.

Green, M. (1986a). Correspondence in apparent motion: defining the heuristics. Proc. Vis. Interface, 337–242.

Green, M. (1986b) What determines correspondence strength in apparent motion? Vision Res., 26: 599–607.

Green, M. (1989) Color correspondence in apparent motion. Percept. Psychophys., 45(1): 15–20.

Grill-Spector, K., Kushnir, T., Edelman, S., Itzchak, Y. and Malach, R. (1998) Cue-invariant activation in object-related areas of the human occipital lobe. Neuron, 21: 191–202.

Grill-Spector, K., Kushnir, T., Edelman, S., Avidan, G., Itzchak, Y. and Malach, R. (1999) Differential processing of objects under various viewing conditions in the human lateral occipital complex. Neuron, 24: 187–203.

Grill-Spector, K., Kushnir, T., Hendler, T. and Malach, R. (2000) The dynamics of object-selective activation correlate with recognition performance in humans. Nat. Neurosci., 3(8): 837–843.

Grill-Spector, K., Kourtzi, Z. and Kanwisher, N. (2001) The lateral occipital complex and its role in object recognition. Vision Res., 41: 1409–1422.

Hassenstein, B. and Reichardt, W. (1956) Systemtheoretische analyse der zeit-, reihenfolgen- and vorzeichenauswertung bei der bewegungsperzeption des russelkafers chlorophanus. Z. Naturforsch., 11b: 513–524.

Haxby, J.V., Gobbini, M.I., Furey, M.L., Ishai, A., Schouten, J.L. and Pietrini, P. (2001) Distributed and overlapping representations of faces and objects in ventral temporal cortex. Science, 293(5539): 2425–2430.

Hikosaka, O., Miyauchi, S. and Shimojo, S. (1991) Focal visual attention produces motion sensation in lines. Investig. Ophthamol. Vis. Sci., 32(4): 176.

Hikosaka, O., Miyauchi, S. and Shimojo, S. (1993a) Focal visual attention produces illusory temporal order and motion sensation. Vision Res., 33(9): 1219–1240.

Hikosaka, O., Miyauchi, S. and Shimojo, S. (1993b) Visual attention revealed by an illusion of motion. Neurosci. Res., 18(1): 11–18.

Hildreth, E.C., Ando, H., Andersen, R.A. and Treue, S. (1995) Recovering three-dimensional structure from motion with surface reconstruction. Vision Res., 35(1): 117–137.

Horowitz, T.S. and Wolfe, J.M. (1998) Visual search has no memory. Nature, 394(6693): 575–577.

Hsieh, P.-J., Caplovitz, G.P. and Tse, P.U. (2005) Illusory rebound motion and the motion continuity heuristic. Vision Res., 45(23): 2972–2985.

Hsieh, P.-J. and Tse, P.U. (2006) Stimulus factors affecting illusory rebound motion. Vision Res., 46(12): 1924–1933.

Hubel, D.H. and Wiesel, T.N. (1965) Receptive fields and functional architecture in two nonstriate visual areas (18 and 19) of the cat. J. Neurophysiol., 28: 229–289.

Hubel, D.H. and Wiesel, T.N. (1968) Receptive fields and functional architecture of monkey striate cortex. J. Physiol., 195: 215–243.

Hulleman, J., te Winkel, W. and Boselie, F. (2000) Concavities as basic features in visual search: evidence from search asymmetries. Percept. Psychophys., 62(1): 162–174.

Humphreys, G.W., Keulers, N. and Donnelly, N. (1994) Parallel visual coding in three dimensions. Perception, 23(4): 453–470.

Johnston, A., McOwan, P.W. and Buxton, H. (1992) A computational model of the analysis of some first-order and second-order motion patterns by simple and complex cells. Proc. Biol. Sci., 250(1329): 297–306.

Kahneman, D., Treisman, A. and Gibbs, B.J. (1992) The reviewing of object files: Object-specific integration of information. Cogn. Psychol., 24: 175–219.

Kanizsa, G. (1951) Sulla polarizzazione del movimento gamma [The polarization of gamma movement]. Arch. Psichol., Neurol. Psichiatr., 3: 224–267.

Kanizsa, G. (1971) Organization in Vision: Essays on Gestalt Perception. Praeger, New York.

Kanwisher, N., Chun, M.M., McDermott, J. and Ledden, P.J. (1996) Functional imagining of human visual recognition. Brain Res. Cogn. Brain Res., 5(1–2): 55–67.

Kenkel, F. (1913) Untersuchungen über den Zusammenhang zwischen Erscheinungsgröße und Erscheinungsbewegung bei einigen sogenannten optischen Täuschungen. Z. Psychol., 67: 358–449.

Kolers, P.A. and Pomerantz, J.R. (1971) Figural change in apparent motion. J. Exp. Psychol., 87: 99–108.

Kourtzi, Z. and Kanwisher, N. (2000b) Cortical regions involved in perceiving object shape. J. Neurosci., 20: 3310–3318.

Kourtzi, Z. and Kanwisher, N. (2001) Representation of perceived object shape by the human lateral occipital complex. Science, 293: 1506–1509.

Kourtzi, Z., Erb, M., Grodd, W. and Bülthoff, H.H. (2003) Representation of the perceived 3-D object shape in the human lateral occipital complex. Cereb. Cortex, 13(9): 911–920.

Kourtzi, Z., Tolias, A.S., Altmann, C.F., Augath, M. and Logothetis, N.K. (2003) Integration of local features into global shapes. Monkey and human FMRI studies. Neuron, 37(2): 333–346.

Koyama, S., Sasaki, Y., Andersen, G.J., Tootell, R.B., Matsuura, M. and Watanabe, T. (2005) Separate processing of different global-motion structures in visual cortex is revealed by FMRI. Curr. Biol., 15(22): 2027–2032.

Kristjansson, A. and Tse, P.U. (2001) Curvature discontinuities are cues for rapid shape analysis. Percept. Psychophys., 63(3): 390–403.

Lamme, V.A., Super, H. and Spekreijse, H. (1998) Feedforward, horizontal, and feedback processing in the visual cortex. Curr. Opin. Neurobiol., 8: 529–535.

Ledgeway, T. and Smith, A.T. (1994) Evidence for separate motion-detecting mechanisms for first- and second-order motion in human vision. Vision Res., 34(20): 2727–2740.

Ledgeway, T. and Smith, A.T. (1995) The perceived speed of second-order motion and its dependence on stimulus contrast. Vision Res., 35(10): 1421–1434.

Lee, T.S., Mumford, D., Romero, R. and Lamme, V.A. (1998) The role of the primary visual cortex in higher level vision. Vision Res., 38(15–16): 2429–2454.

Lelkens, A.M. and Koenderink, J.J. (1984) Illusory motion in visual displays. Vision Res., 24(9): 1083–1090.

Liu, T. and Cooper, L.A. (2003) Explicit and implicit memory for rotating objects. J. Exp. Psychol. Learn. Mem. Cogn., 29: 554–562.

Liu, T., Slotnick, S.D. and Yantis, S. (2004) Human MT+ mediates perceptual filling-in during apparent motion. Neuroimage, 21(4): 1772–1780.

Livingstone, M.S. and Hubel, D.H. (1987) Psychophysical evidence for separate channels for the perception of form, color, movement, and depth. J. Neurosci., 7: 3416–3468.

Lorenceau, J. and Shiffrar, M. (1992) The role of terminators in motion integration across contours. Vision Res., 32: 263–273.

Lu, Z.L. and Sperling, G. (1995) The functional architecture of human visual motion perception. Vision Res., 35(19): 2697–2722.

Lu, Z.L. and Sperling, G. (2001) Three-systems theory of human visual motion perception: review and update. J. Opt. Soc. Am. A Opt. Image. Sci. Vis., 18(9): 2331–2370.

Malach, R., Reppas, J.B., Benson, R.R., Kwong, K.K., Jiang, H., Kennedy, W.A., Ledden, P.J., Brady, T.J., Rosen, B.R. and Tootell, R.B. (1995) Object-related activity revealed by functional magnetic resonance imaging in human occipital cortex. Proc. Natl. Acad. Sci., 92(18): 8135–8139.

Malach, R., Grill-Spector, K., Kushnir, T., Edelman, S. and Itzchak, Y. (1998) Rapid shape adaptation reveals position and size invariance. Neuroimage, 7: S43.

Mareschal, I. and Baker, C.L. (1998) Temporal and spatial response to second-order stimuli in cat A18. J. Neurophysiol., 80: 2811–2823.

Mareschal, I. and Baker, C.L. (1999) Cortical processing of second-order motion. Vis. Neurosci., 16: 527–540.

Marr, D. (1982) Vision. W. H. Freeman and Co., New York.

Mather, G. and West, S. (1993) Evidence for second-order motion detectors. Vision Res., 33(8): 1109–1112.

Mendola, J.D., Dale, A.M., Fischl, B., Liu, A.K. and Tootell, R.B.H. (1999) The representation of real and illusory contours in human cortical visual areas revealed by fMRI. J. Neurosci., 19: 8560–8572.

Mikami, A., Newsome, W.T. and Wurtz, R.H. (1986) Motion selectivity in macaque visual cortex: II. Spatiotemporal range of directional interactions in MT and V1. J. Neurophysiol., 55: 1328–1339.

Moore, C. and Engel, S.A. (2001) Neural response to perception of volume in the lateral occipital complex. Neuron, 29: 277–286.

Moutoussis, K., Keliris, G., Kourtzi, Z. and Logothetis, N. (2005) A binocular rivalry study of motion perception in the human brain. Vision Res., 45(17): 2231–2243.

Muckli, L., Kriegeskorte, N., Lanfermann, H., Zanella, F.E., Singer, W. and Goebel, R. (2002) Apparent motion: event-related functional magnetic resonance imaging of perceptual switches and States. J. Neurosci., 22: RC219.

Murray, S.O., Olshausen, B.A. and Woods, D.L. (2003) Processing shape, motion and three-dimensional shape-from-motion in the human cortex. Cereb. Cortex, 13: 508–516.

Nakayama, K. and Silverman, G.H. (1988) The aperture problem I. Perception of nonrigidity and motion direction in translating sinusoidal lines. Vision Res., 28(6): 739–746.

Navon, D. (1976) Irrelevance of figural identity for resolving ambiguities in apparent motion. J. Exp. Psychol. Hum. Percept. Perform., 2: 130–138.

Nowlan, S.J. and Sejnowski, T.J. (1995) A selection model for motion processing in area MT of primates. J. Neurosci., 15(2): 1195–1214.

O'Keefe, L.P. and Movshon, J.A. (1998) Processing of first- and second-order motion signals by neurons in area MT of the macaque monkey. Vis. Neurosci., 15(2): 305–317.

Olavarria, J.F., DeYoe, E.A., Knierim, J.J., Fox, J.M. and Van Essen, D.C. (1992) Neural responses to visual texture patterns in middle temporal area of the macaque monkey. J. Neurophysiol., 68(1): 164–181.

Pack, C.C., Gartland, A.J. and Born, R.T. (2004) Integration of Contour and terminator signals in visual area MT of alert macaque. J. Neurosci., 24(13): 3268–3280.

Pack, C.C. and Born, R.T. (2001) Temporal dynamics of a neural solution to the aperture problem in visual area MT of macaque brain. Nature, 409(6823): 1040–1042.

Pack, C.C., Livingstone, M.S., Duffy, K.R. and Born, R.T. (2003) End-stopping and the aperture problem: two-dimensional motion signals in macaque V1. Neuron, 39(4): 671–680.

Petersik, J.T. (1995) A comparison of varieties of "second-order" motion. Vision Res., 35(4): 507–517.

Ramachandran, V.S., Ginsburg, A.P. and Anstis, S.M. (1983) Low spatial frequencies dominate apparent motion. Perception, 12: 457–461.

Ramachandran, V.S. and Gregory, R.L. (1978) Does colour provide an input to human motion perception? Nature, 275: 55–56.

Reichardt, W. (1957) Autokorrelationsauswertung als funktionsprinzip des zentralnervensystems. Z. Naturforsch., 12b: 447–457.

Reichardt, W. (1961) Autocorrelation, a principle for the evaluation of sensory information by the central nervous system. In: Rosenblith, W.A. (Ed.), Sensory Communication. Wiley, New York.

Rensink, R.A., O'Regan, J.K. and Clark, J.J. (1997) To see or not to see: the need for attention to perceive changes in scenes. Psychol. Sci., 8: 368–373.

Rensink, R.A. (2000) Visual search for change: a probe into the nature of attentional processing. Vis. Cogn., 7: 345–376.

Schiller, P. (1991) Parallel pathways in the visual system: their role in perception at isoluminance. Neuropsychologia, 29(6): 433–441.

Schira, M.M., Fahle, M., Donner, T.H., Kraft, A. and Brandt, S.A. (2004) Differential contribution of early visual areas to the perceptual process of contour processing. J. Neurophysiol., 91(4): 1716–1721.

Seiffert, A.E. and Cavanagh, P. (1998) Position-based motion perception for color and texture stimuli: Effects of contrast and speed. Vision Res., 39(25): 4172–4185.

Seiffert, A.E., Somers, D.C., Dale, A.M. and Tootell, R.B. (2003) Functional MRI studies of human visual motion perception: texture, luminance, attention and after-effects. Cereb. Cortex, 13(4): 340–349.

Shiffrar, M., Li, X. and Lorenceau, J. (1995) Motion integration across differing image features. Vision Res., 35(15): 2137–2146.

Shimojo, S., Silverman, G.H. and Nakayama, K. (1989) Occlusion and the solution to the aperture problem for motion. Vision Res., 29: 619–626.

Smith, A.T., Greenlee, M.W., Singh, K.D., Kraemer, F.M. and Hennig, J. (1998) The processing of first- and second-order motion in human visual cortex assessed by functional magnetic resonance imaging, fMRI. J. Neurosci., 18(10): 3816–3830.

Solomon, J.A. and Sperling, G. (1994) Full-wave and half-wave rectification in second-order motion perception. Vision Res., 34(17): 2239–2257.

Stelmach, L.B. and Herdman, C.M. (1991) Directed attention and perception of temporal order. J. Exp. Psychol. Hum. Percept. Perform., 17(2): 539–550.

Sternberg, S. and Knoll, R.L. (1973) The perception of temporal order: fundamental issues and a general model. In: Kornblum, S. (Ed.) Attention and Performance, Vol. IV. Academic Press, New York, pp. 629–685.

Stone, J.V. (1999) Object recognition: view-specificity and motion-specificity. Vision Res., 39: 4032–4044.

Stoner, G.R. and Albright, T.D. (1992) Motion coherency rules are form-cue invariant. Vision Res., 32(3): 465–475.

Stoner, G.R. and Albright, T.D. (1996) The interpretation of visual motion: evidence for surface segmentation mechanisms. Vision Res., 36(9): 1291–1310.

Titchener, E.B. (1908) Lecture on the Elementary Psychology of Feeling and Attention. McMillan, New York.

Tootell, R.B., Mendola, J.D., Hadjikhani, N.K., Ledden, P.J., Liu, A.K., Reppas, J.B., Sereno, M.I. and Dale, A.M. (1997) Functional analysis of V3A and related areas in human visual cortex. J. Neurosci., 17(18): 7060–7078.

Treisman, A.M. and Gelade, G.A. (1980) Feature-integration theory of attention. Cognit. Psychol., 12(1): 97–136.

Treisman, A. and Gormican, S. (1988) Feature analysis in early vision: evidence from search asymmetries. Psychol. Rev., 95(1): 15–48.

Troncoso, X.G., Macknik, S.L. and Martinez-Conde, S. (2005) Novel visual illusions related to Vasarely's 'nested squares' show that corner salience varies with corner angle. Perception, 34(4): 409–420.

Tse, P.U. (2006) Neural correlates of transformational apparent motion. Neuroimage, 31(2): 766–773.

Tse, P.U. and Albert, M.K. (1998) Amodal completion in the absence of image tangent discontinuities. Perception, 27(4): 455–464.

Tse, P.U. and Cavanagh, P. (1995) Line motion occurs after surface parsing. Investig. Ophth. Vis. Sci., 36: S417.

Tse, P.U., Cavanagh, P. and Nakayama, K. (1996) The roles of attention in shape change apparent motion. Investig. Ophthalmol. Vis. Sci., 37: S213.

Tse, P.U., Cavanagh, P. and Nakayama, K. (1998) The role of parsing in high-level motion processing. In: Watanabe, T. (Ed.), High-Level Motion Processing: Computational, Neurobiological, and Psychophysical Perspectives. MIT Press, Cambridge, MA, pp. 249–266.

Tse, P.U. and Logothetis, N.K. (2002) The duration of 3-d form analysis in transformational apparent motion. Percept. Psychophys., 64(2): 244–265.

Tse, P.U., Sheinberg, D.L. and Logothetis, N.K. (2003) Attentional enhancement opposite a peripheral flash revealed by change blindness. Psychol. Sci., 14(2): 1–8.

Ullman, S. (1979) The Interpretation of Visual Motion. MIT Press, Cambridge, MA.

Ungerleider, L. and Mishkin, M. (1982) Two cortical visual systems. In: Ingle, D., Goodale, M. and Mansfield, R. (Eds.), Analysis of Visual Behavior. MIT Press, Cambridge, MA, pp. 549–586.

Vaina, L.M., Gryzacz, N.M., Saiviroonporn, P., LeMay, M., Bienfang, D.C. and Conway, A. (2003) Can spatial and temporal motion integration compensate for deficits in local motion mechanisms? Neuroosychologia, 41: 1817–1836.

Vanduffel, W., Fize, D., Peuskens, H., Denys, K., Sunaert, S., Todd, J.T. and Orban, G.A. (2002) Extracting 3D from motion: differences in human and monkey intraparietal cortex. Science, 298: 413–415.

van Santen, J.P.H. and Sperling, G. (1984) Temporal covariance model of human motion perception. J. Opt. Soc. Am. A, 1: 451–473.

Victor, J.D. and Conte, M.M. (1990) Motion mechanisms have only limited access to form information. Vision Res., 30: 289–301.

von der Heydt, R., Peterhans, E. and Baumgartner, G. (1984) Illusory contours and cortical neuron responses. Science, 224(4654): 1260–1262.

von Grünau, M.W. (1986) A motion aftereffect for long-range stroboscopic apparent motion. Percept. Psychophys., 40(1): 31–38.

von Grünau, M. and Faubert, J. (1994) Intraattribute and interattribute motion induction. Perception, 23(8): 913–928.

Wallis, G. and Bülthoff, H. (2001) Effects of temporal association on recognition memory. Proc. Natl. Acad. Sci. USA, 98(8): 4800–4804.

Watson, A.B. and Ahumada Jr., A.J. (1985) Model of human visual-motion sensing. J. Opt. Soc. Am. A, 2: 232–342.

Werkhoven, P., Sperling, G. and Chubb, C. (1993) The dimensionality of texture-defined motion: a single channel theory. Vision Res., 33(4): 463–485.

Wertheimer, M. (1912) Experimentelle Studien über das Sehen von Bewegung. Z. Psychol., 61: 161–265.

Wilson, H.R., Ferrera, V.P. and Yo, C. (1992) A psychophysically motivated model for two-dimensional motion perception. Vis. Neurosci., 9(1): 79–97.

Wolfe, J.M., Yee, A. and Friedman-Hill, S.R. (1992) Curvature is a basic feature for visual search tasks. Perception, 21(4): 465–480.

Zanker, J.M. (1997) Is facilitation responsible for the "motion induction" effect? Vision Res., 37(14): 1953–1959.

Zanker, J.M. and Huepgens, I.S. (1994) Interaction between primary and secondary mechanisms in human motion perception. Vision Res., 34(10): 1255–1266.

Zhou, Y.-X. and Baker, C.L. (1993) A processing stream in mammalian visual cortex neurons for non-Fourier responses. Science, 261: 98–101.

Zhou, Y.-X. and Baker, C.L. (1994) Envelope-responsive neurons in Area 17 and 18 of cat. J. Neurophysiol., 72: 2134–2150.

Zhou, Y.-X. and Baker, C.L. (1996) Spatial properties of envelope responses in Area 17 and 18 of the cat. J. Neurophysiol., 75: 1038–1050.

Zhuo, Y., Zhou, T.G., Rao, H.Y., Wang, J.J., Meng, M., Chen, M., Zhou, C. and Chen, L. (2003) Contributions of the visual ventral pathway to long-range apparent motion. Science, 299: 417–420.

Zucker, S.W., Dobbins, A. and Iverson, L. (1989) Two stages of curve detection suggest two styles of visual computation. Neural Comput., 1: 68–81.

Martinez-Conde, Macknik, Martinez, Alonso & Tse (Eds.)
Progress in Brain Research, Vol. 154
ISSN 0079-6123

CHAPTER 16

Neural basis of shape representation in the primate brain

Anitha Pasupathy*

The Picower Institute for Learning and Memory, RIKEN-MIT Neuroscience Research Center and Department of Brain and Cognitive Sciences, Massachusetts Institute of Technology, Cambridge, MA 02139, USA

Abstract: Visual shape recognition — the ability to recognize a wide variety of shapes regardless of their size, position, view, clutter and ambient lighting — is a remarkable ability essential for complex behavior. In the primate brain, this depends on information processing in a multistage pathway running from primary visual cortex (V1), where cells encode local orientation and spatial frequency information, to the inferotemporal cortex (IT), where cells respond selectively to complex shapes. A fundamental question yet to be answered is how the local orientation signals (in V1) are transformed into selectivity for complex shapes (in IT). To gain insights into the underlying mechanisms we investigated the neural basis of shape representation in area V4, an intermediate stage in this processing hierarchy.

Theoretical considerations and psychophysical evidence suggest that contour features, i.e. angles and curves along an object contour, may serve as the basis of representation at intermediate stages of shape processing. To test this hypothesis we studied the response properties of single units in area V4 of primates. We first demonstrated that V4 neurons show strong systematic tuning for the orientation and acuteness of angles and curves when presented in isolation within the cells' receptive field. Next, we found that responses to complex shapes were dictated by the curvature at a specific boundary location within the shape. Finally, using basis function decoding, we demonstrated that an ensemble of V4 neurons could successfully encode complete shapes as aggregates of boundary fragments. These findings identify curvature as a basis of shape representation in area V4 and provide insights into the neurophysiological basis for the salience of convex curves in shape perception.

Keywords: V4; monkey; electrophysiology; curvature; object-centered position; form perception; object recognition; vision

Introduction

One of the primary attributes of the human visual system is its ability to perceive and recognize objects. Humans can recognize an infinite number of complex shapes rapidly. Such recognition proceeds seemingly effortlessly under a variety of viewing conditions — different viewing angles and

distances, illumination levels, partial occlusion, etc. Such invariant recognition is a computationally challenging problem because, depending upon the viewing conditions and the presence of occluding objects, the image cast by the object on the retina can be dramatically different. Even so, the primate visual system segments the relevant parts of the shape from the scene and then perceives and recognizes the appropriate object with a speed and precision that is unmatched by even the most cutting-edge machine vision systems. Very little,

*Corresponding author. Tel.: +1–617-324-0132;
Fax: +1–617-258-7978; E-mail: anitha@mit.edu

DOI: 10.1016/S0079-6123(06)54016-6

however, is known about the neural basis of such visual shape recognition. Discovering the mechanisms that underlie object recognition will further our understanding about the workings of the primate brain. In addition, knowledge of the underlying brain mechanisms will help construct better automated recognition systems that can perform visual tasks as well as humans.

Ventral visual pathway

The ventral stream or temporal processing pathway in the primate brain is implicated in the processing of object shape and color information (Ungerleider and Mishkin, 1982; Felleman and Van Essen, 1991). The temporal pathway runs from primary visual cortex (V1) to V2 to V4 and into various regions of temporal cortex. This pathway has been worked out in detail in the macaque monkey (Felleman and Van Essen, 1991), and physiological studies in the macaque provide most of our knowledge about shape processing in the primate brain.

Neurophysiological studies addressing the question of shape representation in the ventral visual pathway suggest a hierarchical model for shape processing: successive stages are characterized by larger receptive fields (RFs) and more nonlinear response properties. In primary visual area V1 (the first stage of cortical processing), neurons have small RFs and the responses encode visual stimuli in terms of local orientation and spatial frequency information (Hubel and Weisel, 1959, 1965, 1968; Baizer et al., 1977; Burkhalter and Van Essen, 1986; Hubel and Livingstone, 1987). V1 simple cell responses are usually modeled as a linear weighted sum of the input over space and time (with output nonlinearities) and complex cell responses as a sum of the outputs of a pool of simple cells with similar tuning properties but different positions or phases. Such models fit observed data quite well (for a review see Lennie, 2003).

In the next processing stage, area V2, cells have larger receptive fields — on average the area of a V2 RF is approximately six times that of a V1 RF (based on data from Gattass et al., 1981). V2 neurons encode information about complex stimulus characteristics, in addition to local orientation and frequency information. Many V2 neurons are sensitive to illusory or subjective contours (while cells in V1 are not), and this selectivity is thought to be achieved by pooling from several end-stopped V1 cells (von der Heydt and Peterhans, 1989). There is also some evidence that V2 responses may encode stimulus characteristics such as polarity of angles and curves (Hegde and Van Essen, 2000), texture borders and stereoscopic depth cues (von der Heydt et al., 2000). V2 lesions impair discrimination of shapes defined by higher-order cues but not those defined by luminance cues (Merigan et al., 1993). Thus, while V1 responses primarily encode contours defined by luminance, V2 responses encode contours defined by second-order cues as well (Gallant, 2000).

Several lesion and neurophysiological studies have demonstrated that area V4, the next stage along the temporal processing pathway, plays a crucial role in form perception and recognition. Bilateral V4 lesions result in severe impairment in form discrimination (Heywood and Cowey, 1987). V4 lesions in macaques affect perception of intermediate aspects of stimulus form while sparing stimulus properties explicitly represented in V1 (Schiller and Lee, 1991; Schiller, 1995; Merigan and Phan, 1998). For example, V4 lesions reduce the ability to discriminate the orientation of illusory contours (De Weerd et al., 1996) and to identify borders defined by texture discontinuities (Merigan, 1996). In a human patient with a putative ventral V4 (V4v) lesion, Gallant et al. (2000) reported impairments in the discrimination of illusory contours, glass patterns, curvatures and non-Cartesian gratings.

RFs of V4 neurons are much larger than those in V1 and V2 — on average, at a given eccentricity, a V4 RF is 16 times the area of a V1 RF. In addition, V4 neurons have been reported to have large suppressive surrounds (Desimone et al., 1985). Thus, V4 neurons have access to stimulus information over a large region of the stimulus space and this is thought to serve global perceptual mechanisms, and contribute to the selectivity of V4 neurons to stimulus shape. Some V4 neurons, like complex end-stopped cells in V1 and V2, are selective for stimulus length, width, orientation and spatial frequency (Desimone and Schein, 1987) while others show greater nonlinearity in their responses than V2 neurons. Some cells encode edges explicitly (unlike V1 cells) as suggested by their strong responses to bars and square-wave gratings but not to sinusoidal gratings. Gallant et al. (1993, 1996) demonstrated V4 selectivity for non-Cartesian polar and hyperbolic gratings. The authors suggest that selectivity for polar gratings may mediate perception of curvature. Selectivity for hyperbolic gratings may play an important role in image segmentation, since these cells would respond well to line-crossings, a good source of information about image structure. Kobatake and Tanaka (1994) have demonstrated that many V4 neurons, unlike V2 neurons, respond stronger to complex than simpler stimulus features. Thus, while a lot of studies have demonstrated complex response selectivities in V4 neurons, no coherent, principled representational bases have been identified.

Unlike the areas that precede it, inferotemporal cortex (IT) has no discernible visuotopic organization (Desimone and Gross, 1979). IT neurons have very large RFs that almost always include the center of gaze and frequently cross the vertical meridian into the ipsilateral visual field (Gross et al., 1972). Anterior IT appears to be the last stage in the shape processing pathway, since its efferents project to areas in the temporal and frontal lobe that are not exclusively concerned with vision (Desimone et al., 1985). IT neurons are often selective for complex shapes like faces and hands (Gross et al., 1972; Perrett et al., 1982; Desimone et al., 1984; Tanaka et al., 1991). Several studies have demonstrated that IT neurons, at the single cell, columnar and population levels, encode information about features or parts of complex objects (Schwartz et al., 1983; Fujita et al., 1992; Kobatake and Tanaka, 1994; Wang et al., 1996; Booth and Rolls, 1998). Kobatake and Tanaka (1994) measured the "complexity" of the encoded features by measuring the ratio between best responses to complex and simple stimuli and found that the complexity of encoded shapes increased from V4 to posterior IT to anterior IT. These complex shapes often comprised simpler shapes with a specific spatial relationship between the component parts (Tanaka et al., 1991; Kobatake and Tanaka, 1994). Responses were sensitive to small changes or deletions of the critical attributes of the multipart objects (Desimone et al., 1984; Tanaka, 1993; Kobatake and Tanaka, 1994), but relative responses were insensitive to changes in absolute position (Sato et al., 1980; Schwartz et al., 1983; Ito et al., 1995). Studies about the dependence of IT responses on size, position and cue attributes of an object have demonstrated that single IT neurons can abstract shape properties from widely varying stimulus conditions (Sary et al., 1993).

Investigating shape representation in area V4

The results discussed above suggest that, in the earliest stage of shape processing, responses faithfully encode local stimulus characteristics. At successive levels, neural responses are dictated by larger regions of the visual field and become increasingly nonlinear functions of the stimuli. As a result, neural responses reflect the abstraction of stimulus form, with local characteristics becoming increasing irrelevant in determining the neural response. At the highest stages, responses encode the global shape and are largely invariant to local cue information, position, size, etc. How is this achieved, i.e. how are local orientation and spatial frequency signals in V1 transformed into selectivity for complex shapes in IT? To gain insights into the mechanisms that underlie this transformation of stimulus representation, we need to decipher the neural basis of representation at intermediate stages between V1 and IT. At successive stages of processing we need to discover the relevant stimulus dimensions and the quantitative relationship between those dimensions and the neural response. The relationship between the bases of representation at successive stages will reveal the computations required and will thereby provide insights into the underlying mechanisms. This will also provide clues about how these representational schemes might underlie perception and recognition.

To this end, we sought to discover the bases of shape representation in area V4, an intermediate stage in the ventral visual pathway. As described above, V4 neurons show selectivity for a wide variety of complex stimuli but the specific shape dimensions represented are as yet unknown. We therefore studied V4 neurons with the specific aim of discovering the underlying tuning dimensions. The ideal, most desirable, approach to investigating the basis of shape representation in any neuron would be to study the responses to a large set of complex natural stimuli created by uniformly sampling all of "shape space". Then, analytical approaches with minimal assumptions, such as spike-triggered covariance, could be used to extract the shape dimensions along which the responses vary maximally. However, even a coarse, uniform sampling of shape space would require thousands of stimuli, since object shape varies along a very large number of dimensions. For example, all possible combinations from a 10×10 pixel patch, whose pixels take one of two values (background or preferred color of cell), will result in 2^{100} stimuli. Such an exhaustive search approach would be impractical, since a primary experimental consideration is the length of time, on average 1–2 hours, a single cell can be studied in an awake behaving animal. A pragmatic alternative would be to explore shape space in a directed fashion for principled investigations of a specific hypothesis about shape representation. Depending on the hypothesis, an answer could be arrived at by targeted sampling of the relevant subregion of the extremely large shape space consisting of all possible two-dimensional (2D) shapes. We followed this approach to decipher the bases of representation in area V4. On the basis of theoretical, psychological and physiological studies we identified candidate dimensions for representation in area V4 (see below) and then studied the response characteristics of neurons as a function of the shape dimensions in question.

Contour characteristics as basis of shape representation

What might be a good candidate dimension of shape representation in area V4? To answer this question we turned to theoretical and psychophysical studies. Many modern shape theories and computational models propose a scheme of recognition by hierarchical feature extraction, i.e. objects are first decomposed into simple parts that are pooled at subsequent stages to form higher-order parts. This approach has been motivated in part by physiological findings of selectivity for simpler features in earlier stages (see above). It has also been motivated by the human ability to discern the various parts of an object and their relative positions in addition to recognizing the object as a whole. There is also some psychological evidence for the hypothesis that recognition proceeds by parsing shapes into component parts and then identifying the parts and the spatial relationships between them (Biederman, 1987; Biederman and Cooper, 1991). Finally, from a practical and informational standpoint, such basis representations are advantageous — they provide data compression and are therefore compact and good for further computations.

Feature-based models for visual shape representation vary widely in the number of hierarchical levels and the nature of features or shape primitives extracted at each stage. Beyond local edge orientation, which most models invoke as the first-level feature, shape primitives extracted at intermediate levels are related to the object boundaries (Attneave, 1954; Milner, 1974; Ullman, 1989; Poggio and Edelman, 1990; Dickenson et al., 1992) or to its volume (Biederman, 1987; Pentland, 1989). While boundary-related features (i.e. contour features) include 2D angles and curves and three-dimensional (3D) corners, curved surface patches and indentations, solid or volumetric primitives (generalized cones or geons) are simple 3D shapes such as cylinders, spheres, etc. defined by the orientation of their medial axes and several cross-sectional attributes.

Several theoretical studies have asserted the importance of contour features as a basis of representation: they are high in information content and lead to an economical representation (Attneave, 1954), they are largely invariant to deformation (Besl and Jain, 1985) and can be constructed from local edge orientation and curvature signals (Milner, 1974; Zucker et al., 1989). Further, they form natural parts for constructing more complex representations. Psychological findings imply specialized mechanisms for the perception of contour features: angle perception acuity is higher than that predicted by component line orientation acuity (Chen and Levi, 1996; Heeley and Buchanan-Smith, 1996; Regan et al., 1996), the detection thresholds for curvilinear glass patterns is much lower than that for radial glass patterns (Andrews et al., 1973; Wilson et al., 1997) and detection of curved targets among straight distractors is faster than that for straight targets among curved distractors (Triesman and Gormican, 1988; Wolfe et al., 1992).

Thus, based on theory and psychophysics, contour features seem to be a good choice for the basis of shape representation beyond local orientation. But does the primate visual system actually extract contour features as intermediate-level shape primitives beyond oriented edges, in the process of recognizing visual shape? In this paper we describe experiments that explicitly test this hypothesis. If contour features, i.e. convex projections and concave depressions along the object contour, are indeed extracted in area V4 then one would expect:

a) At least some cells to be strongly and systematically tuned to contour features.
b) Responses to complex shapes to be dictated primarily by specific contour features along the object contour.
c) Population representation of complex shapes, in terms of their component contour features, to be complete and accurate.

Below, we demonstrate that each of these statements is true for area V4. In our experiments, roughly 33% of V4 neurons showed strong systematic tuning for contour features, i.e. angles and curves, presented in isolation within their RFs. (Pasupathy and Connor, 1999). Responses of many V4 neurons to moderately complex shapes are dictated by the curvature of the contour at a specific location relative to object center (Pasupathy and Connor, 2001). Finally, the estimate of V4 population representation, derived from single cell responses and curvature tuning properties, provides a complete and accurate representation of complex 2D contours (Pasupathy and Connor, 2002). These results suggest that contour characteristics (parameterized in terms of contour curvature) at a specific spatial location relative to object center are a basis

for shape representation in area V4. Certainly there are other important shape dimensions encoded in area V4 (see the section "Discussion"). Our results described below demonstrate that contour features are an important dimension.

Tuning for isolated contour features in area V4

If contour features are a basis of representation then we would expect single cells to respond selectively to angles and curves presented in isolation within the cell's RF. This is the first question we addressed with our experiments.

We studied the responses of 152 V4 neurons to a large parametric set of single contour features. Many V4 neurons showed strong systematic tuning to these contour feature stimuli responding preferentially to angles and/or curves oriented in a specific direction (Pasupathy and Connor, 1999). Fig. 1A shows the responses of a V4 neuron to the angles and curves that we used in this study. Each icon shows a stimulus with the background gray level representing the average response rate based on five stimulus repetitions. The stimuli were angles, curves or straight edges, presented in isolation within the center of the RF, in the optimum color for the cell under study (shown here in white) as determined by preliminary tests. Stimulus luminance was constant within the estimated RF boundary and gradually faded into the background gray over a distance equal to estimated RF radius.

The stimulus set included convex, concave and outline angles and curves that pointed in one of eight directions at 45° intervals (orientation 45°, 90°, etc.) and subtended one of three angular extents (acuteness: 45°, 90°, 135°). Four edges and lines were also included for a total of 156 stimuli. Stimuli were flashed for 500 ms each and were separated by an interstimulus interval of 250 ms. A sequence of five stimuli comprised a trial. The entire stimulus set was presented in random order without replacement five times.

For the cell in Fig. 1A, response rates range from 0 (light gray) to 42 ± 2.3 spikes/s (black) as indicated by the scale bar. Responses to acute and right angle convex stimuli pointing in the 135–180° range were strong while the same stimuli drawn as

298

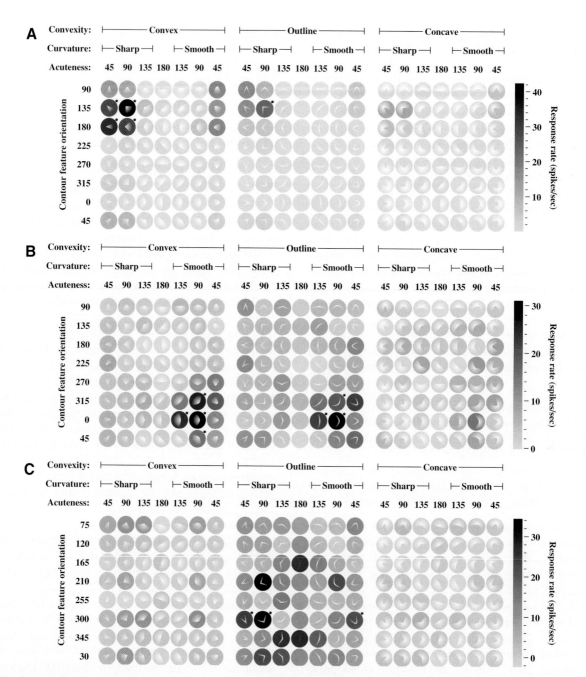

Fig. 1. Examples of V4 responses to contour feature stimuli. Each icon represents a stimulus drawn against a dark background. Circular boundaries were not a part of the actual stimulus display. Each stimulus consists of a straight edge, angle, or curve drawn as a convex projection, outline or concave indentation, centered within the RF. Boundaries are sharp within the RF and gradually fade into the background. Entire extent of fading is not shown here. (A) Example response pattern showing contour feature tuning. Responses of a cell showing strong tuning for sharp angles pointed in the 135–180° range. Background gray levels indicate average responses over five repetitions as per scale bar on the right. (B) Another example of contour feature tuning. Responses show tuning for curves pointed in the 315–360° range. (C) Example response pattern showing orientation tuning. Cell responded strongly to many stimuli that contain a 75° oriented edge or line. Figure originally published in Pasupathy and Connor (1999).

outlines elicited moderate responses. Responses were weaker to smooth (vs. sharp), obtuse (vs. acute) and concave (vs. convex) features. The responses of this cell contribute to a clear, strong, unimodal peak in the orientation × acuteness contour feature space (for detailed analysis see Pasupathy and Connor, 1999). These results cannot be explained in terms of orientation tuning for individual edges, since many of the least effective stimuli (including the straight edges) contain the same edge orientations as the most effective stimuli. This result is strikingly different from the recently reported angle selectivity in area V2 (Ito and Komatsu, 2004). Unlike in V4, angle-selective V2 units responded in comparable amounts to the preferred angle and to one or both of its component end-stopped lines.

A second example of contour feature tuning is shown in Fig. 1B. Convex and outline curves oriented at 315°–360° elicited strong responses from this cell. Responses were strongest to 90° curves but the cell responded well to some obtuse curves as well. Unlike the previous example, this cell did not respond to sharp angle stimuli. Here again, stimuli that drive the cell are clustered together contributing to a single strong peak of high responses in the stimulus space.

Unlike the previous examples, the cell in Fig. 1C did not show clustering of strong responses in the stimulus space. This cell responded well to a variety of angle and curve outline stimuli containing edges oriented near 75° as part of the stimulus. In the preliminary bar orientation test, this cell demonstrated strong orientation tuning with a peak at 75°. The response pattern exhibited by this cell reflects this orientation tuning, since the responses to contour features are primarily dictated by the component orientations of the contour feature stimuli.

Contour feature tuning described above (Figs. 1A, B) cannot be explained in terms of tuning along simpler stimulus dimensions such as edge orientation, contrast polarity or spatial frequency. Component edges of the optimal stimuli also appeared in other stimuli that failed to evoke strong responses from the cell. Secondly, tuning for the orientation of a contour feature was consistent across the three (45°, 90°, 135°) acuteness values despite differences in component edge orientation

and, finally, responses to contour feature stimuli were almost always greater than to single lines or edges. Tuning for spatial frequency or contrast polarity fails to explain contour feature tuning because both optimal and nonoptimal stimuli were composed of similar spatial frequencies and contrast polarities. Further, contour feature tuning was consistent across convex, concave and outline stimuli despite substantial differences in contrast polarity and spatial frequency content. Consistency of tuning across convex, concave and outline stimuli also argue against explanations in terms of area of stimulation or differential surround stimulation hypotheses. Finally, contour feature tuning cannot be explained in terms of hotspots in the RF because tuning profile (rank-order of responses) was similar at different positions in the RF (Pasupathy and Connor, 1999).

Roughly one-third of the entire sample of 152 V4 cells showed strong unimodal peaks (as in Figs. 1A, B) in this orientation × acuteness contour feature space. Most cells responded best to either convex or outline stimuli. This bias towards convex features parallels findings of perceptual significance of convex features reported by several psychophysical studies (see the section "Discussion"). A majority of cells also responded preferentially to 45° (more acute) stimuli but this acuteness bias was weaker than that for convexity. Many cells responded equally well to both angle and curve counterparts while other cells were clearly biased toward either sharp or smooth stimuli. In the case of curves, the angle subtended by the contour feature is analogous to contour curvature — more acute angles correspond to higher curvature. Thus, tuning along the acuteness dimension suggests that curvature may be an important shape dimension encoded by V4 neurons. We explored this further in the next set of experiments.

The results above demonstrate that V4 neurons carry explicit information about the orientation and acuteness of contour features when presented in isolation. This suggests that angles and curves along the object contour could serve as intermediate-level primitives, and that V4 neurons could encode 2D contours in terms of their component angles and curves. To ascertain if they do, in our second set of experiments, we tested how V4 neurons respond when contour features are presented in the context of a closed 2D contour. Do V4 neurons tuned for contour features represent complex shapes in terms of their component contour features, i.e. are the responses to complex shapes dictated by convex and concave features along the shape contour? To answer this question we studied the responses of single V4 neurons to a set of complex shapes created by a systematic combination of convex and concave contour features and the results are discussed below.

Conformation of 2D contours dictate responses of V4 neurons

If contour features are a basis for shape representation in area V4, then responses of V4 neurons to complex-shape stimuli will carry reliable information about the component contour features of the shape in question. To test this question, in the second set of experiments, we studied the responses to complex-shape stimuli of 109 V4 neurons that showed preferential responses to contour features in preliminary tests (Pasupathy and Connor, 2001). A large parametric set of complex shapes, created by systematic variation of curvature along the object contour, was used to study the responses of V4 neurons. An example is shown in Fig. 2.

Each stimulus (shown in white) consisted of a closed shape, presented in the optimal color for the cell, in the center of the RF (represented by the surrounding circle). All parts of all stimuli lay within the estimated RF for the cell under study. The stimulus set consisted of 366 complex shapes with varying combinations of convexities and concavities along the contour. Stimuli had two, three or four convex projections that were separated by 90°, 135° or 180°. Convex projections were sharp points (0° location stimulus 2) or medium convexities (135° location stimulus 2). The intervening contours between the convex projections were rendered as arcs of circles whose curvature and convexity were defined by the angular separation and curvature of the adjoining convex projections. Smooth transitions between contour segments were achieved by rendering the stimuli as piecewise cubic b-splines through the control points associated with the contour segments. Stimuli were presented in eight orientations (*horizontal* within each block) at 45° intervals except for stimuli that were rotationally symmetric. The stimulus set also included two circles with diameters 3/16 and 3/4 of estimated RF diameter. Stimulus presentation protocol was as in experiment above.

For the example in Fig. 2, response rates ranged from -6.3 ± 0.0 (light gray) to 38.1 ± 7.0 spikes/s (black). This cell responded strongly to a wide variety of shapes that had a sharp convexity in the lower left corner of the object (angular position 225°) with an adjoining concavity in the counterclockwise direction. Stimuli with a medium convexity at 225° elicited a moderate response from the cell. For example, compare the responses to stimuli *1* and *3* or stimuli *2* and *4*, which differ in their contour only at angular position 225°. In both cases the stimulus with the sharp convexity (stimuli *1* and *2*) elicited the stronger response. In contrast, stimuli with a broad convexity or a concavity at 225° did not drive the cell. Curvature of the contour from 0° to 90° varied widely across stimuli that evoked strong responses from the cell. The results described so far suggest that the contour curvature at a specific angular position strongly dictates the responses to complex shapes.

To quantify the functional relationship between the contour characteristics and the neural response, we represented each stimulus in a multidimensional shape space in terms of its contour characteristics and then modeled the relationship between the stimulus dimensions and the neuronal responses. Each stimulus in our set can be approximated by 4–8 constant curvature segments that are connected by short segments with variable curvature entirely determined by the adjoining segments. Therefore, each shape can be uniquely represented using a discretized and simplified representation of just the constant curvature segments. These constant curvature segments that comprise the stimuli could be uniquely described just in terms of the angular position relative to center (or tangential orientation) and contour curvature. For a general unconstrained 2D closed contour unique representation would require additional dimensions such as position along the

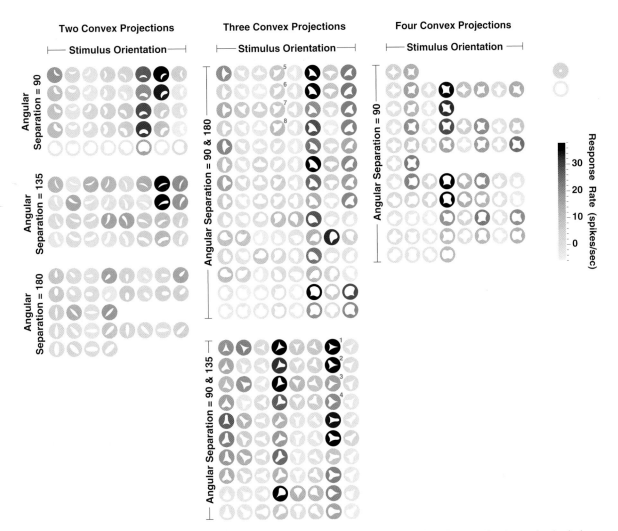

Fig. 2. Example cell from the complex shape test. Within each dark circle, which represents the RF, a complex shape stimulus is drawn in white. The circular boundaries were not an actual part of the visual display. Stimuli were constructed by systematic variation of the convexities and concavities along the contour. Stimuli are divided into groups on the basis of the number of convex projections and the angular separation between convex projections. Background gray level represents average response, to the corresponding stimulus, as per scale bar on the right. Shapes with a sharp convexity pointing to the lower left and an adjoining concavity in the counterclockwise direction elicit strong responses from this cell. Numbers to the upper right of stimuli correspond to references in text. Original figure published in Pasupathy and Connor (2001).

contour, radial position relative to the center of the stimulus and local tangential orientation. Here, since several of these dimensions co-vary, representation in terms of just curvature and angular position is unique.

In this scheme, each stimulus is represented by four to eight points in the curvature by angular position space. For example, stimulus *1* is represented by six points: one for each of the convex

points and the intervening concavities. To quantify the dependence of the neural response on angular position and contour curvature, we modeled the neural response as a 2D Gaussian function (product of two 1D Gaussians with no correlation terms) of contour curvature and angular position. Here, response of a neuron to a stimulus was the maximum of the responses predicted by its component contour segments. The parameters for these

302

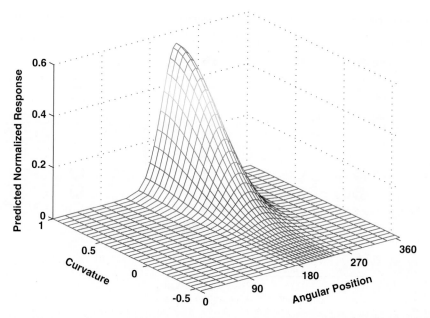

Fig. 3. Best-fitting 2D model for the responses illustrated in Fig. 2. Horizontal axes represent angular position and contour curvature. Vertical axis and surface color represents normalized response predicted by the 2D angular position-curvature model. The Gaussian model had a peak at 229.6° (angular position) and 1.0 (curvature) predicting strong responses for stimuli with sharp convexities pointing to the lower left. Original figure published in Pasupathy and Connor (2001).

curvature × position tuning functions (peak position and standard deviation (SD) for each of the two dimensions and the amplitude of the Gaussian) were estimated by minimizing the sum of squared errors between predicted and observed responses (for further details, see Pasupathy and Connor, 2001).

The tuning surface predicted by the best fitting curvature × position model, for the example in Fig. 2, is shown in Fig. 3. The horizontal axes represent angular position (0°–360°) and contour curvature (−0.31–1.0; negative values represent concavities and positive represent convexities). The z-axis and surface color represent the predicted normalized response. The peak of the Gaussian surface is at contour curvature equal to 1.0 and angular position equal to 229.6°, predicting strongest responses to stimuli with a sharp convex projection at or near the lower left corner of the object. Tuning along the angular position dimension was narrow (SD = 26.7°), implying that a small range of positions (centered at 229.6°) of the sharp convexity evoked strong responses. Predicted responses were close to zero for stimuli with broad convexities and concavities at the lower left. The

goodness of fit of the model, in terms of correlation between observed and predicted values, was 0.70.

The predictions of the 2D model do not accurately reflect the responses of the cell. The model predicted high responses to all stimuli with a sharp convexity at the lower left of the object. However, several stimuli with sharp convex projections at the lower left elicited weak responses from the cell (see stimuli 5–8, Fig. 2). Among the stimuli with a sharp convexity at the lower left, those with a concavity in the counterclockwise direction (~270°) evoked stronger responses than those with a broad convexity in that same location (stimuli 5–8). Thus, an adjoining concavity is essential for a strong response.

To include the influence of the adjoining contour segments, we built a four-dimensional (4D) curvature × position model. In addition to angular position and the corresponding contour curvature, we included curvatures of the adjoining contour segments as independent variables. The 4D model was a product of four 1D Gaussians with no second-order correlation terms. This model had nine parameters: peak position and SD associated with

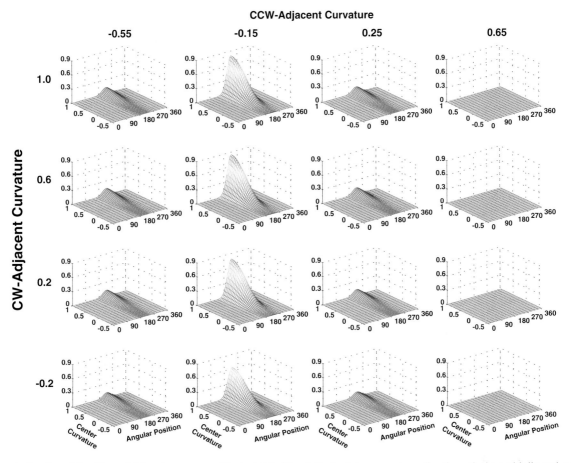

Fig. 4. Results of the 4D model for the responses illustrated in Fig. 2. Each plot represents a slice through the multi-dimensional Gaussian surface. The slices are perpendicular to the CCW and CW curvature dimensions. In the 4×4 grid of plots, horizontal position of the plot represents CCW curvature and vertical position represents CW curvature, as labeled along the top and left, respectively. The horizontal axes of each plot represent angular position (0–360°) and curvature of the central contour segment. The peak of the Gaussian model was located at 230.04° (angular position), 1.0 (curvature of central contour segment), −0.15 (curvature of CCW segment), 0.6 (curvature of CW segment). Model predicts strong responses for shapes with a sharp convexity pointing to the lower left when the adjoining CCW segment is concave. Influence from the CW segment is minimal. CCW: counterclockwise, CW: clockwise. Figure from Pasupathy and Connor (2001).

each of the four dimensions and a parameter for the amplitude of the Gaussian. Fig. 4 illustrates the tuning profile predicted by the 4D model for the example in Fig. 2. This figure shows 16 surface plots, each of which is a slice through the multi-dimensional Gaussian model at different values for the adjoining curvature segments. Horizontal position of each plot represents curvature of the counterclockwise (CCW) contour segment, and vertical position represents curvature of the clockwise (CW) curvature segment. For example, the surface plot in the lower left corner represents the

curvature × position tuning profile for the cell when the adjoining contour segments are both concave (curvatures: CCW = −0.55, CW = −0.2). The horizontal axes for each surface plot represent contour curvature and angular position of the central contour segment, and the z-axis plots predicted normalized response. The peak of the 4D model was at 230.04° and 1.0 along the angular position and central curvature dimensions, similar to the results of the 2D model. In addition, the curvature of the CCW segment had an influence on the predicted response as illustrated by the scaling of the Gaussian peak with CCW curvature (across each row). The tuning profile along the CCW curvature dimension had a peak at broad concavity (curvature = –0.15, second column), and the predicted response dwindled with more convex or concave curvature values. Thus, for stimuli with a sharp convexity at the lower left of the object, the model predicted strong responses when it was flanked by a concavity in the CCW direction and weaker responses when it was flanked by a broad convexity (such as stimuli 5–8, Fig. 2). In contrast, the curvature of the CW segment exerted a weak influence on the response profile as indicated by the almost identical tuning profiles down each column. Tuning along the angular position dimension was narrow (SD = 25.8°), implying that the preferred pattern of contour curvature evoked a strong response if it occupied a small range of angular positions in the lower left of the object.

The SD parameters for the CCW, CW and central curvature dimensions were 0.21, 1.08 and 0.5, respectively. High SD (compared to 1.31, which is the range of curvatures sampled) implies a flat, broad-tuning profile with a small maximum–minimum response difference over the range of curvatures sampled, and therefore a weak dependence of the response on the corresponding dimension. A low SD implies narrow tuning with a large maximum–minimum response difference, and therefore a strong influence on the response. Just as illustrated by the surface plots in Fig. 4, the SD parameters for the 4D model imply a strong dependence of the predicted response on the central and CCW contour segments. The goodness of fit represented by the correlation coefficient was 0.82 for this 4D model.

The coefficient of correlation between observed and predicted responses (r) was used to assess the goodness of fit of the 2D and 4D curvature × position models. For 101/109 cells, the 2D model produced a significant fit (F test, $p < 0.01$) and the median of this distribution was 0.46. Inclusion of the two adjoining curvature segments in the 4D model significantly improved the fit in 93/109 cases (partial F test, $p < 0.01$) and the median of the r distribution increased to 0.57. The model's validity was confirmed by predicting responses to stimuli that were not used to derive the tuning functions.

Alternate hypotheses

Can the response patterns described above be explained in terms of previously described tuning properties of V4 neurons such as edge orientation, bar orientation or contrast polarity? We tested these alternate hypotheses explicitly and found that none explained the results presented here. For each cell, we modeled the responses to complex shapes as a function of edge orientation and contrast polarity and assessed the goodness of fit. We also modeled responses as a function of bar orientation, length and width. Third, we tested whether responses to complex shapes were a function of the mass-based shape of the object by representing each shape in terms of its principal axis orientation and the aspect ratio of the approximating ellipse. For most cells, the predictive power of these models was poor (median $r < 0.3$) and the 2D and 4D curvature × position models provided superior fits to the observed data. The 2D model had as many or fewer parameters than these alternate models. Therefore, the quality of fits was not simply a function of the number of parameters in the model but a true reflection of the cell's tuning along the corresponding stimulus dimensions (such as curvature, edge orientation, etc.).

Another potential explanation is that the apparent shape tuning is due to selective activation of hotspots in the RF. In other words, rather than the shape of the bounding contour, the strength of the response may simply depend on whether certain subregions of the RF are stimulated by the visual stimulus. If this were the case then shape tuning would be dependent on the position of the stimulus within the RF. We investigated position dependence of shape tuning in roughly one-third of the neurons. For each cell, we presented a subset of stimuli at multiple positions within the RF. The subset included at least one stimulus that had the preferred contour segment (as determined by its responses to complex shapes and the angular position-curvature model) as part of its contour and another stimulus that did not. In all we found that responses to all stimuli were modulated as a function of the position of the stimulus in the RF. However, the responses to the preferred stimulus were greater than those to the nonpreferred for all stimulus locations, i.e. while responses decreased in magnitude the preferred features were consistent at all locations within the RF (Pasupathy and Connor, 2001).

The curvature × position models discussed so far test whether complex shape responses were dictated by a specific sector along the object contour. Another possibility is that responses were dictated by two (or more) noncontiguous regions of the contour. We investigated this hypothesis by modeling the neuronal responses to complex shapes as a function of two noncontiguous sections of the contour separated by 180°. We restricted the angular separation to 180°, since curvatures of contour segments closer than 180° were not entirely independent of each other. This 3D model, consisting of one angular position and the curvatures of two contour segments as independent variables, had seven parameters. The performance of the model (median $r = 0.47$) was similar to the 2D curvature × position model, i.e. the second (noncontiguous) contour segment made a negligible contribution to the predictive power of the model. This suggests that the curvature × position models with contiguous contour segments provide a better description of the responses of V4 neurons. A sum of Gaussian's model, where each Gaussian was 4D, further confirmed the negligible influence of other regions of the contour.

As alluded to earlier, in this stimulus set object-centered angular position and tangential orientation for the curvature segments comprising each shape were completely dependent on each other, since the curvature segments were pointed directly away from or towards the object center. With a secondary test, we sought to determine whether the true dependence of the neural response was on object-centered angular position or on orientation of the corresponding curvature segment. We tested cells with a stimulus set (Fig. 5), in which these two factors are partially segregated, since the convex extremities are offset (orthogonal to the main axis) in the CW and CCW directions. If curvature orientation (i.e. the pointing direction of the convex extremities) were the only significant dimension, then tuning patterns should be similar (though scaled in amplitude) across the three blocks of rows in Fig. 5. In fact, the tuning patterns are very different. In the top block, responses are strongest

for the CCW orthogonal offset; in the middle block responses are strongest for the stimuli with no orthogonal offset; in the bottom block, responses are strongest for the CW orthogonal offset. This reversal pattern reflects tuning for angular position: at each orientation, the most effective stimuli are those with convex extremities near 45° relative to the object center. A similar reversal pattern was observed for 23/29 cells. For each of the 29 cells we determined the orthogonal offset position that excited the cell most for each of the three stimulus orientations tested. A linear regression between the orthogonal offset position and stimulus orientation had a negative slope in 23 cases indicating a reversal pattern in the responses similar to the pattern shown here. Also, interaction between orthogonal position and stimulus orientation was significant in all 23 cases that showed selectivity for polar position of the feature relative to the object center. These findings suggest that the position of boundary pattern relative to object center does dictate the neural response in V4.

The experiments described so far demonstrate that the neurons in area V4 are tuned to contour curvature at an object-centered position. This tuning cannot be explained in terms of other previously described properties of V4 neurons. The responses cannot be described in terms of hotspots in the RF. Thus, these results suggest that boundary conformation, which can be parameterized in terms of contour curvature, at a specific position relative to object center, may be a basis of shape representation in area V4. If this were true then we would expect a complete and accurate representation of 2D shape contours in terms of contour curvature in the V4 population response. We tested this hypothesis explicitly and the results are described below.

Ensemble representation of 2D contours in area V4

How complete and accurate is the representation of 2D contours at the population level in area V4? We tested this question by deriving the population representation for each shape in our stimulus set using basis function decoding. As a first approach to population analysis, we focused on the simpler curvature × position 2D domain. All 109 cells that

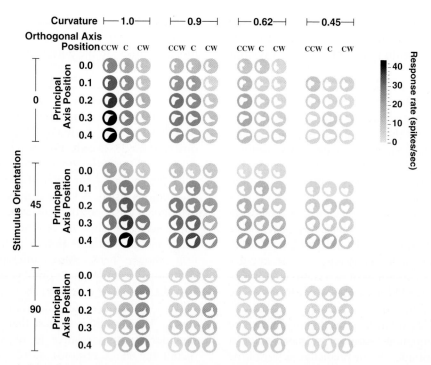

Fig. 5. Example result from the secondary shape test. Stimulus set consists of distorted versions of a teardrop-shaped stimulus. Distortions were created by varying position of the convex point along the principal and orthogonal axes and by varying the curvature of the convex point and the overall stimulus orientation. CCW: counterclockwise; C: center; CW: clockwise. As above, circular disks were not a part of the stimulus display. Background gray levels represent average response rates as per gray scale bar. Responses are strongest to sharp convexities (leftmost vertical block) that are farthest along the principal axis (last row in each horizontal block). Tuning for orthogonal position (horizontal within each block) and stimulus orientation (vertical axis) are interdependent such that responses are strongest for stimuli with the sharp convex point at the upper right corner of the object. For instance, at 0° stimulus orientation, CCW orthogonal position elicits the strongest response, at 45° center elicits the strongest response, and at 90° CW elicits the strongest response. Figure from Pasupathy and Connor (2001).

we recorded from were included in the population analysis.

We used the curvature × position tuning functions (e.g. Fig. 3) to estimate the V4 population response to each shape in our stimulus set. Fig. 6 shows the results of this analysis for an example stimulus, the "squashed raindrop" shape shown at the center. The curvature × position functional representation of this shape is drawn as a white line in the surrounding polar plot. This function has peaks and troughs corresponding to the major features of the shape: a medium convex peak at 0° (right), a concave trough at 45° (upper right), a sharp convex peak at 90° (top), etc. Our analysis was designed to determine whether similar curvature/position information could be decoded from the neural responses.

To derive the population response, we scaled each cell's tuning peak by its response to the shape in question. Thus, each cell "voted" for its preferred boundary fragment with strength proportional to its response rate. The entire set of scaled tuning peaks defined a surface representing coding strength for all combinations of curvature and position. The (Gaussian) smoothed resulting surface is shown as a color image in Fig. 6. Here, color represents the strength of the population representation for the corresponding curvature/position combination. Red represents strong population representation for the corresponding feature, and blue represents weak population representation. The actual curvature function is re-plotted in white for comparison. The population surface contains peaks (red) corresponding to the major boundary features of the

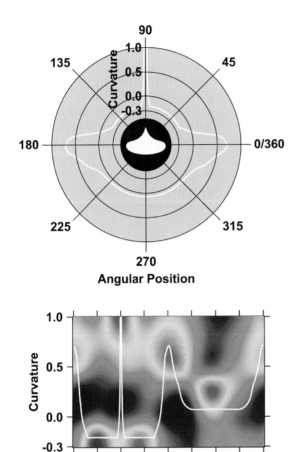

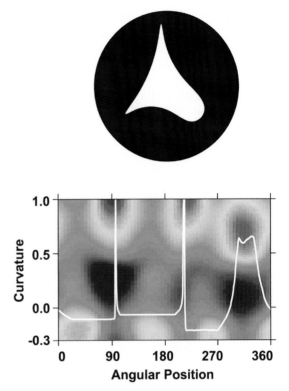

Fig. 6. Estimated population response for an example shape. *Top*: The shape in question is shown in the center. The surrounding white line plots boundary curvature (radial dimension) as a function of angular position (angular dimension) in polar coordinates to highlight the correspondence with boundary features. *Bottom*: Estimated population response across the curvature × position domain (colored surface) with the true curvature function superimposed (white line). *X*-axis represents angular position; *Y*-axis represents curvature. Color scale runs from 0.0 (blue) to 1.0 (red). The peaks and troughs in the curvature function are associated with peaks (red) in the population representation. All the prominent boundary features of the shape are strongly encoded by the V4 population representation. Figure from Pasupathy and Connor (2002).

stimulus: the sharp convexity at 90°, the medium convexities at 0° and 180°, the broad convexity at 270°, and the concavities at 45° and 135°.

A second example is shown in Fig. 7. Again, the population representation contains peaks associated

Fig. 7. Estimated population response for a second example shape. *Top*: The shape. *Bottom*: Estimated V4 population response is shown with the curvature function superimposed in white. Axes and details as in Fig. 6. The three strong peaks (red) at convex curvatures and the somewhat weaker peaks (yellow) at concave curvatures correspond well, in terms of curvature and angular position, with the six prominent boundary features of the shape in question.

with all of the major boundary features of the stimulus: sharp convexity at 90° and 225°, medium convexity at 315°, and the three intervening concavities. We obtained similar results for all 49 shapes in our stimulus set — the V4 population representation had peaks associated with all of the major boundary features for each shape. Thus, the V4 population signal provides a complete representation of 2D shape contours in terms of contour curvature and angular position. To assess the accuracy of the population representation, for each shape we computed the mean absolute difference between the population peaks and the nearest points on the true curvature × position function. For all shapes the errors in representation were small in both dimensions: median difference was 4.04° along the angular position dimension and 0.0704 along the curvature dimension. Thus, the peaks in the population surface corresponded closely to the boundary components in the original shape and therefore provided a complete and accurate representation of the stimuli used here. Thus, the V4 population signal carries all the information needed to represent, perceive and recognize the 2D shape contours used here.

Discussion

Previous studies suggest that area V4 encodes shape in terms of edge orientation, spatial frequency, and bar orientation, length and width (Desimone and Schien, 1987). The results presented here add another pair of dimension to this list — contour curvature and object-centered position. Individual V4 neurons carry information about the boundary conformation at a specific location along the shape contour and the V4 population as a whole encodes a complete and accurate representation of the entire 2D contour in terms of its boundary configuration. With these results we can now successfully describe the responses of many V4 neurons to a more general class of curves; such a description using edge and bar orientation models alone is poor. Thus, these findings extend our understanding of V4 responses to a larger class of 2D shapes.

Our results are consistent with studies demonstrating selectivity for complex shapes with curved parts in area V4 (Gallant et al., 1993; Kobatake and Tanaka, 1994; Gallant et al., 1996; Wilkinson et al., 2000) and IT (Schwartz et al., 1983; Tanaka et al., 1991; Janssen et al., 1999). They also reinforce previous evidence for the extraction of coarse curvature information at earlier levels by end-stopped cells (Heggelund and Hohmann, 1975; Dobbins et al., 1987; Versavel et al., 1990). Recently, Ito and Komatsu (2004) suggested that the process of angle extraction may start but not attain completion in area V2 because angle-selective V2 units respond in comparable amounts to the component end-stopped lines as well. Perhaps then, this process of angle and curve encoding which starts in V2 (or even V1) culminates in the representation of 2D boundaries in terms of their contour features as demonstrated here.

The dimensions used to quantify the encoded boundary features need not have been curvature and angular position — other parameterization schemes that describe contour characteristics might be as or more effective. However, as discussed above, results cannot be explained in terms of mass-based orientation parameters, tuning for oriented bars or lower-level factors such as edge orientation, contrast polarity and spatial frequency. Also, response patterns exhibited by these cells cannot be explained in terms of area of stimulation or by differential surround stimulation. Thus, many V4 neurons carry information about 2D object boundaries in terms of component shape and position. At the population level, this representation is complete and accurate. These results suggest that V4 neurons represent complex 2D contours in a piecewise fashion, encoding information about curvature and other characteristics of a section of the shape boundary.

Implications for perception

Several psychophysical studies have reported high sensitivity in humans for the detection of curved elements in visual displays (Andrews et al., 1973; Triesman and Gormican, 1988; Wolfe et al., 1992; Wilson et al., 1997). Our results demonstrating explicit representation of curvature in area V4 provides a physiological basis for this increased sensitivity to curves. In the experiments presented above, when contour features were presented in isolation and in the context of complex shapes, selectivity for sharp convexities was overrepresented while that for concavities was rare. The distribution of Gaussian tuning peaks along the curvature axis was independent of RF eccentricity and showed a consistent bias toward sharp convexities at all eccentricities. This bias was evident even when curvature was finely sampled (in the secondary fine-scale tuning test). These findings suggest a biased representation of shape in terms of sharp convexities. In the complex-shape study, we sampled curvatures in the range of -0.31 (medium concavity) to 1.0 (sharp convexity). It is possible that sharp concavities are also strongly represented in the cortex and the true bias is in favor of sharp curvatures rather than sharp convexities. However, results from the first experiment with isolated contour features suggested otherwise. In those experiments, the same ranges of concave and convex curvatures were sampled and the results showed a strong bias in favor of convexities. Taken together, these results suggest a bias toward shape representation in terms of sharp convexities.

The overrepresentation of neurons that encode shape in terms of convex projections makes sense from a functional point of view — encoding object shape in terms of sharp convexities leads to a highly efficient and economical shape description, since high curvature regions appear to be rich sources of shape information (Attneave, 1954). The neuronal bias toward convex features may also underlie their perceptual salience. Researchers have found that convex projections, rather than concave indentations, provide the basis for figure/ground interpretations (Kaniza and Gerbino, 1976) and shape similarity judgements (Subirana-Vilanova and Richards, 1996) in human observers. The greater representation of neurons that encode shape in terms of convex projections may underlie the perceptual significance of convex features. Our results provide neurophysiological evidence in support of the "curvature minima" rule (Hoffman and Richards, 1984), which hypothesizes that object segmentation, for the purpose of shape recognition, should occur along boundaries of maximum concavity, producing convex parts. Psychophysical results (Braunstein et al., 1989) in humans support this rule — observers are more likely to recognize parts from a previously viewed object if the parts are convex. Our results suggest that shape representation in the ventral visual pathway proceeds in accordance with the curvature minima rule.

Transformation from local oriented signals to complex shape selectivity

How might tuning for angles and curves in area V4 arise from tuning for local orientation and spatial frequency information available at earlier visual areas? Theorists have proposed that tuning for sharp angles could be achieved by an appropriate combination of end-stopped orientation signals (Milner, 1974; Hummel and Biederman, 1992). For example, the response pattern in Fig. 1A could be achieved by combining the signals from units tuned to a 0°-oriented edge end-stopped at the top and a 90° edge end-stopped at the right, with a preference for a specific contrast direction (brighter to the right). The end result would be a signal related to the presence of a sharp corner pointing to the upper left. The tuning width of the pooled units may then produce the tuning for contour feature orientation and acuteness, and integrating signals from multiple positions could yield position invariance. Transformation from oriented signals to angle representation possibly requires multiple stages of processing, since both in V2 (Ito and Komatsu, 2004) and in V4 there are intermediate units that respond strongly to the preferred angle and its component end-stopped lines.

Cells selective for angles constructed in the above fashion would also respond to curves that contain the appropriate component orientations. Cells that respond preferentially to curves and are tuned to contour curvature may also be derived by integrating an appropriate pattern of local orientation and curvature signals from end-stopped cells in preceding visual areas. Recently Cadieu et al. (2005) have demonstrated that V4-like curvature tuning can be achieved by combining inputs from position-invariant orientation-selective Gabor filters. This is equivalent to pooling appropriate orientation selective units from V1 or V2, followed by nonlinear processing, to derive the necessary curvature-tuned responses. Alternatively, selective responses to curved contours in area V4 may be achieved by integration of V2 signals modeled as linear-nonlinear-linear filters (Wilson, 1999). A third hypothesis suggests that position-independent tuning for corners and curves may be achieved by the processing of local orientation end-stopped signals in the dendrites of V4 neurons (Zucker et al., 1989). Further experiments are required to decide between these possible mechanisms.

At a more global level, the progression from local orientation signals to selectivity for contour conformation supports the proposal of many feature-based shape recognition schemes that suggest contour features as intermediate level shape primitives. Our results imply a coding scheme based on structural description of the shape, consistent with the notion of representation by parts or components. As outlined earlier, in these models shapes are described in terms of the conformations and relative positions (and/or connectivity) of their simpler components (Marr and Nishihara, 1978; Hoffmann and Richards, 1984; Biederman, 1987; Dickenson et al., 1992; Riesenhuber and Poggio, 1999; Edelman and Intrator, 2000). Our findings are consistent with several recent studies that support this idea of a parts-based representation in IT (Tanaka et al., 1991; Wang et al., 2000; Op de Beeck et al., 2001; Tsunoda et al., 2001; Sigala and Logothetis, 2002).

A parts-based representation is advantageous because of its insensitivity to variations in the retinal image of an object and its alphabet-like power to encode an infinite variety of shapes. Most parts-based theories envision a hierarchical progression of parts complexity through a sequence of processing stages. Boundary fragments constitute an appropriate level of parts complexity for an intermediate processing stage like V4. More complex parts are encoded in IT, the next stage in the ventral pathway (Desimone et al., 1984; Tanaka et al., 1991; Tsunoda et al., 2001). In a recent study Brincat and Connor (2004) demonstrated that the responses of posterior IT neurons reflect (linear and nonlinear) integration of information about the characteristics and relative positions of 2–4 contour segments of complex shapes. Further integration in anterior IT could lead to selectivity for more complex features and sparser responses (Tanaka, 1996; Tsunoda et al., 2001; Edelman and Intrator, 2003) or it may culminate in holistic coding for global object shape (Logothetis and Sheinberg, 1996; Booth and Rolls, 1998; Ullman, 1998; Riesenhuber and Poggio, 1999; Baker et al., 2002).

Areas of future investigation

The results presented in this review further our understanding about the basis of 2D contour representation in area V4. Due to practical considerations, our stimuli were restricted to be a small subset of 2D contours — silhouettes with no internal structure — and so our results pertain to the representation of the boundaries of 2D contours. To gain a more complete understanding of 2D contour representation and to extend our results to more naturalistic stimuli further investigations are required.

Our stimuli were 2D contours with convex projections radiating out from the center. In future experiments, these results need to be extended to shapes containing curve segments at other orientations as well. To derive more precise tuning functions of V4 neurons, boundary curvature should be more densely sampled. In this study, stimuli were constructed with a few constant curvature segments. To investigate the dependence of V4 responses on the rate of change or other functions of curvature, responses to stimuli with continuously varying contour curvature needs to be studied. For the stimuli used here, contour representation in terms of curvature and angular position was sufficient (since the stimuli were constructed by combining constant curvature segments) and economical (since it resulted in a small set of points representing each stimulus). For other stimuli, however, representation based on constant curvature segments may be uneconomical (for e.g. stimuli with a high frequency of convex projections along the contour) or insufficient (for e.g. stimuli with long segments of changing curvature). In such cases, it is possible that neurons would encode information in terms of additional stimulus dimensions — a question to be addressed in future studies.

Several other questions need to be addressed — how do V4 neurons respond to more realistic stimuli such as objects with internal structure, simultaneous presentation of contours of multiple objects, stimuli with texture and 3D information or stimuli without explicit contours such as photographic images? In such cases, a different, higher dimensionality would be required to describe the responses of neurons. Finally, only one-third of the neurons that we studied showed systematic tuning for contour characteristics. Amongst the other neurons many were selective for previously described properties such as edge orientation, contrast polarity, oriented bars, etc. A few neurons showed no discernible pattern in their responses or did not respond to any stimuli presented here. These cells may encode shape in terms of other stimulus dimensions not tested here, such as texture patterns (Hanazawa and Komatsu, 2001), 3D stimuli (Hinkle and Connor, 2002; Watanabe et al., 2002; Hegde and Van Essen, 2005; Hinkle and Connor, 2005; Tanabe et al., 2005), kinetic boundaries (Mysore et al., 2006) or other properties that have not been previously discovered in area V4 — all questions to be addressed in future experiments.

Conclusion

Our results demonstrate that area V4 encodes 2D contours as aggregates of boundary fragments, with individual neuron encoding the characteristics of a specific fragment. This result has provided a first step toward deciphering the bases of shape representation in area V4. The general method used here could be extended to investigate the shape dimensions that dictate V4 responses to more naturalistic stimuli. Targeted investigation of shape dimensions, parametric design of stimuli combined with appropriate analytical techniques will help us identify the other bases of representation in the ventral pathway. This will bring us closer to solving the puzzle of how the primate brain parses the 2D representation of the visual world into 3D objects that are perceived and recognized.

Acknowledgements

Work described in this review was done in the laboratory of C. E. Connor at Zanvyl Krieger Mind/Brain Institute, Johns Hopkins University, Baltimore, MD. Ideas presented here were developed in collaboration with C. E. Connor. Many thanks to Mark H. Histed for useful suggestions on the manuscript and Kristin J. MacCully for assistance in its preparation.

References

Andrews, D.P., Butcher, A.K. and Buckley, B.R. (1973) Acuities for spatial arrangement in line figures: human and ideal observers compared. Vision Res., 13: 599–620.

Attneave, F. (1954) Some informational aspects of visual perception. Psychol. Rev., 61: 183–193.

Baker, C.I., Behrmann, M. and Olson, C.R. (2002) Impact of learning on representation of parts and wholes in monkey inferotemporal cortex. Nat. Neurosci., 5: 1210–1216.

Baizer, J.S., Robinson, D.L. and Dow, B.M. (1977) Visual responses of area 18 neurons in awake, behaving monkey. J. Neurophysiol., 40: 1024–1037.

Besl, P.J. and Jain, R.C. (1985) Three dimensional object recognition. Comput. Surv., 17(1): 75–145.

Biederman, I. (1987) Recognition-by-components: a theory of human image understanding. Psychol. Rev., 94: 115–147.

Biederman, I. and Cooper, E.E. (1991) Priming contour-deleted images: evidence for intermediate representations in visual object recognition. Cogn. Psychol., 23(3): 393–419.

Booth, M.C. and Rolls, E.T. (1998) View-invariant representations of familiar objects in the inferior temporal visual cortex. Cereb. Cortex, 8: 510–523.

Braunstein, M.L., Hoffmann, D.D. and Saidpour, A. (1989) Parts of visual objects: an experimental test of the minima rule. Perception, 18: 817–826.

Brincat, S.L. and Connor, C.E. (2004) Underlying principles of visual shape selectivity in posterior inferotemporal cortex. Nat. Neurosci., 7(8): 880–886.

Burkhalter, A. and Van Essen, D.C. (1986) Processing of color, form, and disparity information in visual area VP and V2 of ventral extrastriate cortex in the macaque monkey. J. Neurosci., 6: 2327–2351.

Cadieu, C., Kouh, M., Riesenhuber, M. and Poggio, T. (2005) Shape representation in V4: investigating position-specific tuning for boundary conformation with the standard model of object recognition. Cosyne Abstracts.

Chen, S. and Levi, D.M. (1996) Angle judgment: is the whole the sum of its parts? Vision Res., 36: 1721–1735.

De Weerd, P., Desimone, R. and Ungerleider, L.G. (1996) Cue-dependent deficits in grating orientation discrimination after V4 lesions in macaques. Vis. Neurosci., 13: 529–538.

Desimone, R. and Gross, C.G. (1979) Visual areas in the temporal cortex of the macaque. Brain Res., 178: 363–380.

Desimone, R., Albright, T.D., Gross, C.G. and Bruce, C. (1984) Stimulus-selective properties of inferior temporal neurons in the macaque. J. Neurosci., 4: 2051–2062.

Desimone, R. and Schein, S.J. (1987) Visual properties of neurons in area V4 of the macaque: sensitivity to stimulus form. J. Neurophysiol., 57: 835–868.

Desimone, R., Schien, S.J., Moran, J. and Ungerleider, L.G. (1985) Contour, color and shape analysis beyond the striate cortex. Vision Res., 25: 441–452.

Dickenson, S.J., Pentland, A.P. and Rosenfeld, A. (1992) From volumes to views: an approach to 3-D object recognition. CVGP: Image Understanding, 55: 130–154.

Dobbins, A., Zucker, S.W. and Cynader, M.S. (1987) End-stopped neurons in the visual cortex as a substrate for calculating curvature. Nature, 329: 438–441.

Edelman, S. and Intrator, N. (2000) (Coarse coding of shape fragments) + (retinotopy) approximately = representation of structure. Spat. Vis. 13(2–3): 255–264.

Edelman, S. and Intrator, N. (2003) Towards structural systematicity in distributed, statistically bound visual representations. Cogn. Sci., 27: 73–109.

Felleman, D.J. and Van Essen, D.C. (1991) Distributed hierarchical processing in the primate cerebral cortex. Cereb. Cortex, 1: 1–47.

Fujita, I., Tanaka, K., Ito, M. and Cheng, K. (1992) Columns for visual features of objects in monkey inferotemporal cortex. Nature, 360: 343–346.

Gallant, J.L. (2000) The neural representation of shape. In: DeValois, K.K. and DeValois, R.L. (Eds.), Seeing. Academic Press, San Diego, CA, pp. 311–333.

Gallant, J.L., Braun, J. and Van Essen, D.C. (1993) Selectivity for polar, hyperbolic, and cartesian gratings in macaque visual cortex. Science, 259: 100–103.

Gallant, J.L., Connor, C.E., Rakshit, S., Lewis, J.W. and Van Essen, D.C. (1996) Neural responses to polar, hyperbolic and Cartesian gratings in area V4 of the macaque monkey. J. Neurophysiol., 76: 2718–2739.

Gallant, J.L., Shoup, R.E. and Mazer, J.A. (2000) A human extrastriate area functionally homologous to macaque V4. Neuron, 27: 227–235.

Gattass, R., Gross, C.G. and Sandell, J.H. (1981) Visual Topography of V2 in the macaque. J. Comp. Neurol., 201: 519–539.

Gross, C.G., Rocha-Miranda, C.E. and Bender, D.B. (1972) Visual properties of neurons in inferotemporal cortex of the macaque. J. Neurophysiol., 35: 96–111.

Hanazawa, A. and Komatsu, H. (2001) Influence of the direction of elemental luminance gradients on the responses of V4 cells to textured surfaces. J. Neurosci., 21: 4490–4497.

Heeley, D.W. and Buchanan-Smith, H.M. (1996) Mechanisms specialized for the perception of image geometry. Vision Res., 36: 3607–3627.

Hegde, J. and Van Essen, D.C. (2000) Selectivity for complex shapes in primate visual area V2. J. Neurosci., 20(5): RC61.

Hegde, J. and Van Essen, D.C. (2005) Role of primate visual area V4 in the processing of 3-D shape characteristics defined by disparity. J. Neurophysiol., 94: 2856–2866.

Heggelund, P. and Hohmann, A. (1975) Responses of striate cortical cells to moving edges of different curvatures. Exp. Brain Res., 31: 329–339.

Heywood, C.A. and Cowey, A. (1987) On the role of cortical area V4 in the discrimination of hue and pattern in macaque monkeys. J. Neurosci., 7: 2601–2617.

Hinkle, D.A. and Connor, C.E. (2002) Three-dimensional orientation tuning in macaque area V4. Nat. Neurosci., 5(7): 665–670.

Hinkle, D.A. and Connor, C.E. (2005) Quantitative characterization of disparity tuning in ventral pathway area V4. J. Neurophysiol., 94: 2726–2737.

Hoffmann, D.D. and Richards, W.A. (1984) Parts of recognition. Cognition, 18: 65–96.

Hubel, D.H. and Weisel, T.N. (1959) RFs of single neurones in the cat's striate cortex. J. Physiol. (Lond.), 148: 574–591.

Hubel, D.H. and Weisel, T.N. (1965) RFs and functional architecture in two nonstriate visual areas (18 and 19) of the cat. J. Neurophysiol., 28: 229–289.

Hubel, D.H. and Weisel, T.N. (1968) RFs and functional architecture of monkey striate cortex. J. Physiol. (Lond.), 195: 215–243.

Hubel, D.H. and Livingstone, M.S. (1987) Segregation of form, color, and stereopsis in primate area 18. J. Neurosci., 7: 3378–3415.

Hummel, J.E. and Biederman, I. (1992) Dynamic binding in a neural network for shape recognition. Psychol. Rev., 99: 480–517.

Ito, M. and Komatsu, H. (2004) Representation of angles embedded within contour stimuli in area V2 of macaque monkeys. J. Neurosci., 24(13): 3313–3324.

Ito, M., Tamura, H., Fujita, I. and Tanaka, K. (1995) Size and position invariance of neuronal responses in monkey inferotemporal cortex. J. Neurophysiol., 73(1): 218–226.

Janssen, P., Vogels, R. and Orban, G.A. (1999) Macaque inferior temporal neurons are selective for disparity-defined three-dimensional shapes. Proc. Natl. Acad. Sci. USA, 96: 8217–8222.

Kaniza, G. and Gerbino, W. (1976) Convexity and symmetry in figure-ground organization. In: Henle, M. (Ed.), Art and Artefacts. Springer, New York, pp. 25–32.

Kobatake, E. and Tanaka, K. (1994) Neuronal selectivities to complex object features in the ventral visual pathway of the macaque cerebral cortex. J. Neurophysiol., 71: 856–867.

Lennie, P. (2003) Receptive fields. Curr. Biol., 13(6): R216–R219.

Logothetis, N.K. and Sheinberg, D.L. (1996) Visual object recognition. Annu. Rev. Neurosci., 19: 577–621.

Marr, D. and Nishihara, H.K. (1978) Representation and recognition of the spatial organization of three-dimensional shapes. Proc. R. Soc. Lond B Biol. Sci., 200: 269–294.

Merigan, W.H. (1996) Basic visual capacities and shape discrimination after lesions of extrastriate area V4 in macaques. Vis. Neurosci., 13: 51–60.

Merigan, W.H., Nealy, T.A. and Maunsell, J.H.R. (1993) Visual effects of lesions of cortical area V2 in macaques. J. Neurosci., 13: 3180–3191.

Merigan, W.H. and Phan, H.A. (1998) V4 lesions in macaques affect both single- and multiple-viewpoint shape discriminations. Vis. Neurosci., 15: 359–367.

Milner, P.M. (1974) A model for visual shape recognition. Psychol. Rev., 81: 521–535.

Mysore, S.G., Vogels, R., Raiguel, S.E. and Orban, G.A. (2006) Processing of kinetic boundaries in macaque V4. J. Neurophysiol. 95(3): 1864–1880.

Op de Beeck, H., Wagemans, J. and Vogels, R. (2001) Inferotemporal neurons represent low-dimensional configurations of parameterized shapes. Nat. Neurosci., 4: 1244–1252.

Pasupathy, A. and Connor, C.E. (1999) Responses to contour features in macaque area V4. J.Neurophysiol., 82: 2490–2502.

Pasupathy, A. and Connor, C.E. (2001) Shape representation in area v4: position-specific tuning for boundary conformation. J. Neurophysiol., 86: 2505–2519.

Pasupathy, A. and Connor, C.E. (2002) Population coding of shape in area V4. Nat. Neurosci., 5: 1332–1338.

Pentland, A. (1989) Shape information from shading: a theory about human perception. Spatial Vis., 4(2–3): 165–182.

Perrett, D.I., Rolls, E.T. and Caan, W. (1982) Visual neurones responsive to faces in the monkey temporal cortex. Exp. Brain Res., 47: 329–342.

Poggio, T. and Edelman, S. (1990) A network that learns to recognize three-dimensional objects. Nature, 343: 263–266.

Regan, D., Gray, R. and Hamstra, S.J. (1996) Evidence for a neural mechanism that encodes angles. Vision Res., 36: 323–330.

Riesenhuber, M. and Poggio, T. (1999) Hierarchical models of object recognition in cortex. Nat. Neurosci., 2: 1019–1025.

Sary, G., Vogels, R. and Orban, G.A. (1993) Cue-invariant shape selectivity of macaque inferior temporal neuron. Science, 260: 95–997.

Sato, T., Kawamura, T. and Iwai, E. (1980) Responsiveness of inferotemporal single units to visual pattern stimuli in monkeys performing discrimination. Exp. Brain Res., 38(3): 313–319.

Schiller, P.H. (1995) Effects of lesions in visual cortical area V4 on the recognition of transformed objects. Nature, 376: 342–344.

Schiller, P.H. and Lee, K. (1991) The role of primate extrastriate area V4 in vision. Science, 251: 1251–1253.

Schwartz, E.L., Desimone, R., Albright, T.D. and Gross, C.G. (1983) Shape recognition and inferior temporal neurons. Proc. Natl. Acad. Sci. USA, 80: 5776–5778.

Sigala, N. and Logothetis, N.K. (2002) Visual categorization shapes feature selectivity in the primate temporal cortex. Nature, 415: 318–320.

Subirana-Vilanova, J.B. and Richards, W. (1996) Attentional frames, frame curves and figural boundaries: the outside/inside dilemma. Vision Res., 36: 1493–1501.

Tanabe, S., Doi, T., Umeda, K. and Fujita, I. (2005) Disparity-tuning characteristics of neuronal responses to dynamic random-dot stereograms in macaque visual area V4. J. Neurophysiol., 94: 2683–2699.

Tanaka, K. (1993) Neuronal mechanisms of object recognition. Science, 262: 685–688.

Tanaka, K. (1996) Inferotemporal cortex and object vision. Annu. Rev. Neurosci., 19: 109–139.

Tanaka, K., Saito, H., Fukada, Y. and Moriya, M. (1991) Coding visual images of objects in the inferotemporal cortex of the macaque monkey. J. Neurophysiol., 66: 170–189.

Triesman, A. and Gormican, S. (1988) Feature analysis in early vision: evidence from search asymmetries. Psychol. Rev., 95: 15–48.

Tsunoda, K., Yamane, ., Nishizaki, M. and Tanifuji, M. (2001) Complex objects are represented in macaque inferotemporal cortex by the combination of feature columns. Nat. Neurosci., 4: 832–838.

Ullman, S. (1989) Aligning pictorial descriptions: an approach to object recognition. Cognition, 32: 193–254.

Ullman, S. (1998) Three-dimensional object recognition based on the combination of views. Cognition, 67: 21–44.

Ungerleider, L. and Mishkin, M. (1982) Two cortical visual systems. In: Ingle, D.J., Goodale, M.A. and Mansfield, R.J.W. (Eds.), Analysis of Visual Behavior. MIT Press, Cambridge, MA, pp. 549–586.

Versavel, M., Orban, G.A. and Lagae, L. (1990) Responses of visual cortical neurons to curved stimuli and chevrons. Vision Res., 30: 235–248.

von der Heydt, R. and Peterhans, E. (1989) Mechanisms of contour perception in monkey visual cortex. I. Lines of pattern discontinuity. J. Neurosci., 9: 1731–1748.

von der Heydt, R., Zhou, H. and Friedman, H.S. (2000) Representation of stereoscopic edges in monkey visual cortex. Vision Res., 40(15): 1955–1967.

Wang, Y., Fujita, I. and Murayama, Y. (2000) Neuronal mechanisms of selectivity for object features revealed by blocking inhibition in inferotemporal cortex. Nat. Neurosci., 3: 807–813.

Wang, G., Tanaka, K. and Tanifuji, M. (1996) Optical imaging of functional organization in the monkey inferotemporal cortex. Science, 272: 1665–1668.

Watanabe, M., Tanaka, H., Uki, T. and Fujita, I. (2002) Disparity-selective neurons in area V4 of macaque monkeys. J. Neurophysiol., 87: 1960–1973.

Wilkinson, F., James, T.W., Wilson, H.R., Gati, J.S., Menon, R.S. and Goodale, M.A. (2000) An fMRI study of the selective activation of human extrastriate form vision areas by radial and concentric gratings. Curr. Biol., 10: 1455–1458.

Wilson, H.R. (1999) Non-Fourier cortical processes in texture, form and motion perception. In: Peters, A. and Jones, E.G. (Eds.), Cerebral Cortex: Models of Cortical Circuitry. Plenum, New York, pp. 445–477.

Wilson, H.R., Wilkinson, F. and Asaad, W. (1997) Concentric orientation summation in human form vision. Vision Res., 37: 2325–2330.

Wolfe, J.M., Yee, A. and Friedman-Hill, S.R. (1992) Curvature is a basic feature for visual search tasks. Perception, 21: 465–480.

Zucker, S.W., Dobbins, A. and Iverson, L. (1989) Two stages of curve detection suggest two styles of visual computation. Neural Comput., 1: 68–81.

Martinez-Conde, Macknik, Martinez, Alonso & Tse (Eds.)
Progress in Brain Research, Vol. 154
ISSN 0079-6123

CHAPTER 17

The role of familiarity in the recognition of static and dynamic objects

Isabelle Bülthoff[1,*] and Fiona N. Newell[2]

[1] *Max-Planck-Institut für biologische Kybernetik, Spemannstrasse 38, D 72076 Tübingen, Germany*
[2] *Department of Psychology, Trinity College, University of Dublin, Aras an Phiarsaigh, Dublin 2, Ireland*

Abstract: Although the perception of our world is experienced as effortless, the processes that underlie object recognition in the brain are often difficult to determine. In this chapter, we review the effects of familiarity on the recognition of moving or static objects. In particular, we concentrate on exemplar-level stimuli such as walking humans, unfamiliar objects and faces. We found that the perception of these objects can be affected by their familiarity; for example the learned view of an object or the learned dynamic pattern can influence object perception. Deviations in the viewpoint from the familiar viewpoint, or changes in the temporal pattern of the objects can result in some reduction of efficiency in the perception of the object. Furthermore, more efficient sex categorization and crossmodal matching were found for familiar than for unfamiliar faces. In sum, we find that our perceptual system is organized around familiar events and that perception is most efficient with these learned events.

Keywords: familiarity; faces; dynamic objects; haptic recognition; categorical perception; biological motion

Introduction

The difficulty of object recognition is often not appreciated because our phenomenal experience is that the visual system is very efficient at this task. Fig. 1 is an illustration of some of the problems that the visual system encounters in everyday recognition. It has to achieve what is known as object constancy. Specifically, the system must be able to recognize a particular chair, despite variations in illumination, size, orientation, and shape. Even though shape is the basis of object recognition and its perception has been investigated for decades, there is still an ongoing debate about how objects

are represented in the brain (see e.g., Newell et al., 2005).

One important question about object recognition is how various visual cues, such as motion, disparity, texture, color, and shading, are integrated into a unique object percept. In numerous studies, the role of these cues and their interplay has been investigated thoroughly (for reviews, see Landy et al., 1995; Ernst and Bülthoff, 2004 among many others). Some researchers have reported rather unexpected interactions between visual cues. For example, it has recently been shown that color can influence size perception, in that objects with more saturated colors appear larger than objects with less saturated colors (Ling and Hurlbert, 2004). Other studies have found that shadows can help disambiguate shape perception when shape information is ambiguous (e.g., Bülthoff et al., 1994; Bülthoff and Kersten, in prep). Thus, it seems clear from

*Corresponding author. Tel.: +49-7071-601611;
Fax: +49-7071-601616; E-mail: isabelle.buelthoff@tuebingen.mpg.de

DOI: 10.1016/S0079-6123(06)54017-8

Fig. 1. This image of a complex scene illustrates some of the problems encountered in visual processing when recognizing objects. For example, we can recognize all office chairs in the scene, despite drastic variations in appearance due to different conditions of viewing (e.g., occlusion, orientation, size, location, and illumination).

these studies that distinct visual cues are important in building a robust representation of an object in memory in order to achieve object constancy and efficient object recognition.

The problem of achieving object constancy, however, is further complicated by the fact that most objects can move around the environment. As such, static object information and all associated cues, such as shading and disparity information, change from one moment to the next as the object moves. Consequently, the visual system often has to achieve object constancy despite large changes in the spatial properties of the object.

The purpose of this review is to highlight some of the everyday object recognition tasks the human brain has to solve and to review studies that have helped us better understand how this is achieved. We have decided, in particular, to limit the review to two main tasks: how moving objects are recognized and how very similar exemplar objects, such

as static images of faces, are differentiated. We believe that these types of tasks highlight the complexity and richness of object recognition in the human brain.

Recognizing moving objects

It is well known that dynamic cues can play an important role in object recognition. Humans and even animals can recognize animate objects not only on the basis of their static appearance but also from how they move (Johansson, 1973; Blake, 1993). We can even recognize such stimuli from dynamic information alone in the absence of clear spatial information (Johansson, 1973). Animate objects represented only by bright spots located at their main joints often cannot be identified in static images, but are easily recognized when a series of images is shown in a *biological motion* sequence.

Influence of familiarity on perception of biological motion

In our study (Bülthoff et al., 1997, 1998), point-light walkers similar to those used by Johansson (except for the presence of depth information, see Fig. 2) were employed to test whether the viewpoint-dependent recognition framework (for review, see Bülthoff et al., 1995) could also account for recognition of biological motion. In the viewpoint-dependent recognition framework, static objects are represented as a collection of two-dimensional (2D) views. This framework can be extended to dynamic objects by storing 2D motion traces that are projections of the three-dimensional (3D) trajectories of feature points onto the viewing plane. Dynamic 3D objects are then represented as a collection of several such 2D traces captured from various viewpoints.

If a view-dependent mechanism mediates the recognition of dynamic objects, the familiarity of a view should have a strong influence on recognition performance. That is, recognition should be easiest from viewing positions that have been experienced more often because participants are more likely to have stored internal traces of dynamic objects as seen from these positions. Another prediction is that modifying the depth structure of the stimuli should not affect recognition performance as long as the 2D traces remain unchanged.

All point-light stimuli were derived from one biological motion sequence showing a real human walking. The importance of view familiarity was investigated by measuring the recognition performance of participants viewing the same point-light walker from various viewpoints (Fig. 3). All stimuli were presented either with or without binocular disparity to investigate whether information about the depth structure of the stimuli affected recognition. Participants were asked to report if they recognized a meaningful moving object or saw only random moving points. The recognition rates for stimuli presented with or without binocular disparity indicate a strong viewpoint dependency (Fig. 4). Recognition performance was poor for top views where walking figures were recognized less often than when observed from viewpoints near the equator.

These results support the viewpoint-dependent recognition framework. As the framework predicts, participants' performance should be strongly tied to the familiarity of the viewpoint from which the walker is seen. Furthermore, the results suggest that participants' recognition performance was not based on the use of a viewpoint-invariant internal representation that could have been built during a lifetime of observing humans in motion. Adding depth information to the stimuli did not facilitate recognition. This result suggests that the internal representation used to recognize biological motion

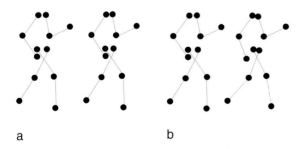

a   b

Fig. 2. Stereogram of undistorted (a) and depth-distorted (b) human walkers represented by dots only. Cross-fusers will notice that the 3D structure is severely distorted in (b) but not in (a). Connecting lines were absent in all experiments. Adapted with kind permission from Bülthoff et al. (1998).

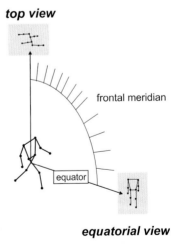

Fig. 3. Viewing positions of a point-light walker used to test view-dependent recognition. The bars along the frontal meridian indicate the viewing positions. The gray screens show the 2D projections of one frame of the animation for the top view and for the equatorial view.

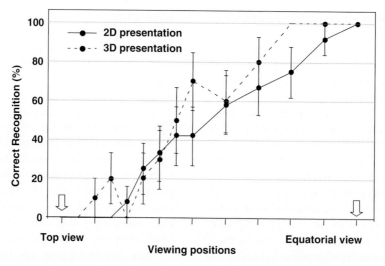

Fig. 4. Recognition performance of a point-light walker plotted as a function of viewpoint. Two-dimensional presentation: point-light displays presented without depth information. Three-dimensional presentation: point-light displays presented with depth information. Adapted with permission from Bülthoff, I. and Bülthoff, H.H. (2003).

is largely 2D, and that viewpoint familiarity is the primary determinant of the results.

As mentioned earlier, if object representations are largely 2D, then distorting the depth structure of the walker should not impair recognition as long as the 2D traces are left unchanged. Further experiments testing recognition of depth-distorted walkers (Fig. 2) supported this prediction. Because of their familiarity with the figure suggested by the 2D motion traces, participants perceived depth-distorted walkers as human walking figures; they did not seem aware that their 3D structures were not congruent with that of a human walker.

Spatiotemporal representations of familiar objects

Although studies of biological motion can reveal how dynamic information alone can affect perception, these studies do not make clear how dynamic information is integrated into an object percept when shape information is available. Some researchers argue that motion offers an alternative route to object perception but that this information is processed independently from shape perception (Kourtzi and Nakayama, 2002). On the other hand, recent studies have found that motion

can often disambiguate shape information when spatial information is reduced (Hill and Johnston, 2001), and that motion information is integrated into the identity of the object (Knappmeyer et al., 2003; Newell et al., 2004).

In a recent study, we have shown that the dynamic properties of an object are learned and integrated into the object percept along with its spatial properties (Newell et al., 2004). In our study, participants were first required to familiarize themselves with a set of novel objects, each with a distinct moving pattern. Thus, the dynamic and static information were both diagnostic of object identity. In our study, we were interested in ascertaining whether or not motion and shape would necessarily be integrated into a spatiotemporal representation of the object. The results suggested that this was indeed the case, but only for object motion that was intrinsic to the object itself. When the familiar intrinsic motion of an object was changed, object categorization performance was significantly reduced. However, manipulating the extrinsic motion information (i.e., the path that the object took in an environment) did not affect performance. More recently, we found that the motion of a novel object's parts as well as the whole object motion can prime the identity of a static version of the object relative to a static view of the object (Setti and Newell, in prep). Taken together, these studies suggest that the familiar movement of an object can be integrated with spatial information into an object's representation in memory that allows for efficient recognition. Moreover, these and other studies on disparity, texture, and color information show that all available visual cues are useful for recognition and suggest that all are integrated in memory as part of an object's representation.

Recognizing static exemplar objects

We now want to turn to a class of stimuli for which the human observer is an expert in terms of shape and texture perception, namely faces. Faces have great social importance because facial information can indicate friend or foe and communicate the emotional state of the individual. Prosopagnosic patients, for example, suffer greatly in their social interactions because of their impaired ability to recognize familiar faces. Compared to many other object classes, faces are all very similar to each other; thus, they are a very homogeneous class. Differentiating one face from another consequently presents a difficult challenge to any perceptual system. Nonetheless, we seem to be able to make facial differentiations effortlessly and without obvious problems. Some theorists argue that efficient face recognition is achieved because the human brain has an innate propensity to process and recognize faces (Grill-Spector et al., 2004; Yovel and Kanwisher, 2004). In this view, faces are a special class of objects. Other researchers, however, have argued that faces are simply a class of object stimuli with which we have a lot of practice. In this view, the phenomena of face recognition are the product of a large accumulation of expertise with these stimuli (Diamond and Carey, 1986; Gauthier et al., 2000). Although this debate is ongoing, this review will not consider whether faces are special or not. Suffice it to say that they are important to our everyday social activities. Instead, our research has focused on the question of how the visual system differentiates between similar objects, such as faces, so that efficient recognition is achieved.

Perceiving familiar face categories

While some object categories are physically very different from each other, (e.g., cars and insects), others are very similar (e.g., male and female faces). Thus, the visual system has to categorize similar objects as the same and highlight differences between different category objects. Generally this is a relatively easy problem to solve because many object categories are intrinsically unique in terms of their spatial properties. However, for object classes that are highly similar, such as male and female faces, the problem becomes more difficult to solve. Recent studies have suggested that the visual system solves this problem through a process known as categorical perception (Harnad, 1987). Thus, objects within a category are perceived as more similar to each other than to objects belonging to another category even if the physical differences between them are equal. For experimental purposes, the hallmarks of categorical perception (CP) are twofold: a CP effect occurs when (1) a sharp change of response occurs at the subjective category boundary in a categorization task, and (2) pairs of stimuli are discriminated more accurately when they straddle the subjective category boundary than when both belong to the same category (even if the physical differences between the pairs are equal). In other words, in CP the peak in discrimination performance occurs at the category boundary defined by the categorization response function.

Recent studies using complex visual stimuli, such as faces, have suggested that CP occurs for categories such as facial expressions (Calder et al., 1996; de Gelder et al., 1997; Young et al., 1997), identity of familiar faces (Beale and Keil, 1995), and identity of familiar objects (Newell and Bülthoff, 2002). There is, however, a suggestion in the literature that effects of categorical perception depend on the familiarity of the object stimuli. For example, some studies have shown that CP effects for unfamiliar or novel stimuli can emerge from short-term learning of category items (Goldstone, 1994; Livingston et al., 1998) and for unfamiliar stimuli, such as face stimuli, learned in the course of an experiment (Levin and Beale, 2000). We have previously reported that the perception of the sex of a face is not automatic but is dependent on the familiarity of the face. In particular, face familiarity affects categorical perception thus allowing for better discrimination and categorization of the sex of faces (Bülthoff and Newell, 2004).

Male and female are well-known facial categories, and our ability to recognize the sex of an unfamiliar face is generally good (Bruce et al., 1993; Wild et al., 2000). Indeed, the well-known model by Bruce and Young (1986) for face processing proposes that the sex of a face can be derived whether the face is familiar or not (Bruce et al., 1987; see also le Gal and Bruce, 2002). Other studies have, however, challenged the notion that sex perception in faces is unrelated to the familiarity of the face (Goshen-Gottstein and Ganel, 2000; Baudoin and Tiberghien, 2002; Rossion, 2002) and argued that familiarity facilitates sex discrimination. In a previous study, we investigated the role of face familiarity on the emergence of CP for the sex of faces. If sex perception is related to facial identity then we expect that CP effects would emerge for familiar faces only. On the other hand, if the sex of a face is information unrelated to its identity, we expect that effects of CP would emerge for all faces, irrespective of familiarity.

A problem with studying sex perception in faces is that two faces differing in sex always differ in identity too; thus, identity is often a confounding variable in these studies. With present media technology and computational methods (e.g., Blanz and Vetter, 1999), one can create face images differing in sex information alone without changing identity. Using the algorithm of Blanz and Vetter (1999), we created sex continua in which the endpoint faces had different sexes but the same facial features. Each of these continua was based on one of six female faces. In each sex set, all features of the original face were transformed (masculinized) in 10% steps into the corresponding male face (for details see Bülthoff and Newell, 2004). Fig. 5 shows six original female faces and their computationally derived corresponding male faces and masculinized faces in-between in 20% steps. With these face stimuli, we could investigate whether the sex of a face is perceived categorically independent of the change of characteristic facial features related to identity.

In our initial experiments, all face stimuli were unfamiliar to the participants. Participants performed a categorization task and a discrimination task. In the categorization task, individual face stimuli were presented and the participant was required to categorize each as either male or female. The discrimination task in our studies typically consisted of a pair-wise face matching (i.e., same or different). The stimuli were presented in blocks according to facial identity in order to promote any existing effects of CP.

Typical categorization and discrimination results are shown in Fig. 6. Importantly, the shape of the categorization function is not step-like, so the first hallmark of CP was not observed. Participants could tell the sex of the endpoint faces, but there was no obvious consensus about the location of the subjective boundary between both sexes that would have been shown up by a step-like function. In the discrimination task, face pairs straddling the sex boundary were not significantly easier to discriminate than within-category pairs. Thus, the categorization and the discrimination results did not exhibit CP for sex in unfamiliar faces.

We expected that sex categories would be well-defined categories requiring no learning, but the results of our initial results suggested that this was not the case. So our next question was whether or not the perception of the sex of a face was dependent on the familiarity of that face.

In our subsequent experiments, participants were trained to categorize all face stimuli by their sex (see Goldstone, 1994; Goldstone et al., 2001 for similar procedures) prior to testing for CP. The training phase consisted of a sex categorization task in which participants were given feedback on the accuracy of their response on each trial and were required to reach a criterion performance before proceeding to the main experiment.

Typical response functions to our sex categorization and discrimination tasks are shown in Fig. 6. The subjective category boundary was very close to the physical sex boundary in the categorization task. The discrimination scores on the same/different task were converted to d' scores. Our statistical analyses revealed that discrimination performance for face pairs straddling the category boundary were significantly better than within-category pairs.

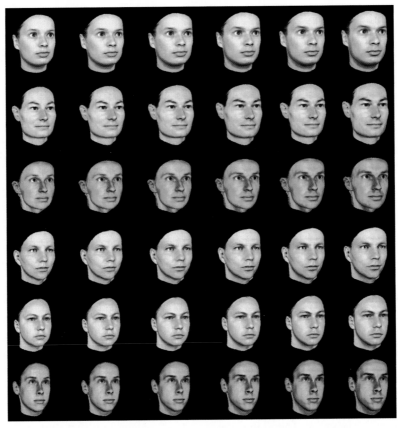

Fig. 5. Sex continua: in each row, the endpoint faces are of different sex but share the same facial features. All leftmost faces are original female faces, and all rightmost faces are computationally derived corresponding male faces. Morphs in-between are shown in 20% steps.

Therefore, our findings exhibited CP for sex after familiarization.

Our results indicate that sex information is available for CP, but only when the faces are familiar. Thus, despite the importance of face perception, sex information present in faces is not naturally perceived categorically. Our findings have implications for functional models of face processing that suggest two independent processing routes, one for facial expression and another for identity (Bruce and Young, 1986): we propose that sex perception is closely linked with the processing of facial identity.

Familiarity and crossmodal face perception

We have recently embarked on a project to investigate the role of familiarity in crossmodal face perception. Our initial studies suggested that familiarity can also facilitate face perception across different modalities (Casey and Newell, 2005). We first investigated whether long-term familiarity has an effect on crossmodal face perception by testing participants' ability to recognize a mask of their own face via touch and vision (Fig. 7). Performance was better for visual self-recognition than for recognition via touch. Thus, despite a lifetime of experience with touching one's own face, a representation of the face is not available in tactile memory. In our subsequent experiments, we trained participants to recognize a set of previously unfamiliar faces via either touch or vision alone. We then conducted a crossmodal, face-matching study in which participants were required to match a face mask sampled with touch with a visual counterpart. Face pairs were either

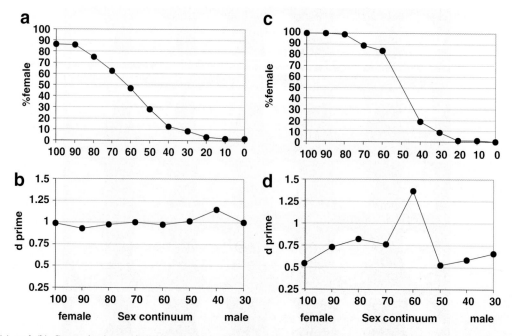

Fig. 6. (a) and (b) Categorization and discrimination performance for unfamiliar face continua. (c) and (d) Categorization and discrimination performance after training. (a) and (c) Mean frequency of "female" responses (%) in the categorization task as a function of the sex of the face stimuli. (b) and (d) Mean discrimination data (d′) as a function of the sex of the face pairs. (b) and (d) Only the most female image of each pair is mentioned on the abscissa (i.e., "100" corresponds to the face pair 100%–70%, etc.). Adapted with kind permission from Bülthoff & Newell (2004). http://www.psypress.co.uk/journals.asp.

familiarized or unfamiliar faces. We found that the matching performance for familiar faces was significantly better than performance for unfamiliar faces. Our findings suggest that familiarity with a face, irrespective of the modality through which it is encoded, benefits face perception by providing a robust representation of the spatial characteristics of that face in memory.

General discussion

Being familiar with objects or categories of objects can drastically change the way we perceive them. Many studies have provided evidence that familiarity evokes different perceptual processes for the purpose of recognition. Familiar views of novel objects are better recognized than less familiar views (e.g., Edelman and Bülthoff, 1992). Furthermore, even very familiar objects are better recognized from more usual viewpoints than from unusual views (Palmer et al., 1981; Newell and Findlay,

1997), suggesting that representations of objects in memory are organized around the more commonly observed aspects of the objects in the environment. In this chapter, we reviewed studies that extend these original findings and reported that familiarity, in most circumstances, can benefit the perception of dynamic point-light displays, moving objects, the sex of faces, and face recognition across different modalities by allowing for better (i.e., faster and more accurate) recognition of these objects.

In the studies reviewed here, the stimuli and tasks were generally familiar to the participants. For example, judging the identity of a person by gait alone is a common task (such as when that person is far away) as is recognizing moving objects. We showed that untrained participants can easily recognize point-light walkers when those figures are shown from viewpoints that they have experienced with real people, but that they cannot identify these walkers from unfamiliar viewing positions (i.e., top views). When seen from unfamiliar viewpoints, moving point-light displays were perceived as

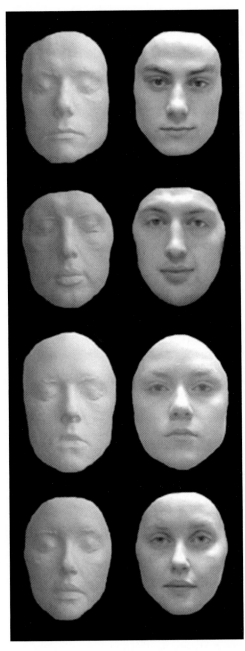

Fig. 7. An image of the face masks used in the crossmodal face recognition experiments. Images of the real faces are shown beside each face mask for illustrative purposes only.

allowed participants to override any depth discrepancies. Participants perceived walking humans when they saw depth-distorted patterns that could not possibly represent normal walking humans. These findings suggest that the 2D projection of the dynamic pattern is primarily used to perceive biological motion. These results were discussed in view of the important question of 2D vs. 3D representations of objects in memory, but here we want to emphasize the influence of familiarity on recognition of familiar dynamic objects. Generally in the real world, the 2D pattern is sufficient for interpreting a perceptual scene; moreover, the 3D representation of that pattern is rarely distorted. Consequently, the observed effects are possibly the result of experience with familiar patterns that allow us to base our perception on assumptions built up during the course of one's lifetime (also referred to as "priors" in Bayes' theorem, 1764).

Similarly, we found that the recognition of novel objects was affected by the familiar motion of the object itself (e.g., whether it wobbled or tumbled) but not by its familiar path or route. The path of the object was useful for recognition when it was the only feature diagnostic of object identity, suggesting that although it was perceived it was not integrated into the object's representation in memory. If we think about moving objects in the real world, objects rarely move along a stereotyped route. Therefore, our results may simply reflect real-world perception where route information is not a familiar cue to object identity and is not integrated into the representation in memory.

Again, categorizing the sex of a face is a task we have done since birth. In the set of experiments reported here, we investigated how the sex of a face is perceived. Participants were confronted with the unusual task of classifying as male or female a set of highly similar faces that varied along the sex dimension only. We claim that evidence of CP emerges only after participants were familiarized with the set of faces and with the task. Our results suggest that our visual system, in the absence of training, would incorrectly treat such similar exemplars as slightly noisy versions of the same face; participants were unable to separate clearly the faces in two categories that would normally allow for effects of CP to emerge. Again, here we want to

random moving patterns and could not be identified. Furthermore, familiar dynamic patterns given by the 2D motion traces of the point-light displays point out the striking influence of familiarity on the perception of sex in faces.

Finally, although the recognition of other faces through touch is not a familiar task to most people, tactile perception of one's own face is a common event. For example, we often feel our own faces for the purposes of grooming, yet our studies showed that despite this experience, recognition of one's own face through touch is not as efficient as through vision. Taken together, these studies suggest that it is familiarity with the stimulus properties *as well as* the task that results in better performance. The benefits of familiarity, we would argue, are therefore likely to be task-specific.

In summary, familiarity with an object or event helps build a robust representation of that object in memory allowing for efficient recognition of objects on the basis of statistical likelihood of the appearance of that object in the natural environment. Some of our findings suggest that the benefit of familiarity with objects seems to be specific to the task at hand and does not generalize to different types of tasks. For example, despite a lifetime of experience with seeing and touching one's own face, only visual recognition of our own face was possible. Without direct investigation of the interplay between task specificity and familiarity, this remains purely speculative. However, several recent studies suggest that the neural coding of objects is influenced by the task, or response contingencies, in that these neurons adapt according to the rules of the task (Duncan, 2001; Freedman et al., 2001). Consequently, it is possible that familiarity is indeed dependent on the task, and that the familiar properties of an object that benefit recognition performance may not influence other types of tasks.

Abbreviations

CP	categorical perception
2D	two-dimensional
3D	three-dimensional

Acknowledgments

This research was partly funded by the HEA: PRTLI fund to the Trinity College Institute of Neuroscience of which FNN is a member. We thank Marty Banks for his comments on an earlier draft of this manuscript.

References

Baudoin, J.Y. and Tiberghien, G. (2002) Sex is a dimension of face recognition. J. Exp. Psychol. Learn. Mem. Cogn., 28: 362–365.

Bayes, T. (1764) An essay toward solving a problem in the doctrine of chances. Philos. Trans. R. Soc. Lond. B Biol. Sci., 53: 370–418.

Beale, J.M. and Keil, F.C. (1995) Categorical effects in the perception of faces. Cognition, 57: 217–239.

Blake, R. (1993) Cats perceive biological motion. Psychol. Sci., 4: 54–57.

Blanz, V. and Vetter, T. (1999) A morphable model for the synthesis of 3D faces. In: SIGGRAPH 99 Conference Proceedings, pp. 187–194.

Bruce, V., Burton, A.M., Hanna, E., Healey, P., Mason, O., Coombe, A., Fright, R. and Linney, A. (1993) Sex discrimination: how do we tell the difference between male and female faces? Perception, 22: 131–152.

Bruce, V., Ellis, H., Gibling, F. and Young, A. (1987) Parallel processing of the sex and familiarity of faces. Can. J. Psychol., 41: 510–520.

Bruce, V. and Young, A.W. (1986) Understanding face recognition. Br. J. Psychol., 77: 305–327.

Bülthoff, I. and Bülthoff, H.H. (2003) Image-based recognition of biological motion, scenes and objects. In: Peterson, M.A. and Rhodes, G. (Eds.), Analytic and Holistic Processes in the Perception of Faces, Objects, and Scenes. Oxford University Press, New York, pp. 146–176.

Bülthoff, I., Bülthoff, H.H. and Sinha, P. (1997) View-based representations for dynamic 3D object recognition. (Technical Report No. 47, http://www.kyb.tuebingen.mpg.de/bu/techr/index.html) Max-Planck-Institut für biologische Kybernetik, Tübingen, Germany.

Bülthoff, I., Bülthoff, H.H. and Sinha, P. (1998) Top-down influences on stereoscopic depth-perception. Nat. Neurosci., 1: 254–257.

Bülthoff, H.H., Edelman, S.Y. and Tarr, M.J. (1995) How are 3-dimensional objects represented in the brain? Cereb. Cortex, 5: 247–260.

Bülthoff, I., Kersten, D. and Bülthoff, H.H. (1994) General lighting can overcome accidental viewing. Invest. Ophthalmol. Vis. Sci., 35: 1741.

Bülthoff, I. and Newell, F.N. (2004) Categorical perception of sex occurs in familiar but not unfamiliar faces. Vis. Cogn., 11: 823–855.

Calder, A.J., Young, A.W., Perrett, D.I., Etcoff, N.L. and Rowland, D.A. (1996) Categorical perception of morphed facial expressions. Vis. Cogn., 3: 81–117.

Casey, S. and Newell, F.N. (2005) The role of long-term and short-term familiarity in visual and haptic face recognition. Exp. Brain Res., 166: 583–591.

de Gelder, B., Teunisse, J. and Benson, P.J. (1997) Categorical perception of facial expressions: categories and their internal structure. Cogn. Emotion, 11: 1–23.

Diamond, R. and Carey, S. (1986) Why faces are and are not special: an effect of expertise. J. Exp. Psychol. Gen., 5: 107–117.

Duncan, J. (2001) An adaptive coding model of neural function in prefrontal cortex. Nat. Rev. Neurosci., 2: 820–829.

Edelman, S. and Bülthoff, H.H. (1992) Orientation dependence in the recognition of familiar and novel views of three-dimensional objects. Vision Res., 32: 2385–23400.

Ernst, M.O. and Bülthoff, H.H. (2004) Merging the senses into a robust percept. Trends Cogn. Sci., 8: 162–169.

Freedman, D.J., Riesenhuber, M., Poggio, T. and Miller, E.K. (2001) Categorical representation of visual stimuli in the primate prefrontal cortex. Science, 291: 312–316.

Gauthier, I., Skudlarski, P., Gore, J.C. and Anderson, A.W. (2000) Expertise for cars and birds recruits brain areas involved in face recognition. Nat. Neurosci., 3: 191–197.

Goldstone, R.L. (1994) Influences of categorization on perceptual discrimination. J. Exp. Psychol. Gen., 123: 178–200.

Goldstone, R.L., Lippa, Y. and Shiffrin, R.M. (2001) Altering object representations through category learning. Cognition, 78: 27–43.

Goshen-Gottstein, Y. and Ganel, T. (2000) Repetition priming for familiar and unfamiliar faces in a sex-judgment task: evidence for a common route for the processing of sex and identity. J. Exp. Psychol. Learn. Mem. Cogn., 26: 1198–1214.

Grill-Spector, K., Knouf, N. and Kanwisher, N. (2004) The fusiform face area subserves face perception, not generic within-category identification. Nat. Neurosci., 7: 555–562.

Harnad, S.R. (Ed.). (1987) Categorical Perception: The Groundwork of Cognition. Cambridge University Press, Cambridge UK.

Hill, H. and Johnston, A. (2001) Categorizing sex and identity from the biological motion of faces. Curr. Biol., 11: 880–885.

Johansson, G. (1973) Visual perception of biological motion and a model of its analysis. Percept. Psychophys., 14: 201–211.

Knappmeyer, B., Thornton, I.M. and Bülthoff, H.H. (2003) The use of facial motion and facial form during the processing of identity. Vision Res., 43: 1921–1936.

Kourtzi, Z. and Nakayama, K. (2002) Distinct mechanisms for the representation of moving and static objects. Vis. Cogn. (Special Issue), 9(1/2): 248–264.

Landy, M.S., Maloney, L.T., Johnston, E.B. and Young, M. (1995) Measurement and modeling of depth cue combination: in defense of weak fusion. Vision Res., 35: 389–412.

Le Gal, P.M. and Bruce, V. (2002) Evaluating the independence of sex and expression in judgements of faces. Percept. Psychophys., 64: 230–243.

Levin, D.T. and Beale, J.M. (2000) Categorical perception occurs in newly learned faces, cross-race faces, and inverted faces. Percept. Psychophys., 62: 386–401.

Ling, Y. and Hurlbert, A. (2004) Colour and size interactions in a real 3D object similarity task. J. Vis., 4: 721–734.

Livingston, K., Andrews, J. and Harnad, S. (1998) Categorical perception effects induced by category learning. J. Exp. Psychol. Learn. Mem. Cogn., 24: 732–753.

Newell, F.N. and Bülthoff, H.H. (2002) Categorical perception of familiar objects. Cognition, 85: 113–143.

Newell, F.N. and Findlay, J.M. (1997) The effect of depth rotation on object identification. Perception, 26: 1231–1257.

Newell, F.N., Sheppard, D.M., Edelman, S. and Shapiro, K.L. (2005) The interaction of shape- and location-based priming in object categorisation: evidence for a hybrid "what + where" representation stage. Vision Res., 45: 2065–2080.

Newell, F.N., Wallraven, C. and Huber, S. (2004) The role of characteristic motion in object categorization. J. Vis., 4: 118–129.

Palmer, S.E., Rosch, E. and Chase, P. (1981). Canonical perspective and the perception of objects. In: Long, J., Baddeley, A. (Eds.), Attention and Performance, Vol. IX. Erlbaum, Hillsdale, NJ, pp. 135–151.

Rossion, B. (2002) Is sex categorisation from faces really parallel to face recognition? Vis. Cogn., 9: 1003–1020.

Wild, H.H., Barrett, S.E., Spence, M.J., O'Toole, A.J., Cheng, Y.D. and Brooke, J. (2000) Recognition and sex categorization of adults' and children's faces in the absence of sex stereotyped cues. J. Exp. Child Psychol., 77: 261–299.

Young, A.W., Rowland, D.A., Calder, A.J., Etcoff, N.L., Seth, A. and Perrett, D.I. (1997) Facial expression megamix: test of dimensionality and category accounts of emotion recognition. Cognition, 63(3): 271–313.

Yovel, G. and Kanwisher, N. (2004) Face perception: domain specific, not process specific. Neuron, 44: 889–898.

Subject Index